Electromagnetic Theory for Electromagnetic Compatibility Engineers

Electromagnetic Theory for Electromagnetic Compatibility Engineers

Tze-Chuen Toh

CRC Press
Taylor & Francis Group
Boca Raton London New York

CRC Press is an imprint of the
Taylor & Francis Group, an **informa** business

CRC Press
Taylor & Francis Group
6000 Broken Sound Parkway NW, Suite 300
Boca Raton, FL 33487-2742

First issued in paperback 2016

Version Date: 20131009

ISBN 13: 978-1-138-03407-5 (pbk)
ISBN 13: 978-1-4665-1815-5 (hbk)

Visit the Taylor & Francis Web site at
http://www.taylorandfrancis.com

and the CRC Press Web site at
http://www.crcpress.com

To my wise and inspirational parents Nancy and Swee-Hock, my endearing wife and muse Linnie, my cherished children by marriage, Courtney, Allison, and Paul . . . and in memory of our cheeky dog Lucky.

Contents

Preface

The irony of a preface is that it is often ignored by the reader; notwithstanding, it is occasionally employed by authors to justify formally the existence of their work. This primer is an outgrowth, and indeed a much expanded version, of a course I gave at Gateway Inc. to help electromagnetic compatibility (EMC) engineers be less dependent upon empirical data when solving puzzling electromagnetic interference (EMI) problems. In short, it is a vehicle to transform practical engineers into theoretical engineers, and in particular, to cultivate academic skills left by the wayside after they have completed their tertiary education. Furthermore, as this course is designed for practical applications, historical developments are bypassed altogether.

This set of lecture notes is written at a level equivalent to that of third- or fourth-year Honors undergraduate study. It is, in essence, a refresher for professional engineers who already have their Bachelor's degree and to reacquaint them with the power of mathematical rigor in solving real-world problems. Conversely, it also serves to introduce undergraduates to the basics of EMC encountered in the technology industry. It is completely self-contained and designed with self-study in mind.

For a condensed course, the recommended topics to cover are Chapters 1, 4, 5, and 6. Having pointed this out, I must nevertheless emphasize that each topic presented herein is essential for EMC engineers to be competent EMC theoreticians. Clearly, in a brief course such as this, it is impossible to cover every aspect of electromagnetic theory. A list of references upon which parts of this course are based is provided for the reader who wishes to delve deeper into topics glossed over for want of space. More importantly, it was written specifically for theoretical physicists and mathematicians new to the field of EMC, signal integrity and RF design.

Electromagnetic theory is the simplest theory among the four known forces of the universe. Indeed, it is the first step toward finding the holy grail of theoretical physics: the grand unification of the four known forces. Maxwell's theory achieved the unification of the electric field and the magnetic field into a single entity called the electromagnetic field. Notwithstanding, its underlying mathematical structure is similar to the other three forces—the weak nuclear force, strong nuclear force, and gravitational force—to wit, gauge theory. Gauge theory provides a common arena for the foundational description of these forces. However, studying gauge theory, fascinating as it is, will take the reader too far afield from the original intent of these notes. Engineers, after all, are practical and down to earth, and hold little interest in the more abstruse mathematical guise of nature.

The primary purpose of this monograph is to integrate theory with practicable engineering applications. As a case in point, it is often difficult to find a monograph on electromagnetic theory that expounds elements of differential transmission line theory, roll-off frequencies, and electrostatic discharge frequently encountered in the industry. That is, the material that professional engineers seek is often not found in standard courses on electrodynamics. On the other hand, although references for differential transmission line theory and electrostatic discharge for professional engineers abound, they often lack mathematical rigor, preferring to cite engineering rules of thumb and observations. It is hoped that this monograph will bridge the gap by providing the needed rigor to engineering applications, and more important, to instill the need for rigorous mathematical science in good engineering practice.

In short, the intent is to appeal to both sets of audiences: to entice the practical engineer to explore some worthwhile mathematical methods, and to reorient the theoretical scientist to apply the theory in the technology industry. For there is much indeed that a professional engineer can profit from in pure academic pursuit to further the cause of innovation and technology advancement.

Finally, SI units are employed throughout this exposition. Although CGS units can still be found in some textbooks on electrodynamics, particularly references prior to the 1970s, I am quite persuaded that the process of converting CGS units back to SI units causes more confusion and frustration than the convenience it purports to impart to the formalism of electromagnetic theory.

Chuen Toh
Lexington, KY

About the Author

Tze-Chuen Toh is a theoretical physicist and a consultant. He received his BSc (Hons I) in mathematics from the University of Queensland in Australia, and his PhD in theoretical physics from the Australian National University. In the past, he worked at Lexmark International Inc. as an electrical engineer, and at Dell Inc. and Gateway Inc. as an electromagnetic compatibility engineer. His research portfolio comprises publications in peer-reviewed scientific journals and patents. His current research interests lie primarily in quantum gravity, gauge theory, quantum computing and information, electrodynamics, and mathematical modeling.

Notations

The following notations are used throughout this monograph.

- The *empty set* is denoted by $\varnothing$.
- Given a set $A \neq \varnothing$, $a \in A$ reads as a is an *element* of (or a *member* of) A.
- The set $A \cup B$ denotes the *union* of sets A and B; $A \cap B$ denotes the *intersection* of sets A and B.
- If A is a set, then $2^A = \{S : S \subseteq A\}$ denotes the set of all subsets of A, that is, the *power set* of A.
- If A and B are nonempty sets, then $A \times B$ is the set of all points* (a,b) such that a belongs to A and b belongs to B. This is written as $A \times B = \{(a,b) : a \in A, b \in B\}$.
- Given two nonempty sets A and B, a *mapping* f.
- $A \to B$ defines a relation between sets A and B whereby f assigns each element $a \in A$ a unique element $b \in B$ called the *value* of f at a. This is written as $f(a) = b$ or $f: a \mapsto b$.
- The sets $\mathbf{R}, \mathbf{R}^2$ denote, respectively, the real line and Euclidean 2-dimensional space. More generally, the Euclidean n-dimensional space $\mathbf{R}^n = \{(x^1, \ldots, x^n) : x^i \in \mathbf{R}, i = 1, \ldots, n\}$.
- A *closed interval* $[a,b] \subset \mathbf{R}$ on the real line is the subset $[a,b] = \{r \in \mathbf{R}: a \leq r \leq b\}$, and an *open interval* $(a,b) \subset \mathbf{R}$ is defined by $(a,b) = \{r \in \mathbf{R}: a < r < b\}$.
- $\forall$ symbolizes *for all* or *for every*.
- $\exists$ symbolizes *there exists*.
- $\Rightarrow$ denotes implication; that is, $P \Rightarrow Q$ reads as *if P then Q*, or equivalently, *P implies Q*.
- $\Leftrightarrow$ denotes equivalence; that is, $P \Leftrightarrow Q$ reads as *P if and only if Q*, or *P* is *equivalent* to *Q*.
- Given a nonempty set A, $B \subseteq A$ denotes that B is a *subset* of A. Recall that $B \subseteq A$ if $\forall x \in B \Rightarrow x \in A$, whereas $B \subset A$ means that B is a *proper* subset of A; that is, $B \subseteq A$ and $B \neq A$.
- $a \notin A$: a does not belong to A.
- $B \not\subset A$: B is not a subset of A.
- Given a nonempty set A, the *boundary* of A is denoted by ∂A.

* Technically known as an *ordered pair*.

- $x \to x_0^+$ denotes $x \to x_0$ such that $x > x_0$; that is, x *tends* toward x_0 in the limit from above.
- $x \to x_0^-$ denotes $x \to x_0$ such that $x < x_0$; that is, x *tends* toward x_0 in the limit from below.
- $a_+(a_-)$ or $a_\pm$ reads a_+ and a_-, respectively.
- In particular, $a_\pm \mp b_\mp$ is short for $a_+ - b_-$ and $a_- + b_+$, respectively.
- $A = \{x{:}\,Q\}$ reads as A is the set of all x such that proposition Q is satisfied.
- $\mathbf{R} = \{x{:} -\infty < x < \infty\} = (-\infty,\infty)$ is the set of *real* numbers.
- $\mathrm{C} = \{x + \mathrm{i}y{:} -\infty < x, y < \infty\}$ is the set of *complex* numbers, where $\mathrm{i} = \sqrt{-1}$.
- $\mathrm{N} = \{1,2,3,\ldots\}$ is the set of *natural* numbers.
- $\mathbf{Z} = \{0,\pm1,\pm2,\ldots\}$ is the set of *integers* $\partial_x = \frac{\partial}{\partial x}$ and $\partial_x^2 = \frac{\partial^2}{\partial x^2}$, and so on.
- $\Delta \equiv \nabla^2$ denotes the *Laplacian operator*. For instance, in 3-dimensional Euclidean space, $\Delta = \partial_x^2 + \partial_y^2 + \partial_z^2$ in rectangular coordinates.
- $\mathbf{R}^n = \left\{(x^1,\ldots,x^n) : -\infty < x^i < \infty \ \forall i = 1,\ldots,n\right\}$ denotes the n-dimensional Euclidean space.
- $(a,b] \equiv \{x{:}\,a < x \le b\}$.
- If $f{:}\,A \to B$ is a function that maps A into B, and $C \subset A$, then the *restriction* of f to the subset C is written as $f|C$ or $f|_C$.
- Given two sets A, B, the *Cartesian product* $A \times B = \{(a,b){:}\,a \in A, b \in B\}$.
- Given a set A and subset $B \subseteq A$, the complement $B^c \equiv A - B = \{a \in A : a \notin B\}$.
- Given a nonempty set A, a *partition* P of A is the collection of subsets $B \subset A$ such that $\forall B, B' \in \mathcal{P}, B \neq B' \Rightarrow B \cap B' = \varnothing$ and $A = \cup_{B \in \mathcal{P}} B$.
- The space $\mathrm{C}(\mathbf{R}^k)$ defines the set of all continuous functions $f : \mathbf{R}^k \to \mathbf{R}$.
- The space $\mathrm{C}^1(\mathbf{R}^k)$ defines the set of all continuous functions $f : \mathbf{R}^k \to \mathbf{R}$ that is continuously one-time differentiable; i.e., $f'(x) = \frac{d}{dx} f(x)$ is continuous on $\mathbf{R}$.
- The space $\mathrm{C}^2(\mathbf{R}^k)$ defines the set of all continuous functions $f : \mathbf{R}^k \to \mathbf{R}$ that is continuously twice differentiable. That is,

$$\frac{\partial^2 f}{\partial x^i \partial x^j} = \frac{\partial^2 f}{\partial x^j \partial x^i} \ \forall i, j = 1,\ldots,k$$

- where $(x^1,\ldots,x^k) \in \mathbf{R}^k$.

1

Brief Review of Maxwell's Theory

In this chapter, Maxwell's theory is briefly reviewed. In particular, it is shown that electromagnetic fields exist and they propagate in space as waves. Historically, the set of Maxwell's equations were not all derived by Maxwell; they were attributed to him because, among other things, he correctly added a displacement current term to Faraday's equation, and predicted the existence of electromagnetic waves from the set of equations that is now known as Maxwell's equations [2,10,11].

As electronic components get smaller, and digital circuits operate in the microwave frequencies, it is imperative that electromagnetic compatibility (EMC) engineers possess a sound understanding of electromagnetic theory in order to root-cause electromagnetic interference issues and to design printed circuit boards effectively with low electromagnetic emissions in mind.

An outline of this chapter runs as follows. Electrostatics is introduced in Section 1.1, followed by a brief overview of steady-state currents in Section 1.2. Maxwell's equations are derived in Section 1.3, and the significance of the equations highlighted. Finally, the existence of electromagnetic waves is illustrated in the last section.

1.1 Electrostatics

It is known empirically that two point charges q_i, $i = 1,2$, exert a force $\boldsymbol{F}$ on each other. This force is the celebrated Coulomb's law:

$$\boldsymbol{F} = \frac{1}{4\pi\varepsilon_0}\frac{q_1 q_2}{r^3}\boldsymbol{r} \tag{1.1}$$

where $\boldsymbol{r}_i = (x_i, y_i, z_i)$ for $i = 1,2$, is the location of charge q_i in the rectangular coordinate system, $r = \sqrt{(x_2 - x_1)^2 + (y_2 - y_1)^2 + (z_2 - z_1)^2}$ and $\boldsymbol{r} = \boldsymbol{r}_2 - \boldsymbol{r}_1$. The constant $\varepsilon_0 = 8.854 \times 10^{-12}$ defines the *electric permittivity of free-space*, in farads per meter (F/m). This fact suggests that a field surrounds a point charge mediating the force between the two charges. Informally then, the *electric field E* generated by a point charge q is defined as follows. Consider a test charge δq in the vicinity of q and let $\boldsymbol{F}$ denote the force between the pair $(q,\delta q)$. Then, in the limit as the test charge $\delta q \to 0$,

$$\boldsymbol{E} \equiv \lim_{\delta q \to 0} \tfrac{1}{\delta q} \boldsymbol{F} = \tfrac{1}{4\pi\varepsilon_0} \tfrac{q}{r^3} \boldsymbol{r}$$

The unit of the electric field is in volts per meter (V/m). A more formal derivation is provided later.

In a more realistic scenario, consider the electric field generated by some charged volume V. First, consider the simpler case wherein finite point charges $q_1,\ldots,q_n$ are embedded in V. Let q_i be located at $\boldsymbol{r}_i$ with respect to some fixed origin where a test charge δq is located. For an arbitrary point $\boldsymbol{r}$ away from the origin, let the electric field generated by the ith charge q_i be $\boldsymbol{E}_i$. Because force is a vector, the superposition principle holds. Thus, it follows from the heuristic definition of the electric field that the electric field at the origin, in the limit as $\delta q \to 0$, is:

$$\boldsymbol{E} \equiv \lim_{\delta q \to 0} \tfrac{1}{\delta q} (\boldsymbol{F}_1 + \cdots + \boldsymbol{F}_n) = \tfrac{1}{4\pi\varepsilon_0} \sum\nolimits_{i=1}^{n} \tfrac{q_i}{r_i^3} \boldsymbol{r}_i \tag{1.2}$$

where $r_i = \sqrt{(x-x_i)^2 + (y-y_i)^2 + (z-z_i)^2} \quad \forall i = 1,\ldots,n$. That is, $\boldsymbol{E} = \sum\nolimits_{i=1}^{n} \boldsymbol{E}_i$.

Indeed, it is seen more formally later that the electric field obeys the superposition principle because the operators defining Maxwell's equations are linear. From (1.2), the extension to a continuous charge distribution in V is obvious. To see this, suppose that V has a volume charge density ρ. Then, heuristically, on setting $\delta q_i = \rho\delta V_i$, where δV is a differential volume element of V, taking the limit carefully as $\delta V_i \to 0$ and $n \to \infty$, such that

$$Q = \lim_{\substack{\delta V_i \to 0 \\ n \to \infty}} \sum_{i=1}^{n} \delta q_i = \int_V \rho(\boldsymbol{r}) \mathrm{d}^3 r$$

is the well-defined total charge on V, it follows that the summation in (1.2) becomes an integral:

$$\boldsymbol{E}(\boldsymbol{r}) = \tfrac{1}{4\pi\varepsilon_0} \int_V \tfrac{\rho(\boldsymbol{r}')}{R'^3} \boldsymbol{R}' \mathrm{d}^3 r' \tag{1.3}$$

where $R' = \sqrt{(x-x')^2 + (y-y')^2 + (z-z')^2}$, for all $\boldsymbol{r}' = (x',y',z') \in V$ and $r = (x,y,z)$.

In the above intuitive approach, the limit $\delta q \to 0$ was invoked in a somewhat cavalier fashion. In reality, it is known that the electric charge cannot be arbitrarily small. To date, the smallest nonzero electric charge is $\frac{1}{3}|e|$,* where e is the

* Subatomic particles called quarks possess a fractional electronic charge composed of integer multiples of one-third that of an electronic charge.

fundamental electronic charge of the electron in Coulomb: $e = -1.602 \times 10^{-19}$ C. However, as this is an exposition of classical physics rather than quantum physics, this subtlety may be ignored for all intents and purposes.

Before formally defining an electric field, consider the example below which is critical in the development of antenna theory; to wit, a static electric dipole. Further details are explored in Chapter 8. An electric dipole is essentially two oppositely charged (point) particles separated by a fixed small distance.

1.1.1 Example

Consider two point charges $(\pm q, d_0)$ separated by a distance $d_0 > 0$. Without loss of generality, suppose that $\pm q$ is located, respectively, at $\boldsymbol{r}_{\pm} = (0,0,\pm\frac{1}{2}d_0)$. Then, the electric field at any point $\boldsymbol{r} = (x,y,z)$ is given by the superposition of the two charges:

$$\boldsymbol{E}(\boldsymbol{r}) = \frac{q}{4\pi\varepsilon_0}\left\{\frac{\boldsymbol{r}-\boldsymbol{r}_+}{|\boldsymbol{r}-\boldsymbol{r}_+|^3} - \frac{\boldsymbol{r}-\boldsymbol{r}_-}{|\boldsymbol{r}-\boldsymbol{r}_-|^3}\right\} = \frac{q}{4\pi\varepsilon_0}\left\{\frac{1}{\left(x^2+y^2+\left(z-\frac{1}{2}d_0\right)^2\right)^{-3/2}}\begin{pmatrix} x \\ y \\ z-\frac{1}{2}d_0 \end{pmatrix}\right.$$

$$\left. - \frac{1}{\left(x^2+y^2+\left(z+\frac{1}{2}d_0\right)^2\right)^{-3/2}}\begin{pmatrix} x \\ y \\ z+\frac{1}{2}d_0 \end{pmatrix}\right\}$$

where the first term is the contribution from $+q$ and the second term is from $-q$. See Figure 1.1.

Now, if $d_0 << r$, then the binomial expansion may be invoked:

$$\frac{1}{|\boldsymbol{r}-\boldsymbol{r}_{\pm}|^3} = \frac{1}{r^3}\left(1 \mp \frac{d_0 z}{r^2}\right)^{-\frac{3}{2}} \approx \frac{1}{r^3}\left(1 \pm \frac{3}{2}\frac{d_0 z}{r^2}\right)$$

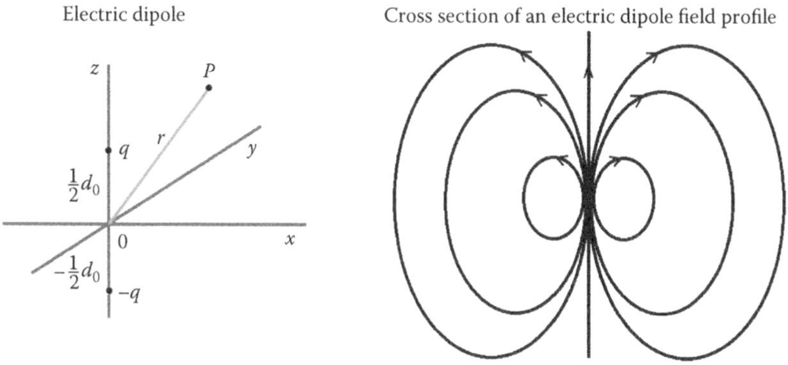

FIGURE 1.1
Electric field generated by an electric dipole.

Whence, substituting the binomial approximation into the above expression yields

$$E(r) = \frac{q}{4\pi\varepsilon_0}\frac{3d_0 z}{r^5}\begin{pmatrix} x \\ y \\ z \end{pmatrix}$$

for the electric field at any point very far away from the pair of point charges. Observe in particular that the electric field is identically zero on the plane $z = 0$ due to the symmetry of the problem, and for $z \neq 0$, it falls of as $\sim \frac{z}{r^4}$ from the pair of charges. □

A closely related quantity to the electric field is the potential field. Intuitively, the electric potential is related to the work done against an electric field in moving a charge, located out in infinity, to some fixed arbitrary position. A more rigorous development is given below.

First, recall that a force $\boldsymbol{F}$ is *conservative* on some domain $\Omega \subseteq \mathbf{R}^3$ if $\exists \psi \in C^2(\Omega)$ such that $\boldsymbol{F} = -\nabla\psi$. Here, the continuously twice differentiable function ψ on Ω is called a *scalar potential*. Geometrically, the gradient $\nabla\psi$ is normal to the surface defined by ψ. Thus, a conservative force is always normal to equipotential surfaces defining the force. By convention, the negative sign is present to indicate that the direction of the force points from a surface of higher potential toward a surface of lower potential. That is, a particle in the presence of a conservative force always traverses from a higher potential to a lower potential surface.

Now, in view of (1.1), it is clear that a conservative electrostatic force acting on a charge q can be formally defined as follows: $\boldsymbol{F} = -q\nabla\varphi \Rightarrow \boldsymbol{E} \equiv -\nabla\varphi$, for some scalar potential φ, where in rectangular coordinates, $\nabla = (\partial_x, \partial_y, \partial_z)$. That is, the electrostatic field is formally defined by the gradient of some scalar potential generating a conservative electrostatic force: $\boldsymbol{E} = -\nabla\varphi$. Thus, the electric field is, by definition, normal to equipotential surfaces, and electric charges move from a higher electric potential surface to a lower electric potential surface.

In this section, only the static case is considered; that is, $\partial_t \boldsymbol{E} = 0$. In hindsight, define the static condition for an electric field by $\nabla \times \boldsymbol{E} = 0$; see Proposition 1.1.2 below, given without proof. Indeed, this stipulation is manifestly apparent in Section 1.3.

1.1.2 Proposition

Suppose an electric field $\boldsymbol{E}$ is defined on $\mathbf{R}^3$ such that $\nabla \times \boldsymbol{E} = 0$. Then, $\exists \varphi \in C^2(\mathbf{R}^3)$ such that $\boldsymbol{E} = -\nabla\varphi$ on $\mathbf{R}^3$, where φ is some scalar potential field. That is, the electric field $\boldsymbol{E}$ is conservative. □

The existence of φ in Proposition 1.1.2 is due to a lemma by Poincaré [1,3]; the exact detail is beyond the scope of this exposition. The potential φ can be recast into a more convenient form. First, recall that a *path* in $\mathbf{R}^3$ is defined by a continuous function $\gamma:[0,1]\to\mathbf{R}^3$. In particular, γ defines a *loop* (or *closed path*) if $\gamma(0) = \gamma(1)$ (*cf.* Appendix A.2 for further details). Second, an immediate consequence of Poincaré's lemma mentioned above (*loc. cit.*) is needed in order to complete the equivalent definition for φ. The corollary is stated below without proof.

1.1.3 Corollary

Let $M \subseteq \mathbf{R}^3$ be a simply connected subspace and let $f: M \to \mathbf{R}$ be any integrable function; that is, $\int_M |f| < \infty$. Then, for any pair of paths $\gamma_1, \gamma_2 : [0,1] \to M$ such that $\gamma_1(0) = \gamma_2(0)$ and $\gamma_1(1) = \gamma_2(1)$, $\int_{\gamma_1} f = \int_{\gamma_2} f$ holds. □

That is, when a space is simply connected, the path integral along any path connecting some fixed pair of endpoints depends solely upon the endpoints and is completely independent of the paths connecting the two endpoints. This leads to the following equivalent definition for an electric potential,

$$\varphi(\boldsymbol{r}) = -\int_{\gamma_\infty} \boldsymbol{E} \cdot \mathrm{d}\boldsymbol{r}' \tag{1.4}$$

where $\boldsymbol{r}$ is the position vector of some arbitrary point $P \in \mathbf{R}^3$ and $\gamma_\infty \subset \mathbf{R}^3$ is a path from ∞ to the point P. Then, the potential is said to be *conservative*. Recall that a path γ_∞ connecting a point at ∞ to a point P is defined by $\gamma_\infty : (0,1] \to \mathbf{R}^3$ such that $\lim_{t\to 0} \gamma_\infty(t) \to \infty$. Finally note that $\mathbf{R}^3$ is simply connected; in contrast, a two-dimensional torus is not simply connected; see Appendix A.2 for details.

The definition for a potential field is well-defined by Corollary 1.1.3. In particular, the *potential difference* $\delta\varphi$ between any two points $P_1, P_2 \in \mathbf{R}^3$ connected by any path $\bar{\gamma}$ such that $\bar{\gamma}(0) = P_1$ and $\bar{\gamma}(1) = P_2$, is given by $\delta\varphi(\boldsymbol{r}) = -\int_{\bar{\gamma}} \boldsymbol{E} \cdot \mathrm{d}\boldsymbol{r}'$. From Corollary 1.1.3, as $\mathbf{R}^3$ is simply connected, this integral is only determined by the two endpoints and not the path connecting them.

1.1.4 Proposition

Suppose $\nabla \times E = 0$ on $\mathbf{R}^3$. Then, $\delta\phi(\boldsymbol{r}) = -\int_\gamma \boldsymbol{E} \cdot \mathrm{d}\boldsymbol{r}' = 0 \; \forall \gamma \subset \mathbf{R}^3$, where γ is any loop in Euclidean 3-space.

Proof

Let $\gamma_+ = \gamma\,|\,[0,\frac{1}{2}]$ be the restriction of γ to the interval $[0,\frac{1}{2}]$ and set $\gamma_- = \gamma\,|\,[\frac{1}{2},1]$. Then, by definition, $\gamma = \gamma_- * \gamma_+$, where

$$\gamma_- * \gamma_+ \equiv \begin{cases} \gamma_+(2t) \text{ for } 0 \le t \le \frac{1}{2}, \\ \gamma_-(2t-1) \text{ for } \frac{1}{2} \le t \le 1. \end{cases}$$

Now, note that if $\Gamma:[0,1]\to\mathbf{R}^3$ is a path, then $\tilde{\Gamma}(t)\equiv\Gamma(1-t)$ defines the reverse orientation of Γ: $\tilde{\Gamma}(0)\equiv\Gamma(1)$ and $\tilde{\Gamma}(1)\equiv\Gamma(0)$. Thus, by Corollary 1.1.3, for any integrable function f on $\mathbf{R}^3$, $\int_\Gamma f=\int_{\Gamma(0)}^{\Gamma(1)} f=-\int_{\Gamma(1)}^{\Gamma(0)} f=-\int_{\tilde{\Gamma}} f$.* Thus, from $\nabla\times\boldsymbol{E}=0\Rightarrow\varphi=-\int_\gamma \boldsymbol{E}\cdot d\boldsymbol{r}'$, it follows immediately that

$$-\int_\gamma \boldsymbol{E}\cdot d\boldsymbol{r}'=-\int_{\gamma_-}\boldsymbol{E}\cdot d\boldsymbol{r}'-\int_{\gamma_+}\boldsymbol{E}\cdot d\boldsymbol{r}'=-\int_{\gamma_-}\boldsymbol{E}\cdot d\boldsymbol{r}'+\int_{\gamma_-}\boldsymbol{E}\cdot d\boldsymbol{r}'=0$$

as required.

In other words, for a simply connected space (i.e., any two paths with the same endpoints can be continuously deformed into each other), the potential difference around a loop is zero. This is a rather critical point to note: indeed, in a space that is not simply connected, for example, a 2-torus, the path integral generally depends upon both the endpoints and the path connecting the endpoints.

More specifically, for a multiply connected space M (*viz.* a nonsimply connected space) $\boldsymbol{E}=-\nabla\varphi$ holds locally in the following sense: $\forall x\in$ M, there exists a neighborhood $N_x\subset M$ of x and a function φ_x on N_x such that $\boldsymbol{E}|N_x=-\nabla\varphi_x$ holds on N_x for each $x\in M$. The collection $\{\varphi_x\}$ thus defines the potential on M. Moreover, it is also manifestly clear in this more general formalism for electrostatics, φ is unique up to an arbitrary constant c: $\varphi\to\varphi+c$ defines the same electric field as $\nabla c=0$.

At a more practical level, imagine measuring the potential difference between two fixed points in a circuit using a voltmeter. If the space within the vicinity of the circuit were multiply connected, then it is conceivable that by merely moving the leads without changing the endpoints of the leads, the voltage reading registered by the meter could change. This clearly renders the voltage measured to be ill-defined. Fortunately, the space within the vicinity of our solar system appears to be simply connected!

1.1.5 Example

Suppose that

$$\boldsymbol{E}=\tfrac{1}{4\pi\varepsilon_0}\tfrac{q}{r^3}\boldsymbol{r}$$

for some charge q, where q is taken to be at the origin in $\mathbf{R}^3$. Then, the potential generated by q is, from Equation (1.4),

$$\varphi=-\tfrac{1}{4\pi\varepsilon_0}\int_\infty^P \tfrac{q}{r'^3}\boldsymbol{r}'\cdot d\boldsymbol{r}'=-\tfrac{1}{4\pi\varepsilon_0}\int_\infty^P \tfrac{q}{r'^2}dr'=\tfrac{1}{4\pi\varepsilon_0}\tfrac{q}{r}$$

* Prove this result without resorting to Lemma 1.1.3. See Exercise 1.5.1(a) for details.

Indeed, it is quite clear that if

$$\boldsymbol{E} = \frac{1}{4\pi\varepsilon_0}\sum_{i=1}^{n}\frac{q_i}{r_i^3}\boldsymbol{r}_i$$

then by the superposition principle,

$$\varphi = \frac{1}{4\pi\varepsilon_0}\sum_{i=1}^{n}\frac{q_i}{r_i}$$

whence, if an extended charged body $V \subset \mathbf{R}^3$ has a charge density ρ, then each differential volume element δV_i has a charge $\delta q_i = \rho\delta V_i$ for $i = 1,\ldots,m$. In the limit as $m\to\infty$ and $\delta V_i \to 0$ such that

$$Q = \lim_{m\to\infty}\sum_{i=1}^{m}\delta q_i \to \int_V \mathrm{d}q = \int_V \rho \mathrm{d}^3 r$$

it follows by construction that

$$\varphi = \frac{1}{4\pi\varepsilon_0}\int_V \frac{\rho(r')}{R'}\mathrm{d}^3 r' \tag{1.5}$$

where $R' = \sqrt{(x-x')^2 + (y-y')^2 + (z-z')^2}$, for all $\boldsymbol{r}' = (x',y',z') \in V$ and $\boldsymbol{r}$ = (x,y,z). See Equation (1.3) above. □

Equation (1.5) thus furnishes the general definition for a potential at an arbitrary point generated by a charged volume (V,ρ).

This section closes with an extremely powerful method for solving electrostatic problems: the *method of images*. This technique is essentially a means to construct a Green's function to solve Poisson's and Laplace's equations in electrostatics (see Appendix A.3). The principle is rather intuitive. Further details are elaborated in Chapter 3.

First, consider a point charge q located a distance z_0 above an infinite, perfect electrical conducting (PEC) plane that is grounded (see Figure 1.2). What is the potential φ at the point $P = (x,y,z)$ above the ground plane? From

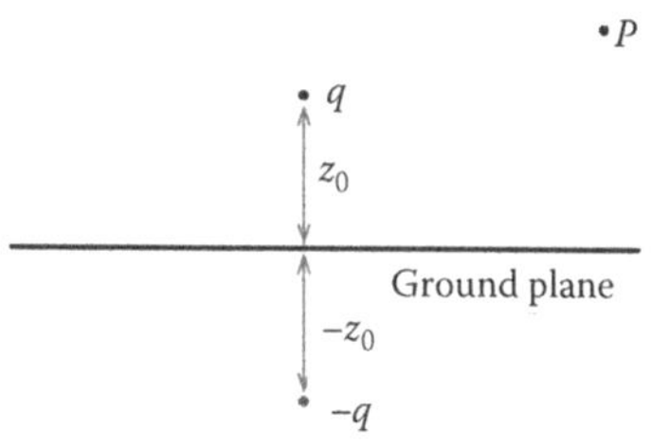

FIGURE 1.2
Point charge and its image below a ground plane.

Appendix A.3, the potential function $\varphi \in C^2(\Omega) \cap C(\bar{\Omega})$ satisfies the following Laplace equation,

$$\Delta\varphi = 0 \text{ on } \Omega \tag{1.6a}$$

subject to the boundary condition

$$\varphi = 0 \text{ on } \partial\Omega \tag{1.6b}$$

where $\Omega = \mathbf{R}_+^3 - \{(0,0,z_0)\}$ and $\mathbf{R}_+^3 = \{(x,y,z) \in \mathbf{R}^3 : z \geq 0\}$ is the upper half-space.

Heuristically, the presence of the charge q induces the opposite charge distribution on the ground plane, via the electric field, such that

a) The resultant induced charge on the plane is equal to $-q$.
b) The potential of the grounded plane is 0.

Then, by construction, $\varphi(\mathbf{r}) = \varphi_+(\mathbf{r} - \mathbf{r}_0) + \varphi_-(\mathbf{r} - \mathbf{r}_0')$ is the potential at $\mathbf{r} = (x,y,z)$, where $\varphi_\pm$ is the potential resulting from $\pm q$. Explicitly, from Example 1.1.5,

$$\varphi(\mathbf{r}) = \frac{1}{4\pi\varepsilon_0}\left\{\frac{q}{\sqrt{x^2+y^2+(z-z_0)^2}} - \frac{q}{\sqrt{x^2+y^2+(z+z_0)^2}}\right\} \tag{1.7}$$

Now, observe that by definition, (b) is satisfied: $\varphi|\partial\Omega = 0$. Furthermore, it is an easy matter to verify that $\Delta\varphi = 0$, where $\Delta = \partial_x^2 + \partial_y^2 + \partial_z^2$. Indeed, this follows at once from $\Delta\varphi_\pm = 0$. This is left as a warm-up exercise for the reader. Because the boundary condition is also satisfied, it follows immediately from the uniqueness theorem of the Laplace equation that (1.7) is the sought-for solution. That is, the potential φ on Ω is defined by (1.7). Finally, note that (1.7) does not apply to $z < 0$, as the *image charge* $-q$ does not really exist below the ground plane. The charge is in fact induced on the surface of the ground plane.

1.1.6 Example

Given two infinitely long, perfect electrical conducting cylinders, consider their cross-section $C_\pm = \{(x,y,0) : x^2 + (y - d_\pm)^2 \leq a_\pm^2\}$, where the axes of the cylinders are parallel to the z-axis, and $a_\pm$ are the radii of the respective cylinders. Suppose the cylinders are held at a potential of $\phi_\pm$. Find the potential at any point external to the cylinders.

The easiest approach to solving this problem is by reducing the charged cylinders to equivalent line charges satisfying the equipotential boundary conditions. Specifically, consider each charged cylinder together with its mirror image across an imaginary boundary initially, and then sum the potentials via superposition.

So, provisionally, consider an equivalent line charge λ_+ located at some distance y_+ (to be determined) from the origin. The distance y_+ is then determined as a function of the cylinder radius and charge via the method of images. For an arbitrary point $\boldsymbol{r} = (x, y)$ away from the line charges, the potential is determined via Equation (1.3) as follows.

$$\boldsymbol{E}_+ = \frac{\lambda_+}{4\pi\varepsilon_0}\int_{-\infty}^{\infty}\frac{1}{\left\{x^2+(y-y_+)^2+(z-z')^2\right\}^{\frac{3}{2}}}\begin{pmatrix} x \\ y-y_+ \\ z-z' \end{pmatrix}\mathrm{d}z'$$

where λ_+ is an infinite line charge density parallel to the z-axis.

Now, observe that

$$\int_{-\infty}^{\infty}\frac{1}{\left\{x^2+(y-y_+)^2+(z-z')^2\right\}^{\frac{3}{2}}}\mathrm{d}z' = -\left.\frac{z-z'}{(x^2+(y-y_+)^2)\sqrt{x^2+(y-y_+)^2+(z-z')^2}}\right|_{-\infty}^{\infty} = \frac{2}{x^2+(y-y_+)^2}$$

$$\int_{-\infty}^{\infty}\frac{z'}{\left\{x^2+(y-y_+)^2+(z-z')^2\right\}^{\frac{3}{2}}}\mathrm{d}z' = \left.\frac{-(x^2+(y-y_+)^2)+zz'-z^2}{(x^2+(y-y_+)^2)\sqrt{x^2+(y-y_+)^2+(z-z')^2}}\right|_{-\infty}^{\infty} = \frac{2z}{x^2+(y-y_+)^2}$$

whence

$$\boldsymbol{E}_+ = \frac{\lambda_+}{2\pi\varepsilon_0}\frac{\boldsymbol{r}_+}{x^2+(y-y_+)^2}$$

where $\boldsymbol{r}_+ = (x, y-y_+, 0)$ defines the electric field in $\mathbf{R}^3 - \{(0, y_+, z) : z \in \mathbf{R}\}$. The electric field $\boldsymbol{E}_-$ follows *mutatis mutandis* via the replacement

$$(\lambda_+, y_+) \to (\lambda_-, y_-): \quad \boldsymbol{E}_- = \frac{\lambda_-}{2\pi\varepsilon_0}\frac{\boldsymbol{r}_-}{x^2+(y-y_-)^2}$$

In particular, inasmuch as the fields are independent of the z-component, in all that follows, we may consider the 2D plane defined by $z = 0$.

Next, appealing to Definition 1.1.2, via polar coordinates

$$\boldsymbol{E}_\pm = \frac{\lambda_\pm}{2\pi\varepsilon_0 r_\pm^2}\mathbf{e}_r$$

and the contribution from λ_+ thus yields

$$\varphi = -\int_\gamma \boldsymbol{E}_+\cdot \mathrm{d}\boldsymbol{r}' = -\frac{\lambda_+}{2\pi\varepsilon_0}\int\frac{\mathrm{d}r}{r} = -\frac{\lambda_+}{2\pi\varepsilon_0}\ln\sqrt{x^2+(y-y_+)^2}+C$$

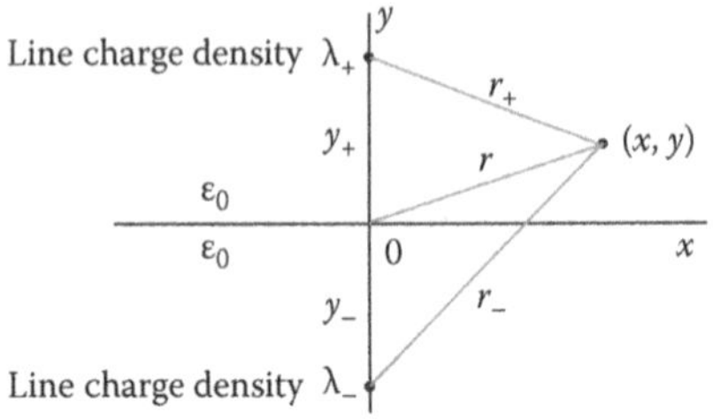

FIGURE 1.3
Potential arising from two infinite line charges.

where C is some arbitrary constant of integration. By symmetry, the resultant potential from the two line charges is:

$$\varphi(r) = -\frac{\lambda_+}{2\pi\varepsilon_0}\ln\frac{\hat{r}}{r_+} - \frac{\lambda_-}{2\pi\varepsilon_0}\ln\frac{\hat{r}}{r_-} \tag{1.8}$$

where $r_\pm = \sqrt{x^2 + (y - y_\pm)^2}$, and some constant reference point $\hat{r}$. See Figure 1.3.

For simplicity, assume provisionally that $\lambda_- = -\lambda_+$: this condition is relaxed later. Then, $\varphi(x,0) = 0\,\forall x \Rightarrow y_- = -y_+$. Furthermore, from (1.8), at any point

$$(x, y) \in C_+,\ \varphi_+ = \frac{\lambda_+}{2\pi\varepsilon_0}\ln\frac{r_+}{r_-}$$

by definition, whence, it follows immediately that

$$\frac{r_+}{r_-} = \text{constant} \Rightarrow \sqrt{\frac{x^2+(y-y_+)^2}{x^2+(y+y_+)^2}} = c$$

for some constant c. However, notice that this expression can be rearranged into the form

$$\left(y - \frac{1+c^2}{1-c^2}y_+\right)^2 + x^2 = \left(\frac{2c}{1-c^2}y_+\right)^2 \tag{1.9}$$

Derive (1.9) as a warm-up exercise; see Exercise 1.5.1(b). Equation (1.9) is nothing but the equation of a circle with

$$\text{radius } \left|\frac{2cy_+}{1-c^2}\right|$$

$$\text{center } \left(\frac{1+c^2}{1-c^2}y_+, 0\right) \in \mathbf{R}^2$$

The circles defined by the above radii and centers describe circles of equipotential. In particular, $\exists c_+$ such that

$$\left|\frac{2c_+y_+}{1-c_+^2}\right| = a_+$$

the equipotential circle coincides with C_+ and

$$d_+ = \frac{1+c_+^2}{1-c_+^2}y_+$$

Thus,

$$d_+^2 = \left(\frac{c_+^2+1}{2c_+}\right)^2\left(\frac{2c_+}{c_+^2-1}y_+\right)^2 = \left(\frac{c_+^2+1}{2c_+}\right)^2 a_+^2$$

Moreover, observe that

$$\left(\frac{1+c_+^2}{1-c_+^2}\right)^2 y_+^2 - y_+^2 = y_+^2\left(\frac{1+c_+^2}{1-c_+^2}+1\right)\left(\frac{1+c_+^2}{1-c_+^2}-1\right) = \left(\frac{2c_+y_+}{1-c_+^2}\right)^2 = a_+^2$$

and hence, yielding

$$d_+^2 = y_+^2 + a_+^2 \tag{1.10}$$

That is, $y_+ = \sqrt{d_+^2 - a_+^2}$.

Now, removing the assumption that $\lambda_- = -\lambda_+$, it is clear by symmetry that $y_- = \sqrt{d_-^2 - a_-^2}$. To proceed with the analysis, observe that on the boundary ∂C_+, $\varphi = \varphi_+$. Hence, choosing the point $(0, d_+ - a_+) \in \partial C_+$,

$$\varphi_+ = -\frac{\lambda_+}{2\pi\varepsilon_0}\ln|d_+ - a_+ - y_+| - \frac{\lambda_-}{2\pi\varepsilon_0}\ln|d_+ - a_+ - y_-| \tag{1.11a}$$

and likewise, $\varphi \mid C_- = \varphi_-$ implies choosing the point $(0, d_- - a_-) \in \partial C_-$ yields

$$\varphi_- = -\frac{\lambda_+}{2\pi\varepsilon_0}\ln|d_- - a_- - y_+| - \frac{\lambda_-}{2\pi\varepsilon_0}\ln|d_- - a_- - y_-| \tag{1.11b}$$

Hence, solving for $\lambda_\pm$ simultaneously via (1.11) gives

$$\lambda_+ = \left(\varphi_- - \varphi_+\frac{\beta_-}{\beta_+}\right)\left(\alpha_- + \alpha_+\frac{\beta_-}{\beta_+}\right)^{-1}$$

$$\lambda_- = \frac{\varphi_+}{\beta_+} - \frac{\alpha_+}{\beta_+}\left(\varphi_- - \varphi_+\frac{\beta_-}{\beta_+}\right)\left(\alpha_- + \alpha_+\frac{\beta_-}{\beta_+}\right)^{-1}$$

where

$$\alpha_{\pm} = -\tfrac{1}{2\pi\varepsilon_0}\ln\left|d_{\pm} - a_{\pm} - y_{+}\right| \text{ and } \beta_{\pm} = -\tfrac{1}{2\pi\varepsilon_0}\ln\left|d_{\pm} - a_{\pm} - y_{-}\right| \qquad \square$$

1.1.7 Remark

The case wherein the electric permittivity ε of a dielectric such that $\varepsilon > \varepsilon_0$ can be found in Reference [14] by applying the method of transformation in a circle [7].

1.2 Magnetostatics

Along a vein similar to the definition for an electric field, the magnetic field can be defined as the field generated by an element of current. More formally, given a conductor (M, σ), where σ (in units of $\Omega^{-1}\text{m}^{-1}$) is the electrical conductivity, if a constant electric field $\boldsymbol{E}$ is applied across ∂M, it will generate a flow of electrons within M defined by $\boldsymbol{J} = \rho_e \boldsymbol{v}_e$. Here, $\boldsymbol{J}$ defines the *current density*, ρ_e is the electron charge density, and $\boldsymbol{v}_e$ is the average drift velocity of the electrons. The current density is related to the electric field via *Ohm's law*

$$\boldsymbol{J} = \sigma \boldsymbol{E} \tag{1.12}$$

This defines the *conduction* current density. Furthermore, if there exist free mobile charges of density ρ in the medium moving at an average velocity $\boldsymbol{v}$, then the conduction current density also contains a *convection* current density term: $\rho\boldsymbol{v}$. That is, $\boldsymbol{J} = \sigma\boldsymbol{E} + \rho\boldsymbol{v}$. In most media of interest, this convection term is zero; on the other hand, this term is nonzero in cases such as the vacuum tube, wherein the primary conduction current density comprises that of convection current density, simply because $\sigma = 0$ in vacuum.

1.2.1 Remark

If M is suspended in a constant electric field, then electrons will migrate toward $\partial M_- \subset \partial M$, causing it to be negatively charged, whereas its complement $\partial M_+ = \partial M - \partial M_-$ becomes positively charged. Because M is initially uncharged, its total charge must remain zero. Hence, the charges will redistribute on ∂M such that the induced electric field within M generated by the pair $(\partial M_-, \partial M_+)$ cancels out the applied electric field. At which point, the flow of electrons will cease as there is a zero electric field within M. As an immediate corollary static electric field cannot sustain a current in a conductor.

Throughout the book, references made to a cross section of a conductor and an axis of a conductor are made. As such, their formal definitions are provided for future reference. Let $M \subset \mathbf{R}^3$ such that $\exists \gamma \subset M$ a simple path and a collection $\mathrm{C} = \{\Sigma_x \subset M : x \in \gamma\}$ of compact subsets satisfying $\Sigma_x \cap \Sigma_{x'} = \varnothing \; \forall x, x' \in \gamma,\, x \neq x'$, and $M = \cup_{x \in \gamma} \Sigma_x$ and the tangent vector along γ at X is normal to Σ_x. Then $\Sigma \in \mathrm{C}$ is called a *cross-section* of M and M is called a *closed tubular neighborhood* of γ. If each $\Sigma \in \mathrm{C}$ is open with compact closure, then M is called an *open tubular neighborhood* of γ. Finally, $\gamma \subset M$ is said to be the *axis* of the tubular neighborhood M if for each cross-section $\Sigma \subset M$ with $\{x\} = \gamma \cap \Sigma$, $B_2(\hat{r}', x') \subseteq B_2(\hat{r}, x) \; \forall x' \in \Sigma$, where the maximal radius $\hat{R} = \max\{R > 0 : B_2(R, u) \subseteq \Sigma, u \in \Sigma\}$, is the largest radius such that the two-dimensional disc $B_2(R, u)$ is contained in Σ, with $B_2(r, u) = B(r, u) \cap \{\mathbf{R}^2 + u\}$,

$$B(r, u) = \left\{ y \in \mathbf{R}^3 : |y - u| < r \right\}$$

is a three-dimensional disc in $\mathbf{R}^3$, and $\mathbf{R}^2 + u$ is the *hyperplane* in $\mathbf{R}^3$ translated by u: $\{y' + u \in \mathbf{R}^3 : y \in \mathbf{R}^2\}$. It is easy to see that the axis of a cylinder coincides with the above more general definition.

1.2.2 Definition

Given a conductor (M, σ), suppose a current density J flows across a cross-section $\Sigma \subset M$ of M. Then, the *current* I flowing through M is defined by

$$I = \int_{\Sigma} J \cdot n \mathrm{d}^2 x$$

where n is a normal vector field on Σ.

1.2.3 Definition

Suppose J is defined on (M, σ). Then, a *vector potential* A generated by J is defined on $\mathbf{R}^3$ by

$$A(r) = \tfrac{\mu_0}{4\pi} \int_M \tfrac{J(r')}{R} \mathrm{d}^3 x$$

where $R = |r - r'| = \sqrt{(x - x')^2 + (y - y')^2 + (z - z')^2} \; \forall (x', y', z') \in M$ and $(x, y, z) \in \mathbf{R}^3$. Here, $\mu_0 = 4\pi \times 10^{-7}$ H/m (in Henry/meter) denotes the *magnetic permeability* of free-space. Moreover, the *magnetic field density* (or to be more technically precise, the *magnetic flux density*) B on $\mathbf{R}^3$ is given by $B = \nabla \times A$.

Note in passing that the definition given above for a magnetic field is motivated by the Lorentz force law. Explicitly, in the absence of an electric field, a charge particle moving at some fixed velocity v in a static magnetic field B

obeys the *Lorentz force* law: $\boldsymbol{F} = -q\boldsymbol{v} \times \boldsymbol{B}$. In the presence of an electric field Lorentz's law becomes, via Coulomb's law, $\boldsymbol{F} = q\boldsymbol{E} - q\boldsymbol{v} \times \boldsymbol{B}$. Thus, considering for simplicity, the absence of an electric field, the magnetic field density $\boldsymbol{B}$ may be defined to be the field acting on a test charge such that it satisfies the Lorentz force law. More specifically, if the Lorentz force is generated by some potential field, then, given that $\nabla \cdot \boldsymbol{B} = 0$ empirically (see Section 1.3 for further details), it follows that there exists some vector potential A such that $\boldsymbol{B} = \nabla \times \boldsymbol{A}$; refer to Appendix A.1. This completes the justification for Definition 1.2.3.

In many instances, evaluating the magnetic field via Definition 1.2.3 is easier than using the conventional definition often encountered in first-year electromagnetics courses. This is given below for completeness. Suppose γ is a loop and I a current circulating around γ. Then, the magnetic flux density $\boldsymbol{B}$ on $\mathbf{R}^3$ is defined by

$$\boldsymbol{B}(\boldsymbol{r}) = \tfrac{\mu_0}{4\pi} \oint_\gamma \tfrac{I(\boldsymbol{r}')}{R^2} \mathrm{d}\boldsymbol{l} \times \boldsymbol{e}_R \tag{1.13a}$$

where

$$R = |\boldsymbol{r} - \boldsymbol{r}'| = \sqrt{(x - x')^2 + (y - y')^2 + (z - z')^2},\ \boldsymbol{e}_R = \tfrac{\boldsymbol{R}}{R},\ \boldsymbol{r}' \in \gamma \text{ and } \boldsymbol{r} \in \mathbf{R}^3$$

More generally, Equation (1.13a) can be extended to a volume in the following fashion. Suppose $M \subset \mathbf{R}^3$ has a current density $\boldsymbol{J}$ flowing across its cross-section $\Sigma \subset M$. Then, the magnetic field at $\boldsymbol{r} \in \mathbf{R}^3$ resulting from the triple $(M, \boldsymbol{J}, \Sigma)$ is given by

$$\boldsymbol{B}(\boldsymbol{r}) = \tfrac{\mu_0}{4\pi} \int_\Sigma \tfrac{\boldsymbol{J}(\boldsymbol{r}')}{R^2} \times \mathbf{e}_R \mathrm{d}^2 x \tag{1.13b}$$

The magnetic field developed above did not depend on time: $\partial_t \boldsymbol{B} = 0$. That is, only the static scenario was considered. In view of Definition 1.2.3, this is equivalent to $\partial_t \boldsymbol{A} = 0$ and hence $\partial_t \boldsymbol{J} = 0$. Thus a static magnetic field is generated by a constant current.

Reflecting on the treatment of electrostatics in Section 1.1, wherein the equivalent case of $\partial_t \boldsymbol{E} = 0$ was defined by $\nabla \times \boldsymbol{E} = 0$, is there a dual criterion wherein $\nabla \times \boldsymbol{B}$ defines a magnetostatic scenario? This question is addressed in Section 1.3. Finally, this section concludes with an application of Equation (1.13a).

1.2.4 Example

Let $C = \{(x, y, 0) \in \mathbf{R}^3 : x^2 + y^2 = a^2\}$ define a circle on the (x,y)-plane and suppose that a constant current I is circulating around C; see Figure 1.4. Then, given any point $\boldsymbol{r} \in \mathbf{R}^3 - C$, the magnetic field density at $\boldsymbol{r}$ can be determined via Equation (1.13a) as follows.

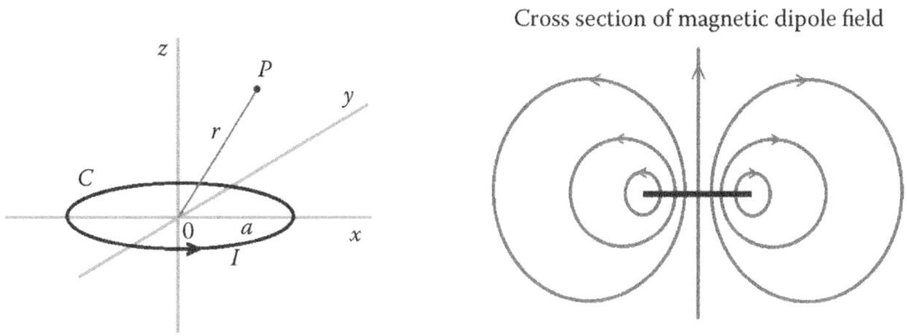

FIGURE 1.4
Magnetic dipole moment generated by a loop current I.

First, observe that

$$\frac{\mathbf{e}_R}{R^2} = (x + a\sin\theta, y - a\cos\theta, z)\{(x + a\sin\theta)^2 + (y - a\cos\theta)^2 + z^2\}^{-\frac{3}{2}}$$

Next, from $\mathrm{d}\boldsymbol{l} = a(-\sin\theta, \cos\theta, 0)\mathrm{d}\theta$,

$$\mathrm{d}\boldsymbol{l} \times \frac{\mathbf{e}_R}{R^2} = \frac{1}{\{(x-a\cos\theta)^2+(y-a\sin\theta)^2+z^2\}^{\frac{3}{2}}} \begin{vmatrix} \mathbf{e}_x & \mathbf{e}_y & \mathbf{e}_z \\ -\sin\theta & \cos\theta & 0 \\ x - a\cos\theta & y - a\sin\theta & z \end{vmatrix} a\mathrm{d}\theta$$

$$= \frac{a}{\{(x-a\cos\theta)^2+(y-a\sin\theta)^2+z^2\}^{\frac{3}{2}}} \begin{pmatrix} z\cos\theta \\ z\sin\theta \\ -(y - a\sin\theta)\sin\theta - (x - a\cos\theta)\cos\theta \end{pmatrix} \mathrm{d}\theta.$$

Thus,

$$B_x = \frac{\mu_0 I z}{4\pi a^3} \int_0^{2\pi} \frac{\cos\theta\,\mathrm{d}\theta}{\left\{\left(\frac{x}{a}-\cos\theta\right)^2+\left(\frac{y}{a}-\sin\theta\right)^2+\left(\frac{z}{a}\right)^2\right\}^{\frac{3}{2}}}$$

$$B_y = \frac{\mu_0 I z}{4\pi a^3} \int_0^{2\pi} \frac{\sin\theta\,\mathrm{d}\theta}{\left\{\left(\frac{x}{a}-\cos\theta\right)^2+\left(\frac{y}{a}-\sin\theta\right)^2+\left(\frac{z}{a}\right)^2\right\}^{\frac{3}{2}}}$$

$$B_z = -\frac{\mu_0 I}{4\pi a^3} \int_0^{2\pi} \frac{(y-a\sin\theta)\sin\theta+(x-a\cos\theta)\cos\theta}{\left\{\left(\frac{x}{a}-\cos\theta\right)^2+\left(\frac{y}{a}-\sin\theta\right)^2+\left(\frac{z}{a}\right)^2\right\}^{\frac{3}{2}}} \mathrm{d}\theta$$

Now, suppose that $r \gg a$. Then, on setting

$$\xi = \sqrt{\left(\tfrac{x}{a} - \cos\theta\right)^2 + \left(\tfrac{y}{a} - \sin\theta\right)^2 + \left(\tfrac{z}{a}\right)^2} \text{ and } \xi_0 = \sqrt{\left(\tfrac{x}{a}\right)^2 + \left(\tfrac{y}{a}\right)^2 + \left(\tfrac{z}{a}\right)^2} = \left(\tfrac{r}{a}\right)$$

it follows that up to first order in sin θ and cos θ,

$$\xi \approx \xi_0 \sqrt{1 + \frac{2}{\xi_0^2}\left(\frac{x}{a}\cos\theta + \frac{y}{a}\sin\theta\right)}$$

whence, by appealing to the binomial expansion,

$$\frac{1}{\xi^3} \approx \frac{1}{\xi_0^3}\left\{1 + \frac{3}{\xi_0^2}\left(\frac{x}{a}\cos\theta + \frac{y}{a}\sin\theta\right)\right\}$$

Substituting this approximation into the integrals for B_x, B_y and B_z yields

$$B_x = \frac{3}{4}\frac{\mu_0 I\, xz}{a^3\xi_0^5} = \frac{3}{4}\frac{\mu_0 I\, a^2}{r^3}\frac{xz}{r^2} \tag{1.14a}$$

$$B_y = \frac{3}{4}\frac{\mu_0 I\, yz}{a^3\xi_0^5} = \frac{3}{4}\frac{\mu_0 I\, a^2}{r^3}\frac{yz}{r^2} \tag{1.14b}$$

$$B_z = -\frac{3}{4}\frac{\mu_0 I}{a^3\xi_0^3}\left\{\frac{3x^2}{r^2} + \frac{3y^2}{r^2} - 2\right\}a^2 = -\frac{3}{4}\frac{\mu_0 I a^2}{r^3}\left\{\frac{3x^2}{r^2} + \frac{3y^2}{r^2} - 2\right\} \tag{1.14c}$$

Indeed, observe from Equation (1.14) that for $r >> a$, the magnetic field is directly proportional to the area of the current loop. □

1.2.5 Remark

The above example defines a *magnetic dipole*: to wit, a simple loop with a non-zero current circulating around the loop (*cf.* an electric dipole of Example 1.1.1). Further details can be found in Chapter 8.

1.3 Maxwell's Equations

Maxwell's electromagnetic theory is in fact the first step toward field unification. It unifies the electric field and the magnetic field into a single entity, giving rise to the electromagnetic field.* In all that follows, unless stated explicitly, rectangular coordinates $\mathbf{r} = (x, y, z)$ are chosen.

* The question regarding the existence of the electric and the magnetic fields is an interesting one. Although an electromagnetic field exists globally, it cannot be decomposed into electric and magnetic fields if space–time cannot be split into space and time. The various ways of splitting up flat space–time (a property known as *foliation*) yields the *Lorentz* symmetry group—this forms the basis for Einstein's theory of special relativity. In particular, the constancy of the speed of light in any inertial frame is indeed a consequence of the Lorentz symmetry of flat space–time; see, for example, Reference [12].

Let $\boldsymbol{E}$ be an electric field and $\boldsymbol{B}$ some magnetic field defined on $(\mathbf{R}^3, \varepsilon, \mu, \sigma)$, where ε denotes the electric permittivity, μ the magnetic permeability, and σ the electric conductivity of the medium. Free charges present in the medium are defined by the charge density ρ. Then, *Maxwell's equations* comprise the following,

$$\nabla \times \boldsymbol{E}(\boldsymbol{r},t) = -\partial_t \boldsymbol{B}(\boldsymbol{r},t) \tag{1.15}$$

$$\nabla \cdot \boldsymbol{B}(\boldsymbol{r},t) = 0 \tag{1.16}$$

$$\nabla \times \boldsymbol{B}(\boldsymbol{r},t) = \mu \boldsymbol{J}(\boldsymbol{r},t) + \mu\varepsilon\,\partial_t \boldsymbol{E}(\boldsymbol{r},t) \tag{1.17}$$

$$\nabla \cdot \boldsymbol{E}(\boldsymbol{r},t) = \tfrac{\rho}{\varepsilon} \tag{1.18}$$

where $\boldsymbol{J} \equiv \sigma \boldsymbol{E}$. That is, the entire theory of classical electromagnetism involves finding the solution $(\boldsymbol{E}, \boldsymbol{B})$ to Maxwell's equations while satisfying the appropriate boundary conditions.

As a side comment, define the *magnetic field intensity* $\boldsymbol{H}$ by $\boldsymbol{B} = \mu \boldsymbol{H}$ and the *electric displacement* by $\boldsymbol{D} = \varepsilon \boldsymbol{E}$. Then, the set of Maxwell's equations is often presented as

$$\nabla \times \boldsymbol{E}(\boldsymbol{r},t) = -\partial_t \boldsymbol{B}(\boldsymbol{r},t) \tag{1.15$'$}$$

$$\nabla \times \boldsymbol{H}(\boldsymbol{r},t) = \boldsymbol{J}(\boldsymbol{r},t) + \partial_t \boldsymbol{D}(\boldsymbol{r},t) \tag{1.16$'$}$$

$$\nabla \cdot \boldsymbol{D}(\boldsymbol{r},t) = \rho \tag{1.17$'$}$$

$$\nabla \cdot \boldsymbol{B}(\boldsymbol{r},t) = 0 \tag{1.18$'$}$$

The reason for presenting Maxwell's equations as (1.15)–(1.18) is elaborated in Remark 1.3.2.

Here, the medium is assumed to be homogeneous and isotropic; that is, μ,σ,ε are independent of the *x*-, *y*-, *z*-variables and direction. For simplicity, the constants μ,σ,ε herein are also assumed to be independent of frequency and temperature. The frequency dependency of ε and σ is fully explored via the Kramers-Kronig relations in Chapter 9.

Equation (1.15) is known as *Faraday's law*; it shows how a time-varying magnetic field couples with an electric field. Intuitively, the time variation of the magnetic field generates the spatial variation in the electric field. Strictly speaking, this is not a technically precise statement (*cf.* Section 8.1 for the correct interpretation). Equation (1.16) shows how a time-varying electric field couples with a magnetic field. Intuitively, the time variation of the electric field generates the spatial variation of the magnetic field.

1.3.1 Theorem

Suppose the pair $(\boldsymbol{E},\boldsymbol{B})$ is a solution to Maxwell's equations on a simply connected, open subset $\Omega \subseteq \mathbf{R}^3$ satisfying the appropriate boundary conditions on $\partial\Omega$. Then, the most general expression for the electric field is given by $\boldsymbol{E} = -\nabla\varphi - \partial_t \boldsymbol{A}$, where $\boldsymbol{A}$ satisfies $\boldsymbol{B} = \nabla \times \boldsymbol{A}$ and φ is some conservative scalar potential field on Ω. □

Proof

From (1.15), $\nabla \times \boldsymbol{E} = -\partial_t \nabla \times \boldsymbol{A} \Rightarrow 0 = \nabla \times (\boldsymbol{E} + \partial_t \boldsymbol{A})$. Set $\tilde{\boldsymbol{E}} = \boldsymbol{E} + \partial_t \boldsymbol{A}$. Then, Ω is simply connected implies $\exists \varphi \in C^2(\Omega)$ such that $\tilde{\boldsymbol{E}} = -\nabla\varphi$ on Ω. Whence, $\boldsymbol{E} = -\nabla\varphi - \partial_t \boldsymbol{A}$, as required. □

It is obvious from Theorem 1.3.1 that the general expression for an electric field comprises a static part and a time-varying part. In particular, the vector potential contributes to the time variation of the electric field and the scalar potential contributes to the conservative aspect of the electric field.

1.3.2 Corollary

Let $(\boldsymbol{E},\boldsymbol{B})$ be a solution of Maxwell's equations. Then, $\partial_t \boldsymbol{E}(\boldsymbol{r},t) = 0 = \partial_t \boldsymbol{B}(\boldsymbol{r},t)$ if and only if $\nabla \times \boldsymbol{E}(\boldsymbol{r},t) = 0$ and $\nabla \times \boldsymbol{B}(\boldsymbol{r},t) = \mu \boldsymbol{J}(\boldsymbol{r},t)$.

Proof

Suppose $\partial_t \boldsymbol{E} = 0 = \partial_t \boldsymbol{B}$. Then, $\partial_t \boldsymbol{B} = 0 \Rightarrow \partial_t \boldsymbol{A} = 0$ as $\boldsymbol{B} = \nabla \times \boldsymbol{A}$ by definition, whence, Equation (1.15) reduces to $\nabla \times \boldsymbol{E}(\boldsymbol{r},t) = 0$, and from $\partial_t \boldsymbol{E} = 0 \Rightarrow \partial_t \boldsymbol{D} = 0$, it follows at once that Equation (1.17) reduces to $\nabla \times \boldsymbol{B}(\boldsymbol{r},t) = \mu \boldsymbol{J}(\boldsymbol{r},t)$. Conversely, $\nabla \times \boldsymbol{E} = 0 \Rightarrow \partial_t \boldsymbol{A} = 0$ from Theorem 1.3.1, and hence, $\partial_t \boldsymbol{B}(\boldsymbol{r},t) = 0$. Likewise, $\nabla \times \boldsymbol{B}(\boldsymbol{r},t) = \mu \boldsymbol{J}(\boldsymbol{r},t) \Rightarrow \partial_t \boldsymbol{E} = 0$, as asserted. □

Thus, the static condition is precisely equivalent to the case wherein the electric field and magnetic field are decoupled. In particular, the definition of electrostatics defined in Section 1.1 by the *irrotational* criterion $\nabla \times \boldsymbol{E}(\boldsymbol{r},t) = 0$ is completely justified in view of Corollary 1.3.2.

Furthermore, it ought to be pointed out that the pair $(\varphi, \boldsymbol{A})$ defining $(\boldsymbol{E},\boldsymbol{B})$ is not unique. Indeed, it is unique up to some arbitrary, differentiable, function $f = f(\boldsymbol{r},t)$ as follows: $\varphi \to \varphi' = \varphi + \partial_t f$ and $\boldsymbol{A} \to \boldsymbol{A}' = \boldsymbol{A} - \nabla f$. This can be easily seen via direct substitution:

$$\boldsymbol{E}' = -\nabla\varphi' - \partial_t \boldsymbol{A}' = -\nabla\varphi - \partial_t \nabla f - \partial_t \boldsymbol{A} + \partial_t \nabla f = -\nabla\varphi - \partial_t \boldsymbol{A} = \boldsymbol{E}$$

$$\boldsymbol{B}' = \nabla \times \boldsymbol{A}' = \nabla \times \boldsymbol{A} - \nabla \times \nabla f = \nabla \times \boldsymbol{A} = \boldsymbol{B}$$

as $\nabla \times \nabla f = 0$ on $\mathbf{R}^3$ for any twice differentiable function f on $\mathbf{R}^3$. This property is known as *gauge transformation* and Maxwell's theory is said to be *invariant* under gauge transformation. In particular, fixing the choice of $(\varphi, \boldsymbol{A})$ is called *gauge fixing*. Fixing the choice of gauge will in no way affect the physical significance of the problem. Thus, there is freedom to choose a gauge to simplify solving Maxwell's equations.

1.3.3 Remark

Equations (1.15) and (1.16) are called the first pair of Maxwell's equations, and (1.17) and (1.18) are called the second pair of Maxwell's equations. This is because in the language of differential geometry, the pair (1.15) and (1.16) can be expressed as a single equation, and the second pair (1.17) and (1.18) can be expressed as the dual of the former equation. For those readers interested in understanding the geometric formalism of Maxwell's theory, consult References [1,3,8,9].

The first term of Equation (1.16) is the *displacement* term governed by bound charges in the medium; it gives rise to the phenomenon of the displacement current present in capacitors. The second term, as mentioned in Section 1.2, is the conduction term: $\boldsymbol{J}(\boldsymbol{r},t) = \sigma \boldsymbol{E}(\boldsymbol{r},t)$ is the current density flowing across the medium surrounding the conductors, and it may also include a convection term; see Equation (1.12). If the medium has zero conductivity, that is, $\sigma = 0$, then Equation (1.16) is dominated purely by the displacement current term.

Equation (1.17) is *Gauss' law*. It states that the electric field is generated by the presence of a charge density ρ. In contrast, Equation (1.18) states that the magnetic field is not generated by magnetic charges: there are no magnetic monopoles.* This equation is based purely upon the observation that magnetic monopoles have never been found in nature.

1.3.4 Theorem (Charge Conservation)

The electric charge is conserved in $\mathbf{R}^3$. That is,

$$\nabla \cdot \boldsymbol{J} + \partial_t \rho = 0 \tag{1.19}$$

Proof

From (1.16), let Σ be the boundary of some compact volume M in $\mathbf{R}^3$. Then, by Corollary A.1.2,

$$0 = \oiint_{\Sigma} \nabla \times \boldsymbol{H} \cdot \boldsymbol{n} dS = \varepsilon \partial_t \oiint_{\Sigma} \boldsymbol{E} \cdot \boldsymbol{n} dS + \sigma \oiint_{\Sigma} \boldsymbol{E} \cdot \boldsymbol{n} dS$$

* This is a rather lively topic that is open to much theoretical debate: once again, its existence depends largely upon the topological structure of space–time (*cf.*, e.g., References [1,3,8,9]). From a mathematical standpoint, magnetic monopoles exist in space–times that have holes. An even deeper consequence for the existence of magnetic monopoles is that electric charges must be quantized (i.e., discrete)!

Because, via the divergence theorem,

$$\oiint_{\Sigma} \boldsymbol{E} \cdot \boldsymbol{n} dS = \oiiint_{M} \nabla \cdot \boldsymbol{E} dV = \tfrac{1}{\varepsilon} \oiiint_{M} \rho dV$$

it follows immediately that

$$0 = \partial_t \oiiint_{M} \rho dV + \sigma \oiiint_{M} \nabla \cdot \boldsymbol{E} dV \equiv \partial_t \oiiint_{M} \rho dV + \oiiint_{M} \nabla \cdot \boldsymbol{J} dV,$$

whence $\nabla \cdot \boldsymbol{J} + \partial_t \rho = 0$ as claimed. □

Equation (1.19) is known as the *continuity equation* for electric charges. It shows that the electric charge is conserved in the following manner. Because the divergence $\nabla \cdot \boldsymbol{J}$ of the current density $\boldsymbol{J}$ denotes the net loss of charges, if there is a zero net loss of charge within a compact volume M, that is, $\nabla \cdot \boldsymbol{J} = 0$, then by (1.19), the time variation of the charge density $\partial_t \rho = 0$. That is, the electric charge is conserved because $\partial_t \rho = 0$ implies that ρ is constant in time.

Conversely, observe that Theorem 1.3.1 implies that $\nabla \cdot \boldsymbol{J} = 0$ is precisely the condition for steady-state current. To see this, it suffices to note that $\partial_t \rho = 0$ implies that the charge remains constant in time, and hence, the current density cannot vary with time. Indeed, a consequence of the continuity equation is the following. From Equation (1.18), $\nabla \cdot \boldsymbol{J}$ can be expressed as

$$\nabla \cdot \boldsymbol{J} = \nabla \cdot \sigma \boldsymbol{E} = \tfrac{\sigma}{\varepsilon} \rho$$

because the electric conductivity σ is assumed constant. Hence, from Equation (1.19),

$$\tfrac{\sigma}{\varepsilon} \rho + \partial_t \rho = 0 \Rightarrow \rho = \rho_0 e^{-(\sigma/\varepsilon)t}$$

where ρ_0 is the initial charge density within a compact volume M.

Notice the exponential decay of charges with respect to time within the volume M. This means in particular that any charges placed within a conductor will quickly diffuse onto the surface (i.e., the boundary) of the conductor in accordance with the exponential decay rate given by

$$\rho(\boldsymbol{r}, t) = \rho_0(\boldsymbol{r}, t) e^{-(\sigma/\varepsilon)t} \tag{1.20}$$

Physically, the charges are distributed on the surface such that the resultant electric field within the conductor is zero.

Two consequences are discernable from the continuity equation.

(a) For a perfect dielectric where the conductivity $\sigma = 0$, $\rho = \rho_0$.

That is, the charges will remain where they are; they are bound charges. On the other hand, if no charges are added to the dielectric, that is, the dielectric remains uncharged, then an electric field will polarize the molecules into electric dipoles.

(b) For a perfect conductor, that is, $\sigma \to \infty$, the charges diffuse very rapidly onto the surface of the conductor such that the resultant field within the conductor is zero.

That is, the charges are mobile. Another way of seeing this is as follows. By definition, the current density within the conductor (or any medium) is given by $\boldsymbol{J} = \sigma \boldsymbol{E}$. Inasmuch as physically, $\|\boldsymbol{J}\| < \infty$, it follows that $\boldsymbol{E} \to 0$ in the limit as $\sigma \to \infty$ such that the current density will continue to remain finite. This fact gives rise to the statement that a perfect conductor has a zero electric field inside it. There is a zero electric field inside a perfect conductor, therefore it follows that all the charges must reside on the surface of a perfect conductor and be distributed in such a manner that the fields cancel within the conductor (viz., $\boldsymbol{E} \equiv 0$).

By appealing to Faraday's law, it can also be shown that the *magnetic flux* $\Psi = \int_\Sigma \boldsymbol{B} \cdot dx^2$ is conserved—that is the reason why $\boldsymbol{B}$ is called the magnetic flux density. That is, the sum of the magnetic flux entering and exiting a compact surface is zero. This in turn implies that magnetic flux forms closed loops: i.e., no magnetic monopoles.

1.4 Electromagnetic Waves

First, consider a stationary electric charge in space. It was shown above that $\nabla \times E = 0$ defines an *electrostatic field* because of the absence of a time-varying field. Consequently, Maxwell's equations reduce to:

$$\nabla \times \boldsymbol{E} = 0 \tag{1.21}$$

$$\nabla \cdot \boldsymbol{B} = 0 \tag{1.22}$$

$$\nabla \times \boldsymbol{B} = \mu \boldsymbol{J} \tag{1.23}$$

$$\nabla \cdot \boldsymbol{E} = \tfrac{\rho}{\varepsilon} \tag{1.24}$$

where, for convenience, the boundary conditions are left unspecified for the moment.

To recapitulate Section 1.3, Equation (1.21) implies that $\partial_t \boldsymbol{H} = 0$; that is, the electric field does not generate a time-varying magnetic field. The electric field is decoupled from the magnetic field. Similarly, from Equation (1.22), the magnetic field is decoupled from a time-varying electric field: the magnetic field does not generate a time-varying electric field.

From Equation (1.23), if the electric charges are stationary, then $\mathbf{J} = 0$ and hence, $\nabla \times \boldsymbol{B} = 0$. However, this does not imply that $\boldsymbol{B} = 0$. What makes $\boldsymbol{B} = 0$ is Equation (1.22): the absence of a magnetic charge. Hence, in this universe, $\nabla \times \boldsymbol{B} = 0$ implies that $\boldsymbol{B} = 0$ due to the absence of magnetic monopoles.

Recall that the full Maxwell's equations state that a time-varying electric (magnetic) field generates a time-varying magnetic (electric) field, which in turn generates an electric (magnetic) field and so forth. From this standpoint, it is intuitively clear how an electromagnetic wave can propagate in free space: it is self-sustaining, wherein each field generates the other field in space–time, provided both the fields are time-varying. Hence, it is once again intuitively clear that in the case of static fields, there can be no propagation of electromagnetic waves.

To study the existence of electromagnetic waves, consider the full Maxwell's equations, where, for simplicity, the homogeneous and isotropic medium is assumed to be charge-free. As always, in the absence of any particular symmetry, rectangular coordinates are used for simplicity. Finally, for convenience, it is tacitly assumed that ε and μ are scalars.

From Equation (1.15), where $\{\mathbf{e}_x, \mathbf{e}_y, \mathbf{e}_z\}$ denotes the standard basis in rectangular coordinates,

$$\nabla \times \boldsymbol{E} = \begin{vmatrix} \mathbf{e}_x & \mathbf{e}_y & \mathbf{e}_z \\ \partial_x & \partial_y & \partial_z \\ E_x & E_y & E_z \end{vmatrix} = \begin{pmatrix} \partial_y E_z - \partial_z E_y \\ -\partial_x E_z + \partial_z E_x \\ \partial_x E_y - \partial_y E_x \end{pmatrix} = -\begin{pmatrix} \partial_t B_x \\ \partial_t B_y \\ \partial_t B_z \end{pmatrix} \tag{1.25}$$

From Equation (1.17),

$$\nabla \times \boldsymbol{B} = \begin{vmatrix} \mathbf{e}_x & \mathbf{e}_y & \mathbf{e}_z \\ \partial_x & \partial_y & \partial_z \\ B_x & B_y & B_z \end{vmatrix} = \begin{pmatrix} \partial_y B_z - \partial_z B_y \\ -\partial_x B_z + \partial_z B_x \\ \partial_x B_y - \partial_y B_x \end{pmatrix} = \begin{pmatrix} \varepsilon \partial_t E_x + \sigma E_x \\ \varepsilon \partial_t E_y + \sigma E_y \\ \varepsilon \partial_t E_z + \sigma E_z \end{pmatrix} \tag{1.26}$$

Because the wave is propagating in a charge-free space, Equation (1.18) becomes $\nabla \cdot \boldsymbol{E} = 0$. This leads to

$$\partial_x E_x + \partial_y E_y + \partial_z E_z = 0 \tag{1.27}$$

Likewise, Equation (1.18) leads to $\partial_x B_x + \partial_y B_y + \partial_z B_z = 0$.

Recall that a wave ψ defined in $\mathbf{R}^4$ satisfies the *wave equation*:

$$\Delta\psi = \alpha \partial_t^2 \psi + \beta \partial_t \psi \tag{1.28}$$

where $\frac{1}{\sqrt{\alpha}}$ is the speed of the propagating wave and β is the loss present in the medium. Some comments regarding Equation (1.28) are now due.

- If the medium is lossless, $\beta = 0$, and the wave equation reduces to

$$\Delta\psi = \alpha\,\partial_t^2\,\psi$$

- If $\left|\alpha\,\partial_t^2\,\psi\right| << \left|\beta\,\partial_t\,\psi\right|$ in $\mathbf{R}^4$, then Equation (1.28) reduces to

$$\Delta\psi \approx \beta\,\partial_t\,\psi$$

This is precisely the *diffusion equation,* also known as the *heat equation.* Under this condition, there are no waves propagating in the medium. This equation can also be solved using separation of variables if the configuration possesses nice symmetries. If the medium is very lossy, that is, $|\beta| >> |\alpha|$ such that $\left|\alpha\,\partial_t^2\,\psi\right| << \left|\beta\,\partial_t\,\psi\right|$ is satisfied, then no waves will propagate through the medium.

In fact, the astute reader will quickly observe that because $\left|\partial_t^2\,\psi\right| < \infty$ holds physically for all times, it follows that $\left|\alpha\,\partial_t^2\,\psi\right| << \left|\beta\,\partial_t\,\psi\right|$ is satisfied for any $|\beta| < \infty$ if $\alpha \to 0$. This means that the effect of any changes made to the wave will spread almost instantaneously throughout the entire space. This superficially appears to be in violation of causality; food for thought here!

The wave equation for the z-component is worked out explicitly below. The derivation for the other two components as well as for the magnetic field follows *mutatis mutandis.* To begin, differentiating the first line of Equation (1.25) with respect to y and using the last row of (1.26) yield:

$$\partial_y^2 E_z - \partial_y\,\partial_z E_y = -\mu\,\partial_t\,\partial_y H_x = \mu\,\partial_t(\varepsilon\,\partial_t E_z + \sigma E_z - \partial_x H_y)$$

Next, differentiating the second line of Equation (1.26) with respect to x and using the last row of (1.26) yield:

$$-\partial_x^2 E_z + \partial_x\,\partial_z E_x = -\mu\,\partial_t\,\partial_x H_y = -\mu\,\partial_t(-\varepsilon\,\partial_t E_z - \sigma E_z + \partial_y H_x)$$

From Equation (1.27),

$$\partial_z E_z = -\partial_x E_x - \partial_y E_y$$

Taking the difference of the first two equations yields:

$$\partial_x^2 E_z + \partial_y^2 E_z - \partial_z(\partial_x E_x + \partial_y E_y) = -\partial_x H_y + \partial_y H_x$$

Using (1.27) and the last row of (1.26) yields:

$$\partial_x^2 E_z + \partial_y^2 E_z + \partial_z^2 E_z = \mu\varepsilon\,\partial_t^2 E_z + \mu\sigma\,\partial_t E_z$$

This expression is precisely the wave equation given by (1.28):

$$\Delta E_z = \mu\varepsilon\,\partial_t^2 E_z + \mu\sigma\,\partial_t E_z$$

Going through the same procedure yields the identical form for the other components of the electric field and the magnetic field. Explicitly, the fields are defined by the Helmholtz equations:

$$-\Delta \boldsymbol{E} + \mu\varepsilon\,\partial_t^2 \boldsymbol{E} + \mu\sigma\,\partial_t \boldsymbol{E} = 0 \tag{1.29}$$

$$-\Delta \boldsymbol{B} + \mu\varepsilon\,\partial_t^2 \boldsymbol{B} + \mu\sigma\,\partial_t \boldsymbol{B} = 0 \tag{1.30}$$

From Equation (1.28), it is evident that electromagnetic waves propagate in any medium at the speed $v = \frac{1}{\sqrt{\mu\varepsilon}}$, where ε, μ depend upon the medium. Diffusion through the medium is related directly to its conductivity and permeability: $\mu\sigma$. In particular, for a perfect—that is, lossless—dielectric, $\sigma \equiv 0$, and Equation (1.29) reduces in a lossless medium to the wave equation:

$$-\Delta \boldsymbol{E} + \mu\varepsilon\,\partial_t^2 \boldsymbol{E} = 0. \tag{1.31}$$

For a medium with a very high conductivity, (1.29) reduces to the diffusion equation. Consequently, electromagnetic waves cannot propagate inside a good conductor. They will diffuse in a good conductor the way heat diffuses through a medium. Technically, the electric field will still propagate inside a good conductor as waves; however, the waves will be severely attenuated. This, in turn, may be approximated by the diffusion equation (*cf.* the discussion carried out above). All the comments just made about the electric field also apply to the time-varying magnetic field. In summary, Maxwell's equations predict the existence of electromagnetic waves.

Transverse electric and magnetic (TEM) waves possess a number of interesting properties that are outlined below. TEM waves are electromagnetic waves with no longitudinal (also called *axial*) field components. That is, if the direction of propagation is along the z-axis, then the z-components of both the electric and magnetic fields are identically zero. In all that follows, it is assumed that the waves are propagating along the z-axis.

For TEM waves, Equation (1.25) reduces to (1.30)

$$\nabla \times \boldsymbol{E} = \begin{vmatrix} \mathbf{e}_x & \mathbf{e}_y & \mathbf{e}_z \\ \partial_x & \partial_y & \partial_z \\ E_x & E_y & 0 \end{vmatrix} = \begin{pmatrix} -\partial_z E_y \\ \partial_z E_x \\ \partial_x E_y - \partial_y E_x \end{pmatrix} = \begin{pmatrix} -\partial_t B_x \\ -\partial_t B_y \\ 0 \end{pmatrix}$$

and (1.26) reduces to (1.31)

$$\nabla\times\boldsymbol{B}=\begin{vmatrix}\mathbf{e}_x & \mathbf{e}_y & \mathbf{e}_z\\ \partial_x & \partial_y & \partial_z\\ B_x & B_y & 0\end{vmatrix}=\begin{pmatrix}-\partial_z B_y\\ \partial_z B_x\\ \partial_x B_y-\partial_y B_x\end{pmatrix}=\mu\begin{pmatrix}\varepsilon\partial_t E_x+\sigma E_x\\ \varepsilon\partial_t E_y+\sigma E_y\\ 0\end{pmatrix}$$

First, consider the last row in Equation (1.31): $\partial_x E_y-\partial_y E_x=0$. Observe that this can be expressed in the form:

$$\nabla_\perp\times\boldsymbol{E}_\perp\equiv\begin{vmatrix}\mathbf{e}_x & \mathbf{e}_y & \mathbf{e}_z\\ \partial_x & \partial_y & 0\\ E_x & E_y & 0\end{vmatrix}=\begin{pmatrix}0\\ 0\\ \partial_x E_y-\partial_y E_x\end{pmatrix}=0 \tag{1.32}$$

The notation $\nabla_\perp\equiv(\partial_x,\partial_y,0)$ defines the transverse (i.e., normal to the direction of propagation) component of ∇; likewise, $\boldsymbol{E}_\perp\equiv(E_x,E_y,0)$ defines the transverse electric field components. In general, $\nabla=\nabla_\perp+\boldsymbol{e}_z\partial_z$ and $\boldsymbol{E}=\boldsymbol{E}_\perp+\boldsymbol{e}_z E_z$ (in rectangular or cylindrical coordinates; *cf.* Appendix A.1).

There are a number of implications to be drawn from Equation (1.32). The first is the following. Recall that $\nabla_\perp\times\boldsymbol{E}_\perp=0$ implies the existence of some potential function φ such that $\boldsymbol{E}_\perp=-\nabla_\perp\varphi$. That is, from (1.32), the transverse components of a TEM wave are static! Indeed, from Equations (1.30) and (1.32), it is clear that $\partial_t B_x=0=\partial_t B_y$: this is precisely the static condition. Secondly, can a single conductor (such as a waveguide cavity) sustain TEM wave propagation? This clearly has implications in the design of high-speed digital circuits.

1.4.1 Theorem

Let $\Omega\subset\mathbf{R}^3$ be a perfect conductor that is compact and connected. Then Ω cannot support a TEM wave propagation.

Proof

By (1.32), $\exists\varphi$ on Ω such that $\Delta_\perp\varphi=0$ is subject to $\varphi\,|\,\partial\Omega=\varphi_0$, for some constant φ_0. Because φ is analytic on Ω, by the maximum modulus principle [5], its minimum and maximum must lie on $\partial\Omega$. However, $\varphi=\varphi_0$ is constant on $\partial\Omega$; hence, φ must be constant on Ω. Thus, as it is a perfect conductor Ω cannot sustain a TEM wave, as claimed. □

1.4.2 Corollary

Let $\Omega=S\times\mathbf{R}$ be a perfect conductor such that $S\subset\mathbf{R}^2$ is compact and connected. Then, a TEM wave cannot propagate along Ω.

Proof

Suppose Ω can support a TEM wave. Then, the restriction to any connected compact subset $\Omega' \subset \Omega$ must also support a TEM wave, yielding a contradiction by Theorem 1.4.1. □

Thus, at least two conductors are required to support a TEM wave. In particular, a TEM wave incident on a waveguide will be transformed into a transverse electric (TE) or a transverse magnetic (TM) wave as it propagates along the waveguide. Note that a TEM wave can only exist along a perfect electrical conductor: $\sigma \to \infty$. When $\sigma < \infty$, there exists a small longitudinal component of electric field $\mathbf{e}_z E_z$ near the surface of the conductor in order to overcome the ohmic loss resulting from the finite conductivity. Hence, good conductors can at best sustain an approximate TEM wave known as a *quasi-TEM* wave. Only perfect conductors can sustain TEM waves.

Returning to Equation (1.30), differentiating the first two rows with respect to z and using the first two rows of (1.31), the following wave equation ensues,

$$\partial_z^2 \boldsymbol{E}_\perp = \mu\varepsilon \partial_t^2 \boldsymbol{E}_\perp + \mu\sigma \partial_t \boldsymbol{E}_\perp \tag{1.33}$$

To solve this equation, try the following solution: $\psi = A(x,y)e^{-\gamma z}e^{i\omega t}$, where $\omega = 2\pi f$ is the angular frequency of the propagating wave, with f (in Hertz) being the frequency of propagation. Substituting into (1.33) yields

$$\gamma^2 A(x,y)e^{-\gamma z}e^{i\omega t} = -\omega^2\mu\varepsilon A(x,y)e^{-\gamma z}e^{i\omega t} + i\omega\mu\sigma A(x,y)e^{-\gamma z}e^{i\omega t}$$

Verify this as a simple exercise (see Exercise 1.5.2). Hence, this immediately leads to the following expression:

$$\gamma^2 = -\omega^2\mu\varepsilon + i\omega\mu\sigma \tag{1.34}$$

Set $\gamma = \alpha + i\beta$. Then, the evaluation of α, β is given as follows. First, observe that

$$\alpha^2 - \beta^2 + i2\alpha\beta = \gamma^2 = -\omega^2\mu\varepsilon + i\omega\mu\sigma$$

implies:

$$\alpha^2 - \beta^2 = -\omega^2\mu\varepsilon$$

$$2\alpha\beta = \omega\mu\sigma$$

Next, noting that $(\alpha^2+\beta^2)^2=(\alpha^2-\beta^2)^2+4\alpha^2\beta^2=(\omega^2\mu\varepsilon)^2+(\omega\mu\sigma)^2$, it is thus clear that $\alpha^2+\beta^2=\sqrt{(\omega^2\mu\varepsilon)^2+(\omega\mu\sigma)^2}$. Finally, adding and subtracting $\alpha^2+\beta^2$ and $\alpha^2-\beta^2$ from each other yield

$$\alpha=\tfrac{1}{\sqrt{2}}\left(\sqrt{(\omega^2\mu\varepsilon)^2+(\omega\mu\sigma)^2}-\omega^2\mu\varepsilon\right)^{\frac{1}{2}}=\omega\sqrt{\tfrac{\mu\varepsilon}{2}}\left(\sqrt{1+\left(\tfrac{\sigma}{\omega\varepsilon}\right)^2}-1\right)^{\frac{1}{2}}$$

and

$$\beta=\tfrac{1}{\sqrt{2}}\left(\sqrt{(\omega^2\mu\varepsilon)^2+(\omega\mu\sigma)^2}+\omega^2\mu\varepsilon\right)^{\frac{1}{2}}=\omega\sqrt{\tfrac{\mu\varepsilon}{2}}\left(\sqrt{1+\left(\tfrac{\sigma}{\omega\varepsilon}\right)^2}+1\right)^{\frac{1}{2}}$$

whence,

$$\gamma=\tfrac{1}{\sqrt{2}}\omega\sqrt{\mu\varepsilon}\left(\sqrt{1+\left(\tfrac{\sigma}{\omega\varepsilon}\right)^2}-1\right)^{\frac{1}{2}}+i\tfrac{1}{\sqrt{2}}\omega\sqrt{\mu\varepsilon}\left(\sqrt{1+\left(\tfrac{\sigma}{\omega\varepsilon}\right)^2}+1\right)^{\frac{1}{2}} \tag{1.35}$$

The solution of (1.33) is thus

$$\psi(x,y,z,t)=A(x,y)e^{-\frac{1}{\sqrt{2}}\omega\sqrt{\mu\varepsilon}\left(\sqrt{1+\left(\frac{\sigma}{\omega\varepsilon}\right)^2}-1\right)^{\frac{1}{2}}z}e^{i\left\{\omega t-\frac{1}{\sqrt{2}}\omega\sqrt{\mu\varepsilon}\left(\sqrt{1+\left(\frac{\sigma}{\omega\varepsilon}\right)^2}+1\right)^{\frac{1}{2}}z\right\}} \tag{1.36}$$

for a wave propagating in the +z-direction.

From $\gamma=\alpha+i\beta$ and Equation (1.36), it is clear from (1.35) that α corresponds to the loss associated with the conductivity of the medium. This is because $e^{-\alpha z}$ is real and $\alpha\geq 0$ implies that it decays exponentially along the direction of propagation. On the other hand, $e^{-i\beta z}$ is the oscillating (sinusoidal) portion of the wave. This is due to the presence of the imaginary factor $i=\sqrt{-1}$. To see this, it will suffice to recall Euler's rule: $e^{i\theta}=\cos\theta+i\sin\theta$. Thus, Equation (1.36) corresponds to a sinusoidal wave propagating along the *z*-direction whose amplitude decays exponentially. In particular, the negative sign in front of $i\beta z$ indicates that the wave is propagating forward along the +z-direction.

As an aside, recall that the one-dimensional wave equation—that is, one that involves only one space variable—differs from all other partial differential equations in the sense that it can be solved without imposing any boundary conditions. However, there is a price to pay: the partial differential equation has infinitely many solutions.

Explicitly, given a one-dimensional wave equation $\partial_z^2\psi=\mu\varepsilon\partial_t^2\psi$ in vacuum, any solution of the form

$$\psi=f\left(z-\tfrac{1}{\sqrt{\mu\varepsilon}}t\right)+g\left(z+\tfrac{1}{\sqrt{\mu\varepsilon}}t\right)$$

where f, g are any functions that are twice differentiable with respect to z and t, is a solution of the wave equation, as can be easily verified by direct substitution (*cf.* Exercise 1.5.3). This is known as the *D'Alembert wave solution* [4,6].

Returning to Equation (1.36), can the form of $\boldsymbol{A}$ be determined further? First, recall from (1.33) that $\boldsymbol{A} = \boldsymbol{A}(x,y)$ must be a solution satisfying $\boldsymbol{E}_\perp = -\nabla_\perp \varphi$. Because the wave is propagating in a charge-free region, $\nabla_\perp \cdot \boldsymbol{E}_\perp = 0 \Rightarrow \nabla_\perp^2 \varphi = 0$, which is Laplace's equation. Hence, the final form for $\boldsymbol{A} = \boldsymbol{A}(x,y)$ necessarily depends on the boundary conditions imposed on $\nabla_\perp^2 \varphi = 0$. Show that this is indeed the case; see Exercise 1.5.4.

In summary, a TEM wave in some domain $M \times \mathbf{R} \subset \mathbf{R}^4$ has the solution:

$$\boldsymbol{E}(x,y,z,t) = \boldsymbol{E}_\perp(x,y) e^{-\frac{1}{\sqrt{2}}\omega\sqrt{\mu\varepsilon}\left(\sqrt{1+\left(\frac{\sigma}{\omega\varepsilon}\right)^2}-1\right)^{-\frac{1}{2}} z} e^{i\left\{\omega t - \frac{1}{\sqrt{2}}\omega\sqrt{\mu\varepsilon}\left(\sqrt{1+\left(\frac{\sigma}{\omega\varepsilon}\right)^2}+1\right)^{\frac{1}{2}} z\right\}} \tag{1.37}$$

where $\boldsymbol{E}_\perp(x,y) = -\nabla\varphi(x,y)$ for some twice-differentiable function $\varphi \in C^2(M) \cap C(\bar{M})$ that depends upon the boundary conditions on ∂M wherein the wave is propagating. Indeed, the method of images outlined in Section 1.1 can be applied to solve for $\boldsymbol{E}_\perp(x,y) = -\nabla\varphi(x,y)$. Inasmuch as Equation (1.37) is just the solution for a propagating plane wave, it follows that a TEM wave is an electromagnetic plane wave.

From (1.35), Equation (1.37) can be written more simply as

$$\boldsymbol{E}(x,y,z,t) = \boldsymbol{E}_\perp(x,y) e^{-\alpha z} e^{i(\omega t - \beta z)} \tag{1.38}$$

This is the solution for a TEM wave propagating in an unbounded medium in the +z-direction, where $\boldsymbol{E}_\perp(x,y) \to 0$ in the limit as $x \to \infty$ or $y \to \infty$. Observe that when the conductivity is zero (i.e., in a perfect dielectric) there is zero attenuation of the plane waves: $\alpha = 0$ when $\sigma = 0$. In particular, when

a) $\frac{\sigma}{\omega\varepsilon} << 1, \alpha \approx \frac{1}{\sqrt{2}}\omega\sqrt{\mu\varepsilon}\left(1+\frac{1}{2}\left(\frac{\sigma}{\omega\varepsilon}\right)^2 - 1\right)^{\frac{1}{2}} \approx \frac{1}{2}\sigma\sqrt{\frac{\mu}{\varepsilon}}$

b) $\frac{\sigma}{\omega\varepsilon} << 1, \ \beta \approx \frac{1}{\sqrt{2}}\omega\sqrt{\mu\varepsilon}\left(1+\frac{1}{2}\left(\frac{\sigma}{\omega\varepsilon}\right)^2 + 1\right)^{\frac{1}{2}} \approx \frac{1}{\sqrt{2}}\omega\sqrt{\mu\varepsilon}\,(1+1)^{\frac{1}{2}} = \omega\sqrt{\mu\varepsilon}$

Thus, when the conduction term is small, the displacement term dominates. Hence, for a good dielectric,

$$\boldsymbol{E}(x,y,z,t) \approx \boldsymbol{E}_\perp(x,y) e^{-\frac{1}{2}\sigma\sqrt{\frac{\mu}{\varepsilon}}} e^{i(\omega t - \beta z)} \approx \boldsymbol{E}_\perp(x,y) e^{i(\omega t - \beta z)}$$

In summary, a *good dielectric* is defined by the condition that $\frac{\sigma}{\omega\varepsilon} << 1$, and the waves are thus approximately plane waves. Conversely, in a good conductor, TEM waves are attenuated exponentially. In particular, no waves can propagate in a perfect conductor as $\sigma \to \infty \Rightarrow \alpha \to \infty$ So, for $\frac{\sigma}{\omega\varepsilon} >> 1$,

$$\alpha \approx \tfrac{1}{\sqrt{2}}\omega\sqrt{\mu\varepsilon}\sqrt{\tfrac{\sigma}{\omega\varepsilon}} = \sqrt{\tfrac{\mu\omega\varepsilon}{2}} \equiv \tfrac{1}{\delta} \text{ and } \beta \approx \tfrac{1}{\sqrt{2}}\omega\sqrt{\mu\varepsilon}\sqrt{\tfrac{\sigma}{\omega\varepsilon}} = \tfrac{1}{\delta}$$

The quantity $\delta = \sqrt{\frac{2}{\mu\omega\sigma}}$ is called the *skin depth* of a conductor. In other words, at a depth $z = \delta$ from the surface of a conductor, the field falls off to e^{-1} of its original magnitude.

Notice that for very good conductors, $\alpha \approx \frac{1}{\delta} \approx \beta$. Hence, for a good conductor,

$$\boldsymbol{E}(x,y,z,t) \approx \boldsymbol{E}_{\perp}(x,y)e^{-\frac{z}{\delta}}e^{i\left(\omega t-\frac{z}{\delta}\right)}$$

Thus, a *good conductor* is defined by the condition that $\frac{\sigma}{\omega\varepsilon} >> 1$ and the electromagnetic field propagates within the conductor via diffusion. It is also interesting to note that TEM waves are attenuated exponentially when the magnetic permeability of a medium is high. This follows directly from Equation (1.35):

$$\alpha = \omega\sqrt{\tfrac{\mu\varepsilon}{2}}\left(\sqrt{1+\left(\tfrac{\sigma}{\omega\varepsilon}\right)^2}-1\right)^{\frac{1}{2}} \to \infty \text{ as } \mu \to \infty$$

That is, electromagnetic fields do not propagate in a high-magnetic permeability medium. Finally, verify directly from (1.35) that for $\varepsilon >> \frac{\sigma}{\omega}$, $\alpha \approx \frac{1}{2}\sigma\sqrt{\frac{\mu}{\varepsilon}}$. In particular, $\varepsilon \to \infty \Rightarrow \alpha \to 0$. However, observe in this instance that the field will undergo very rapid oscillations as $\beta \approx \omega\sqrt{\mu\varepsilon} \to \infty$. Physically, the field becomes ill-defined and hence will cease to propagate within the medium (*cf.* the phenomenon within a good conductor wherein the field also ceases to propagate).

To complete the analysis, the magnetic field propagation is determined. For simplicity, set $\boldsymbol{E} = \boldsymbol{E}_{\perp}e^{-\gamma z}e^{i\omega t}$, where $\gamma = \alpha + i\beta$. Second, observe from Equation (1.31) that

$$\mathbf{e}_z \times (\nabla \times \boldsymbol{E}) = \begin{vmatrix} \mathbf{e}_x & \mathbf{e}_y & \mathbf{e}_z \\ 0 & 0 & 1 \\ -\partial_z E_y & \partial_z E_x & 0 \end{vmatrix} = \begin{pmatrix} -\partial_z E_x \\ -\partial_z E_y \\ 0 \end{pmatrix} = \begin{pmatrix} -\partial_t B_x \\ -\partial_t B_y \\ 0 \end{pmatrix}$$

Hence, it is clear for TEM waves that

$$\partial_t \boldsymbol{B} = -\mathbf{e}_z \times (\nabla \times \boldsymbol{E}) \tag{1.39}$$

That is, $\boldsymbol{B} = -\mathbf{e}_z \times \int (\nabla \times \boldsymbol{E})\mathrm{d}t$. This yields

$$\boldsymbol{B} = -\mathbf{e}_z \times \int (\nabla \times \boldsymbol{E})\mathrm{d}t = -\mathbf{e}_z \times \nabla \times \boldsymbol{E}_\perp \int \mathrm{e}^{-\gamma z} \mathrm{e}^{\mathrm{i}\omega t} \mathrm{d}t = \tfrac{\mathrm{i}}{\omega} \mathbf{e}_z \times \nabla \times \boldsymbol{E} \tag{1.40}$$

Explicitly,

$$\mathbf{e}_z \times (\nabla \times \boldsymbol{E}) = \begin{vmatrix} \mathbf{e}_x & \mathbf{e}_y & \mathbf{e}_z \\ 0 & 0 & 1 \\ -\partial_z E_y & \partial_z E_x & 0 \end{vmatrix} = \begin{pmatrix} -\partial_z E_x \\ -\partial_z E_y \\ 0 \end{pmatrix} = \mathbf{e}_z \times \boldsymbol{E}_\perp \partial_z \mathrm{e}^{\mathrm{i}(\omega t + \mathrm{i}\gamma z)} = -\gamma \mathbf{e}_z \times \boldsymbol{E}_\perp \mathrm{e}^{\mathrm{i}(\omega t + \mathrm{i}\gamma z)}$$

Hence,

$$\boldsymbol{B} = -\tfrac{\mathrm{i}\gamma}{\omega} \mathbf{e}_z \times \boldsymbol{E}_\perp \mathrm{e}^{\mathrm{i}(\omega t - \gamma z)} \equiv -\tfrac{\mathrm{i}\gamma}{\omega} \mathbf{e}_z \times \boldsymbol{E} \tag{1.41}$$

The magnetic field is thus completely defined in terms of the electric field. In particular, all the comments made for the electric field above also apply to the magnetic field with one important proviso: that the magnetic field be nonstatic, as must be the case for the TEM mode.

The above results are briefly summarized below.

- In a perfect dielectric, there is no attenuation for a time-varying magnetic field.
- In a perfect conductor, there is no time-varying magnetic field propagation as the electric field is zero. In particular, a time-varying magnetic field is attenuated exponentially. It is critical to note that the derivations were made with time-varying magnetic fields. This condition need not hold for static magnetic fields.
- Increasing the electric permittivity of the medium wherein the waves propagate decreases the attenuation; however, this leads to "infinite" oscillation of the field, rendering the field ill-defined.
- Last but not least, the magnetic permeability also introduces an exponential attenuation.

Furthermore, observe trivially that for a TEM wave, the electric and magnetic fields possess similar attenuation properties; although, perhaps, this fact is not surprising inasmuch as the electric field and magnetic field are part of the same physical field called the electromagnetic field.

In short, a good conductor is as effective a shield for a propagating electric field as it is for a propagating magnetic field of a TEM wave *normal* to the surface of a conductor. This follows very clearly from the fact that the transverse fields incident normally on a conductor are attenuated exponentially by the skin depth $\delta = \sqrt{\frac{2}{\mu\omega\sigma}}$.

Finally, for a TEM wave propagating in a medium (ε,μ) such that $\frac{\sigma}{\omega\varepsilon} << 1$, $\beta \approx \omega\sqrt{\mu\varepsilon}$ and in particular, $\lim_{\sigma\to 0}\gamma = i\beta$; whence, (1.41) becomes

$$\boldsymbol{B} = \tfrac{\beta}{\omega}\mathbf{e}_z \times \boldsymbol{E} = \tfrac{\mu}{\eta}\mathbf{e}_z \times \boldsymbol{E} \tag{1.42}$$

where

$$\eta = \tfrac{\omega\mu}{\beta} = \sqrt{\tfrac{\mu}{\varepsilon}}$$

This is called the *TEM wave impedance*. In a vacuum, the TEM impedance $\eta_0 = \sqrt{\frac{\mu_0}{\varepsilon_0}}$. Observe that the TEM impedance η is independent of frequency in the lossless case where $\sigma = 0$; in particular, it is real.

1.4.3 Remark

For a general dielectric medium with loss (i.e., $\sigma > 0$) the TEM impedance is given by

$$\eta = -\tfrac{\omega\mu}{i\gamma} = \tfrac{\omega\mu}{\alpha^2+\beta^2}(\beta + i\alpha)$$

In particular, for $\frac{\sigma}{\omega\varepsilon} << 1$, it can be shown (*cf.* Exercise 1.5.5) that

$$\eta \approx \sqrt{\tfrac{\mu}{\varepsilon}}\,\tfrac{1+i(\sigma/2\omega\varepsilon)}{1+(\sigma/2\omega\varepsilon)^2}$$

Finally, this chapter closes with a brief description of the power transferred by an electromagnetic wave. First, recall that given an electromagnetic field $(\boldsymbol{E},\boldsymbol{B})$ propagating in a dielectric medium, the *Poynting vector* $\boldsymbol{S} = \boldsymbol{E}\times\boldsymbol{H}$ defines the power flow per unit area, with units of W/m^2. In short, it represents the power density of an electromagnetic field. Next, recall that the energy density of the electric field is given by $\frac{1}{2}\boldsymbol{D}\cdot\boldsymbol{E}^* = \frac{1}{2}\varepsilon|\boldsymbol{E}|^2$, the energy density of the magnetic field is given by $\frac{1}{2}\boldsymbol{B}\cdot\boldsymbol{H}^* = \frac{1}{2}\mu|\boldsymbol{H}|^2$, and lastly, the ohmic loss in a dielectric medium is given by the ohmic power density $\boldsymbol{J}\cdot\boldsymbol{E}^* = \sigma|\boldsymbol{E}|^2$. Then, the following result holds.

1.4.4 Theorem (Poynting)

Suppose an electromagnetic field $(\boldsymbol{E},\boldsymbol{B})$ is propagating in $(\mathbf{R}^3,\mu,\varepsilon,\sigma)$. If $\Omega\subset\mathbf{R}^3$ is a compact domain, then

$$0 = \oint_{\partial\Omega}\boldsymbol{S}\cdot\boldsymbol{n}\,\mathrm{d}^2r + \tfrac{1}{2}\partial_t\int_\Omega\left\{\varepsilon|\boldsymbol{E}|^2 + \mu|\boldsymbol{H}|^2\right\}\mathrm{d}^3r + \int_\Omega\sigma|\boldsymbol{E}|^2\mathrm{d}^3r \qquad \square$$

The result states that the power leaving a compact surface is equal to the integral of the Poynting vector over the compact surface. This is essentially the principle of energy conservation.

1.5 Worked Problems

1.5.1 Exercise

(a) Given a path $\gamma = \gamma(t)$, $t \in [0,1]$, show that $\int_{\gamma} f(z(t))\mathrm{d}t = -\int_{\tilde{\gamma}} f(z(t))\mathrm{d}t$

(b) Establish Equation (1.9).

Solution

(a) Set $\tau = 1 - t$. Then, $\mathrm{d}\tau = -\mathrm{d}t$ and hence,

$$\int_{\tilde{\gamma}} f(z)\mathrm{d}z = \int_0^1 f(\tilde{\gamma}(t))\tfrac{\mathrm{d}}{\mathrm{d}t}\tilde{\gamma}(t)\mathrm{d}t = \int_1^0 f(\gamma(\tau))(-\dot{\gamma}(\tau))(-\mathrm{d}\tau) = -\int_0^1 f(\gamma(s))\dot{\gamma}(s)\mathrm{d}s$$

However, this is precisely $-\int_{\gamma} f(z)\mathrm{d}z$ as s is just a dummy variable.

(b) From

$$\frac{x^2+(y-y_+)^2}{x^2+(y+y_+)^2} = c^2$$

expanding $(y \pm y_+)^2 = y^2 \pm 2yy_+ + y_+^2$ yields

$$0 = x^2(1-c^2) + y^2(1-c^2) - 2yy_+(1+c^2) + y_+^2(1-c^2) \quad \Leftrightarrow$$

$$0 = x^2 + \left(y - y_+\tfrac{1+c^2}{1-c^2}\right)^2 - \left(y_+\tfrac{1+c^2}{1-c^2}\right)^2 + y_+^2 \quad \Leftrightarrow$$

$$x^2 + \left(y - y_+\tfrac{1+c^2}{1-c^2}\right)^2 = \left(y_+\tfrac{1+c^2}{1-c^2}\right)^2 - y_+^2 = y_+^2\left\{\left(\tfrac{1+c^2}{1-c^2}\right)^2 - 1\right\} = \left(y_+\tfrac{2c}{1-c^2}\right)^2$$

via

$$\left(\tfrac{1+c^2}{1-c^2}\right)^2 - 1 = \left(\tfrac{1+c^2}{1-c^2} - 1\right)\left(\tfrac{1+c^2}{1-c^2} + 1\right)$$

as required. □

1.5.2 Exercise

Show that

$$\gamma^2 A(x,y)e^{-\gamma z}e^{i\omega t} = -\omega^2\mu\varepsilon A(x,y)e^{-\gamma z}e^{i\omega t} + i\omega\mu\sigma A(x,y)e^{-\gamma z}e^{i\omega t}$$

Solution

From Equation (1.33), set $E_\perp = Ae^{-\gamma z}e^{i\omega t}$. Then, $\partial_z E_\perp = -\gamma Ae^{-\gamma z}e^{i\omega t}$ and hence, $\partial_z^2 E_\perp = \gamma^2 Ae^{-\gamma z}e^{i\omega t}$. Likewise, $\partial_t E_\perp = i\omega Ae^{-\gamma z}e^{i\omega t} \Rightarrow \partial_t^2 E_\perp = -\omega^2 Ae^{-\gamma z}e^{i\omega t}$. Substituting these into (1.33) yields the claim. □

1.5.3 Exercise

Find a general solution to the D'Alembert wave equation $\partial_z^2\psi = \mu\varepsilon\,\partial_t^2\psi$ on $\mathbf{R}^3$.

Solution

Set $\xi = z + ct$ and $\zeta = z - ct$, where $c = \frac{1}{\sqrt{\varepsilon\mu}}$. Let $\psi(z,t) = f(\xi) + g(\zeta)$, for arbitrary twice differentiable functions f, g on $\mathbf{R}^3$. Then, $\partial_z\psi = \partial_\xi f\,\partial_z\xi + \partial_\zeta g\,\partial_z\zeta$. That is, $\partial_z\psi = \partial_\xi f + \partial_\zeta g$ and hence, $\partial_z^2\psi = \partial_\xi^2 f + \partial_\zeta^2 g$. Similarly, $\partial_t\psi = \partial_\xi f\,\partial_t\xi + \partial_\zeta g\,\partial_t\zeta \Rightarrow \partial_t\psi = -c\,\partial_\xi f + c\,\partial_\zeta g \Rightarrow \partial_z^2\psi = c^2(\partial_\xi^2 f + \partial_\zeta^2 g)$; whence, $\mu\varepsilon\,\partial_t^2\psi = \partial_\xi^2 f + \partial_\zeta^2 g = \partial_z^2\psi$, as required. □

1.5.4 Exercise

For a TEM wave, show that $A = A(x,y)$ necessarily depends on the boundary conditions imposed on $\nabla_\perp^2\varphi = 0$.

Solution

From Equation (1.33), set $E = A(x,y)e^{-\gamma z}e^{i\omega t}$. Then, $E_\perp = -\nabla_\perp\varphi \Rightarrow \exists\tilde{\varphi}$ twice differentiable such that $\varphi = \tilde{\varphi}(x,y)e^{-\gamma z}e^{i\omega t}$; whence, $A = -\nabla_\perp\tilde{\varphi}$ by construction and in particular, $\Delta_\perp\varphi = 0 \Rightarrow \Delta_\perp\tilde{\varphi} = 0$. Because the unique solution $\tilde{\varphi}$ to Laplace's equation, if it should exist, depends upon the boundary conditions imposed, the assertion follows. □

1.5.5 Exercise

Establish the approximation

$$\eta \approx \sqrt{\frac{\mu}{\varepsilon}}\,\frac{1 + i(\sigma/2\omega\varepsilon)}{1 + (\sigma/2\omega\varepsilon)^2}$$

given in Remark 1.43.

Solution

For $\frac{\sigma}{\omega\varepsilon} << 1$, via (1.35), it was established above that $\alpha \approx \frac{\sigma}{2}\sqrt{\frac{\mu}{\varepsilon}}$ and $\beta \approx \omega\sqrt{\mu\varepsilon}$. Thus, from (1.43), $\alpha^2 + \beta^2 \approx \frac{1}{4}\sigma^2 \frac{\mu}{\varepsilon} + \omega^2\mu\varepsilon = \omega^2\mu\varepsilon\left\{1 + \left(\frac{\sigma}{2\omega\varepsilon}\right)^2\right\}$ and $\beta + i\alpha = \omega\sqrt{\mu\varepsilon}\left\{1 + i\frac{\sigma}{2\omega\varepsilon}\right\}$. Substituting these quantities into

$$\eta = \frac{\omega\mu}{\alpha^2 + \beta^2}(\beta + i\alpha)$$

leads to the approximation.

References

1. Baez, J. and Munian, J. 1994. *Gauge Fields, Knots and Gravity*. Singapore: World Scientific.
2. Cheng, D. 1992. *Field and Electromagnetics*. Reading, MA: Addison-Wesley.
3. Choquet-Bruhat, Y., DeWitt-Morette, C., and Dillard-Bleick, M. 1982. *Analysis, Manifolds and Physics, Part I: Basics*. Amsterdam: North-Holland.
4. Chester, C. 1971. *Techniques in Partial Differential Equations*. New York: McGraw-Hill.
5. Churchill, R. and Brown, J. 1990. *Complex Variables and Applications*. New York: McGraw-Hill.
6. Farlow, S. 1993. *Partial Differential Equations for Scientists and Engineers*. New York: Dover.
7. Lin, W. and Jin, H. 1990. Analytic solutions to the electrostatic problems of two dielectric spheres. *J. Appl. Phys.* 67(3): 1160–1166.
8. Nakahara, M. 2003. *Geometry, Topology and Physics*. Bristol, UK: IOP.
9. Nash, C. and Sen, S. 1983. *Topology and Geometry for Physicists*. New York: Academic Press.
10. Neff, Jr., H. 1981. *Basic Electromagnetic Fields*. New York: Harper & Row.
11. Plonsey, R. and Collin, R. 1961. *Principles and Applications of Electromagnetic Fields*. New York: McGraw-Hill.
12. Schröder, U. 1990. *Special Relativity*. LNP 33. Singapore: World Scientific.
13. Stratton, J. 1941. *Electromagnetic Theory*. New York: McGraw-Hill.
14. Toh, T.-C. 2011. Potential generated by rotating charged cylinders. *Prog. Electromagn. Res. B*. 33: 239–256.

2

Fourier Transform and Roll-Off Frequency

An informal survey of Fourier analysis is presented here to motivate practical applications in electromagnetic interference (EMI), even though this topic is too vast for this chapter to do it proper justice. In particular, the subject matter is presented to provide insight into the origin of harmonics, transients, and methods to suppress harmonics in high-speed digital circuits.

The chapter concludes with a pithy overview of some elementary properties of filter networks. An understanding of filter circuits clearly plays an important role in EMC mitigation and signal integrity in general. Readers who are interested in an in-depth exposition on Fourier analysis and its applications will benefit greatly from References [1,3,7,14,15]. The application of complex theory to filter theory can be found in References [12,13].

2.1 Fourier Series

Periodic functions can be expressed as an infinite sum of sine and cosine functions. Recall that a function $f:\mathbf{R}\to\mathbf{R}$ is *periodic* with *period* $T>0$ if $f(t+T)=f(t)$ for all $t\in\mathbf{R}$. The function is said to be *even* if $f(t)=f(-t)$ $\forall t\in\mathbf{R}$, and *odd* if $f(t)=-f(t)$ for all $t\in\mathbf{R}$.

Note that the concept of even and odd symmetry of a function depends upon the origin chosen. To elaborate further: consider the function $f(t)=\sin t$. Clearly, choosing the origin to be at $t=0$, $f(-t)=\sin(-t)=-\sin t=-f(t)$ $\forall t\in\mathbf{R}$. Hence, f is odd. However, suppose the origin along the t-axis were translated to $t=-\frac{\pi}{2}$; that is, $f(t)\to F(t)\equiv f(t+\frac{\pi}{2})$. Then, by definition, $F(t)=\sin(t+\frac{\pi}{2})=\cos t$ is an even function as $\cos t=\cos(-t)$ $\forall t\in\mathbf{R}$. Thus, the odd/even symmetry of a function is a relative concept and not an absolute one. By fixing the origin, the concept then becomes an absolute one.

2.1.1 Definition

Given a function $f:\mathbf{R}\to\mathbf{R}$, it is said to be of *bounded variation* on $[a,b]\subset\mathbf{R}$, if $\exists M>0$ such that $\sup_{P\in\mathcal{P}}\Sigma_P\left|f(x_i)-f(x_{i+1})\right|\le M$, where $\mathcal{P}$ is the set of all finite partition $P=\{x_i : a\le x_i<x_{i+1}\le b, i=1,\ldots,|P|\}$ of $[a,b]$, and the *cardinality* $|P|$ is the number of elements in P.

2.1.2 Theorem (Fourier)

Let $f: \mathbf{R} \to \mathbf{R}$ be a periodic function with period $T > 0$ satisfying (i) $\int_0^T |f(t)| dt < \infty$, (ii) f is piecewise continuous on $[0, T]$, and (iii) f is of bounded variation on $[0, T]$. Then, $\tilde{f}(t) = \frac{1}{2}(f(t^+) + f(t^-))$ on **R**, where

$$\tilde{f}(t) = \sum_{n=0}^{\infty} \{a_n \cos n\omega t + b_n \sin n\omega t\} \tag{2.1}$$

$\omega \equiv \frac{2\pi}{T}$ and $a_n, b_n \in \mathbf{R}$ $\forall n = 0,1,2, \ldots$, are constants called the *Fourier coefficients* of f defined by

$$a_n = \tfrac{2}{T}\int_0^T f(t)\cos n\omega t dt \quad \text{for} \quad n > 0, \tag{2.2a}$$

$$b_n = \tfrac{2}{T}\int_0^T f(t)\sin n\omega t dt \quad \text{for} \quad n > 0, \tag{2.2b}$$

$$a_0 = \tfrac{1}{T}\int_0^T f(t) dt \tag{2.2c}$$

□

2.1.3 Corollary

Suppose that $f: \mathbf{R} \to \mathbf{R}$ is an *aperiodic* function—that is, nonperiodic—such that (i) $f(t) = 0$ $\forall t \notin (a,b)$, for some $-\infty < a < b < \infty$, (ii) $\int_a^b |f| < \infty$, and (iii) f is of bounded variation on $[a, b]$. Then, the Fourier expansion defined by Equation (2.1) exists for f.

Proof

Define a periodic function $\tilde{f}: \mathbf{R} \to \mathbf{R}$ as follows. First, set $\tilde{f} = f$ on $[a, b]$. Now, observe that given $[\alpha, \beta] \subset \mathbf{R}$, there exists a homeomorphism $\tau: [\alpha, \beta] \to [a, b]$ defined by

$$\tau(t) = \tfrac{b-a}{\beta-\alpha}(t - \alpha) + a$$

called a *reparametrization*. Moreover, given any $t > b$, $\exists i > 0$ such that $b + (i-1)a \leq t \leq b + ia$, set $\alpha_+ = b + (i-1)a$ and $\beta_+ = b + ia$. Likewise, for any $t < a$, $\exists i > 0$ such that $(i+1)a - ib \leq t \leq ia - (i-1)b$. So, set $\alpha_- = (i+1)a - ib$ and $\beta_- = ia - (i-1)b$, and define

$$\tau_\pm(t) = \tfrac{b-a}{\beta_\pm - \alpha_\pm}(t - \alpha_\pm) + a$$

Thus, define

$$\tilde{f}(t)=\begin{cases} f(\tau_-(t)) \text{ for } t<a, \\ f(t) \text{ for } a\le t\le b, \\ f(\tau_+(t)) \text{ for } t>b. \end{cases}$$

Then, this is the desired extension of f to the real line. In particular, observe that the extension need not be continuous at $t=\alpha_\pm$ or $t=\beta_\pm$. Clearly if $f(a)=f(b)$, then $\tilde{f}$ is continuous on the real line. See Figure 2.1.

By construction, $\tilde{f}$ is periodic with period $T = b - a$ and it satisfies the conditions (i) and (ii) of Theorem 2.1.2. Thus, the conclusion of Theorem 2.1.2 applies. □

2.1.4 Remark

By appealing to Euler's formula, to wit, $e^{i\varphi}=\cos\phi+i\sin\phi$, (2.1) can be expressed in complex Fourier coefficients by

$$\tilde{f}(t)=\sum_{n=-\infty}^{\infty} c_n e^{in\omega t} \tag{2.3}$$

where

$$c_n=\tfrac{1}{T}\int_{-T/2}^{T/2} f(t)e^{-in\omega t}\,dt$$

for all integer n. In particular, it will be left as an easy exercise to verify that

$$c_n=\tfrac{1}{2}(a_n-ib_n),\ n>0 \tag{2.4a}$$

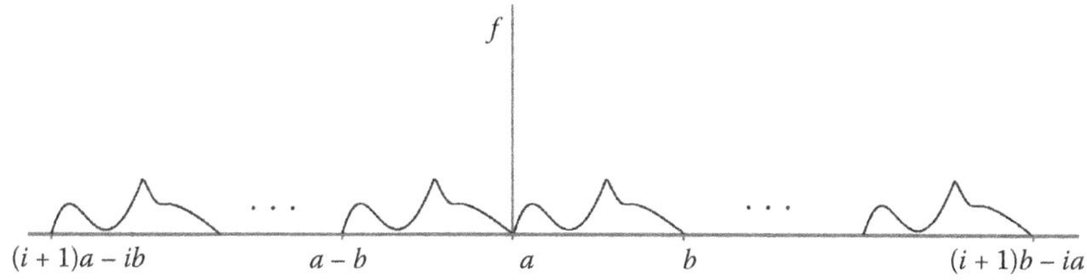

FIGURE 2.1
Extending a continuous function defined on the closed interval [a,b].

$$c_{-n} = \tfrac{1}{2}(a_n + ib_n),\ n > 0 \tag{2.4b}$$

$$c_0 = a_0 \tag{2.4c}$$

See Exercise 2.5.1 for more details.

The pair (a_n, b_n) is called the nth *harmonic of* f. Specifically, the *discrete spectrum* of f is defined by the quantity $\sqrt{a_n^2 + b_n^2}$. This then determines the magnitude of the n-harmonic of f. In the EMC world, the n-harmonic refers to the frequency $\frac{n\omega}{2\pi}$, and Equation (2.1) then yields the magnitude of the n-harmonic. That is, (2.1) determines the strength of the emissions, and if they exceed regulatory limits, knowing the harmonic provides a means to target the source of the emission.

2.1.5 Example

Consider the following square wave defined by

$$f = \begin{cases} 1 & \text{for } 0 \le t < \tfrac{1}{2}\tau, \\ 0 & \text{for } \tfrac{1}{2}\tau \le t < \tau, \end{cases}$$

where $\tau > 0$ is the period of the square wave. Then, from (2.1), and setting $\omega = \frac{2\pi}{\tau}$,

$$a_0 = \tfrac{1}{\tau}\int_0^{\frac{1}{2}\tau} dt = \tfrac{1}{2}$$

$$a_n = \tfrac{2}{\tau}\int_0^{\frac{1}{2}\tau} \cos n\omega t\, dt = 0, \quad \text{for} \quad n = 1, 2, \ldots,$$

$$b_n = \tfrac{2}{\tau}\int_0^{\frac{1}{2}\tau} \sin n\omega t\, dt = -\tfrac{1}{n\pi}(\cos n\pi - 1) = \tfrac{2}{(2n-1)\pi}, \quad \text{for} \quad n = 1, 2, \ldots.$$

That is,

$$f(t) = \tfrac{1}{2} + \sum_{n=1}^{\infty} \tfrac{2}{(2n-1)\pi} \sin(2n-1)\omega t$$

is the Fourier decomposition for a square wave. A plot of f, where $n = 200$ and period $\tau = 1$ s, is given below. The phenomenon of ringing indicated in the plot is related to discontinuity of the Fourier series; more details are provided later.

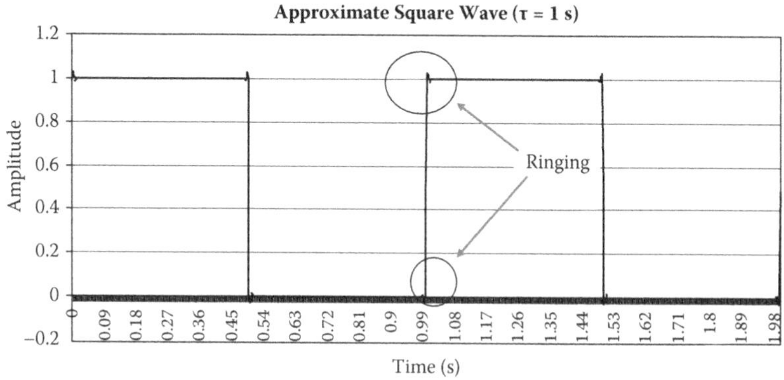

2.1.6 Example

Consider a triangular wave defined as

$$f = \begin{cases} \frac{2}{\tau}t & \text{for } 0 \le t \le \frac{1}{2}\tau, \\ -\frac{2}{\tau}t + 2 & \text{for } \frac{1}{2}\tau \le t \le \tau, \end{cases}$$

where $\tau > 0$ is the period of the triangular wave. Then, from Equation (2.1),

$$a_0 = \frac{1}{\tau}\int_0^{\frac{1}{2}\tau} \frac{2}{\tau}t\,dt - \frac{1}{\tau}\int_{\frac{1}{2}\tau}^{\tau} \frac{2}{\tau}t\,dt + \frac{2}{\tau}\int_{\frac{1}{2}\tau}^{\tau} dt = 1$$

$$a_n = \left(\frac{2}{\tau}\right)^2 \int_0^{\frac{1}{2}\tau} t\cos n\omega t\,dt + \frac{2}{\tau}\left\{-\frac{2}{\tau}\int_{\frac{1}{2}\tau}^{\tau} t\cos n\omega t\,dt + 2\int_{\frac{1}{2}\tau}^{\tau} \cos n\omega t\,dt\right\} \Rightarrow$$

$$a_n = \begin{cases} -\frac{4}{\pi^2 n^2} & \text{for } n \text{ odd,} \\ 0 & \text{for } n \text{ even.} \end{cases}$$

Likewise,

$$b_n = \left(\frac{2}{\tau}\right)^2 \int_0^{\frac{1}{2}\tau} t\sin n\omega t\,dt + \frac{2}{\tau}\left\{-\frac{2}{\tau}\int_{\frac{1}{2}\tau}^{\tau} t\sin n\omega t\,dt + 2\int_{\frac{1}{2}\tau}^{\tau} \sin n\omega t\,dt\right\} = 0\ \forall n$$

Hence,

$$f(t) = 1 - \frac{4}{\pi^2}\sum_{n=1}^{\infty} \frac{1}{(2n-1)^2}\cos(2n-1)\omega t$$

The plot of f is obtained by setting $n = 200$ and $\tau = 1$ s.

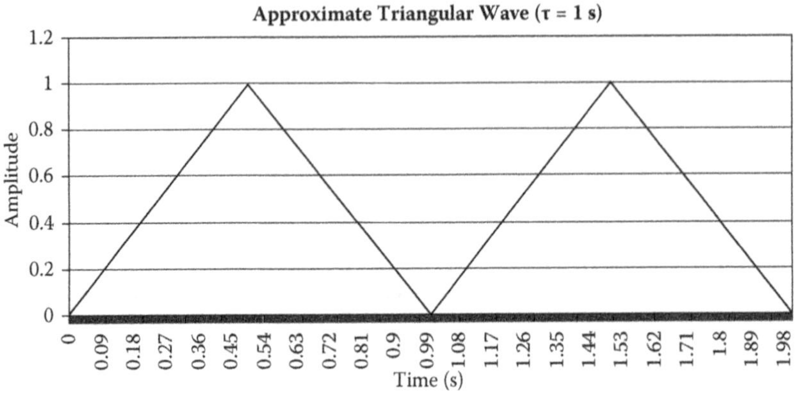

Observe from Example 2.1.5 that there is ringing occurring at $t = 0.5$ s of the square wave, where the function transitions discontinuously. This phenomenon is called *Gibb's phenomenon.* This can be intuitively seen from Theorem 2.1.2, where $\tilde{f}(t_i) = \frac{1}{2}(f(t_i^+) + f(t_i^-))$ at the points t_i of discontinuities: $\tilde{f}(t_i) \neq f(t_i)$. This is because the rate of change as the function approaches the discontinuity changes rapidly.

Indeed, this has physical implications for EMC engineers: ringing happens when there is a sharp transition in signal level. For instance, ringing is apparent if a signal trace is inductive. This can be seen heuristically as follows. When a line is inductive, and the signal is decomposed into its Fourier series, the higher harmonics are truncated, and the amplitudes for the lower harmonics are much larger than those of higher harmonics. On the other hand, for a capacitive line, ringing is absent as the transitions at the edges are smoothed out: a capacitive line truncates low harmonics. Refer to Section 2.4 for more details.

2.2 Fourier Transform

Intuitively—or perhaps not!—an arbitrary function that vanishes sufficiently rapidly away from the origin so that its integral over the real line is finite also has a Fourier decomposition. In some instances, it might not be possible to Fourier-decompose the function into a countably infinite sum of sines and cosines. Instead, the summation is uncountably infinite and hence gives rise to the Fourier integral.

2.2.1 Theorem

Suppose that $f: \mathbf{R} \to \mathbf{R}$ is a function such that

$$\int_{-\infty}^{\infty} |f(t)|^2 dt < \infty$$

Then, the integral

$$F(\omega) = \tfrac{1}{2\pi}\int_{-\infty}^{\infty} e^{-i\omega t} f(t)dt \tag{2.5}$$

exists and its inverse is given by

$$f(t) = \int_{-\infty}^{\infty} e^{i\omega t} F(\omega)d\omega \tag{2.6}$$

□

2.2.2 Remark

It is crucial to note that if the condition $\int_{-\infty}^{\infty} |f(t)|^2 dt < \infty$ were relaxed to $\int_{-\infty}^{\infty} |f(t)| dt < \infty$, the inverse Fourier transform might not exist.

Equations (2.5) and (2.6) constitute a *Fourier transform pair* and (2.5) is called the *Fourier transform* of *f*, and (2.6) is called the *inverse Fourier transform* of *F*. Therefore what is the relationship between a Fourier series and a Fourier transform? To see this, consider the coefficients of (2.6): $c_n = \tfrac{1}{T}\int_{-T/2}^{T/2} f(t)e^{-in\omega t}dt$, and recall that $\omega = \tfrac{2\pi}{T}$. Clearly, in the limit as $T \to \infty, \omega \to 0$; and for an aperiodic function, $T \to \infty$. Hence, for very large *T*, ω may be expressed as an infinitesimal quantity: $\omega \to \delta\omega$.

Secondly, as $\omega \to \delta\omega$, the discrete spectrum $n\omega$ becomes a continuous non-zero spectrum $\bar{\omega}$: that is, $n\omega \to \bar{\omega}$ in the limit as $\omega \to 0$ and $n \to \infty$ such that the quantity $0 < \bar{\omega} < \infty$. Then, taking the limit as the pair $T,n \to \infty$, the quantity $c_n T \to \int_{-\infty}^{\infty} f(t)e^{-i\bar{\omega}t}dt < \infty$. This limiting process yields, informally, the Fourier transform.

Conversely, rewriting Equation (2.3) as $f(t) = \sum_{n=-\infty}^{\infty} c_n T e^{in\omega t} \tfrac{1}{T}$, in the limiting process as $T \to \infty$, and noting from above that $\lim_{T,n\to\infty} c_n T \to F(\omega)$ converges to the Fourier transform of *f*, $\tfrac{2\pi}{T} \to d\bar{\omega}$ and the summation becomes an integral:

$$\tfrac{1}{2\pi}\sum_{n=-\infty}^{\infty} c_n T e^{in\omega t} \tfrac{2\pi}{T} \to \tfrac{1}{2\pi}\int_{-\infty}^{\infty} e^{i\bar{\omega}t} F(\bar{\omega})d\bar{\omega}$$

yielding the inverse Fourier transform. Thus, informally, a Fourier transform takes a function and transforms it into another function.* See Exercise 2.5.1 for details. It is a linear operator that maps a certain set of functions into another set of functions. For more details on linear operators, refer to Appendix A.1. For the present, it suffices to recall that linearity means that both scaling $T(af) = aT(f), a \in \mathbb{C}$, and superposition $T\left(\sum_i f_i\right) = \sum_i T(f_i)$, are preserved under the mapping.

* Mathematically, a Fourier transform is a linear isomorphism on a Hilbert space of square integrable functions (*cf.* Remark 2.2.2). This factoid is for readers interested in exploring the wonderful world of Hilbert spaces and their various applications in the field of engineering and the theory of partial differential equations; see Appendix A.4 for details.

These two properties—scaling and the superposition principle—make linear systems a lot easier to study than nonlinear systems. In other words, a scaled solution of a linear system is again a solution of the system, and the sum of solutions of a linear system is again a solution of the system. These conclusions are generally false for nonlinear systems.

From an engineering perspective, a system is linear if its response to an input function is linear. As a concrete example, let an operator T describe the response of a system under an input voltage (function) v. Explicitly, let $v \mapsto v' = T(v)$. If v_1, v_2 are two voltages simultaneously applied into the input terminal of the system in question, then the resultant output voltage (i.e., response of the system to the input voltages) satisfies

$$v_1 + v_2 \xrightarrow{T} v_1' + v_2' = T(v_1) + T(v_2)$$

if and only if T is linear. Thus, the Fourier transform can be employed to describe linear systems and it cannot, by definition, be used to describe nonlinear systems. The following result from Fourier theory has great applications in the physical sciences.

2.2.3 Theorem (Parseval)

Let $f: \mathbf{R} \to \mathbf{R}$ be a function such that $\int_{-\infty}^{\infty} |f(t)|^2 \, dt < \infty$. Then, $\frac{1}{2\pi}\int_{-\infty}^{\infty} |F(\omega)|^2 \, d\omega = \int_{-\infty}^{\infty} f(t)^2 \, dt$. □

This is known as the energy integral, and its proof can be found in any standard reference on Fourier analysis. The integrand $|F(\omega)|^2$ has units of joules per Hz. Intuitively then, $\frac{1}{2\pi}\int_a^{a+\delta a} |F(\omega)|^2 \, d\omega$ represents the energy content of the wave f in the frequency range $[a, a + \delta a]$. The physical significance of this is considered in the next section.

This section closes with an example regarding an application of the Fourier transform to analyze Maxwell's equations and a brief remark on the Laplace transform, as it is essentially a natural extension of the Fourier transform. It is clear from the definition of the Fourier transform that even some simple functions do not possess "proper" Fourier transforms (without resorting to distribution theory). For instance, the unit function 1, $e^{i\omega t}$ and $\cos t$, do not possess Fourier transforms that are functions in the strict sense:

$$1 \leftrightarrow 2\pi\delta(\omega): \quad F(\omega) = \frac{1}{2\pi}\int_{-\infty}^{\infty} 1e^{-i\omega t} \, dt \equiv 2\pi\delta(\omega)$$

$$e^{it} \leftrightarrow 2\pi\delta(\omega - 1): \quad F(\omega) = \frac{1}{2\pi}\int_{-\infty}^{\infty} e^{it}e^{-i\omega t} \, dt = \frac{1}{2\pi}\int_{-\infty}^{\infty} e^{-i(\omega-1)t} \, dt = 2\pi\delta(\omega - 1)$$

$$\cos t \leftrightarrow \pi(\delta(\omega - 1) + \delta(\omega + 1))$$

Verify this in Exercise 2.5.2.

These situations are clearly not very satisfactory at all. Indeed, the Laplace transform will rectify these unpleasant scenarios. Heuristically, without going into the development of Laplace transform theory, this can be easily accomplished by the following replacement: $i\omega \to s = \sigma + i\omega$, where σ is real. The Laplace transform pair is defined by

$$\begin{cases} F(s) = \int_0^\infty f(t)e^{-st}\,dt \\ f(t) = \frac{1}{2\pi i}\int_{\sigma - i\infty}^{\sigma + i\infty} F(s)e^{st}\,ds \end{cases} \tag{2.7}$$

2.2.4 Theorem

Suppose $f: \mathbf{R} \to \mathbf{R}$ is piecewise continuous and satisfies (a) f is of bounded variation for all compact interval $[a,b] \subset \mathbf{R}$, and (b) $\exists \alpha, M, \tau > 0$ constants such that $e^{-\alpha t}|f(t)| < M\ \forall t > \tau$. Then, the Laplace transform $F = F(s)$ of $f = f(t)$ exists whenever $\Re e(s) > \inf\{\alpha : e^{-\alpha t}|f(t)| < M\ \forall t > \tau\}$. □

As a final comment, Fourier and Laplace transforms are often employed to solve ordinary differential equations and partial differential equations. By way of example, consider the one-dimensional diffusion equation $\partial_t \psi = \alpha^2 \partial_x^2 \psi\ \forall (x,t) \in \mathbf{R} \times [0,\infty)$ subject to the initial condition $\psi(x,0) = g(x)$ on $\mathbf{R}$, where g is square-integrable.

By definition, $f(t) \leftrightarrow F(\omega) \Rightarrow \partial_t F(\omega) \equiv 0$. In particular,

$$\partial_t(f(t)e^{-i\omega t}) = e^{-i\omega t}\partial_t f(t) - i\omega f(t)e^{-i\omega t} \Rightarrow \mathcal{F}[\partial_t f] = i\omega \mathcal{F}[f]$$

where $\mathcal{F}[f]$ denotes the Fourier transform of f for convenience. Indeed, on setting $h = \partial_t f$, it follows immediately that $\mathcal{F}[\partial_t^2 f] = \mathcal{F}[\partial_t h] = i\omega\mathcal{F}[h] = -\omega^2\mathcal{F}[f]$. Hence, the Fourier transform with respect to the x variable yields $\partial_t \Psi(\xi,t) = -\alpha^2\xi^2\Psi(\xi,t)$, where $\Psi(\xi,t) = \frac{1}{2\pi}\int_{-\infty}^{\infty}\psi(x,t)e^{-i\xi x}dx$ and ξ is an arbitrary parameter. Because ξ is a parameter, eliminating it as a variable makes the result more transparent: $\frac{d}{dt}\Psi(t) = -\alpha^2\xi^2\Psi(t)$. Thus, the Fourier transform reduces the partial differential equation to an ordinary differential equation $\frac{d}{dt}\Psi(t) = -\alpha^2\xi^2\Psi(t)$ satisfying $\Psi(0) = \mathcal{F}[g](\xi)$. The solution Ψ can be easily solved and finding the inverse Fourier transform yields the desired result, $\psi = \mathcal{F}^{-1}[\Psi]$.

The astute reader will notice immediately that the Fourier transform operated on the x variable instead of the t variable. Indeed, the Laplace transform is generally used to transform the time variable instead of the space variable. This is because (i) the time variable typically lies in the open interval $[0,\infty)$

[*cf.* Equation (2.7)] and (ii) it takes care of the initial conditions. See, for example, References [12,13,15] for further details. This section concludes with a final example.

2.2.5 Example

The Fourier transform can be employed to gain further insight into Maxwell's equations [4,10]. Assuming the fields are time harmonic, the Fourier transform of the electric field $E(r,t)$ is defined by

$$\tilde{E}(k,\omega) = \iint E(r,t)e^{-i(k\cdot r-\omega t)}d^3r dt$$

The inverse is $E(r,t) = \frac{1}{(2\pi)^4}\iint \tilde{E}(k,\omega)e^{i(k\cdot r-\omega t)}d^3k d\omega$. Here, $k = \frac{2\pi}{\lambda}$ defines the *wave number* and λ is the wavelength. Then, by definition, noting that $\partial_\xi e^{i(k\cdot r-\omega t)} = ik_\xi e^{i(k\cdot r-\omega t)}$ for $\xi = x, y, z$, that is, $\partial_\xi \leftrightarrow ik_\xi$, and $\partial_t e^{i(k\cdot r-\omega t)} = -i\omega e^{i(k\cdot r-\omega t)} \Rightarrow \partial_t \leftrightarrow -i\omega$, it is easy to see that

$$\begin{cases} \nabla\times E(r,t) = ik\times E(r,t) \\ \nabla\cdot E(r,t) = ik\cdot E(r,t) \\ \nabla\times B(r,t) = ik\times B(r,t) \\ \nabla\cdot B(r,t) = ik\cdot B(r,t) \end{cases}$$

and $\partial_t(E(r,t), B(r,t)) = -i\omega(E(r,t), B(r,t))$.

In particular, taking the Fourier transform of the scalar and vector potential,

$$\tilde{E}(k,\omega) = -ik\tilde{\varphi}(k,\omega) + i\omega\tilde{A}(k,\omega)$$

$$\tilde{B}(k,\omega) = ik\times\tilde{A}(k,\omega)$$

Whence, Maxwell's equations can be rewritten as

$$\begin{cases} k\times\tilde{E}(k,\omega) = \omega\tilde{B}(k,\omega) \\ k\cdot\tilde{B}(k,\omega) = 0 \\ ik\times\tilde{B}(k,\omega) = \mu\tilde{J} - i\omega\mu\varepsilon\tilde{E}(k,\omega) \\ ik\cdot\varepsilon\tilde{E}(k,\omega) = \tilde{\rho}(k,\omega) \end{cases}$$

Observe that k is the direction of the wave propagation. Hence, defining $\mathbf{e}_k = \frac{1}{|k|}k$, it follows at once that $\tilde{E} = \tilde{E}_\perp + \tilde{E}_\parallel$, where $\tilde{E}_\perp = (\mathbf{e}_k\times\tilde{E})\times\mathbf{e}_k$

and $\tilde{E}_{\parallel} = (\mathbf{e}_k \cdot \tilde{E})\mathbf{e}_k$. Then, recalling that $\nabla \leftrightarrow i\boldsymbol{k}$, it is quite clear that $\boldsymbol{k} \cdot \tilde{\boldsymbol{E}} = \boldsymbol{k} \cdot \tilde{\boldsymbol{E}}_{\parallel} \Rightarrow \nabla \cdot \tilde{\boldsymbol{E}}_{\perp} = 0$, and $\boldsymbol{k} \times \tilde{\boldsymbol{E}} = \boldsymbol{k} \times \tilde{\boldsymbol{E}}_{\perp} \Rightarrow \nabla \times \tilde{\boldsymbol{E}}_{\parallel} = 0$. Physically, this means that the longitudinal component does not affect the rotation of the electric field and the transverse component of the electric field does not affect the divergence of the electric field.

An immediate consequence is the following:

- *Coulomb gauge:* $\boldsymbol{k} \cdot \tilde{\boldsymbol{A}} = 0 \Rightarrow \boldsymbol{k} \cdot \tilde{\boldsymbol{A}}_{\parallel} = 0 \Rightarrow \tilde{\boldsymbol{A}}_{\parallel} = 0$. That is, the longitudinal component is identically zero in the Coulomb gauge.
- Charge conservation: $i\omega\tilde{\rho} = \boldsymbol{k} \cdot \tilde{\boldsymbol{J}} = \boldsymbol{k} \cdot \tilde{\boldsymbol{J}}_{\parallel}$. That is, charge conservation is solely influenced by the longitudinal component of the current density.

Finally, to conclude this example, via Maxwell's equation (1.17),

$$\nabla \times \nabla \times \boldsymbol{A} = \mu \boldsymbol{J} - \mu\varepsilon \partial_t (\nabla\varphi + \partial_t \boldsymbol{A})$$

the vector identity $\nabla \times \nabla \times \boldsymbol{A} = \nabla(\nabla \cdot \boldsymbol{A}) - \Delta \boldsymbol{A}$ leads to the wave equation:

$$-\Delta \boldsymbol{A} + \nabla(\nabla \cdot \boldsymbol{A}) + \mu\varepsilon \partial_t^2 \boldsymbol{A} = \mu \boldsymbol{J} - \mu\varepsilon \nabla \partial_t \varphi$$

However, instead of invoking the *Lorentz gauge* $\nabla \cdot \boldsymbol{A} = -\mu\varepsilon \partial_t \varphi$, continue to use the Coulomb gauge. From Poisson's equation, $\varepsilon\nabla \cdot \nabla\varphi = \rho \Rightarrow \varepsilon\nabla \cdot \partial_t \nabla\varphi = \partial_t \rho = \nabla \cdot \boldsymbol{J}$ and hence,

$$\nabla \cdot (\varepsilon \partial_t \nabla\varphi) = \nabla \cdot \boldsymbol{J} \Rightarrow i\omega\varepsilon\tilde{\varphi}\boldsymbol{k} = \tilde{\boldsymbol{J}}_{\parallel}$$

whence substituting this into the wave equation for the vector potential yields

$$-k^2 \tilde{\boldsymbol{A}} - \omega^2 \mu\varepsilon \tilde{\boldsymbol{A}} = \mu \tilde{\boldsymbol{J}} - \mu \tilde{\boldsymbol{J}}_{\parallel} = \mu \tilde{\boldsymbol{J}}_{\perp}$$

In particular, taking the Fourier transform gives $-\Delta \boldsymbol{A} + \mu\varepsilon \partial_t \boldsymbol{A} = -\mu \boldsymbol{J}_{\perp}$. Thus, the transverse component of the electric current density contributes to the vector potential whereas the longitudinal component contributes to the scalar potential. Finally, from $\boldsymbol{E} = -\nabla\varphi - \partial_t \boldsymbol{A}$,

$$\tilde{\boldsymbol{E}} = -i\tilde{\varphi}\boldsymbol{k} + i\omega\tilde{\boldsymbol{A}} \Rightarrow \tilde{\boldsymbol{E}}_{\parallel} = -i\tilde{\varphi}\boldsymbol{k} \quad \text{and} \quad \tilde{\boldsymbol{E}}_{\perp} = i\omega\tilde{\boldsymbol{A}}_{\perp}$$

as $\tilde{\boldsymbol{A}}_{\parallel} = 0$. In other words, the longitudinal component of the electric field is the result of the scalar potential and the transverse component of the electric field is the result of the vector potential. □

2.3 Roll-Off Frequency

The Fourier transform of a function f is related to the energy content of the function as follows. $|F(i\omega)|^2\, d\omega$ defines the energy content in the infinitesimal frequency bandwidth $d\omega$. This fact becomes important to EMC engineers in the qualitative assessment of the energy contents of radio frequency harmonics [11]. Indeed, the source of electromagnetic interference is often mitigated by suppressing the harmonic with the highest energy content.

In summary, the total energy content of a function is given by $\frac{1}{2\pi}\int_{-\infty}^{\infty}|F(\omega)|^2\, d\omega$. By Theorem 2.2.3, $|f(t)|^2\, dt$ defines the energy content of the function within the infinitesimal time interval dt. In particular, $\int_{-\infty}^{\infty}|f(t)|^2\, dt$ defines the total energy (i.e., spectral energy) content of the function by Parseval's theorem.

In fact, by plotting the graph of $|F(\omega)|$ as a function of ω, a piecewise linear upper bound that envelops the graph $|F(\omega)|$ defines the roll-off frequency [5,11]. In this section, the foundational concepts are developed instead of approaching the subject via the upper bound envelope approximation typically cited in the EMC literature.

2.3.1 Example

Consider the function $f(t) = e^{-t^2}$ on $\mathbf{R}$. Its Fourier transform is

$$F(\omega) = \int_{-\infty}^{\infty} e^{-t^2} e^{-i\omega t}\, dt = \sqrt{\pi} e^{-\left(\frac{\omega}{2}\right)^2}$$

Thus, the spectrum of f is $|F(\omega)| = \sqrt{\pi} e^{-\left(\frac{\omega}{2}\right)^2}$ and the graph of the spectrum is given below.

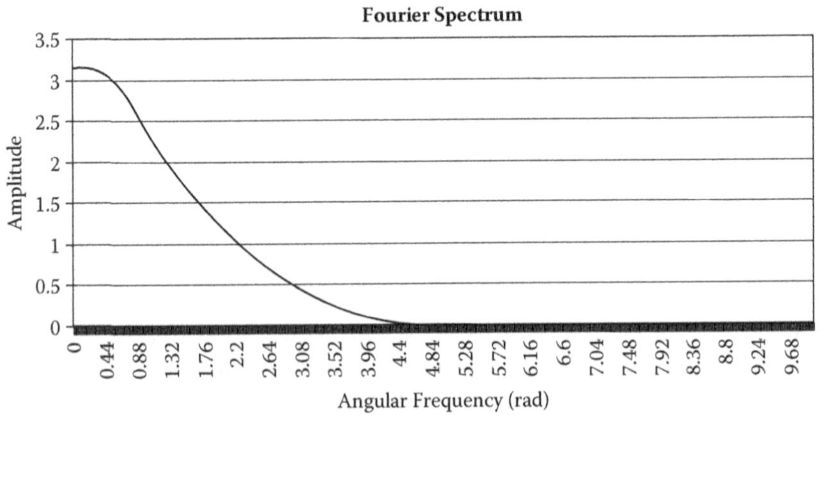

The following example is particularly relevant to EMC engineers: the spectrum of a square wave. A square waveform is an idealized digital wave generated for data transmission. A more cogent example to consider is that of a trapezoidal wave.

2.3.2 Example

Consider a square pulse defined by

$$f(t) = \begin{cases} 1 \text{ for } 0 \le t \le \tau \\ 0 \quad \text{for } t \ge \tau \end{cases}$$

Then, $F(\omega) = \int_0^\tau e^{-i\omega t} dt = \frac{i}{2}(e^{-i\omega\tau} - 1) = e^{-i\omega\tau}\frac{2}{\omega}\sin\frac{\omega\tau}{2} = \tau e^{-i\omega\tau}\mathrm{sinc}\frac{\omega\tau}{2}$, where $\mathrm{sinc}\varphi = \frac{\sin\varphi}{\varphi}$. Hence, $|F(\omega)| = |\tau\,\mathrm{sinc}\frac{\omega\tau}{2}|$. A plot for $\tau = 1$ is given below.

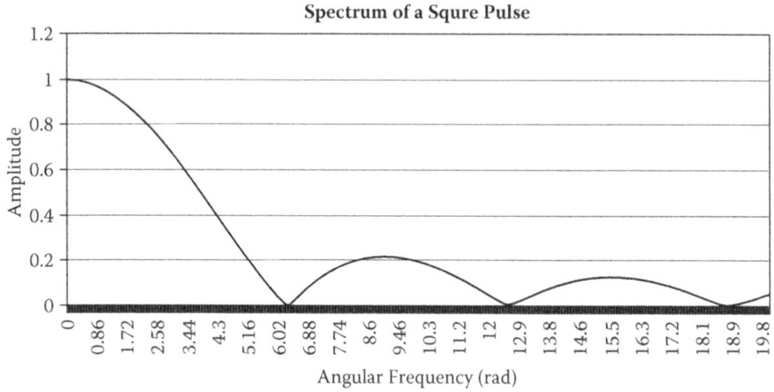

It is clear from the plot that the majority of the spectral energy lies in the lower harmonics. □

2.3.3 Proposition

Let $f : \mathbf{R} \to \mathbf{R}$ be a continuous trapezoidal pulse of period τ, rise time τ_+ and fall time τ_- defined by

$$f(t) = \begin{cases} \frac{A}{\tau_+}t & \text{for } 0 \le t \le \tau_+ \\ A & \text{for } \tau_+ \le t \le \tau\text{-}\tau_- \\ \frac{A}{\tau_-}(\tau - t) & \text{for } \tau\text{-}\tau_- \le t \le \tau \\ 0 & \text{for } t \ge \tau \end{cases} \tag{2.8}$$

Then, $0 < \tau_\pm << \tau \Rightarrow |F(\omega)| \approx \frac{1}{2}A(\tau_+ - \tau_-)$.

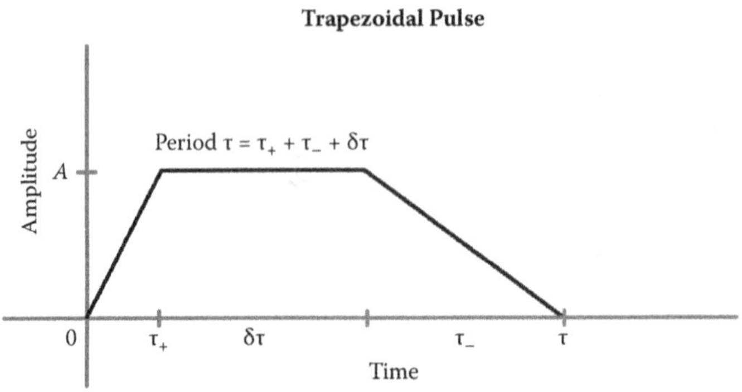

FIGURE 2.2
A trapezoidal pulse of period τ.

Proof

Consider the waveform defined by Equation (2.8) depicted in Figure 2.2.
The Fourier transform of f is given by

$$F(\omega)=\int_{-\infty}^{\infty} f(t)\mathrm{e}^{-\mathrm{i}\omega t}\,\mathrm{d}t=\tfrac{A}{\tau_+}\int_0^{\tau_+} t\mathrm{e}^{-\mathrm{i}\omega t}\,\mathrm{d}t+A\int_{\tau_+}^{\tau-\tau_-}\mathrm{e}^{-\mathrm{i}\omega t}\,\mathrm{d}t+\tfrac{A}{\tau_-}\int_{\tau-\tau_-}^{\tau}\left(\tfrac{\tau+\tau_++\tau_-}{2}-t\right)\mathrm{e}^{-\mathrm{i}\omega t}\,\mathrm{d}t$$

Now, noting that $\omega\tau=2\pi\Rightarrow \mathrm{e}^{\pm \mathrm{i}2\pi}=1$, evaluating the integral term by term yields:

$$\tfrac{A}{\tau_+}\int_0^{\tau_+} t\mathrm{e}^{-\mathrm{i}\omega t}\,\mathrm{d}t=\tfrac{A}{\tau_+\omega^2}\mathrm{e}^{-\mathrm{i}\omega\tau_+}\left\{1+\mathrm{i}\omega\tau_+-\mathrm{e}^{\mathrm{i}\omega\tau_+}\right\}$$

$$A\int_{\tau_+}^{\tau-\tau_-}\mathrm{e}^{-\mathrm{i}\omega t}\,\mathrm{d}t=\mathrm{i}\tfrac{A}{\omega}\{\mathrm{e}^{\mathrm{i}\omega\tau_-}-\mathrm{e}^{-\mathrm{i}\omega\tau_+}\}$$

$$\tfrac{A}{\tau_-}\int_{\tau-\tau_-}^{\tau}(\tau-t)\mathrm{e}^{-\mathrm{i}\omega t}\,\mathrm{d}t=\tfrac{4\pi A}{\omega^2\tau_-}\mathrm{e}^{\mathrm{i}\frac{1}{2}\omega\tau_-}\sin\tfrac{\omega\tau_-}{2}+\mathrm{i}\tfrac{A}{\omega^2\tau_-}\left\{2\mathrm{e}^{\mathrm{i}\frac{1}{2}\omega\tau_-}\sin\tfrac{\omega\tau_-}{2}(1+2\pi\mathrm{i})-\omega\tau_-\mathrm{e}^{\mathrm{i}\omega\tau_-}\right\}$$

whence appealing to (a) $\sin\varphi=\frac{1}{2\mathrm{i}}(\mathrm{e}^{\mathrm{i}\varphi}-\mathrm{e}^{-\mathrm{i}\varphi})$, and (b) $1-\mathrm{e}^{\mathrm{i}\varphi}=\mathrm{e}^{\mathrm{i}\frac{\varphi}{2}}(\mathrm{e}^{-\mathrm{i}\frac{\varphi}{2}}-\mathrm{e}^{\mathrm{i}\frac{\varphi}{2}})\Rightarrow$ $1-\mathrm{e}^{\mathrm{i}\varphi}=-2\mathrm{i}\mathrm{e}^{\mathrm{i}\frac{\varphi}{2}}\sin\frac{\varphi}{2}$, it follows at once that

$$F(\omega)=\mathrm{i}\tfrac{2A}{\omega^2}\left\{\tfrac{1}{\tau_-}\mathrm{e}^{\mathrm{i}\frac{1}{2}\omega\tau_-}\sin\tfrac{\omega\tau_-}{2}-\tfrac{1}{\tau_+}\mathrm{e}^{-\mathrm{i}\frac{1}{2}\omega\tau_+}\sin\tfrac{\omega\tau_+}{2}\right\} \tag{2.9}$$

Equation (2.9) can be expressed equivalently as

$$F(\omega)=\mathrm{i}\tfrac{A}{\omega}\left\{\mathrm{e}^{\mathrm{i}\frac{1}{2}\omega\tau_-}\operatorname{sinc}\tfrac{\omega\tau_-}{2}-\mathrm{e}^{-\mathrm{i}\frac{1}{2}\omega\tau_+}\operatorname{sinc}\tfrac{\omega\tau_+}{2}\right\} \tag{2.10}$$

So, suppose that $\tau_{\pm} << \tau$. Then, $\tau_{+} + \tau_{-} << \tau$ and hence, via the binomial approximation,

$$e^{-i\frac{1}{2}\omega\tau_{\pm}} \approx 1 - i\tfrac{1}{2}\omega\tau_{\pm} + o\left(\left(\tfrac{\omega\tau_{\pm}}{2}\right)^2\right) \quad \text{and} \quad \mathrm{sinc}\left(\tfrac{\omega\tau_{\pm}}{2}\right) \approx 1 - \tfrac{1}{3!}\left(\tfrac{\omega\tau_{\pm}}{2}\right)^2 + o\left(\left(\tfrac{\omega\tau_{\pm}}{2}\right)^4\right)$$

and hence,

$$F(\omega) \approx \tfrac{A}{2}(\tau_{+} - \tau_{-})\left\{1 - \tfrac{i}{3!}\tfrac{\omega}{2}(\tau_{+} + \tau_{-})\right\}$$

to first order in $\frac{1}{2}\omega(\tau_{+} + \tau_{-})$. That is, $|F(\omega)| \approx \frac{1}{2}A(\tau_{-} + \tau_{+})$, as required. Fill in the details in Exercise 2.5.3. □

2.3.4 Corollary

Suppose $0 < \tau_{+} << \tau_{-} \sim \tau$, then

$$|F(\omega)| \approx \tfrac{A}{\omega}\left\{1 + \tfrac{1}{2}\left(\tfrac{\omega\tau_{+}}{2}\right)^2\right\}$$

Proof

From Equation (2.10), and $\tau_{-} \sim \tau \Rightarrow \sin\omega\tau_{-} \approx 0$. Hence,

$$\begin{aligned} F(\omega) &\approx -i\tfrac{A}{\omega}e^{-i\frac{1}{2}\omega\tau_{+}}\,\mathrm{sinc}\left(\tfrac{\omega\tau_{+}}{2}\right) \\ &\approx -i\tfrac{A}{\omega}\left\{\left(1 - i\tfrac{\omega\tau_{+}}{2}\right)\left(1 - \tfrac{1}{3!}\left(\tfrac{\omega\tau_{+}}{2}\right)^2\right)\right\} \\ &\approx -i\tfrac{A}{\omega}\left\{1 - \tfrac{1}{3!}\left(\tfrac{\omega\tau_{+}}{2}\right)^2 - i\tfrac{\omega\tau_{+}}{2}\right\} \end{aligned}$$

implies immediately that

$$|F(\omega)| \approx \tfrac{A}{\omega}\left\{1 + \tfrac{1}{2}\left(\tfrac{\omega\tau_{+}}{2}\right)^2\right\}$$

as claimed. □

In summary, it is clear from the above results that the spectral energy of radiation emissions is essentially dictated by the fast rise and fall times: that is, whether the rise/fall time is much less than the period of the digital signal. Hence, from an electromagnetic interference perspective, fast rise/fall times have the potential to radiate strongly, indeed, to the extent of violating the upper bounds for radiation emissions set by various international regulatory agencies. Reducing the rise time is a common practice in reducing radio emissions in the EMC industry.*

2.4 Frequency Response and Filter Theory: A Primer

The study of system response in terms of the time variable t (known also as the *time domain*) can be equivalently studied in terms of the frequency variable ω (also known as the *frequency domain*) via Fourier transform. Oftentimes, transforming a problem into the frequency domain simplifies the problem tremendously. The time domain solution is then obtained from the frequency domain solution via the inverse Fourier transform.

By way of motivation, consider the generalization of the frequency variable ω by expressing it as a complex frequency $s = \sigma + i\omega$, where $\sigma, \omega \in \mathbf{R}$. Next, consider an exponentially damped sinusoidal voltage given by $v(t) = V_0 e^{\sigma t} \cos(\omega t + \theta)$, where $V_0 \in \mathbf{R}$ is the magnitude of the voltage, $\omega = 2\pi f$ the angular frequency and θ is an arbitrary phase.

- For $\sigma < 0$, $e^{\sigma t} \to 0$ in the limit as $t \to \infty$; that is, $v(t)$ decays exponentially in time (exponentially damped).
- For $\sigma > 0$, $e^{\sigma t} \to \infty$ in the limit as $t \to \infty$; this is not physically realizable, and hence if $\sigma \neq 0$, it is always chosen to be negative. It represents the loss as the voltage wave propagates through a lossy medium.

2.4.1 Example

Given the damped sinusoidal voltage $v(t) = V_0 e^{\sigma t} \cos(\omega t + \theta)$, show that this can be expressed as $v(t) = \Re e(V_0 e^{st} e^{i\theta})$. To see that this is indeed the case, it suffices to appeal to Euler's formula: $e^{i\theta} = \cos\theta + i\sin\theta$. Then, it is evident that

$$e^{st}e^{i\theta} = e^{st+i\theta} = e^{\sigma t + i(\omega t+\theta)} = e^{\sigma t}e^{i(\omega t+\theta)} = e^{\sigma t}(\cos(\omega t+\theta) + i\sin(\omega t+\theta))$$

whence $\Re e(e^{st}e^{i\theta}) = e^{\sigma t}\cos(\omega t+\theta) \Rightarrow v(t) = \Re e(V_m e^{st}e^{i\theta})$, as claimed. □

* However, EMC engineers are also aware that reducing the rise/fall times too much will result in compromising the signal integrity of the digital device in question. Indeed, an ingenious workaround to this problem (typically found in clock generators in integrated circuits) is the implementation of spread spectrum [8,9].

Example 2.4.1 leads to the following generalization. Physically measurable quantities such as current and voltage can be generalized to complex quantities. There are two reasons for making the generalization. First, only physically measurable quantities are real; however, the imaginary component encodes information regarding phase and loss. More on this is shown in Chapter 4.

Before proceeding, a small mathematical detour is made below for sinusoidal waves. Let $f = f(x, t)$ be a complex function. Then, f is said to be *time harmonic* (or a *steady state sinusoidal* function of t) if there exists a complex function $F = F(x, \theta)$ such that $f(x,t) = F(x,\theta)e^{i\omega t}$, for some phase $\theta \in \mathbf{R}$. Thus, there is a clear bijection between f and F; in particular, the time variable t can be factored from the original function f. This leads to the following definition.

2.4.2 Definition

The *phasor transform* of a time harmonic function $f(x,t) = F(x,\theta)e^{i\omega t}$ is a mapping $P : f \mapsto F$ defined by $P[f](x,\theta) = f(x,t)e^{-i\omega t}$. The inverse P^{-1} exists for time-harmonic functions and is trivially defined by $P^{-1}[F](x,t) = F(x,\theta)e^{i\omega t}$.

Notice that the positive sign in $e^{i\omega t}$ is merely an arbitrary choice. It is equally valid to use $e^{-i\omega t}$ in the definition of time harmonicity as long as consistency is maintained. The positive sign was chosen to be consistent with the definition of the Fourier transform.

2.4.3 Lemma

The phasor transform P satisfies the following properties:

(a) linearity: $P[af + bg] = aP[f] + bP[g]$ for all time-harmonic f, g and scalars $a, b \in C$:

(b) $P[\partial_t f](x,\theta) = i\omega P[f](x,\theta)$

(c) $P[\partial_x f](x,\theta) = \partial_x P[f](x,\theta)$

for some fixed phase $\theta \in \mathbf{R}$.

Proof

Linearity is clear: $P[af + bg] = af(x,t)e^{-i\omega t} + bg(x,t)e^{-i\omega t} = aP[f] + bP[g]$, where f, g are any two steady-state sinusoidal functions and a, b are constants. Property (b) is also easily established:

$$P[\partial_t f](x,\theta) = P[\partial_t(F(x,\theta)e^{i\omega t})] = P[F(x,\theta)\tfrac{d}{dt}e^{i\omega t}] = P[i\omega F(x,\theta)e^{i\omega t}] \equiv i\omega P[f](x,\theta)$$

and finally, to prove (c), it is enough to observe that

$$P[\partial_x f](x,\theta) = \partial_x(f(x,t)e^{-i\omega t}) = \partial_x F(x,\theta) \equiv \partial_x P[f](x,\theta) \qquad \square$$

Throughout the analysis, a steady-state sinusoidal propagating wave is assumed for simplicity. The justification for a sinusoidal assumption is somewhat obvious: (a) most physical waveforms have Fourier decompositions; (b) the superposition principal can be applied to the propagating waves; (c) the phasor transform can be applied to convert the time dependency in partial differential equations into steady-state differential equations. More generally, Fourier and Laplace transforms are often used to solve differential equations, as pointed out in Section 2.2.

Suppose that a time harmonic function $f = f(x,t)$ is twice differentiable with respect to x and t. Then, it is very easy to see the following.

$$\mathrm{P}[\partial_t^2 f] = \mathrm{P}[\partial_t(\partial_t f)] = \mathrm{i}\omega\mathrm{P}[\partial_t f] = -\omega^2\mathrm{P}[f]$$

and

$$\mathrm{P}[\partial_x^2 f](x,\omega) = \partial_x^2(F(x,\omega)\mathrm{e}^{-\mathrm{i}\omega t}) \equiv \partial_x^2\mathrm{P}[f](x,\omega)$$

Thus, consider a simple wave equation $\partial_x^2 f(x,t) = \alpha^2\,\partial_t^2 f(x,t)$. Phasor transforming the wave equation yields $\mathrm{P}[\partial_x^2 f] = \mathrm{P}[\alpha^2\,\partial_t^2 f] = -\omega^2\alpha^2\mathrm{P}[f]$. Thus, the partial differential equation is converted into a second-order ordinary differential equation (with respect to x): $\partial_x^2 F(x,\omega) = -\omega^2\alpha^2 F(x,\omega)$.

Hence, for steady-state sinusoidal waves, it is convenient to apply phasor transform to convert a time domain solution into a frequency domain solution via the following rules.

- $\frac{\partial}{\partial t} \to \mathrm{i}\omega$ under the phasor transform.
- $\frac{\partial^2}{\partial t^2} \to -\omega^2$ under the phasor transform.

Furthermore, ∂_x and P *commute*; that is, $\mathrm{P}[\partial_x f] = \partial_x\mathrm{P}[f]$ for any steady-state sinusoidal function f.

Finally, define a complex voltage by $v(t) = \tilde{V}\mathrm{e}^{st}$, where $\tilde{V} = V_0\mathrm{e}^{\mathrm{i}\theta}$ for some real constants V_0, θ. This is known as the *phasor representation* for the time-harmonic voltage. In the above notation, the impedance of an inductor is represented by $X_L = sL$, and that of a capacitor is represented by $X_C = \frac{1}{sC}$. The motivation for using phasor representation is to eliminate time dependency, as pointed out above. Another example is given to illustrate this point.

Consider a voltage induced across an inductor: $v(t) = L\frac{di(t)}{dt}$. Substituting $i(t) = \tilde{I}\mathrm{e}^{st}$ yields $\frac{di(t)}{dt} = s\tilde{I}\mathrm{e}^{st} = si(t)$. That is, $v(t) = sLi(t) \Rightarrow \tilde{V} = sL\tilde{I}$. Because the time variable has been eliminated, the expression is solely a function of frequency; the time domain expression has thus been transformed into the frequency domain expression. Likewise, the voltage induced across a capacitor is $v(t) = \frac{1}{C}\int i(t)dt$. Substituting $i(t) = \tilde{I}\mathrm{e}^{st}$ yields $\int i(t)dt = \frac{1}{s}\tilde{I}\mathrm{e}^{st} = \frac{1}{s}i(t)$. That is, $v(t) = \frac{1}{sC}i(t) \Rightarrow \tilde{V} = \frac{1}{sC}\tilde{I}$.

2.4.4 Example

Consider a simple series RL-circuit driven by some voltage $\tilde{v}(t) = \tilde{V}(\theta)e^{st}$, where $\tilde{V}(\theta) = V_0 e^{i\theta}$ and $V_0, \theta \in \mathbf{R}$ are fixed. Here, the system is defined by the following ordinary differential equation $Ri(t) + L\frac{d}{dt}i(t) = v(t)$. By Lemma 2.4.3, via phasor representation,

$$Ri(t) + L\tfrac{d}{dt}i(t) = v(t) \Leftrightarrow R\tilde{I}(\phi) + i\omega L\tilde{I}(\phi) = \tilde{V}(\theta)$$

for some fixed phase ϕ, whence, $\tilde{I}(\phi) = \frac{\tilde{V}(\theta)}{R + i\omega L}$.

In order to evaluate ϕ, it suffices to note that $\tilde{I}(\phi) = \frac{\tilde{V}(\theta)}{R + i\omega L} = I_0 e^{i\phi}$, for some real I_0. Therefore

$$\frac{\tilde{V}(\theta)}{R + i\omega L} = V_0 e^{i\theta} \frac{R - i\omega L}{R^2 + \omega^2 L^2} = V_0 e^{i\theta} \frac{1}{\sqrt{R^2 + \omega^2 L^2}} e^{i\theta'} \equiv I_0 e^{i\phi}$$

where $\theta' = \arctan\frac{-\omega L}{R}$, $\phi = \theta + \theta'$ and $I_0 = \frac{V_0}{\sqrt{R^2 + \omega^2 L^2}}$. Thus, in time domain representation,

$$i(t) = \tilde{I}(\phi)e^{i\omega t} = \frac{V_0}{\sqrt{R^2 + \omega^2 L^2}} e^{i\left(\theta - \arctan\frac{\omega L}{R} + \omega t\right)}$$

yielding the solution to the differential equation governing the circuit. In particular, the current measured by an ammeter is $\Re e(i(t)) = \frac{V_0}{\sqrt{R^2 + \omega^2 L^2}} \cos(\theta - \arctan\frac{\omega L}{R} + \omega t)$.

□

In general, particularly for forcing functions that are not necessarily time harmonic, the differential equation governing the linear system is solved via Laplace transform by converting the time domain into a complex frequency domain. This leads to the concept of a transfer function. Informally, a *transfer function* represents the response of a linear system to an input function. Let $H = H(s)$ characterize a linear system response in the frequency domain. Then, some examples of transfer functions are:

- $\frac{\tilde{V}_1(s)}{\tilde{V}_0(s)} = H(s)$, where $\tilde{V}_0$ represents the input voltage and $\tilde{V}_1$ represents the output voltage.
- $\frac{\tilde{V}(s)}{\tilde{I}(s)} = H(s)$; here, $H(s)$ represents the impedance of the system.
- $\frac{\tilde{I}(s)}{\tilde{V}(s)} = H(s)$; here, $H(s)$ represents the admittance of the system.

Recall that the admittance $Y(s)$ is defined as the inverse of the impedance $Z(s)$: $Y(s) = 1/Z(s)$.

Before proceeding further, some formal definitions are given for completeness. A circuit (network) is often said to be linear if it comprises resistors, capacitors, and inductors, that is, if $I \mapsto V$ is a linear invertible mapping. If the properties of the network do not vary with time, it is time invariant. And finally, if the network comprises passive linear elements, then it satisfies the reciprocity theorems (cf. Section 6.1) and hence the network is said to be a reciprocal network.

2.4.5 Definition

A *network* $\mathcal{N} = \{(E_i, \hat{C}_{ij})\}$ comprises circuit elements E_i and conductors $\hat{C}_{ij}$ connecting the pair (E_i, E_j).

a) It is *linear* if each E_i can be represented by an invertible linear operator L_i such that $L_i[I(z,t)] = V(z,t)$, where the pair (V, I) is the voltage and current associated with E_i.
b) The invertible linear operator L_i is *time invariant* if $L_i[I(z,t+\tau)] = V(x,t+\tau)\ \forall\tau$. More precisely, if T_τ is *a time translation* operator defined by $T_\tau[\xi(z,t)] = \xi(z,t+\tau)$, then L_i is time invariant if $L_i T_\tau = T_\tau L_i\ \forall\tau$.

Some examples are given below for clarification. Consider a simple resistor R. By Ohm's law, $V = RI$; that is, the linear operator is just multiplication by R. Next, consider an ideal capacitor C. Then, this is represented by the following linear operator: $V = \frac{1}{C}\int I(t')dt'$ (recall that the integral operator is a linear operator). Finally, consider an ideal inductor L. This is represented by the operator: $V = -L\frac{dI}{dt}$; clearly, the operator $\frac{d}{dt}$ is linear and its inverse is the integral operator.

2.4.6 Definition

Let $H : \mathbf{C} \to \mathbf{C}$ be a complex function. Then, it is called a *rational function* if it can be expressed as

$$H(s) = A\frac{(s-s_1)(s-s_2)\cdots(s-s_n)}{(s-s_{n+1})(s-s_{n+2})\cdots(s-s_{n+m})}$$

where $A, s_1, \ldots, s_{n+m} \in \mathbf{C}$ are constants, for some $n, m \in \mathbf{N}$. The complex numbers $s_1, \ldots, s_n$ are called the *zeros* of H and $s_{n+1}, \ldots, s_{n+m}$ are called the *poles* of H.

2.4.7 Remark

Let $h = h(t)$ be time harmonic. Then, $s = i\omega$ is purely imaginary and $H(s)H(-s) = |H(s)|^2$. Studying the poles and zeros of the transfer function H is important in the design of filters for signal integrity and suppressing electromagnetic interference without affecting signal quality.

This chapter concludes with a qualitative sketch of the natural response of a linear system response in terms of the poles of its transfer function. Recall that in general, $s = \sigma + i\omega$, for some real σ, ω. The plot of $|H(s)|$ as a function of s can be analyzed as follows. First, set $\sigma = 0$ and plot $|H(i\omega)|$ as a function of ω. Next, set $\omega = 0$ and plot $|H(\sigma)|$ as a function of σ. The phase of H is determined as follows: set $H(i\omega) = A(\omega) + iB(\omega)$, where $A(\omega)$, $B(\omega) \in \mathbf{R}$. Then, the phase of $H(i\omega)$ on the complex plane is, by definition, $\arg H(i\omega) = \arctan \frac{B(\omega)}{A(\omega)}$.

A physical interpretation of the poles of a transfer function $H = H(s)$ is given below:

- An input function (also known as a *forcing function*) operating at any one of the poles will result in an infinite response by the system. An example of a forcing function is an input voltage.
- A *natural response* is defined by the response of a system when the input terminals are shorted, that is, when the forcing function is set to zero. An example is the natural resonance of a series RC-circuit.

On the other hand, the zeros of a transfer function determine the magnitude and phase of the response in time domain.

As a concrete example, consider a transfer function

$$H(s) = \frac{K(s)}{(s-p_1)\cdots(s-p_m)}$$

where $K = K(s)$ encodes the zeros $z_1, \dots, z_n$ of the system. Now, by taking the partial fraction expansion, this can be expressed as

$$H(s) = \frac{K(s)}{(s-p_1)\cdots(s-p_m)} = \frac{K_1}{s-p_1} + \cdots + \frac{K_m}{s-p_m} \tag{2.11}$$

for some constants $K_1, \dots, K_m \in \mathbf{C}$ associated with $K(s)$. It is thus obvious from Equation (2.11) that the zeros, via $\{K_1, \dots, K_m\}$, determine the magnitude and phase of H. More specifically, converting (2.11) back to the time domain via inverse Laplace transform yields

$$h(t) = K_1 e^{p_1 t} + \cdots + K_m e^{p_m t} \tag{2.12}$$

The above comments are now apparent: as $|e^{i\alpha}| = 1 \; \forall \alpha \in \mathbf{R}$, it follows at once that $K_i \in \mathbf{C} \Rightarrow K_i = |K_i| e^{i\phi}$, for some phase ϕ. Thus, $\{K_i\}$ encodes the phase and magnitude information, as asserted.

Next, consider the output voltage resulting from a natural response of the system. Given the poles $s_{n+1}, \dots, s_{n+m}$ of the transfer function $H(s)$, the output voltage has the form $v(t) = A_1 e^{s_{n+1} t} + A_2 e^{s_{n+2} t} + \cdots + A_m e^{s_{n+m} t}$, where $A_1, \dots, A_m$ are constants that depend on the initial conditions of the system. Thus, by studying the frequency response of a linear system in the frequency domain, a qualitative solution in the time domain can be obtained via the poles of the transfer function. See Exercise 2.5.4 for a concrete example.

A transfer function $H = H(s)$ is also called a *response function*. Suppose that

$$H(s) = \frac{P_1}{s-p_1} + \cdots + \frac{P_n}{s-p_n} + \frac{Q_1}{s-q_1} + \cdots + \frac{Q_m}{s-q_m}$$

where P_i, Q_j for $i = 1, \ldots, n$ and $j = 1, \ldots, m$, are constants, $\{p_i\}$ are natural frequencies and $\{q_j\}$ are the forcing frequencies. That is, the response function can be decomposed as

$$H(s) = H_1(s) + H_2(s)$$

where $H_1(s) = \frac{P_1}{s-p_1} + \cdots + \frac{P_n}{s-p_n}$ is the natural response and $H_2(s) = \frac{Q_1}{s-q_1} + \cdots + \frac{Q_m}{s-q_m}$ is the forcing response. Taking the inverse Laplace transform yields

$$h(t) = h_1(t) + h_2(t)$$

where $h_1(t) = P_1 e^{p_1 t} + \cdots + P_n e^{p_n t}$ and $h_2(t) = Q_1 e^{q_1 t} + \cdots + Q_m e^{q_m t}$ are, respectively, the natural part and the forcing part of the system in the time domain.

The poles of the system determine the natural response. The poles of the excitation applied to the system determine the forced response. Now, observe that the *steady-state* response of the system is defined by $h(t)|_{t>T}$, where $T >> 0$ is some arbitrarily large number. In particular, if $\Re e(p_i) > 0$, then $\lim_{t\to\infty} h_1(t) \to \infty$ and hence the system is unstable; there is no steady-state response. Thus, in order for steady state to exist, it is necessary that the natural poles $\{p_i\}$ lie on the left-hand side of the complex plane; that is, $\Re e(p_i) < 0 \; \forall i$. Because $\Re e(p_i) < 0 \Rightarrow \lim_{t\to\infty} h_1(t) \to 0$, it follows that $h(t) \approx h_2(t)$ for large $t > 0$. Hence, in the study of the steady-state response of a system, only the forced response need be considered.

2.4.8 Example

Consider the LC-network defined [5] as

$$Z(s) = H \frac{(s^2+\omega_1^2)(s^2+\omega_3^2)\cdots(s^2+\omega_{2n+1}^2)}{s(s^2+\omega_2^2)(s^2+\omega_4^2)\cdots(s^2+\omega_{2n}^2)} \tag{2.13}$$

where $0 < \omega_i < \omega_{i+1} \; \forall i = 1, \ldots, 2n$. Then, via partial fraction expansion, this can be reduced to

$$Z(s) = sH + \frac{K_0}{s} + \frac{sK_1}{s^2+\omega_2^2} + \frac{sK_2}{s^2+\omega_4^2} \cdots + \frac{sK_n}{s^2+\omega_{2n}^2} \tag{2.14}$$

where $K_i > 0 \; \forall i = 0, 1, \ldots, n$. Clearly,

$$\lim_{s\to 0} sZ(s) = \lim_{s\to 0}\left\{s^2 H + K_0 + \cdots + \frac{s^2 K_n}{s^2+\omega_{2n}^2}\right\} = K_0$$

and likewise,

$$\lim_{s\to i\omega_i} \frac{s^2+\omega_{2i}^2}{s} Z(s) = \lim_{s\to 0}\left\{H(s^2+\omega_{2i}^2) + \cdots + K_i + \cdots + K_n \frac{s^2+\omega_{2i}^2}{s^2+\omega_{2n}^2}\right\} = K_i, \text{ for } i = 1, \ldots, n.$$

Now, observe that

$$\frac{sK_i}{s^2+\omega_{2i}^2} = \left\{\frac{s^2+\omega_{2i}^2}{sK_i}\right\}^{-1} = \left\{\frac{s}{K_i} + \frac{1}{sK_i/\omega_{2i}^2}\right\}^{-1}$$

where $sH \leftrightarrow$ inductance H, $\frac{K_0}{s} \leftrightarrow$ capacitance $\frac{1}{K_0}$. Hence,

$$\left\{\frac{s}{K_i} + \frac{1}{sK_i/\omega_{2i}^2}\right\}^{-1} = \frac{1}{s\frac{1}{K_i} + \frac{1}{s(K_i/\omega_{2i}^2)}} \Rightarrow \frac{1}{K_i} \leftrightarrow \text{capacitor and } \frac{K_i}{\omega_{2i}^2} \leftrightarrow \text{inductor}$$

connected in parallel, as it has the form $\left\{\frac{1}{sC+\frac{1}{sL}}\right\}^{-1}$ corresponding to the impedance of a parallel LC circuit. The first term corresponds to an inductor, the second term to a capacitor, and the subsequent terms are parallel LC circuits, all of which are connected in series, as indicated by the summation in Equation (2.13). Hence, (2.13) has the circuit representation depicted in Figure 2.3 and it is called the *LC Foster I network*.

By inspecting (2.13), the poles are $\{0, \pm i\omega_{2k}, \infty : i = 1, \ldots n\}$ and the zeros are $\{\pm i\omega_{2k-1} : k = 1, \ldots, n\}$. Thus, all the finite poles and zeros lie on the imaginary axis, as expected, for by construction, the resistance of the network is zero, and also by definition, the poles and zeros alternate with a singularity at the

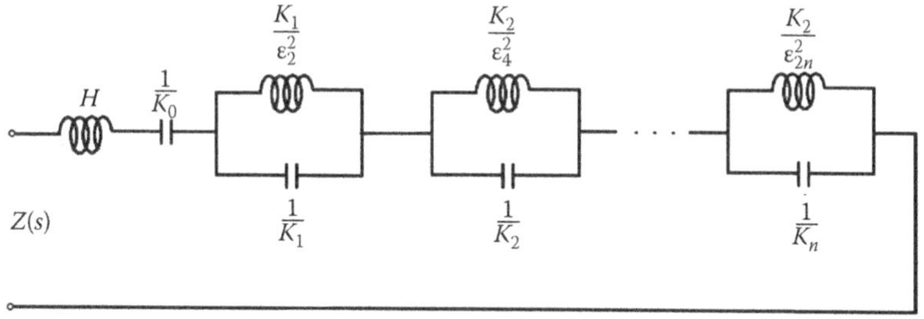

FIGURE 2.3
A realization of a general *LC* Foster I network.

origin. Moreover, it is also clear that (i) $\lim_{s\to\infty} Z(s) = \infty$ if and only if a series inductor is present, and (ii) $\lim_{s\to 0} Z(s) = \infty$ if and only if a series capacitor is present.

To conclude this example, observe that if Equation (2.13) has the following form instead:

$$Z(s) = H\frac{s(s^2+\omega_2^2)(s^2+\omega_4^2)\cdots(s^2+\omega_{2n}^2)}{(s^2+\omega_1^2)(s^2+\omega_3^2)\cdots(s^2+\omega_{2n-1}^2)} \tag{2.15}$$

where $0 < \omega_i < \omega_{i+1}\ \forall i = 1, \ldots, 2n-1$, then a similar analysis leads to:

(a) A zero instead of a singularity at the origin.
(b) The alternating poles and zeros pattern is the reverse of Equation (2.13); that is, the poles of (2.13) correspond to the zeros of (2.15) and vice versa along the imaginary axis. □

2.4.9 Example

Consider the LC Foster I network defined by (2.14) above. Its admittance is $Y(s) = \frac{1}{Z(s)}$. Once again, proceeding as in Example 2.4.8 via partial fraction expansion:

$$Y(s) = sH + \frac{K_0}{s} + \frac{sK_1}{s^2+\omega_1^2} + \frac{sK_2}{s^2+\omega_2^2} + \cdots + \frac{sK_n}{s^2+\omega_n^2} \tag{2.16}$$

Noting by construction that $\frac{1}{Z_i} \equiv \frac{sK_i}{s^2+\omega_i^2}$, it follows immediately from Example 2.4.8 that

$$Z_i = \frac{s}{K_i} + \frac{1}{s\frac{K_i}{\omega_i^2}}$$

for each $i = 1, \ldots, n$. That is, each Z_i is a series LC circuit of inductance $\frac{1}{K_i}$ and capacitance $\frac{K_i}{\omega_i^2}$. Thus, Equation (2.16) has the physical representation illustrated in Figure 2.4.

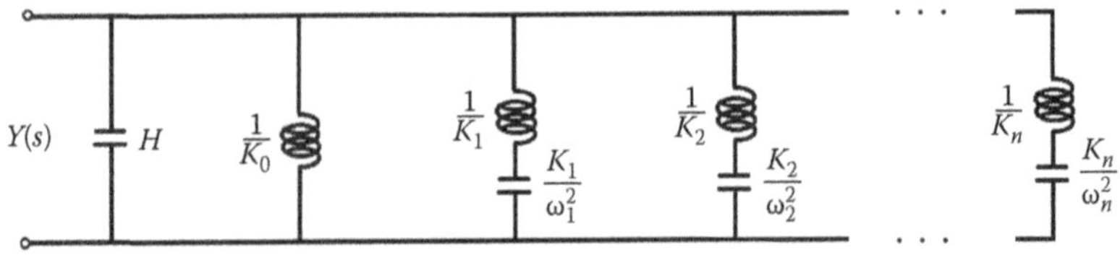

FIGURE 2.4
A realization of a general *LC* Foster II network.

This network is called the *LC Foster II network*. The analysis follows that of Example 2.4.8 *mutatis mutandis*. In particular, by inspecting Figure 2.3, it is evident that $\lim_{s\to 0} Y(s) = \infty \Rightarrow K_0 > 0$ and the capacitor H is absent. On the other hand, $\lim_{s\to\infty} Y(s) = 0 \Rightarrow H > 0$ and the inductor $\frac{1}{K_0}$ is absent. □

2.4.10 Example

Consider the LC Cauer network defined [2] in Figure 2.5.

Set $\Delta_{n-1} = Z_{n-1} + Z_n = Z_{n-1} + \frac{1}{Y_n}$ and $\Delta_{n-2} = Y_{n-2} + \frac{1}{\Delta_{n-1}}$. Then, by induction, it is clear from Figure 2.4, that

$$\Delta_k = \begin{cases} Z_k + \frac{1}{\Delta_{k+1}} & \text{for } k \text{ odd,} \\ Y_k + \frac{1}{\Delta_{k+1}} & \text{for } k \text{ even,} \end{cases}$$

where by construction, n is even. Observe that by definition, Δ_{2k} defines admittance whereas Δ_{2k+1} defines impedance for $k = 1, \ldots n-1$.

The impedance Z is thus given by the following continued fraction expansion,

$$Z = Z_1 + \cfrac{1}{Y_2 + \cfrac{1}{Z_3 + \cfrac{1}{Y_4 + \ddots \cfrac{1}{\Delta_{n-1}}}}}. \tag{2.17}$$

Then, the *LC Cauer I network* is defined by

$$Z_k = \begin{cases} sL & \text{for } k \text{ odd,} \\ \frac{1}{sC} & \text{for } k \text{ even.} \end{cases}$$

Two basic properties of the LC Cauer I network can be easily deduced from Equation (2.17):

(a) $\lim_{s\to\infty} |Z(s)| \to \infty \Leftrightarrow Z_1 = sL \neq 0$

(b) $\lim_{s\to\infty} |Z(s)| \to 0 \Leftrightarrow Z_1 = sL \equiv 0$ and $Z_2 = \frac{1}{sC} \neq 0$

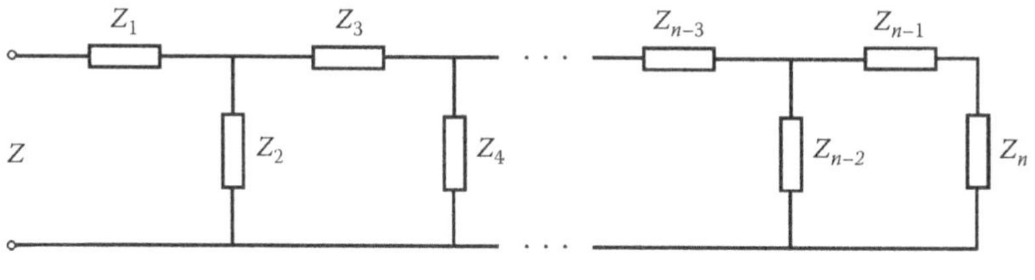

FIGURE 2.5
A general LC Cauer network (also known as a ladder network).

The LC Cauer II network is defined by

$$Z_k = \begin{cases} \frac{1}{sC} \text{ for } k \text{ odd,} \\ sL \text{ for } k \text{ even.} \end{cases}$$

a) $\lim_{s\to 0}|Z(s)| \to \infty \Leftrightarrow Z_1 = \frac{1}{sC} \neq 0$

b) $\lim_{s\to 0}|Z(s)| \to 0 \Leftrightarrow Z_1 = \frac{1}{sC} \equiv 0$ and $Z_2 = sL \neq 0$

The RC Foster I and II networks and Cauer I and II networks can likewise be constructed and studied. In particular, a general filter is often the synthesis of Foster and Cauer networks. These considerations are not pursued here. See Exercises 2.5.5–2.5.7 for some concrete examples.

2.5 Worked Problems

2.5.1 Exercise

Given the Fourier expansion $\tilde{f}(t) = \Sigma_{n=0}^{\infty}\{a_n \cos n\omega t + b_n \sin n\omega t\}$, (i) show that it can be expressed as $\tilde{f}(t) = \Sigma_{n=-\infty}^{\infty} c_n e^{in\omega t}$, where the coefficients are given by Equation (2.4); (ii) formally deduce Fourier transform from the Fourier expansion.

Solution

(i) From Euler's formula, $e^{it} = \cos t + i \sin t$, it follows that

$$\tilde{f}(t) = \sum_{n=0}^{\infty}\{a_n \cos n\omega t + b_n \sin n\omega t\}$$

$$= \sum_{n=0}^{\infty} \tfrac{1}{2} a_n (e^{in\omega t} + e^{-in\omega t}) - \sum_{n=0}^{\infty} \tfrac{1}{2} i b_n (e^{in\omega t} - e^{-in\omega t})$$

$$= \sum_{n=0}^{\infty} \tfrac{a_n - ib_n}{2} e^{in\omega t} + \sum_{n=0}^{\infty} \tfrac{a_n + ib_n}{2} e^{-in\omega t}$$

$$\equiv \sum_{n=-\infty}^{\infty} c_n e^{in\omega t}$$

where $c_0 = a_0$, $c_n = \frac{1}{2}(a_n - ib_n)$ and $c_{-n} = \frac{1}{2}(a_n + ib_n)\ \forall n > 0$.
(ii) From (i) and the fact that

$$c_n = \frac{1}{T}\int_{-T/2}^{T/2} \tilde{f}(t)e^{-in\omega t}\, dt$$

it follows by direct substitution that

$$\tilde{f}(t) = \sum_{-\infty}^{\infty} \frac{1}{T}\left\{\int_{-T/2}^{T/2} \tilde{f}(t)e^{-in\omega t}\, dt\right\}e^{in\omega t}$$

Now, on setting $\omega_n = n\omega$ and $\delta\omega = \frac{2\pi}{T}$, it clearly follows that

$$\lim_{T,n\to\infty} \sum_{-\infty}^{\infty} \frac{1}{T}\left\{\int_{-T/2}^{T/2} \tilde{f}(t)e^{-i\omega_n t}\, dt\right\}e^{i\omega_n t} \to \frac{1}{2\pi}\int_{-\infty}^{\infty} F(\bar{\omega})e^{i\bar{\omega}t}\, \delta\bar{\omega}$$

whence $\tilde{f}(t) = \frac{1}{2\pi}\int_{-\infty}^{\infty} F(\omega)e^{i\omega_n t}d\omega$, as required. □

2.5.2 Exercise

Verify the Fourier transform pair: $\cos t \leftrightarrow \pi(\delta(\omega - 1) + \delta(\omega + 1))$.

Solution

First, set $f(t) = \cos t$ and recall that $\frac{1}{2\pi}\int_{-\infty}^{\infty} e^{-i\omega t}dt = 2\pi\delta(\omega)$. Then, from $\cos t = \frac{1}{2}(e^{it} + e^{-it})$, it follows that

$$F(i\omega) = \frac{1}{2}\left\{\int_{-\infty}^{\infty} e^{it}e^{-i\omega t}\, dt + \int_{-\infty}^{\infty} e^{-it}e^{-i\omega t}\, dt\right\} = \frac{1}{2}\left\{\int_{-\infty}^{\infty} e^{-i(\omega-1)t}\, dt + \int_{-\infty}^{\infty} e^{-i(\omega+1)t}\, dt\right\}$$

$$= \pi\big(\delta(\omega - 1) + \delta(\omega + 1)\big).$$

2.5.3 Exercise

Derive Equation (2.12) and hence establish (2.14).

Solution

$$\frac{A}{\tau_+}\int_0^{\tau_+} te^{-i\omega t}\, dt = \frac{A}{\tau_+\omega^2}\left\{e^{-i\omega\tau_+}(1 + i\omega\tau_+) - 1\right\} = \frac{A}{\tau_+\omega^2}e^{-i\frac{1}{2}\omega\tau_+}\left\{e^{-i\frac{1}{2}\omega\tau_+}(1 + i\omega\tau_+) - e^{i\frac{1}{2}\omega\tau_+}\right\}$$

From $\sin\phi = \frac{1}{2i} i(e^{i\phi} - e^{-i\phi})$, it follows that

$$\frac{A}{\tau_+\omega^2} e^{-i\frac{1}{2}\omega\tau_+}\left\{e^{-i\frac{1}{2}\omega\tau_+}(1+i\omega\tau_+) - e^{i\frac{1}{2}\omega\tau_+}\right\}$$

$$= \frac{A}{\tau_+\omega^2} e^{-i\frac{1}{2}\omega\tau_+}\left\{-2i\sin\frac{\omega\tau_+}{2} + i\omega\tau_+ e^{i\frac{1}{2}\omega\tau_+}\right\}$$

$$= -i\frac{2A}{\omega^2\tau_+} e^{-i\frac{1}{2}\omega\tau_+}\sin\frac{\omega\tau_+}{2} + i\frac{A}{\omega} e^{i\omega\tau_+}$$

Likewise,

$$\frac{A\tau}{\tau_-}\int_{\tau-\tau_-}^{\tau} e^{-i\omega t}\, dt = i\frac{A}{\omega}\frac{\tau}{\tau_-}(1 - e^{i\omega\tau_-}) = i\frac{2\pi A}{\omega^2\tau_-} e^{i\frac{1}{2}\omega\tau_-}\left(e^{-i\frac{1}{2}\omega\tau_-} - e^{i\frac{1}{2}\omega\tau_-}\right)$$

$$= \frac{4\pi A}{\omega^2\tau_-} e^{i\frac{1}{2}\omega\tau_-}\sin\frac{\omega\tau_-}{2}$$

Finally,

$$-\frac{A}{\tau_-}\int_{\tau-\tau_-}^{\tau} t e^{-i\omega t}\, dt = -\frac{A}{\omega^2\tau_-}\left\{1 + 2\pi i - e^{i\omega\tau_-} - 2\pi i e^{i\omega\tau_-} + i\omega\tau_- e^{i\omega\tau_-}\right\}$$

$$= -\frac{A}{\omega^2\tau_-}\left\{-i2e^{i\frac{1}{2}\omega\tau_-}\sin\frac{\omega\tau_-}{2} + 4\pi e^{i\frac{1}{2}\omega\tau_-}\sin\frac{\omega\tau_-}{2} + i\omega\tau_- e^{i\omega\tau_-}\right\}$$

$$= i\frac{A}{\omega^2\tau_-}\left\{2e^{i\frac{1}{2}\omega\tau_-}\sin\frac{\omega\tau_-}{2}(1 + 2\pi i) - \omega\tau_- e^{i\omega\tau_-}\right\}$$

$$= i\frac{2A}{\omega^2\tau_-} e^{i\frac{1}{2}\omega\tau_-}\sin\frac{\omega\tau_-}{2} - \frac{4\pi A}{\omega^2\tau_-} e^{i\frac{1}{2}\omega\tau_-}\sin\frac{\omega\tau_-}{2} - i\frac{A}{\omega} e^{i\omega\tau_-}$$

From this,

$$F(\omega) = -i\frac{2A}{\omega^2\tau_+} e^{-i\frac{1}{2}\omega\tau_+}\sin\frac{\omega\tau_+}{2} + \frac{4\pi A}{\omega^2\tau_-} e^{i\frac{1}{2}\omega\tau_-}\sin\frac{\omega\tau_-}{2} + i\frac{2A}{\omega^2\tau_-} e^{-i\frac{1}{2}\omega\tau_-}\sin\frac{\omega\tau_-}{2}(1 + 2\pi i)$$

$$= i\frac{2A}{\omega^2}\left\{\frac{1}{\tau_-} e^{i\frac{1}{2}\omega\tau_-}\sin\frac{\omega\tau_-}{2} - \frac{1}{\tau_+} e^{-i\frac{1}{2}\omega\tau_+}\sin\frac{\omega\tau_+}{2}\right\}$$

$$= i\frac{A}{\omega}\left\{e^{i\frac{1}{2}\omega\tau_-}\operatorname{sinc}\frac{\omega\tau_-}{2} - e^{-i\frac{1}{2}\omega\tau_+}\operatorname{sinc}\frac{\omega\tau_+}{2}\right\}$$

The conclusion of the proof thus follows accordingly. □

2.5.4 Exercise

Find the transfer function $H(s) = \frac{V_0(s)}{V_i(s)}$ for the circuit illustrated in Figure 2.6, where V_i is the input voltage and V_0 is the output voltage.

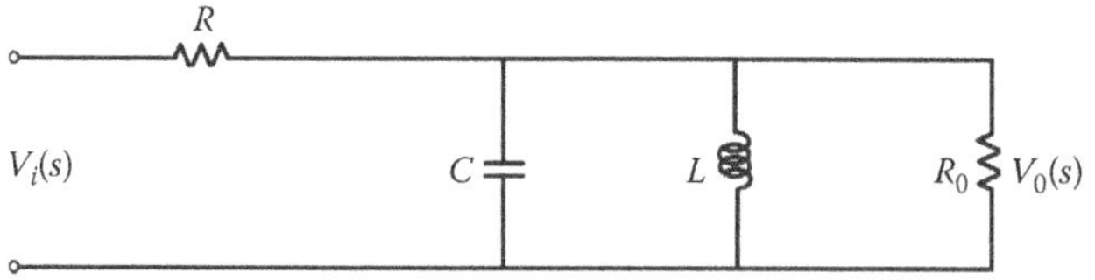

FIGURE 2.6
A simple *RLC* filter network.

Solution

The admittance of the *RLC* network is

$$Y' = sC + \frac{1}{sL} + \frac{1}{R_0} \Rightarrow Z' = \frac{1}{Y'} = \frac{sL}{s^2LC+\alpha}$$

where $\alpha = 1 + \frac{1}{R_0}$. Hence, the total impedance $Z = R + \frac{sL}{s^2LC+s\frac{L}{R_0}+1}$. From the total current $I = \frac{V_i}{Z}$, it follows that $V_0 = V_i - IR = V_i\left(1 - \frac{R}{Z}\right)$ and hence,

$$H(s) = 1 - \frac{R}{Z} = \frac{s\beta}{s\beta+s^2\alpha+R}$$

where $\alpha = RLC$ and $\beta = L\left(1+\frac{R}{R_0}\right)$. Performing partial fraction expansion,

$$H(s) = \frac{s\beta}{s\beta+s^2\alpha+R} = \frac{A}{s+\omega_+} + \frac{B}{s+\omega_-}$$

where

$$\omega_\pm = -\frac{1+\frac{R}{R_0}}{2RC} \pm \frac{1}{2LC}\sqrt{\left(1+\frac{R}{R_0}\right)^2 - 4LC}$$

are the roots of the denominator $s^2\alpha + s\beta + R = 0$,

$$s\beta = A(s+\omega_-) + B(s+\omega_+)$$

To evaluate for the constants *A*,*B*, set $s = -\omega_+$. Then, $A = \frac{\omega_+ L}{\omega_+ - \omega_-}$. Likewise, setting $s = -\omega_-$ yields $B = -\frac{\omega_- L}{\omega_+ - \omega_-}$.

Hence, the poles of the transfer function *H* are $\{\omega_+, \omega_-\}$ and the zero of *H* is clearly at the origin $s = 0$ whenever $R \neq 0$. Finally, observe that if $2\sqrt{LC} > 1 + \frac{R}{R_0}$, $R > 0 \Rightarrow \omega_\pm \in \mathbf{C}$, which are not purely imaginary numbers. On the other hand, for $2\sqrt{LC} \leq 1 + \frac{R}{R_0} \Rightarrow \omega_\pm \in \mathbf{R}$ and hence, the poles and zeros all lie on the real line. □

2.5.5 Exercise

Construct a general low-pass filter to allow all frequencies $0 \le \omega \le \omega_0$ to pass through with very little attenuation.

Solution

Consider a general transfer function H for a filter: $|H(\omega)|^2 = \frac{P(\omega^2)}{Q(\omega^2)}$, where P, Q are polynomial functions of ω^2, and $0 \le |H(\omega)|^2 \le 1$. Ideally, $|H(\omega)|^2 = 1 - u(\omega^2 - \omega_0^2)$, where

$$u(x) = \begin{cases} 1 & \text{for } x > 0, \\ 0 & \text{for } x \le 0, \end{cases}$$

is the *Heavyside* function. Hence, $Q(\omega^2) \approx P(\omega^2)$ for $\omega \le \omega_0$ and $Q(\omega^2) >> P(\omega^2)$ for $\omega > \omega_0$. Thus, set $Q(\omega^2) = P(\omega^2) + P'(\omega^2)$, where $P'(\omega^2) \approx 0$ for $\omega \le \omega_0$ and $P'(\omega^2) >> 0$ for $\omega > \omega_0$. Then, $|H(\omega)|^2 = \frac{1}{1+h(\omega^2)}$, where $h(\omega^2) = \frac{P'(\omega^2)}{P(\omega^2)}$. It thus remains to determine the functional form of h.

It is obvious that by setting $h = \left(\frac{\omega}{\omega_0}\right)^{2n}$, then $h << 1$ for $n >> 1$ whenever $\omega < \omega_0$. In particular, that filter is *maximally flat* at $\omega = 0$: $|H(0)|^2 = 1$. This choice of h is called the *Butterworth approximation.* To determine the property of H, make the substitution: $\omega^2 = -s^2$. Then, by definition, $H(s)H(-s) = |H(s)|^2$, and rewriting $\varepsilon^2 = \frac{1}{\omega_0^n}$, it follows that $|H(s)|^2 = \frac{1}{1+\varepsilon^2(-s^2)^n}$.

Now, the poles of $H(s)$ lie precisely on the curve $1 + \varepsilon^2(-s^2)^n = 0$. However, this curve is precisely the equation of a circle in **C** of radius $\frac{1}{\varepsilon^{2/n}}$. To see this, it suffices to note that

$$1 + \varepsilon^2(-s^2)^n = 0 \Leftrightarrow (-s^2)^n = \frac{1}{\varepsilon^{2/n}}(-1)^{1/n} = \frac{1}{\varepsilon^{2/n}} e^{\,i\pi/n} \Rightarrow s = \frac{1}{\varepsilon^{2/n}} e^{i\frac{\pi}{2}\frac{n+1}{n}}$$

That is,

$$se^{-i\frac{\pi}{2}\frac{n+1}{n}} = \frac{1}{\varepsilon^{2/n}}$$

Thus, the poles lie on the circle of radius $\frac{1}{\varepsilon^{2/n}}$. And in particular, from $s = \sigma + i\omega$, it is clear that for physically realizable filters, the poles must lie on the circle intersecting the left-hand complex plane; that is, $\sigma < 0$. □

2.5.6 Exercise

Construct a general high-pass filter to allow all frequencies $\omega \ge \omega_0 >> 0$ to pass through with very little attenuation.

Solution

Now, observe trivially that $\frac{1}{s} \to 0$ as $s \to \infty$. That is, if $H = H(s)$ is a low-pass filter, then $G(s)=1/H(s)$ represents a high-pass filter. Hence, from Exercise 2.5.5, make the following transformation $s \to 1/s$. Then, it is clear that $\hat{H}(s) \equiv H(1/s)$ defines a high-pass filter. That is, using Exercise 2.5.5,

$$\hat{H}(s) = \frac{(-s^2)^n}{(-s^2)^n + \varepsilon^2}$$

More generally, for a low-pass filter $G(s) = \frac{P(s)}{Q(s)}$, where $P(s) = \Sigma_{i=0}^{n} a_i s^i$ and $Q(s) = \Sigma_{i=0}^{m} b_i s^i$, with $m > n$. Then, $s \to \frac{1}{s} \Rightarrow P\left(\frac{1}{s}\right) = \frac{1}{s^n} \Sigma_{i=0}^{n} a_i s^{n-i}$ and $Q\left(\frac{1}{s}\right) = \frac{1}{s^m} \Sigma_{i=0}^{m} b_i s^{m-i}$. This leads to the following high-pass filter:

$$\hat{G}(s) = G\left(\frac{1}{s}\right) = s^{m-n} \frac{a_0 s^n + \cdots + a_n}{b_0 s^m + \cdots + b_n}$$

The physical realization is not difficult to construct. Indeed, it is enough to observe that $s \to 1/s$ transforms an inductive reactance to a capacitive reactance and vice versa. Hence, replacing inductors with capacitors and vice versa in a low-pass filter immediately leads to a high-pass filter. □

References

1. Brown, J. and Churchill, R. 2011. *Fourier Series and Boundary Value Problems*. New York: McGraw Hill.
2. Campbell, G. 1922. Physical theory of the electric wave-filter. *Bell Syst. Tech. J.*, **1**(2): 1–32.
3. Carslaw, H. 1921. *Introduction to the Theory of Fourier's Series and Integrals*. London: Macmillian.
4. Dressel, M. and Grüner, G. 2002. *Electrodynamics of Solids: Optical Properties of Electrons in Matter*. Cambridge: Cambridge University Press, UK.
5. Duff, W. 1988. *Fundamentals of Electromagnetic Compatibility (Handbook Series on Electromagnetic Interference and Compatibility)* Vol. 1. Gainesville, VA: Interference Control Technologies Inc.
6. Foster, R. 1924. A reactance theorem. *Bell Syst. Tech. J.*, **3**(2): 259– 267.
7. Hanna, R. and Rowland, J. 1990. *Fourier Series, Transforms and Boundary Value Problems*. New York: Wiley-Interscience.
8. Hardin, K., Fessler, J., and Bush, D. 1994. Spread spectrum clock generation for the reduction of radiated emissions. In *Proceedings of the IEEE Int. Symp. on Electromagn. Compat.*, 227–231.

9. Hardin, K., Fessler, J., and Bush, D. 1995. A study of the interference potential of spread spectrum clock generation techniques. In *Proceedings of the IEEE Int. Symp. on Electromagn. Compat.*, 624–629.
10. Jackson, J. 1962. *Classical Electrodynamics*. New York: John Wiley & Sons.
11. Johnson, H. and Graham, M. 1993. *High-Speed Digital Design*. Upper Saddle River, NJ: Prentice Hall.
12. Hayt, W. and Kemmerly, Jr., J. 1978. *Engineering Circuit Analysis*. Sydney: McGraw Hill.
13. LePage, W. 1961. *Complex Variables and the Laplace Transform for Engineers*. New York: Dover.
14. Reed, M. and Simon, B. 1980. *Methods of Modern Mathematical Physics*. Vol. I: *Functional Analysis*. New York: Academic Press.
15. Wylie, C., Jr. 1960. *Advanced Engineering Mathematics*. New York: McGraw-Hill.

3

Boundary Value Problems in Electrostatics

No exposition on electrodynamics is complete without delving into some basic boundary value problems encountered in electrostatics. Indeed, neither would the exposition be complete if a cursory glimpse of multipole theory were absent [1,5–8]. The former is crucial to EMC engineers in developing an intuitive feel for real-world problems, and how simplifying a complicated scenario via a toy model can greatly help resolve electromagnetic interference problems. The latter is useful in understanding the basis for various rules of thumb employed by EMC engineers. Unfortunately, as is often the case, sometimes EMC engineers apply these rules with reckless abandon without being cognizant of the origins of the rules.

Finally, the power of the method of images is developed further in this chapter. The technique is particularly useful for solving many problems encountered by EMC engineers. In particular, for 2-dimensional problems, utilizing techniques in complex variables [2,10] also come in very handy, and EMC engineers are encouraged to review the theory of analytic functions to solve two-dimensional Laplace equations encountered in electrostatics and magnetostatics.

3.1 Electromagnetic Boundary Conditions

A brief summary of electromagnetic boundary conditions is collected here for ease of reference. The derivations can be found in Section A.3 of the Appendix. These conditions are utilized in subsequent sections to solve boundary value problems. By way of establishing some notations, let $\Omega_{\pm} \subset \mathbf{R}^3$ be two connected open sets such that $\partial\Omega_0 = \Omega_+ \cap \Omega_-$ is a 2-dimensional surface. Given the pair $(\Omega_{\pm}, \varepsilon_{\pm}, \mu_{\pm}, \sigma_{\pm})$, where $\varepsilon_{\pm}$ is the electric permittivity, $\mu_{\pm}$ is the magnetic permeability, and $\sigma_{\pm}$ is the conductivity on $\Omega_{\pm}$ consider an electric field $\boldsymbol{E}_-$ in Ω_- incident on $\partial\Omega_0$, and the resultant transmitted field $\boldsymbol{E}_+$ in Ω_+. Finally, the unit normal vector field $\boldsymbol{n}_{\pm}$ on $\partial\Omega_0$ is defined to be directed into $\Omega_{\mp}$ and let ρ_0 denote the free surface charge density and $\boldsymbol{J}_0$ the surface current density on $\partial\Omega_0$.

3.1.1 Theorem

Suppose $\sigma_\pm = 0$ and $\rho_0 = 0$. Then, on setting $D_\pm = \varepsilon_\pm E_\pm$, the following conditions hold on $\partial\Omega_0$:

$$\text{(a)}\ \boldsymbol{n}_+ \cdot (\boldsymbol{D}_- - \boldsymbol{D}_+) = 0$$

$$\text{(b)}\ \boldsymbol{n}_+ \times (\boldsymbol{E}_- - \boldsymbol{E}_+) = 0 \qquad \square$$

Condition 3.1.1(a) states that for lossless dielectrics, the normal component of the electric displacement field is continuous across the boundary; by definition then, the normal component of the electric field across the interface must be discontinuous. In contrast, 3.1.1(b) asserts that the tangential component of the electric field across the boundary interface is continuous. A similar interpretation holds for the subsequent results stated below.

3.1.2 Corollary

Suppose $\sigma_\pm \neq 0$, and let $J_\pm = \sigma_\pm E$ be a non-time–varying current density in $\Omega_\pm$. Then, on $\partial\Omega_0$, $\rho_0 \neq 0$ and

$$\text{(a)}\ \boldsymbol{n}_+ \cdot (\boldsymbol{D}_- - \boldsymbol{D}_+) = \rho_0$$

$$\text{(b)}\ \boldsymbol{n}_+ \cdot (\boldsymbol{J}_- - \boldsymbol{J}_+) = 0$$

$$\text{(c)}\ \boldsymbol{n}_+ \times \left(\tfrac{1}{\sigma_-}\boldsymbol{J}_- - \tfrac{1}{\sigma_+}\boldsymbol{J}_+\right) = 0$$

$$\text{(d)}\ \sigma_+ \to \infty \Rightarrow \boldsymbol{n}_+ \times \boldsymbol{E}_- = 0 \quad \text{and} \quad \boldsymbol{n}_+ \cdot \boldsymbol{D}_- = \rho_0 \qquad \square$$

3.1.3 Theorem

Let $\sigma_\pm = 0$ and $J_0 = 0$. Then, the following conditions hold on $\partial\Omega_0$.

$$\text{(a)}\ \boldsymbol{n}_+ \cdot (\boldsymbol{B}_- - \boldsymbol{B}_+) = 0$$

$$\text{(b)}\ \boldsymbol{n}_+ \times \left(\tfrac{1}{\mu_-}\boldsymbol{B}_- - \tfrac{1}{\mu_+}\boldsymbol{B}_+\right) = 0 \qquad \square$$

3.1.4 Corollary

Suppose $\sigma_\pm \neq 0$. Then, on $\partial\Omega_0$, $J_0 \neq 0$ and

$$\text{(a)}\ \boldsymbol{n}_+ \times \left(\tfrac{1}{\mu_-}\boldsymbol{B}_- - \tfrac{1}{\mu_+}\boldsymbol{B}_+\right) = \boldsymbol{J}_0$$

$$\text{(b)}\ \sigma_+ \to \infty \Rightarrow \boldsymbol{n}_+ \times \tfrac{1}{\mu_-}\boldsymbol{B}_- = \boldsymbol{J}_0 \quad \text{and} \quad \boldsymbol{n}_+ \cdot \boldsymbol{D}_- = 0 \qquad \square$$

3.1.5 Remark

A surface charge density ρ_0 and a surface current density J_0 are idealizations that do not truly exist across a boundary interface. More precisely, $\exists\delta > 0$ sufficiently small, and some small open neighborhood $N_\delta \subset \Omega_+ \cup \Omega_-$ such that $\partial\Omega_0 \subset N_\delta$ and $N_\delta \subseteq \bigcup_{x\in\partial\Omega_0} B_\delta(x)$, wherein a differential volume charge density and volume current density exist in N_δ, instead of an idealized surface charge and current densities on $\partial\Omega_0$. Notwithstanding, for convenience, surface charge and current densities are used without further ado. Technically, they only exist when Ω_+ is a perfect conductor.

As a final comment, observe from Theorem 3.1.1 and Corollary 3.1.2 that for lossy dielectrics, at the interface between two media, the continuity requirement is dependent upon the respective conductivities of the media and not on the electric permittivities under steady-state conditions. Indeed, for lossy media, the electric permittivities determine the charge density accumulated at the interface. Note that for general lossy dielectric media, there exists a transient response, depending upon the charge relaxation times before the steady-state condition is attained.

Recall from Equation (1.20) that charges placed in a dielectric will quickly diffuse to the surface according to $\rho(r,t) = \rho_0(r,t)e^{-\frac{t}{\tau}}$, where $\tau \equiv \frac{\varepsilon}{\sigma}$ is called the *charge relaxation time* of the dielectric. More specifically, for time $t \ll \tau$ (charge relaxation time), the dielectric constants determine the electric field profile. Once charge density accumulates on the interface, the electrical conductivities of the media dictate how the electric field behaves in the media; more details follow shortly.

Indeed, it is evident that for an ideal (lossless) dielectric, free charges injected into the dielectric will remain where they are: they will not diffuse onto the boundary as $\tau \to \infty$. On the contrary, for good conductors, $0 < \tau \ll 1$ and hence, charges injected into a conductor will diffuse very rapidly onto the boundary of the conductor. The response of the fields within a lossy dielectric for $\tau > 0$ clearly fall under two categories: (i) transient response when $t \le \tau$, and (ii) steady-state response when $t \gg \tau$. Transient response is usually complicated to solve whereas the steady state is somewhat easier. Examples of steady-state response are given below.

3.1.6 Corollary

Consider a domain $\Omega \subset \mathbf{R}^3$ such that $\Omega = (\Omega_+, \varepsilon_+, \sigma_+) \cup (\Omega_-, \varepsilon_-, \sigma_-)$, $\Omega_+ \cap \Omega_- = \varnothing$, and $\Gamma = \bar{\Omega}_+ \cap \bar{\Omega}_- \neq \varnothing$ is the boundary interface between the two media. Finally, set $\partial\Omega = \Gamma_+ \cup \Gamma_-$, where $\partial\Omega_\pm = \Gamma_\pm \cup \Gamma$. Suppose some fixed potential φ is applied at the boundary of Ω:

$$\varphi = \begin{cases} V_+ \text{ on } \partial\Omega_+ \\ V_- \text{ on } \partial\Omega_- \end{cases}$$

Then, the charge ρ on Γ satisfies $\rho = (\tau_- - \tau_+)J_{+,\perp}$, where $\tau_\pm = \frac{\varepsilon_\pm}{\sigma_\pm}$ is the respective charge relaxation time in $\Omega_\pm$ and $J_{+,\perp} = |J_+ \cdot n_+|$ is the normal component of the current density.

Proof

By definition, $\rho = n_+ \cdot (D_- - D_+) = n_+ \cdot (\varepsilon_- E_- - \varepsilon_+ E_+) = n_+ \cdot (-\varepsilon_- \nabla\varphi_- + \varepsilon_+ \nabla\varphi_+)$ on Γ. However, by Corollary 3.1.2, $0 = n_+ \cdot (J_- - J_+) = n_+ \cdot (-\sigma_- \nabla\varphi_- + \sigma_+ \nabla\varphi_+)$ on Γ implies that $n_+ \cdot J_- = n_+ \cdot J_+ \Rightarrow \partial_{n_+} \varphi_- = \frac{\sigma_+}{\sigma_-} \partial_{n_+} \varphi_+$ and hence,

$$\rho = n_+ \cdot \nabla\varphi_+ \left(-\varepsilon_- \tfrac{\sigma_+}{\sigma_-} + \varepsilon_+\right) = \left(-\tfrac{\varepsilon_-}{\sigma_-} + \tfrac{\varepsilon_+}{\sigma_+}\right)\sigma_+ n_+ \cdot \nabla\varphi_+$$

Now, by construction, $n_+ \cdot J_+ = -J_{+,\perp} \Rightarrow \rho = (\tau_+ - \tau_-)J_{+,\perp}$, as claimed. □

It is thus clear from the above result that $\rho = 0 \Leftrightarrow \tau_+ = \tau_-$ in lossy media. That is, lossy media appear to behave like lossless dielectric media whenever their respective charge relaxation times should coincide.

3.1.7 Remark

As a concrete example, consider the domain $\Omega = (\Omega_+, \varepsilon_+, \sigma_+) \cup (\Omega_-, \varepsilon_-, \sigma_-)$, where $\Omega_- = (0,c) \times (0,a]$ and $\Omega_+ = (0,c) \times [a,b)$, for some $0 < a < b$. Suppose the boundary conditions are:

$$\varphi = \begin{cases} V_0 & \text{for } y = b, \\ 0 & \text{for } y = 0, \end{cases} \quad \text{and} \quad \partial_x \varphi = \begin{cases} 0 & \text{for } x = a \\ 0 & \text{for } x = 0 \end{cases}$$

Consider the Laplace solution $\Delta\varphi_0 = 0$ on Ω for the case wherein $\sigma_\pm = 0$, that is, the lossless case. Finally, let $\Delta\varphi_\infty = 0$ on Ω be the solution for the steady-state case. Then, for

$$0 \le t << \min\left\{\tfrac{\varepsilon_+}{\sigma_+}, \tfrac{\varepsilon_-}{\sigma_-}\right\}$$

the general solution $\varphi \approx \varphi_0$, whereas for

$$t >> \max\left\{\tfrac{\varepsilon_+}{\sigma_+}, \tfrac{\varepsilon_-}{\sigma_-}\right\}$$

$\varphi \approx \varphi_\infty$. The solution that lies between the two extremes is technically a transient solution, and it defines a smooth homotopy between φ_0 and φ_∞ as the charge density build-up reaches a maximum on the interface at $y = a$ if

$\frac{\varepsilon_+}{\sigma_+} \neq \frac{\varepsilon_-}{\sigma_-}$. The transient solution can be solved numerically by following coupled equations subject to the above boundary conditions:

$$\begin{cases} -\Delta\varphi + \mu_0\varepsilon\,\partial_t^2\varphi = \frac{\rho}{\varepsilon} \\ \partial_t\rho = \sigma\Delta\varphi - \sigma\mu_0\varepsilon\,\partial_t^2\varphi \end{cases}$$

This is easily derived via Gauss' law and the charge conservation equation, and exploiting the Lorentz gauge: $\nabla\cdot \boldsymbol{A} + \mu\varepsilon\,\partial_t\varphi = 0$. Indeed, the above equation yields the complete solution.

Explicitly, invoking Gauss: $\varepsilon\nabla\cdot \boldsymbol{E} = \rho \Rightarrow -\varepsilon\Delta\varphi - \varepsilon\,\partial_t\nabla\cdot \boldsymbol{A} = \rho$. Moreover, via charge continuity: $\partial_t\rho = -\nabla\cdot \boldsymbol{J} = \sigma\Delta\varphi + \sigma\,\partial_t\nabla\cdot \boldsymbol{A}$. Finally, applying the Lorentz gauge yields the pair of partial differential equations.

Indeed, it is quite clear from the above discussion that even if the relaxation times are the same, $\frac{\varepsilon_+}{\sigma_+} = \frac{\varepsilon_-}{\sigma_-}$, the field will nevertheless deform from a pure dielectric solution to that of a steady-state conductive solution. The difference is the absence of charge density on the media interface.

3.2 Image Theory Revisited

In Chapter 1, image theory was introduced to solve some electrostatic problems. In this section, this method is developed in some depth to help EMC engineers apply the technique to product development and research. The material herein is organized in a series of examples with various methods demonstrated for ease of reference.

3.2.1 Example

As a first example, the method of inversion is utilized to find the potential induced by a point charge outside a grounded conducting solid sphere. Let $B(a) = \{(x, y, z) \in \mathbf{R}^3 : x^2 + y^2 + z^2 < a^2\}$ be a perfect, electrical, conducting solid sphere located at the origin, and a point charge $Q \neq 0$ located at $r_Q = (x_Q, 0, 0)$ without any loss of generality, where $x_Q > a$. See Figure 3.1.

To begin, consider some image charge Q' in $B(a)$ located at $r'_Q = (x'_Q, 0, 0)$ such that the potential $\varphi \mid \overline{B}(a) = 0$. Set

$$\varphi(r) = \frac{1}{4\pi\varepsilon_0}\left\{\frac{Q}{r} + \frac{Q'}{r'}\right\}$$

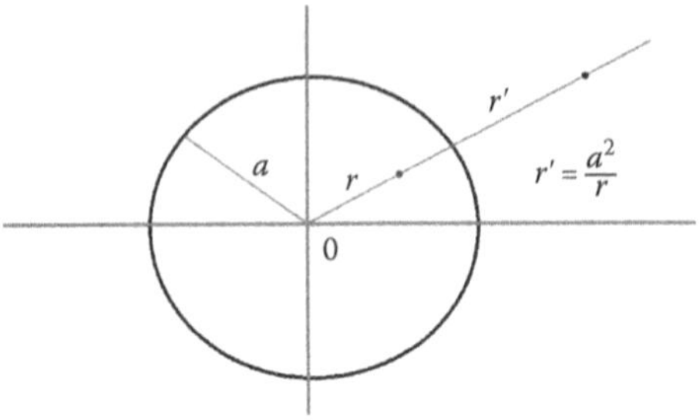

FIGURE 3.1
Transformation of inversion in a sphere.

where

$$r = \|\boldsymbol{r} - \boldsymbol{r}_Q\| \quad \text{and} \quad r' = \|\boldsymbol{r}' - \boldsymbol{r}_Q\|$$

for arbitrary point $\boldsymbol{r} = (x, y, z)$. Then,

$$\varphi \mid \partial B(a) = 0 \Rightarrow \tfrac{Q}{r} + \tfrac{Q'}{r'} = 0 \ \forall \boldsymbol{r} \in \partial B(a)$$

Does the pair $(Q', \boldsymbol{r}')$ exist?

Define a mapping $\zeta : \bar{B}(a) \rightarrow \mathbf{R}^3 - B(a)$ by $\boldsymbol{r} \mapsto \boldsymbol{r}' \equiv \frac{a^2}{r^2}\boldsymbol{r}$, called the *inversion in a sphere*. In short, ζ maps the origin into a point at infinity. Very briefly, the properties [2,4,9] may be summarized as follows: (a) planes are mapped into spheres tangent at the origin of the inversion, (b) spheres are mapped into spheres, (c) points within the circle are mapped into points outside the circle, and (d) points on the circle are mapped onto themselves. Some insight into this mapping can be found in Exercise 3.6.1.

Under the inversion transformation, $\exists\, (Q', \boldsymbol{r}')$ such that $\frac{Q}{r} + \frac{Q'}{r'} = 0 \ \forall \boldsymbol{r} \in \partial B(a)$. Thus,

$$Q' = -\tfrac{Q}{a} r_Q' = -Q \tfrac{a}{r_Q}$$

is the image charge within the sphere such that $\varphi | \partial B(a) = 0$. Thus, for an arbitrary point $\boldsymbol{r} \in \mathbf{R}^3 - B(a)$, the potential is given by

$$\varphi(\boldsymbol{r}) = \tfrac{1}{4\pi\varepsilon_0} \left\{ \tfrac{Q}{R} + \tfrac{Q'}{R'} \right\}$$

where

$$R = \sqrt{(x - x_Q)^2 + y^2 + z^2} \quad \text{and} \quad R' = \sqrt{(x - x_Q')^2 + y^2 + z^2}$$

In spherical coordinates, given an arbitrary point $(r, \phi, \theta) \in \mathbf{R}^3 - B(a)$, under the inversion mapping, $(r, \theta, \phi) \mapsto \left(\frac{a^2}{r}, \theta, \phi\right)$. Let ϑ denote the angle between $\boldsymbol{r}$

and $\boldsymbol{r}_Q = (r_Q, 0, \frac{\pi}{2})$. Then, $R = \sqrt{r^2 + r_Q^2 - 2rr_Q \cos\vartheta}$ and $R' = \sqrt{r^2 + \frac{a^4}{r_Q^2} - 2r\frac{a^2}{r_Q}\cos\vartheta}$. In particular, executing a minor algebraic manipulation leads to

$$R' = \frac{a}{r_Q}\sqrt{\left(\frac{r_Q r}{a}\right)^2 + a^2 - 2r_Q r\cos\vartheta}$$

and hence,

$$\varphi(\boldsymbol{r},\vartheta) = \frac{Q}{4\pi\varepsilon_0}\left\{\frac{1}{\sqrt{r^2+r_Q^2-2r_Q r\cos\vartheta}} - \frac{1}{\sqrt{\left(\frac{Qr}{a}\right)^2+a^2-2r_Q r\cos\vartheta}}\right\} \tag{3.1}$$

on $\mathbf{R}^3 - B(a)$, where $\cos\vartheta = \frac{\boldsymbol{r}\cdot\boldsymbol{r}_Q}{r_Q r} = \sin\phi\cos\theta$. Show this in Exercise 3.6.3.

Finally, from Equation (3.1), it is clear that the Green's function for a unit charge external to a grounded conducting sphere is

$$G(\boldsymbol{r},\boldsymbol{r}_Q) = \frac{1}{\sqrt{r^2+r_Q^2-2r_Q r\cos\vartheta}} - \frac{1}{\sqrt{\left(\frac{r_Q r}{a}\right)^2+a^2-2r_Q r\cos\vartheta}} \tag{3.2}$$

In particular, if $\boldsymbol{r}_Q = (r_Q, \theta_Q, \phi_Q)$, then $\cos\vartheta = \sin\phi\sin\phi_Q\cos(\theta-\theta_Q) + \cos\phi\cos\phi_Q$. See Exercise 3.6.2. □

3.2.2 Example

Suppose the above conducting solid sphere $\overline{B}(a)$ is charged at some constant potential φ_0. What is the resultant potential in $\mathbf{R}^3 - B(a)$?

First, it is enough to observe that as $\overline{B}(a)$ is a perfect conductor; it follows that $\varphi \mid \overline{B}(a) = \varphi_0$, and hence, may represent φ_0 by some charge q located at the center of $\overline{B}(a)$. Now, observe further that as the external charge Q induces an image charge Q' in $B(a)$, it follows that the replacement charge $q \to q - Q'$ must be made so that the resultant charge in $B(a)$ in the presence of Q is q.

Then, by the superposition principle, it follows immediately that on $\mathbf{R}^3 - B(a)$,

$$\varphi(\boldsymbol{r},\vartheta) = \frac{Q}{4\pi\varepsilon_0}\left\{\frac{1}{\sqrt{r^2+r_Q^2-2r_Q r\cos\vartheta}} - \frac{1}{\sqrt{\left(\frac{r_Q r}{a}\right)^2+a^2-2r_Q r\cos\vartheta}}\right\} + \frac{1}{4\pi\varepsilon_0}\frac{q-Q'}{r}$$

At $r = a$,

$$\varphi(\boldsymbol{r},\vartheta) = \frac{Q}{4\pi\varepsilon_0}\times 0 + \frac{1}{4\pi\varepsilon_0}\frac{q-Q'}{r} = \varphi_0$$

Hence, $q - Q' = 4\pi\varepsilon_0\varphi_0 a$; and substituting this into the above equation yields

$$\varphi(r,\vartheta) = \frac{Q}{4\pi\varepsilon_0}\left\{\frac{1}{\sqrt{r^2+r_Q^2-2r_Q r\cos\vartheta}} - \frac{1}{\sqrt{\left(\frac{r_Q r}{a}\right)^2+a^2-2r_Q r\cos\vartheta}}\right\} + \varphi_0\frac{a}{r} \tag{3.3}$$

As a quick verification, it is seen that by construction, $\{\cdots\}|_{r=a}=0$ and hence, $\varphi(a,\vartheta)=\varphi_0\ \forall\vartheta$, as required. □

3.2.3 Example

Consider the space $\Omega = \{(x,y,z)\in\mathbf{R}^3 : 0<y<a\}$, for some constant $a > 0$, and suppose a point charge Q is located at $r_0=(x_0,y_0,c)\in\Omega$. Suppose $\partial\Omega = \partial\mathbf{R}^3_+\cup\partial\Omega_a$ are perfect electrical conductors, where $\partial\Omega_a=\{(x,y,a): x,y\in\mathbf{R}\}$. Solve the Dirichlet boundary value problem:

$$\begin{cases} -\Delta\varphi = \frac{1}{\varepsilon_0}Q\delta^3(r-r_0) & \text{on } \Omega \\ \varphi = 0 & \text{on } \partial\Omega \end{cases}$$

Now, as $\lim\limits_{x,y\to\pm\infty}\varphi\to 0$, it follows that [10, p. 226] if a Green's function G for the boundary value problem can be determined, then the solution is given by

$$\varphi(r) = \int_\Omega \frac{Q}{\varepsilon_0}\delta^3(r-r')G(r,r')d^3r' = \frac{Q}{\varepsilon_0}G(r,r_0)$$

To find the Green's function for Ω, it suffices to consider the space $\Omega' = \mathbf{R}^2\times(0,1)$ and construct a Green's function G' on Ω'. Toward this end, consider Figure 3.2.

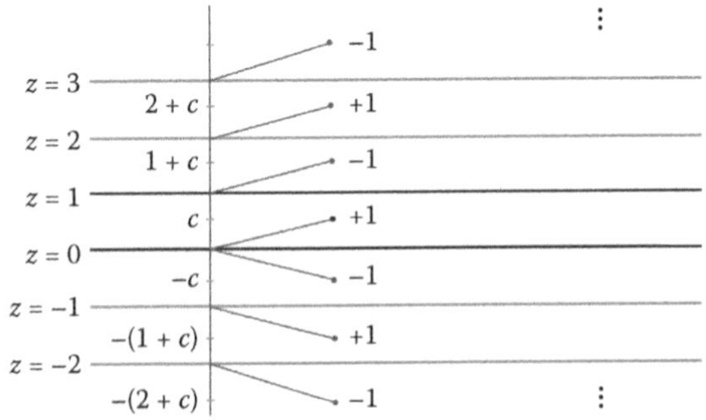

FIGURE 3.2
Infinite sequence of image charges induced by a unit charge.

Referring to Figure 3.2, set a unit charge at $r=(x,y,c)$, and consider an image charge placed at $(x,y,-c)$, that is, the reflection across the x–y plane $\mathbf{E}_0=\mathbf{R}^2$. The presence of the image charge preserves the boundary condition on $\mathbf{E}_0$. Next, consider the reflection across the affine plane $\mathbf{E}_1=\mathbf{R}^2+\{(0,0,1)\}$ independent of $\mathbf{E}_0$ at $(x,y,2-c)$. Placing an image charge at $(x,y,2-c)$ preserves the boundary condition at $\mathbf{E}_1$. However, the presence of an image charge at $(x,y,2-c)$ breaks the boundary condition on $\mathbf{E}_0$; hence, a third image charge must be placed at $(x,y,c-2)$. Likewise, to preserve the boundary condition of $\mathbf{E}_1$ as the result of the second image charge across $\mathbf{E}_0$, a fourth image charge must be placed at $(x,y,2+c)$. By induction, the process continues indefinitely, yielding a sequence of image charges at the following locations:

$$r_k=(x,y,2k+c) \text{ and } \bar{r}_k=(x,y,2k-c)\,\forall k\in\mathbf{Z}$$

See Figure 3.2.

Thus, on setting

$$\delta r_k=|\boldsymbol{r}'-\boldsymbol{r}_k|=\sqrt{(x'-x)^2+(y'-y)^2+(z'-c-2k)^2}$$

and

$$\delta \bar{r}_k=|\boldsymbol{r}'-\bar{\boldsymbol{r}}_k|=\sqrt{(x'-x)^2+(y'-y)^2+(z'+c-2k)^2}$$

and observing that for any fixed $z'\in\mathbf{Z}$ and $c\in(0,1)$,

$$\{|z'+c+2k|:k\in\mathbf{Z}\}=\{|z'-c+2k|:k\in\mathbf{Z}\}\Rightarrow\sum_{k\in\mathbf{Z}}\left\{\tfrac{1}{\delta r_k}-\tfrac{1}{\delta\bar{r}_k}\right\}=0$$

This suggests at once that

$$G'(\boldsymbol{r},\boldsymbol{r}')=\tfrac{1}{4\pi}\sum_{k\in\mathbf{Z}}\left\{\tfrac{1}{\delta r_k}-\tfrac{1}{\delta\bar{r}_k}\right\}$$

is the sought-for Green's function on Ω'. Moreover, as

$$\frac{1}{4\pi}\frac{1}{r}$$

is the Green's function for a point charge, that is, $-\Delta\frac{1}{r}=4\pi\delta(r)$, and $G'(\boldsymbol{r},\boldsymbol{r}')$ satisfies the boundary conditions at $z'=0,1$, the uniqueness of Poisson's equation implies immediately that $G'(\boldsymbol{r},\boldsymbol{r}')$ is the required Green's function on Ω'.

The generalization to Ω is trivial. Transform $G'(\boldsymbol{r},\boldsymbol{r}') \to G(\boldsymbol{r},\boldsymbol{r}')$ under the transformation $2k \to 2ka$. Then, clearly,

$$\{|c+2ka| : k \in \mathbf{Z}\} = \{|-c+2ka| : k \in \mathbf{Z}\} \Rightarrow \sum_{k\in\mathbf{Z}} \left\{ \tfrac{1}{\delta R_k} - \tfrac{1}{\delta \bar{R}_k} \right\} = 0 \tag{3.4}$$

where

$$\delta R_k = |\boldsymbol{r}' - \boldsymbol{R}_k| = \sqrt{(x'-x)^2 + (y'-y)^2 + (z'-c-2ka)^2}$$

$$\delta \bar{R}_k = |\boldsymbol{r}' - \bar{\boldsymbol{R}}_k| = \sqrt{(x'-x)^2 + (y'-y)^2 + (z'+c-2ka)^2}$$

with $z' = 0$. Finally, note that when $z' = a$, Equation (3.4) is once again satisfied: the construction would fail if $k \in \mathbf{N}$. The required Green's function on Ω is thus

$$G'(\boldsymbol{r},\boldsymbol{r}') = \frac{1}{4\pi} \sum_{k\in\mathrm{Z}} \left\{ \tfrac{1}{\delta R_k} - \tfrac{1}{\delta \bar{R}_k} \right\} \tag{3.5}$$

□

This section ends with two more examples regarding the application of the method of images: determine the potential resulting from a charged cylinder over a ground plane and a charged conductive sphere over a lossless dielectric half-space.

3.2.4 Example

Consider a cross-section of an infinitely long conducting cylinder C of radius $a > 0$ over a conducting ground plane $\partial\mathbf{R}_+^2 = \{(x,y) \in \mathbf{R}^2 : y = 0\}$: $C = \{(x,y) \in \mathbf{R}^2 : x^2 + (y-y_0)^2 \le a\}$, where $y_0 > a > 0$, and set $\Omega = \mathbf{R}_+^2 - C$. By construction, the center of the cylinder is $(0, y_0)$. Suppose that the cylinder is set at a potential $\varphi = \varphi_0$, determine the potential in Ω.

Consider a line charge λ located at $(0, y_+) \in C$, where the pair (λ, y_+) are to be determined. By the method of images, consider a line charge density $-\lambda$ located at $y = (0, -y_+)$. Recall that the potential in $\mathbf{R}_+^2 - \{(0, y_+)\}$ is given by

$$\varphi(x,y) = -\frac{\lambda}{2\pi\varepsilon_0} \ln \sqrt{\tfrac{x^2+(y-y_+)^2}{x^2+(y+y_+)^2}} \tag{3.6}$$

To see Equation (3.6), consider, for simplicity, a unit charge at the center of a disk $B_r(0) = \{(x,y) : x^2 + y^2 \le r^2\}$ of radius r. Invoking Gauss' law, $\nabla \cdot \boldsymbol{E} = \frac{\rho}{\varepsilon_0} \Rightarrow -\Delta\psi = \frac{\delta(x)}{\varepsilon_0}$. By Stokes' theorem,

$$-\int_{B_r(0)} \Delta\psi \mathrm{d}^2x = -\int_{\partial B_r(0)} \nabla\psi \cdot \mathbf{e}_r \mathrm{d}^2x = -\int_0^{2\pi} \partial_r \psi r \,\mathrm{d}\theta = \int \tfrac{\delta(x)}{\varepsilon_0} \mathrm{d}x = \tfrac{1}{\varepsilon_0}$$

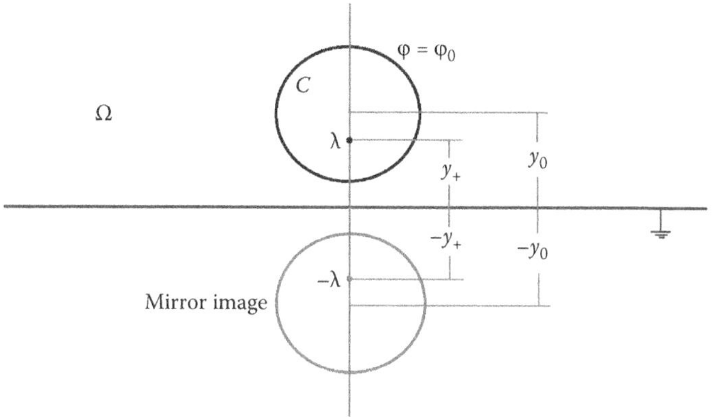

FIGURE 3.3
Potential generated by a charged infinitely long conducting cylinder.

whence

$$-\partial_r \psi = \frac{1}{2\pi\varepsilon_0 r} \Rightarrow \psi = -\frac{1}{2\pi\varepsilon_0} \ln r$$

and the potential resulting from two line charges yields

$$\varphi(x,y) = -\frac{\lambda}{2\pi\varepsilon_0} \ln \sqrt{x^2 + (y - y_+)^2} + \frac{\lambda}{2\pi\varepsilon_0} \ln \sqrt{x^2 + (y + y_+)^2}$$

establishing Equation (3.6). See Figure 3.3 for an intuitive explanation.

Observe that as C is a perfect conductor, it forms a surface of equipotential. Hence, it suffices to determine surfaces of equipotential defined by the line charge density. From (3.6), let $S(V) \subset \mathbf{R}^2_+$ define an equipotential surface such that

$$V = -\frac{\lambda}{2\pi\varepsilon_0} \ln \sqrt{\frac{x^2 + (y - y_+)^2}{x^2 + (y + y_+)^2}}$$

for all $(x,y) \in S(V)$, where V is a constant potential. Set $K = e^{-2\pi\varepsilon_0 V/\lambda}$. Then, by definition,

$$K^2 = \frac{x^2 + (y - y_+)^2}{x^2 + (y + y_+)^2} \Leftrightarrow x^2 + y^2 + 2yy_+ \frac{K^2+1}{K^2-1} + y_+^2 = 0$$

Noting that

$$\left(y + y_+ \frac{K^2+1}{K^2-1}\right)^2 = y^2 + 2yy_+ \frac{K^2+1}{K^2-1} + \left(y_+ \frac{K^2+1}{K^2-1}\right)^2$$

it follows clearly that the above equation reduces, after some algebraic manipulation, to

$$x^2 + \left(y + y_+ \tfrac{K^2+1}{K^2-1}\right)^2 = \left(y_+ \tfrac{K^2+1}{K^2-1}\right)^2 - y_+^2 = \left(\tfrac{2K}{K^2-1} y_+\right)^2 \quad (3.7)$$

However, this is nothing but the equation of a circle centered at

$$\left(0, -\tfrac{K^2+1}{K^2-1} y_+\right)$$

below the ground plane and via reflection, at

$$\left(0, \tfrac{K^2+1}{K^2-1} y_+\right)$$

above the ground plane. Set

$$y_0 = \tfrac{K^2+1}{K^2-1} y_+$$

Then $(0, y_0)$ is the location of the line charge above the ground plane such that C is at equipotential.

It thus remains to determine $\lambda = \lambda(\varphi_0)$. From Equation (3.7), it follows immediately that $a^2 = \left(\tfrac{2K}{K^2-1} y_+\right)^2$. Now, observe that

$$a^2 + y_+^2 = y_+^2 \left\{\tfrac{4K^2+(K^2-1)^2}{(K^2-1)^2}\right\} = y_+^2 \left(\tfrac{K^2+1}{K^2-1}\right)^2 = y_0^2$$

Hence, $y_+ = \sqrt{y_0^2 - a^2}$. Next, set $\alpha = \frac{y_0}{y_+}$. Then,

$$\alpha = \tfrac{K^2+1}{K^2-1} \Rightarrow K^2 = \tfrac{\alpha+1}{\alpha-1} \Rightarrow K = \sqrt{\tfrac{\alpha+1}{\alpha-1}}$$

whence

$$K = e^{-2\pi\varepsilon_0 V/\lambda} \Rightarrow \lambda = -\frac{2\pi\varepsilon_0\varphi_0}{\ln\sqrt{\frac{\alpha+1}{\alpha-1}}} \Rightarrow \varphi = -\frac{\varphi_0}{\ln\sqrt{\frac{\alpha+1}{\alpha-1}}} \ln\sqrt{\frac{x^2+(y-y_+)^2}{x^2+(y+y_+)^2}}$$

as required. □

3.2.5 Example

Let $\mathbf{R}_-^3 = \{(x,y,z) \in \mathbf{R}^3 : z \le 0\}$ denote an infinite dielectric half-space with electric permittivity ε_- and $\mathbf{R}_+^3 = \mathbf{R} - \mathbf{R}_-^3$ denote a pure dielectric medium of electric permittivity ε_+. Suppose $D_a(r_0) = \{(x, y, z) \in \Omega : x^2 + y^2 + (z - z_0)^2 \le a^2\}$ is a charged PEC sphere of charge Q, the center of which is $r_0 = (0, 0, z_0)$ above the dielectric plane, where $z_0 > a > 0$. Determine the potential field in $\Omega = \mathbf{R}_+^3 - D_a(r_0)$.

Because $D_a(r_0)$ is a perfect conductor, its surface is at equipotential. Hence, suppose without loss of generality that $\exists Q_0$ is such that V_0 is the potential on the surface of $D_a(r_0)$. Then, by Exercise 3.5.8,

$$\exists Q_1 = -\tfrac{\varepsilon_- - \varepsilon_+}{\varepsilon_- + \varepsilon_+} Q_0$$

located at $r_1 = (0, 0, z_1)$, with $z_1 = -z_0$, such that the potential on $\partial\mathbf{R}_-^3$ is zero. However, the presence of Q_1 violates the equipotential condition on $\partial D_a(r_0)$. Hence, appealing to Example 3.2.1, $\exists Q_2 = -\frac{a}{2z_0} Q_1$ at $r_2 = (0, 0, z_2)$, where $z_2 = z_0 - \frac{a}{2z_0}$, such that $\partial D_a(r_0)$ is once again an equipotential surface.

It is clear by now that the presence of Q_2 breaks the equipotential condition on $\partial\mathbf{R}_-^3$. Thus, a charge

$$Q_3 = -\tfrac{\varepsilon_- - \varepsilon_+}{\varepsilon_- + \varepsilon_+} Q_2$$

must be placed at $r_3 = (0, 0, z_3)$, where $z_3 = -z_2$, such that the equipotential condition on $\partial\mathbf{R}_-^3$ is restored. As this in turn violates the equipotential condition on $\partial D_a(r_0)$, another fictitious charge

$$Q_4 = \frac{a}{2z_4 + a^2\{2z_0 - a^2/(2x_0)\}^{-1}} Q_3 \quad \text{at} \quad r_4 = (0, 0, z_4)$$

where

$$z_4 = 2z_0 - \frac{a^2}{2z_0 - a^2/(2z_0)}$$

For notational convenience, set

$$\alpha = \tfrac{\varepsilon_- - \varepsilon_+}{\varepsilon_- + \varepsilon_+}, d_0 = 2z_0 \quad \text{and} \quad d_1 = d_0 - \tfrac{a^2}{d_0}$$

Then, $Q_1 = -\alpha Q_0$, $Q_2 = -\frac{a}{d_0} Q_1$, $Q_3 = -\alpha Q_2$, and $Q_4 = -\frac{a}{d_1} Q_3$, where $d_2 = d_0 - \frac{a^2}{d_1}$, with $z_2 = d_1$, $z_3 = -z_2$, and $z_4 = d_2$. By induction, it is clear that

$$Q_{2k-1} = -\alpha Q_{2k-2} \quad \text{and} \quad Q_{2k} = -\tfrac{a}{d_{2k-1}} Q_{2k-1} \text{ for all } k = 1, 2, \ldots$$

$$z_{2k-1} = -z_{2k-2} \quad \text{and} \quad z_{2k} = d_{2k-1} \text{ for all } k = 1, 2, \ldots$$

and $Q = \Sigma_{k \ge 0} Q_{2k}$ by construction.

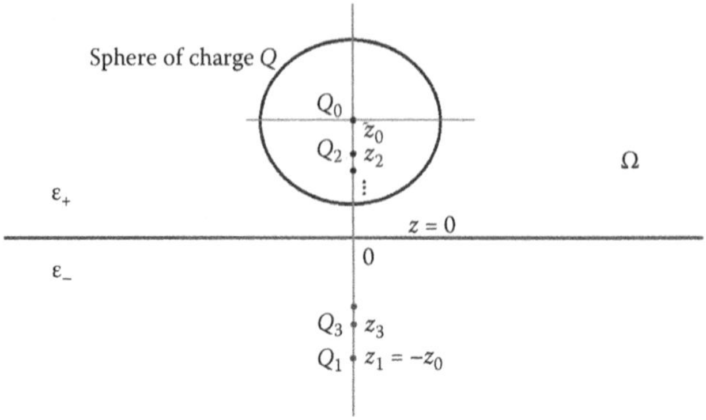

FIGURE 3.4
Method of images applied to a charged conducting sphere over a dielectric plane.

Indeed, further simplification can be achieved as

$$
\begin{aligned}
Q_{2k+1} &= -\alpha Q_{2k} = -\alpha\left(-\frac{a}{d_{2k-1}}\right)Q_{2k-1} = \cdots = (-1)^{2k+1}\alpha^{k+1}\frac{a^k}{d_{2k-1}d_{2k-2}\cdots d_0}Q_0 \\
&= -\frac{\alpha^{k+1}a^k}{d_{2k-1}d_{2k-2}\cdots d_0}Q_0
\end{aligned}
$$

$$
\begin{aligned}
Q_{2k} &= -\frac{a}{d_{2k-1}}Q_{2k-1} = -\alpha\left(-\frac{a}{d_{2k-1}}\right)Q_{2k-1} = \cdots = (-1)^{2k}\alpha^{k}\frac{a^k}{d_{2k-1}d_{2k-2}\cdots d_1}Q_0 \\
&= \frac{\alpha^{k}a^k}{d_{2k-1}d_{2k-2}\cdots d_0}Q_0
\end{aligned}
$$

Thus, given an arbitrary point $\boldsymbol{r} = (x,y,z) \in \Omega$, by the superposition principle, the electric potential field φ on Ω is given by the summation of all the charges

$$
\begin{aligned}
\varphi &= \frac{1}{4\pi\varepsilon_+}\sum_{k\geq 0}\left\{\frac{Q_{2k}}{\sqrt{x^2+y^2+(z-z_{2k})^2}} + \frac{Q_{2k+1}}{\sqrt{x^2+y^2+(z-z_{2k+1})^2}}\right\} \\
&= \frac{Q_0}{4\pi\varepsilon_+}\sum_{k\geq 0}\frac{\alpha^k a^k}{d_{2k-1}d_{2k-2}\cdots d_0}\left\{\frac{1}{\sqrt{x^2+y^2+(z-z_{2k})^2}} - \frac{\alpha}{\sqrt{x^2+y^2+(z-z_{2k+1})^2}}\right\}
\end{aligned}
$$

See Figure 3.4 for the mathematical representation of fictitious charges. □

3.3 Multipole Expansion

It is clear by now that the scalar potential resulting from a localized static charge density ρ in $\mathbf{R}^3$ satisfies the Poisson equation $-\Delta\varphi(\boldsymbol{r}) = \frac{\rho(\boldsymbol{r})}{\varepsilon}$; see Exercise 3.6.5. From Section A.4, it is seen that the Green's function for a unit point charge is $G(\boldsymbol{r}-\boldsymbol{r}')=\frac{1}{4\pi}\frac{1}{|\boldsymbol{r}-\boldsymbol{r}'|}$. Now, in $\mathbf{R}^3$, $\lim_{r\to\infty}\rho(\boldsymbol{r})=0$, whence the solution to Poisson's equation is

$$\varphi(\boldsymbol{r})=\frac{1}{4\pi\varepsilon_0}\int\frac{\rho(\boldsymbol{r}')}{|\boldsymbol{r}-\boldsymbol{r}'|}\mathrm{d}^3\boldsymbol{r}' \tag{3.8}$$

in free space, where a charge density $\rho:\mathbf{R}^3\to\mathbf{R}$ with nonempty compact support, $\Omega_\rho=\mathrm{supp}(\rho)$ is assumed in all that follows.

Now, observe that for $r>>r'$, the denominator $\frac{1}{|\boldsymbol{r}-\boldsymbol{r}'|}$ can be Taylor expanded as follows. Set $f(r)=\frac{1}{|\boldsymbol{r}-\boldsymbol{r}'|}$. Then, $|\boldsymbol{r}-\boldsymbol{r}'|=\sqrt{r^2+r'^2-2r'r\cos\theta}$, where $\cos\theta=\frac{\boldsymbol{r}\cdot\boldsymbol{r}'}{rr'}$. This can be rewritten as

$$|\boldsymbol{r}-\boldsymbol{r}'|=r\sqrt{1+\left(\tfrac{r'}{r}\right)^2-2\tfrac{r'}{r}\cos\theta}\equiv r\sqrt{1+\xi}$$

where $\xi=\left(\frac{r'}{r}\right)^2-2\frac{r'}{r}\cos\theta$ and $|\xi|<<1$, whence, Taylor expanding about 1 yields

$$\{1+\xi\}^{-\frac{1}{2}}=1-\tfrac{1}{2}\xi+\tfrac{3}{8}\xi^2-\tfrac{5}{16}\xi^3+o(\xi^4)$$

That is,

$$f(\boldsymbol{r})=\tfrac{1}{r}\left\{1+\tfrac{r'}{r}\cos\theta+\left(\tfrac{r'}{r}\right)^2\tfrac{3\cos^2\theta-1}{2}+o\left(\left(\tfrac{r'}{r}\right)^3\right)\right\}$$

However, this is just the *spherical harmonic expansion* [5,8,9]:

$$f(\boldsymbol{r})=\tfrac{1}{r}\sum_{n=0}^{\infty}P_n(\cos\theta)\left(\tfrac{r'}{r}\right)^n \tag{3.9}$$

for $r>r'$, where

$$P_n(x)=\sum_{k=0}^{[n]}\frac{(-1)^k(2n-k)!}{2^n k!(n-k)!(n-2k)!}x^{n-2k}$$

defines the *Legendre polynomials* of order n, with

$$[n]=\begin{cases}\frac{n}{2}, & \text{for } n \text{ even}\\ \frac{n-1}{2}, & \text{for } n \text{ odd}\end{cases}$$

It can be expressed more compactly via the *Rodrigues' formula*

$$P_n(x) = \frac{1}{2^n n!}\frac{d^n}{dx^n}(x^2-1)^n$$

The first few terms are evaluated as

$$P_0(x) = 1$$

$$P_1(x) = x$$

$$P_2(x) = \tfrac{1}{2}(3x^2-1)$$

$$P_3(x) = \tfrac{1}{2}(5x^3-3x),\ldots$$

For completeness, consider the scenario wherein $r < r'$; physically, this corresponds precisely to the case where the point is located within the charge body Ω_ρ. It can be shown that

$$f(\boldsymbol{r}) = \tfrac{1}{r'}\sum_{n=0}^{\infty} P_n(\cos\theta)\left(\tfrac{r}{r'}\right)^n \tag{3.10}$$

Together, Equations (3.9) and (3.10) are often expressed in a more compact fashion as

$$f(\boldsymbol{r}) = \tfrac{1}{r_>}\sum_{n=0}^{\infty} P_n(\cos\theta)\left(\tfrac{r_<}{r_>}\right)^n \tag{3.11}$$

where $r_> = \max\{r, r'\}$ and $r_< = \min\{r, r'\}$.

3.3.1 Definition

The multipole expansion of $\varphi(\boldsymbol{r}) = \int_{\Omega_\rho} \frac{\rho(\boldsymbol{r}')}{|\boldsymbol{r}-\boldsymbol{r}'|} d^3\boldsymbol{r}'$, for $r > r'$, is defined by

$$\varphi(\boldsymbol{r}) = \tfrac{1}{4\pi\varepsilon r}\sum_{n=0}^{\infty}\int_{\Omega_\rho} \rho(\boldsymbol{r}')P_n(\cos\theta)\left(\tfrac{r'}{r}\right)^n d^3\boldsymbol{r}' \tag{3.12}$$

The first term corresponds to an electric *monopole*, the second term a *dipole*, the third term a *quadrupole*, and so forth.

3.3.2 Remark

Notice that whereas the monopole falls off as $\frac{1}{r}$, the dipole falls off as $\frac{1}{r^2}$, the quadrupole falls off as $\frac{1}{r^3}$, and so on. It is thus obvious that for an arbitrary compact source Ω_ρ, if the distance r away from the source satisfies $r \gg \text{diam}(\Omega_\rho)$, where $\text{diam}(\Omega_\rho)$ denote the diameter of Ω_ρ (i.e., the largest side of Ω_ρ or diagonal, whichever is larger), and is defined by $\text{diam}(\Omega_\rho) = \max\{\|\boldsymbol{x}-\boldsymbol{x}'\| : \boldsymbol{x}, \boldsymbol{x}' \in \Omega_\rho\}$, then Ω_ρ may be approximated by a point source plus a small correction term from the dipole contribution that rapidly becomes negligible for very large r. On the other hand, close to the charge distribution Ω_p, the higher-order poles dominate, and in particular, it is incorrect to approximate the extended source by a point source. This is evident from Examples 3.2.1 and 3.2.5.

From the definition of the magnetic potential

$$A(\boldsymbol{r}) = \frac{\mu_0}{4\pi}\int \frac{J(\boldsymbol{r}')}{|\boldsymbol{r}-\boldsymbol{r}'|}\mathrm{d}^3\boldsymbol{r}'$$

it follows from the above discussion *mutatis mutandis* that multipole expansion for the magnetic field $B = \nabla \times A$ for $r > r'$:

$$A(\boldsymbol{r}) = \frac{\mu_0}{4\pi r}\sum_{n=0}^{\infty}\int J(\boldsymbol{r}')P_n(\cos\theta)\left(\frac{r'}{r}\right)^n \mathrm{d}^3\boldsymbol{r}' \tag{3.13}$$

Finally, to complete the discussion, consider the special case wherein the charge (or current) is distributed on the surface of some compact set S. This leads to the introduction of *spherical harmonics* $Y_{nm}(\theta,\phi)$ and they are defined as follows. First, the *associated Legendre polynomials* are given by

$$P_n^m(x) = (-1)^m(1-x^2)^{\frac{m}{2}}\frac{\mathrm{d}^{n+m}}{\mathrm{d}x^{n+m}}(x^2-1)^n$$

An important property satisfied by the polynomials is the *orthogonality relation*:

$$\int_{-1}^{1} P_k^m(x)P_n^m(x)\mathrm{d}x = \frac{2}{2n+1}\frac{(n+m)!}{(n-m)!}\delta_{kn}$$

and

$$\delta_{ij} = \begin{cases} 1 \text{ for } i = j \\ 0 \text{ for } i \neq j \end{cases}$$

is the *Kronecker-delta* function. Then, the normalized spherical harmonics are defined by

$$Y_{nm}(\theta,\phi) = \sqrt{\frac{2n+1}{4\pi}\frac{(n-m)!}{(n+m)!}}P_n^m(\cos\theta)\mathrm{e}^{\mathrm{i}m\varphi} \quad \text{and} \quad Y_{n,-m}(\theta,\phi) = (-1)^m Y_{nm}^*(\theta,\phi)$$

satisfying the orthogonality relation $\int_0^{2\pi}\mathrm{d}\phi\int_0^{\pi}\sin\theta\mathrm{d}\theta Y_{n'm'}^*(\theta,\phi)Y_{nm}(\theta,\phi) = \delta_{n'n}\delta_{m'm}$. In particular, one has the completeness relation:

$$\sum_{n=0}^{\infty}\sum_{m=-n}^{n} Y_{nm}^*(\theta',\phi')Y_{nm}(\theta,\phi) = \delta(\phi-\phi')\delta(\cos\theta - \cos\theta')$$

As mentioned before, an immediate application of spherical harmonics is to expand a function $f:\partial S \to R$ defined on the boundary of some solid sphere $S \subset \mathbf{R}^3$:

$$f(\theta,\phi) = \sum_{n=0}^{\infty}\sum_{m=-n}^{n} A_{nm}Y_{nm}(\theta,\phi)$$

where the coefficients A_{nm} are defined via the orthogonal relation as

$$A_{nm} = \int_0^{2\pi} d\phi \int_0^{\pi} \sin\theta d\theta Y^*_{n'm'}(\theta,\phi) f(\theta,\phi)$$

Here, f could represent a boundary value or surface charge. This concludes a brief sketch on the topic of spherical harmonics.

3.4 Steady-State Currents

Recall from the conservation of charge that $\partial_t \rho + \nabla \cdot J = 0$ in some domain Ω. Hence, when the charge density is constant, that is, $\partial_t \rho = 0 \Rightarrow \nabla \cdot J = 0$ on Ω, this defines the condition for steady-state current. The intuitive notion behind this definition is clear: when a constant current flows through a wire, $\frac{d}{dt} I = 0$ and hence, the current density is independent of time.

3.4.1 Example

Consider two rectangles $(\Omega_\pm, \varepsilon_\pm, \sigma_\pm)$, where $\Omega_+ = (0,a)\times(b_-,b_+)$, $\Omega_- = (0,a)\times(0,b_-)$, and $\varepsilon_\pm$ are the respective electric permittivities, and $\sigma_\pm$ the respective conductivities. Set $\Omega = \Omega_- \cup \Omega_+ \cup \Gamma$, where $\Gamma = \{(x,b_-): 0 < x < a\}$. Find the potential $\varphi \in C^2(\Omega) \cap C^0(\bar{\Omega})$ satisfying the following Dirichlet boundary value problem:

$$\Delta\varphi = 0 \text{ on } \Omega \tag{3.14}$$

$$\varphi = \begin{cases} V_+ \text{ for } y = b_+ \\ V_- \text{ for } y = b_- \\ 0 \text{ for } x = 0 \\ 0 \text{ for } x = a \end{cases} \tag{3.15}$$

First, observe that under steady-state conditions, $J_+ \cdot \mathbf{e}_y = J_- \cdot \mathbf{e}_y$ on Γ, where $J_\pm = \sigma_\pm E_\pm$ is the current density defined on $\Omega_\pm$ and $E_\pm$ is the electric field on $\Omega_\pm$. Set $\varphi_\pm$ to be the potential on $\Omega_\pm$. Then, on Γ, $\sigma_- \partial_y \varphi_- = \sigma_+ \partial_y \varphi_+$. From Exercise 3.6.4, the general solution on $\Omega_\pm$ is given by

$$\varphi_\pm(x,y) = \sum_{n>0} \left\{ \alpha_n^\pm \cosh \lambda_n y + \beta_n^\pm \sinh \lambda_n y \right\} \sin \lambda_n x \tag{3.16}$$

where $\lambda_n = \frac{\pi n}{a}$, $n = 1, 2, \ldots$

Now, the coefficients $\alpha_n^\pm, \beta_n^\pm$ are evaluated by imposing Equation (3.15) on (3.16). Proceeding systematically, consider $y = b$ on Ω_+:

$$V_+ = \varphi_+(x,b) = \sum_{n>0} \left\{ \alpha_n^+ \cosh \lambda_n b + \beta_n^+ \sinh \lambda_n b \right\} \sin \lambda_n x$$

Thus, multiplying both sides by $\sin \lambda_m x$ and integrating along $(0, a)$ yields

$$\alpha_n^+ \cosh \lambda_n b + \beta_n^+ \sinh \lambda_n b = V_+ \int_0^a \sin \lambda_n x dx = \begin{cases} \frac{4V_+}{n\pi} & \text{for } n \text{ odd} \\ 0 & \text{for } n \text{ even} \end{cases} \tag{3.17a}$$

Likewise, for $y = 0$ on Ω_-:

$$\alpha_n^- \cosh \lambda_n b = V_- \int_0^a \sin \lambda_n x dx = \begin{cases} \frac{4V_-}{n\pi} & \text{for } n \text{ odd} \\ 0 & \text{for } n \text{ even} \end{cases} \tag{3.17b}$$

Furthermore, from the continuity of the potential across a boundary,

$$\left\{ \alpha_n^+ \cosh \lambda_n a + \beta_n^+ \sinh \lambda_n a \right\} \sin \lambda_n x = \left\{ \alpha_n^- \cosh \lambda_n a + \beta_n^- \sinh \lambda_n a \right\} \sin \lambda_n x \; \forall x$$

That is,

$$\alpha_n^+ \cosh \lambda_n a + \beta_n^+ \sinh \lambda_n a = \alpha_n^- \cosh \lambda_n a + \beta_n^- \sinh \lambda_n a \tag{3.17c}$$

and from the continuity of the normal component of the current density,

$$\frac{\sigma_+}{\sigma_-} \left\{ \alpha_n^+ \sinh \lambda_n a + \beta_n^+ \cosh \lambda_n a \right\} = \alpha_n^- \sinh \lambda_n a + \beta_n^- \cosh \lambda_n a \tag{3.17d}$$

Thus, via Equation (3.17), the coefficients can be determined. After some tedious but trivial routine manipulations, the coefficients to (3.16) are:

$$\alpha^+_{2n-1} = \frac{4}{(2n-1)\pi}\left\{V_+\left(1-\frac{\sigma_+}{\sigma_-}\right)\coth\lambda_{2n-1}a - \frac{2V_-}{\sinh 2\lambda_{2n-1}b}\right\}\times$$

$$\left\{\frac{\sigma_+}{\sigma_-}\left(\sinh\lambda_{2n-1}a - \cosh\lambda_{2n-1}a\coth\lambda_{2n-1}b\right) - \cosh\lambda_{2n-1}a\left(\coth\lambda_{2n-1}a - \coth\lambda_{2n-1}b\right)\right\}^{-1}$$

$$\beta^+_{2n-1} = \frac{4V_+}{(2n-1)\pi}\frac{1}{\sinh\lambda_{2n-1}b} - \alpha^+_{2n-1}\coth\lambda_{2n-1}b$$

$$\beta^-_{2n-1} = \alpha^+_{2n-1}\left(\coth\lambda_{2n-1}a - \coth\lambda_{2n-1}b\right) + \frac{4}{(2n-1)\pi}\left(\frac{V_+}{\sinh\lambda_{2n-1}b} - \frac{V_-\coth\lambda_{2n-1}a}{\cosh\lambda_{2n-1}b}\right)$$

$$\alpha^-_{2n-1} = \frac{4V_-}{(2n-1)\pi}\frac{1}{\cosh\lambda_{2n-1}b}$$

In particular, the solution on Ω is given by

$$\varphi(x,y) = \begin{cases} \sum_{n>0}\left(\alpha^+_{2n-1}\cosh\lambda_{2n-1}y + \beta^+_{2n-1}\sinh\lambda_{2n-1}y\right)\sin\lambda_{2n-1}x & \text{on } \bar{\Omega}_+ \\ \sum_{n>0}\left(\alpha^-_{2n-1}\cosh\lambda_{2n-1}y + \beta^-_{2n-1}\sinh\lambda_{2n-1}y\right)\sin\lambda_{2n-1}x & \text{on } \bar{\Omega}_- \end{cases} \quad \square$$

Observe from Example 3.4.1 that when dielectric media are lossy, at the boundary interface between the two media, in the steady-state limit, the potential is dictated by the conductivities of the media instead of the electric permittivities. It is only when the dielectric media are lossless that the electric permittivities dictate how the field behaves (*cf.* Remark 3.1.7 for details). This section concludes with a final example.

3.4.2 Example

Consider a composite annulus $\Omega = (\Omega_+, \varepsilon_+, \sigma_+) \cup (\Omega_-, \varepsilon_-, \sigma_-)$, where $\varepsilon_\pm$ are the respective electric permittivities and $\sigma_\pm$ the respective conductivities in $\Omega_\pm$,

$$\Omega_+ = \{(x,y) \in \mathbf{R}^2 : a^2 < x^2 + y^2 < b^2\}$$

$$\Omega_- = \{(x,y) \in \mathbf{R}^2 : b^2 < x^2 + y^2 < c^2\}$$

Moreover, set $\Gamma = \partial\Omega_+ \cap \partial\Omega_-$ and $\Gamma_\pm = \partial\Omega_\pm - \Gamma$. Determine the electric potential φ on Ω if $\varphi = \varphi_\pm$ on $\Gamma_\pm$, where $\varphi_\pm$ are constants.

Inasmuch as the field is static, it must satisfy Laplace's equation $\Delta\varphi = 0$ subject to the boundary condition at the interface Γ from Corollary 3.1.2. Employing cylindrical coordinates, where

$$\Delta\varphi = \tfrac{1}{r}\partial_r(r\,\partial_r\varphi) + \tfrac{1}{r^2}\partial^2_\theta\varphi + \partial^2_z\varphi$$

$\partial_z \equiv 0$ and the angular symmetry of the problem, that is, $\varphi(r,\theta) = \varphi(r,\theta')\forall\theta,\theta'$, reduces Laplace's equation to

$$\Delta\varphi = \tfrac{1}{r}\partial_r(r\,\partial_r\varphi) = 0 \Rightarrow r\,\partial_r\varphi = k$$

for some constant k. Whence, $\varphi = k\int\frac{dr}{r} \Rightarrow \varphi = k\ln r + k'$, where k' is a constant of integration. Thus, the general solution is $\varphi = a + b\ln r$.

Now, let $\varphi_\pm = a_\pm + b_\pm \ln r$ on $\Omega_\pm$, subject to the following boundary condition at the interface: (i) $0 = \boldsymbol{n}_- \cdot (\boldsymbol{J}_+ - \boldsymbol{J}_-) \Rightarrow \sigma_+\partial_r\varphi_+ = \sigma_-\partial_r\varphi_-$ on Γ, and (ii) $\lim_{r\to b^-}\varphi_+(r) = \lim_{r\to b^+}\varphi_-(r)$ on Γ. Imposing the Dirichlet boundary conditions leads to:

$$\varphi_0 = \varphi_+(a) = a_+ + b_+\ln a \quad \text{and} \quad 0 = \varphi_-(c) = a_- + b_-\ln c$$

Thus, $a_- = -b_-\ln c$ and $a_+ = \varphi_0 - b_+\ln a$. Because there are four unknown variables to evaluate, two more equations are required. Those are supplied by conditions (i) and (ii). From (i), $\partial_r\varphi_\pm = \frac{b_\pm}{r} \Rightarrow \sigma_+ b_+ = \sigma_- b_-$, and the continuity relation (ii) yields $a_+ + b_+\ln b = a_- + b_-\ln b$.

Direct substitution into (ii) yields $\varphi_0 + b_+\ln\frac{b}{a} = b_-\ln\frac{b}{c}$ and hence, on setting $\gamma = \sigma_+\ln\frac{b}{c} + \sigma_-\ln\frac{a}{b}$, it follows that $b_+ = \varphi_0\frac{\sigma_-}{\gamma}$, $b_- = \varphi_0\frac{\sigma_+}{\gamma}$, $a_- = -\varphi_0\frac{\sigma_+}{\gamma}\ln c$, and $a_+ = \frac{\varphi_0}{\gamma}\left(\sigma_+\ln\frac{b}{c} - \sigma_-\ln b\right)$. Thus, $\varphi_+ = a_+ + b_+\ln r = \frac{\varphi_0}{\gamma}\left(\sigma_+\ln\frac{b}{c} + \sigma_-\ln\frac{r}{b}\right)$. It is easy to see that $\varphi_+(a) = \varphi_0$, as expected by construction. Similarly, $\varphi_- = a_- + b_-\ln r = \varphi_0\frac{\sigma_+}{\gamma}\ln\frac{r}{c}$. As a sanity check, $\varphi_-(c) \equiv 0$ and $\varphi_+(b) = \varphi_-(b)$, as expected. Thus,

$$\varphi = \begin{cases} \varphi_+ \text{ on } \Omega_+ \\ \varphi_- \text{ on } \Omega_- \end{cases}$$

is the required solution for Ω. □

Once again, from the above example, the field is determined by the conductivities of the media, as opposed to pure dielectric media, wherein the electric permittivities determine the field profile. On the other hand, the electric permittivities determine the charge density on Γ. Explicitly, the line charge density is $\rho = \varepsilon_+\partial_r\varphi_+ - \varepsilon_-\partial_r\varphi_-$ on Γ. The current density is trivially given by

$$\boldsymbol{J} = \begin{cases} -\sigma_+\nabla\varphi_+ \text{ on } \Omega_+ \\ -\sigma_-\nabla\varphi_- \text{ on } \Omega_- \end{cases}$$

Finally, recall that if Ω were a lossless composite dielectric, then $\rho|_\Gamma \equiv 0$ and φ is obtained by replacing $\sigma_\pm$ with $\varepsilon_\pm$. Clearly in this instance, the current density is zero.

3.5 Duality

It is clear by inspecting Maxwell's equations that they can be rendered symmetric by the introduction of a fictitious magnetic charge density ς and its associated magnetic current density j; to wit,

$$\nabla \cdot \boldsymbol{B} = \varsigma \tag{3.18a}$$

$$\nabla \times \boldsymbol{E} = -\partial_t \boldsymbol{B} - j \tag{3.18b}$$

whence following the proof of Theorem 1.3.1 *mutatis mutandis* yields the equivalent magnetic charge conservation:

$$\nabla \cdot j + \partial_t \varsigma = 0 \tag{3.19}$$

The symmetrized extensions of Maxwell's equations are given below for ease of reference:

$$\nabla \times \boldsymbol{E}(\boldsymbol{r},t) = -j(\boldsymbol{r},t) - \partial_t \boldsymbol{B}(\boldsymbol{r},t) \tag{3.20a}$$

$$\nabla \cdot \boldsymbol{B}(\boldsymbol{r},t) = \varsigma \tag{3.20b}$$

$$\nabla \times \boldsymbol{H}(\boldsymbol{r},t) = \boldsymbol{J}(\boldsymbol{r},t) + \partial_t \boldsymbol{D}(\boldsymbol{r},t) \tag{3.20c}$$

$$\nabla \cdot \boldsymbol{D}(\boldsymbol{r},t) = \rho \tag{3.20d}$$

where $\boldsymbol{D} = \varepsilon \boldsymbol{E}$ and $\boldsymbol{B} = \mu \boldsymbol{H}$. Maxwell's equations are rewritten as (3.20) to display the symmetry between them.

Now, observe that transforming Equation (3.20a) to (3.20c) requires the replacement:

$$\boldsymbol{E} \to \boldsymbol{H},\ j \to -\boldsymbol{J},\ \boldsymbol{B} \to -\boldsymbol{D}$$

In particular, this implies that (3.20b) becomes $\varsigma = \nabla \cdot \boldsymbol{B}(\boldsymbol{r},t) \to \nabla \cdot (-\boldsymbol{D}(\boldsymbol{r},t)) = -\rho$, and hence, transforming (3.20b) to (3.20d) requires the replacement:

$$\varsigma \to -\rho$$

Finally, transforming from (3.20c) to (3.20a) yields:

$$\boldsymbol{H} \to -\boldsymbol{E},\ \boldsymbol{J} \to j,\ \boldsymbol{D} \to \boldsymbol{B}$$

Likewise, transforming (3.20d) to (3.20b) leads to the replacement:

$$\rho \to \varsigma$$

In summary, the former four pairs constitute the required transformation transforming the first pair of Maxwell's equations into the second pair of Maxwell's equations, and vice versa for the latter pairs.

It is clear that the analogy can be carried out further by defining magnetic conductivity σ_m as $j = \frac{\sigma_m}{\mu} \boldsymbol{B}$. The four fictitious quantities (ς, σ_m, j) lead in a natural way to magnetic boundary conditions presented in Section 3.1 *mutatis mutandis*:

$$\boldsymbol{n}_+ \cdot (\boldsymbol{B}_- - \boldsymbol{B}_+) = \varsigma \tag{3.21a}$$

$$\boldsymbol{n}_+ \times (\boldsymbol{E}_- - \boldsymbol{E}_+) = -j \tag{3.21b}$$

For perfect electric conductors (PEC) and perfect magnetic conductors (PMC), the boundary conditions are summarized as

$$\begin{cases} \boldsymbol{n}_+ \times \boldsymbol{E}_+ = 0 \\ \boldsymbol{n}_+ \times \boldsymbol{B}_+ = \mu \boldsymbol{J} \end{cases} \tag{PEC}$$

$$\begin{cases} \boldsymbol{n}_+ \times \boldsymbol{B}_+ = 0 \\ \boldsymbol{n}_+ \times \boldsymbol{E}_+ = -j \end{cases} \tag{PMC}$$

Note in passing that PMC is an idealized condition and it does not exist; however, it is a useful condition to impose when solving radiation problems.

3.5.1 Theorem (Lorentz reciprocity)

Given some open subset $(\Omega, \varepsilon, \mu) \subseteq \mathbf{R}^3$, suppose $(S_\pm, J_\pm, j_\pm)$ are two time-varying current sources on some compact $S_\pm \subset \Omega$. Set $\Omega_0 = \Omega - (S_+ \cup S_-)$. If the current sources are time harmonic, then

$$\oiint_{\partial\Omega_0} \boldsymbol{E}_+ \times \boldsymbol{B}_- \cdot \boldsymbol{n} \mathrm{d}^2 r = \oiint_{\partial\Omega_0} \boldsymbol{E}_- \times \boldsymbol{B}_+ \cdot \boldsymbol{n} \mathrm{d}^2 r$$

Proof

Now, from Maxwell's equations, the source generates $(\boldsymbol{E}_\pm, \boldsymbol{B}_\pm)$ via

$$\nabla \times \boldsymbol{E}_\pm = -\partial_t \boldsymbol{B}_\pm - j_\pm \quad \text{and} \quad \nabla \times \boldsymbol{B}_\pm = \mu \partial_t \boldsymbol{E}_\pm + \mu \boldsymbol{J}_\pm$$

Next, consider the pair $E_+ \times B_-$ and $E_- \times B_+$ and motivated by the vector identity [3]

$$\nabla \cdot (E_+ \times B_-) = (\nabla \times E_+) \cdot B_- - (\nabla \times B_-) \cdot E_+$$

it follows that

$$\begin{aligned}
&\nabla \cdot (E_+ \times B_- - E_- \times B_+) \\
&= (\nabla \times E_+) \cdot B_- - (\nabla \times B_-) \cdot E_+ - (\nabla \times E_-) \cdot B_+ + (\nabla \times B_+) \cdot E_- \\
&= -\partial_t B_+ \cdot B_- - j_+ \cdot B_- - \mu\, \partial_t E_- \cdot E_+ - \mu J_- \cdot E_+ + \partial_t B_- \cdot B_+ + j_- \cdot B_+ + \mu\, \partial_t E_+ \cdot E_- + \mu J_+ \cdot E_-
\end{aligned}$$

However, by assumption, the sources are time harmonic; hence, the fields generated are time harmonic.* In particular, $\partial_t(E_\pm, B_\pm) = i\omega(E_\pm, B_\pm)$ implies that the ∂_t-terms cancel, yielding

$$\nabla \cdot (E_+ \times B_- - E_- \times B_+) = j_- \cdot B_+ - j_+ \cdot B_- + \mu J_+ \cdot E_- - \mu J_- \cdot E_+$$

Finally, appealing to the divergence theorem,

$$\oiiint_{\Omega_0} \nabla \cdot (E_+ \times B_- - E_- \times B_+) \mathrm{d}^3 r = \oiint_{\partial\Omega_0} (E_+ \times B_- - E_- \times B_+) \cdot n \mathrm{d}^2 r$$

where n is the unit, outward, normal vector field on $\partial\Omega_0$,

$$\oiiint_{\Omega_0} (j_- \cdot B_+ - j_+ \cdot B_- + \mu J_+ \cdot E_- - \mu J_- \cdot E_+) \mathrm{d}^3 r = 0$$

by construction implies immediately that

$$\oiint_{\partial\Omega_0} (E_+ \times B_- - E_- \times B_+) \cdot n \mathrm{d}^2 r = 0$$

□

3.5.2 Corollary

Given some open, bounded, subset $(\Omega, \varepsilon, \mu) \subseteq \mathbf{R}^3$, suppose $(S_\pm, J_\pm, j_\pm)$ are two time-varying current sources on some compact $S_\pm \subset \Omega$. If $\partial\Omega$ satisfies either the PEC or PMC boundary condition, then

$$\iiint_\Omega (J_+ \cdot E_- - \tfrac{1}{\mu} j_+ \cdot B_-) \mathrm{d}^3 r = \iiint_\Omega (J_- \cdot E_+ - \tfrac{1}{\mu} j_- \cdot B_+) \mathrm{d}^3 r$$

* See Exercise 3.6.7.

Proof

From the proof of Theorem 3.5.1, it suffices to show that

$$\iint_{\partial\Omega}(\boldsymbol{E}_+\times\boldsymbol{B}_- - \boldsymbol{E}_-\times\boldsymbol{B}_+)\cdot\boldsymbol{n}\mathrm{d}^2r = 0$$

Now, noting [3] the vector identity $\boldsymbol{A}\times\boldsymbol{B}\cdot\boldsymbol{C} = \boldsymbol{A}\cdot\boldsymbol{B}\times\boldsymbol{C}$, it follows that

$$(\boldsymbol{E}_+\times\boldsymbol{B}_- - \boldsymbol{E}_-\times\boldsymbol{B}_+)\cdot\boldsymbol{n} = \boldsymbol{n}\times\boldsymbol{E}_+\cdot\boldsymbol{B}_- - \boldsymbol{n}\times\boldsymbol{E}_-\cdot\boldsymbol{B}_+ = 0$$

for PEC: $\boldsymbol{n}\times\boldsymbol{E}_\pm = 0$. Conversely, applying the vector identity again,

$$\boldsymbol{n}\times\boldsymbol{E}_+\cdot\boldsymbol{B}_- - \boldsymbol{n}\times\boldsymbol{E}_-\cdot\boldsymbol{B}_+ = -\boldsymbol{n}\times\boldsymbol{B}_-\cdot\boldsymbol{E}_+ + \boldsymbol{n}\times\boldsymbol{B}_+\cdot\boldsymbol{E}_- = 0$$

for PMC: $\boldsymbol{n}\times\boldsymbol{B}_\pm = 0$, and the result thus follows. □

As an interesting application of the reciprocity theorem, consider an open subspace $\Omega\subset\mathbf{R}^3$ bounded by a metal chassis $\partial\Omega$. Suppose also that there exists a current density J_+ induced on some compact subset $K_+\subset\Omega$, where K_+ is a PEC, and some current density $\boldsymbol{J}_-$ on a compact source $K_-\subset\Omega$. Then, from Corollary 3.5.2, setting $j_\pm = 0$ yields $\iiint_\Omega \boldsymbol{J}_+\cdot\boldsymbol{E}_-\mathrm{d}^3r = \iiint_\Omega \boldsymbol{J}_-\cdot\boldsymbol{E}_+\mathrm{d}^3r$.

However, $\iiint_\Omega \boldsymbol{J}_+\cdot\boldsymbol{E}_-\mathrm{d}^3r = \iiint_{K_+}\boldsymbol{J}_+\cdot\boldsymbol{E}_-\mathrm{d}^3r = 0$ as $\boldsymbol{E}_-\,|\,K_+ = 0$ and hence, $\iiint_\Omega \boldsymbol{J}_-\cdot\boldsymbol{E}_+\mathrm{d}^3r = 0$. Furthermore, noting that as (J_-, K_-) is arbitrary, it follows at once that $\iiint_\Omega \boldsymbol{J}_-\cdot\boldsymbol{E}_+\mathrm{d}^3r = 0 \Rightarrow \boldsymbol{E}_+\equiv 0$ on Ω. That is, induced current on a PEC does not radiate; this is because the electric field generated on the PEC precisely cancels out the incident electric field. From an EMC perspective, this is an interesting example. It demonstrates that conductors within a chassis do not radiate from currents induced upon them. In particular, the walls of a chassis do not reradiate from surface current densities induced on them by Corollary 3.5.2.

3.6 Worked Problems

3.6.1 Exercise

Prove that the *inversion in a circle* $\varsigma:\overline{B}(a)\to\mathbf{R}^2 - B(a)$ by $r\mapsto r'\equiv\frac{a^2}{r}$ such that $\varsigma\,|\,\partial B(a) = 1_{\partial B(a)}$, is conformal about any deleted neighborhood of 0, and hence, deduce that the inversion in a sphere mapping is also conformal.

Solution

Recall that a mapping is *conformal* in some neighborhood $N \subset \mathbf{C}$ if it is analytic and its derivative is nowhere zero in N. As $\mathbf{R}^2 \cong \mathbf{C}$ under the canonical homeomorphism $h:(x,y) \mapsto x+iy$, it suffices to consider the mapping $\zeta:\mathbf{C} \to \mathbf{C}$ defined by $\zeta = \frac{1}{z}$. However, this mapping is clearly conformal away from the origin: $\zeta' = -\frac{1}{z^2} \neq 0$ on $\mathbf{C} - \{0\}$. Hence, the circle inversion mapping is conformal.

Regarding the inversion in a sphere, it suffices to note that each cross-section of the fixed sphere is a circle. As the inversion in a sphere is merely the mapping restricted to the respective (circular) cross-section, it follows immediately that the mapping must also be conformal, as claimed. □

3.6.2 Exercise

Given $\boldsymbol{r} = (r,\theta,\phi)$ and $\boldsymbol{r}_Q = (r_Q, 0, \frac{\pi}{2})$, let ϑ denote the angle between $\boldsymbol{r}$ and $\boldsymbol{r}_Q$. Show that $\cos\vartheta = \sin\phi\cos\theta$. Hence, deduce the result for $\boldsymbol{r}_Q = (r_Q, \theta_Q, \phi_Q)$.

Solution

In rectangular coordinates, $x = r \sin\phi \cos\theta$, $y = r \sin\phi \sin\theta$, $z = r\cos\phi$. By definition, $\cos\vartheta = \frac{\boldsymbol{r}\cdot\boldsymbol{r}_Q}{r_Q r}$. Hence, without loss of generality, we may set $r = 1 = r_Q$. Then,

$$\boldsymbol{r}\cdot\boldsymbol{r}_Q = x x_Q + y\cdot 0 + z\cdot 0 = \sin\phi\cos\theta$$

and the result thus follows. To complete the proof, it is enough to note that

$$\boldsymbol{r}\cdot\boldsymbol{r}_Q = \sin\phi\cos\theta\sin\phi_Q\cos\theta_Q + \sin\phi\sin\theta\sin\phi_Q\sin\theta_Q + \cos\phi\cos\phi_Q$$

and invoking the identity $\cos(a + b) = \cos a \cos b - \sin a \sin b$, the conclusion thus follows. □

3.6.3 Exercise

Consider two infinite strips $(\Omega_\pm, \varepsilon_\pm)$, where $\Omega_- = \mathbf{R}\times(0,a)$ and $\Omega_+ = \mathbf{R}\times(a,b)$, with $\varepsilon_\pm$ being the respective electric permittivities and $\varepsilon_+ \neq \varepsilon_-$. Find the potential φ on $\Omega = \bar{\Omega}_- \cup \bar{\Omega}_+$, if

$$\varphi = \begin{cases} \varphi_0 \text{ if } y = b \\ 0 \text{ if } y = 0 \end{cases} \tag{3.22}$$

Deduce the explicit expression for the electric field on Ω.

Solution

First, define $\partial\Omega_0 = \{(x,a): -\infty < x < \infty\}$, $\partial\Omega_- = \{(x,0): -\infty < x < \infty\}$, and $\partial\Omega_+ = \{(x,b): -\infty < x < \infty\}$. Then, φ satisfies the Laplace equation $\Delta\varphi = 0$ subject to the boundary conditions (3.22),

$$\lim_{y\to a^-} \varphi(x,y) = \lim_{y\to a^+} \varphi(x,y) \quad \text{and} \quad \lim_{y\to a^-} \varepsilon_- \partial_y \varphi(x,y) = \lim_{y\to a^+} \varepsilon_+ \partial_y \varphi(x,y)$$

The uniformity of the potential along the x-axis suggests that there is no variation along the x-axis: $\partial_x \varphi = 0$ on Ω, whence,

$$\Delta\varphi = \partial_x^2 \varphi + \partial_y^2 \varphi = \partial_y^2 \varphi = 0 \Rightarrow \varphi = Cy + D$$

for some constants C, D.

Furthermore, as $\varepsilon_+ \neq \varepsilon_-$, set $\varphi_\pm = C_\pm y + D_\pm$ on $\Omega_\pm$. Then, appealing to the boundary conditions,

$$\varphi_0 = C_+ b + D_+ \Rightarrow D_+ = \varphi_0 - C_+ b \quad \text{and} \quad 0 = C_- 0 + D_- \Rightarrow D_- = 0$$

Also, the two continuity conditions yield

$$C_+ a + D_+ = C_- a \Rightarrow \varphi_0 + C_+(a-b) = C_- a$$

$$\varepsilon_- C_- = \varepsilon_+ C_+ \Rightarrow C_- = \tfrac{\varepsilon_+}{\varepsilon_-} C_+$$

whence $C_+ \left\{\left(\tfrac{\varepsilon_+}{\varepsilon_-} - 1\right)a + b\right\} = \varphi_0 \Rightarrow C_+ = \varphi_0 \left\{\left(\tfrac{\varepsilon_+}{\varepsilon_-} - 1\right)a + b\right\}^{-1}$ and hence,

$$\varphi = \begin{cases} \varphi_0 \left\{1 - \dfrac{b-y}{\left(\frac{\varepsilon_+}{\varepsilon_-}-1\right)a+b}\right\} & \text{on } \Omega_+ \\ \varphi_0 \dfrac{\varepsilon_+}{\varepsilon_-} \dfrac{y}{\left(\frac{\varepsilon_+}{\varepsilon_-}-1\right)a+b} & \text{on } \Omega_- \end{cases}$$

Finally, from $\boldsymbol{E} - \nabla\phi$, it follows clearly that

$$\boldsymbol{E} = \begin{cases} \dfrac{\varphi_0}{\left(\frac{\varepsilon_+}{\varepsilon_-}-1\right)a+b} \mathbf{e}_y & \text{on } \Omega_+ \\ \dfrac{\varepsilon_+}{\varepsilon_-} \dfrac{\varphi_0}{\left(\frac{\varepsilon_+}{\varepsilon_-}-1\right)a+b} \mathbf{e}_y & \text{on } \Omega_- \end{cases}$$

As a quick sanity check, it can be easily seen that when $\varepsilon_+ = \varepsilon_-$, $\varphi = \varphi_0 \frac{y}{b}$ and $\boldsymbol{E} = \frac{\varphi_0}{b}$, as expected. □

3.6.4 Exercise

Given (Ω, ε), where $\Omega = (0,a) \times (0,b)$, suppose

$$\varphi = \begin{cases} V_+ & \text{for } y = b \\ V_- & \text{for } y = 0 \\ 0 & \text{for } x = 0, a \end{cases} \tag{3.23}$$

Find the potential φ on Ω. How does ε affect φ?

Solution

The solution $\varphi \in C^2(\Omega) \cap C(\bar{\Omega})$ is given by the Dirichlet boundary value problem $\Delta\varphi = 0$ on Ω satisfying boundary conditions (3.23). The simple geometry of the domain and the simple boundary conditions lead one to attempt to solve Laplace's equation via the separation of variables. So, set $\varphi(x,y) = \Phi(x)\Theta(y)$. Then,

$$0 = \Delta\varphi = \Theta(y)\partial_x^2\Phi(x) + \Phi(x)\partial_y^2\Theta(y) \Rightarrow \frac{\partial_x^2\Phi}{\Phi} = -\frac{\partial_y^2\Theta}{\Theta} \equiv -\lambda^2$$

for some constant $\lambda \in \mathbf{R}$. The negative sign in front of λ^2 was chosen because of the periodic boundary condition $\varphi(0,y) = 0 = \varphi(0,a)\ \forall y$. The general solution is thus $\Phi(x) = a\ cos\ \lambda x + b\ sin\ \lambda x$.

A solution satisfying the boundary condition is $a = 1$, $b = 0$ and $\Phi(x) = sin\ \lambda x$, where $\lambda = \frac{\pi n}{a}$, $n = 0, 1, 2, \ldots$. Likewise, the general solution satisfying $\frac{d^2}{dy^2}\Theta - \lambda^2\Theta = 0$ is given by

$$\Theta(y) = c\ cosh\ \lambda y + d\ sinh\ \lambda y$$

Hence, $\varphi = (\alpha \cosh \lambda y + \beta \sinh \lambda y)\sin \lambda x$ satisfies $\Delta\varphi = 0$. As this holds for arbitrary integer n, the linearity of the Laplacian operator Δ implies that the general solution is given by $\varphi(x,y) = \Sigma_{n\in\mathbf{Z}}(\alpha_n \cosh \lambda_n y + \beta_n \sinh \lambda_n y)\sin \lambda_n x$, where $\lambda_n \equiv \frac{\pi n}{a}$ to display the explicit dependence of n, and the pair of coefficients (α_n, β_n) are determined via the remaining boundary conditions. Furthermore, as $\varphi \equiv 0$ for $n = 0$, it follows that $n \in \mathbf{N}$.

Explicitly, $V_+ = \varphi(x,b) = \Sigma_{n\in\mathbf{N}}(\alpha_n \cosh \lambda_n b + \beta_n \sinh \lambda_n b)\sin \lambda_n x$. Whence, multiplying both sides by $\sin \lambda_m x$ and integrating yields

$$\alpha_n \cosh \lambda_n b + \beta_n \sinh \lambda_n b = \frac{2V_+}{a}\int_0^a \sin \lambda_n x dx = \begin{cases} \frac{4V_+}{n\pi} & \text{for } n \text{ odd} \\ 0 & \text{for } n \text{ even} \end{cases}$$

Likewise,

$$V_- = \varphi(x,0) = \sum_n \alpha_n \sin \lambda_n x \Rightarrow \alpha_n = \begin{cases} \frac{4V_-}{n\pi} & \text{for } n \text{ odd} \\ 0 & \text{for } n \text{ even} \end{cases}$$

Thus,

$$\beta_{2n-1} = \tfrac{4}{(2n-1)\pi} \tfrac{V_+ - V_- \cosh \lambda_{2n-1} b}{\sinh \lambda_{2n-1} b} \quad \forall n = 1, 2, \ldots$$

and the general solution is

$$\varphi(x,y) = \tfrac{4}{\pi} \sum_n \tfrac{1}{2n-1} \left(\cosh \lambda_{2n-1} y + \tfrac{V_+ - V_- \cosh \lambda_{2n-1} b}{\sinh \lambda_{2n-1} b} \sinh \lambda_{2n-1} y \right) \sin \lambda_{2n-1} x$$

Lastly, note that φ is independent of ε in this example. In general, this is not the case if ε varies in Ω as the boundary condition at the interface wherein ε changes in Ω must be satisfied by the solution. □

3.6.5 Exercise

Show that a localized static charge density ρ in $\mathbf{R}^3$ satisfies the Poisson equation $-\Delta\varphi(\boldsymbol{r}) = \frac{\rho(\boldsymbol{r})}{\varepsilon}$. And hence, establish that the Poisson equation $-\Delta\varphi = f$ describes electrostatics in general. In particular, deduce that steady-state conditions are also satisfied by Laplace's equation.

Solution

From Gauss' law, $\nabla \cdot \boldsymbol{E} = \frac{\rho}{\varepsilon}$; substituting $\boldsymbol{E} = -\nabla\varphi$ yields $-\Delta\varphi(\boldsymbol{r}) = \frac{\rho(\boldsymbol{r})}{\varepsilon}$, as required. Finally, set $f = \frac{\rho}{\varepsilon}$, and the assertion is established. Next, to establish the last assertion, consider the charge conservation relation $\partial_t \rho = -\nabla \cdot \boldsymbol{J}$. Under steady-state conditions, $\partial_t \rho = 0 = -\nabla \cdot \boldsymbol{J} \Rightarrow 0 = -\nabla \cdot (-\sigma\nabla\varphi) \Rightarrow -\Delta\varphi = 0$, as required. □

3.6.6 Exercise

Suppose $\boldsymbol{J}(\boldsymbol{r},t)$ is some current density defined on a compact conductor $\Omega \subset \mathbf{R}^3$ above a ground plane $\partial\mathbf{R}^3_+$. Show that its mirror image is $-\boldsymbol{J}(\overline{\boldsymbol{r}},t)$, where $\overline{\boldsymbol{r}} = (x,y,-z)$ with $\boldsymbol{r} = (x,y,z)$.

Solution

By definition, $\boldsymbol{J}(\boldsymbol{r},t) = \rho(\boldsymbol{r},t)v(\boldsymbol{r})$, where $\rho(\boldsymbol{r},t)$ is the charge density defined on Ω and v is the average velocity of the charge density circulating on Ω. By definition, under the mirror image transformation (i.e., reflection on $\partial\mathbf{R}^3_+$), $\boldsymbol{r} = (x,y,z) \to \overline{\boldsymbol{r}} \equiv (x,y,-z)$, the image of $\rho(\boldsymbol{r},t) \to -\rho(\overline{\boldsymbol{r}},t)$ and $v(\boldsymbol{r},t) \to v(\overline{\boldsymbol{r}},t)$. Hence, the image current is $-\rho(\overline{\boldsymbol{r}},t)v(\overline{\boldsymbol{r}},t) = -\boldsymbol{J}(\overline{\boldsymbol{r}},t)$, as claimed. □

3.6.7 Exercise

Consider a segment of an annulus defined by

$$\Omega = \left\{(x,y) \in \mathbf{R}^2 : a^2 < x^2 + y^2 < b^2\right\} \cap \Delta$$

where $\Delta = \{(x,y) \in \mathbf{R}^2 : x = b\cos\theta, y = b\sin\theta, 0 < \theta_0 < \theta < \theta_1\}$ is a wedge of a disk of radius b. Let $\partial\Omega_a$ $(\partial\Omega_b)$ denote the inner (outer) radial boundary of Ω, and $\Gamma_0(\Gamma_1)$ denote the boundary of Ω that subtends the x-axis by an angle $\theta_0(\theta_1)$. If

$$\varphi = \begin{cases} V_- \text{ on } \Gamma_0 \\ V_+ \text{ on } \Gamma_1 \end{cases}$$

such that $\partial_n \varphi = 0$ on $\partial\Omega_a \cup \partial\Omega_b$, where $\boldsymbol{n}$ is the outward pointing unit normal vector field on $\partial\Omega_a \cup \partial\Omega_b$, what is the potential in Ω? Hint: consider the conformal transformation $w = \mathrm{e}^z$ in the complex plane, with Ω embedded in $\mathbf{C}$ and utilize the result from Exercise 3.6.4.

Solution

Set $z = x + iy$ and $w = \mathrm{e}^z$. Then, $w = re^{i\theta} \Rightarrow r = \mathrm{e}^x$ and $\theta = y$ (modulo 2π). From this, it is clear that the point $(c,y) \mapsto (\mathrm{e}^c, \theta)$. Specifically, if $\theta_0 < y < \theta_1$ and $c = a$, then the mapping

$$\{(a,y) : \theta_0 < y < \theta_1\} \to \partial\Omega_a$$

is a bijection. Likewise, let $a < x < b$, and set $y = \theta_0$. Then, the mapping

$$\{(x,\theta_0) : a < x < b\} \to \Gamma_0$$

is also a bijection under the conformal mapping.

Thus, under the conformal map $w = \mathrm{e}^z$, vertical lines are mapped into angular arcs of the complex plane, and horizontal lines are mapped into the radial lines of the complex planes; see Figure 3.5 for details.

Because the Dirichlet problem for the Laplace equation is invariant under a conformal transformation, it suffices to transform Ω onto the rectangular domain (via the inverse conformal transformation), solve the Laplace equation, and then transform the solution back into the original domain Ω.

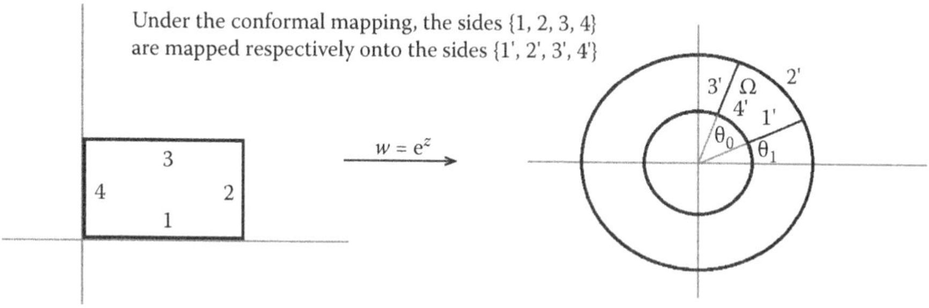

FIGURE 3.5
Mapping under a conformal transformation.

The solution to the rectangular domain $R = (0,a) \times (0,b)$ is precisely that of Exercise 3.6.4. Because boundary conditions transform invariantly under a conformal transformation, it follows that the potential ψ defined on R subject to the following boundary conditions,

$$\psi = \begin{cases} V_- & \text{on } x = 0 \\ V_+ & \text{on } x = a \end{cases}$$

and $\partial_y \psi = 0$ on $\{y = 0\} \cup \{y = b\}$, is given by

$$\varphi(x,y) = \tfrac{4}{\pi}\sum_n \tfrac{1}{2n-1}\left(\cosh \lambda_{2n-1}\, y + \tfrac{V_+ - V_- \cosh \lambda_{2n-1} b}{\sinh \lambda_{2n-1} b} \sinh \lambda_{2n-1} y\right) \sin \lambda_{2n-1} x$$

Now, the inverse conformal transformation yields $x = \ln r$ and $y = \theta$. Hence, via the composition of maps, $\varphi(r,\theta) \equiv \psi \circ w^{-1}$, the desired potential on Ω is

$$\psi(r,\theta) = \tfrac{4}{\pi}\sum_n \tfrac{1}{2n-1}\left(\cosh \lambda_{2n-1}\theta + \tfrac{V_+ - V_- \cosh \lambda_{2n-1} b}{\sinh \lambda_{2n-1} b} \sinh \lambda_{2n-1}\theta\right) \sin\left(\lambda_{2n-1} \ln r\right)$$

□

3.6.8 Exercise

Suppose a point charge Q is located at $(0,0,z_0) \in \mathbf{R}^3_+$ above a pure dielectric half-space $(\mathbf{R}^3_-, \varepsilon)$, where ε is the electric permittivity of the dielectric medium. Determine the electric potential φ in $\Omega = \left(\mathbf{R}^3_+ - \{(0,0,z_0)\}, \varepsilon_0\right)$, where ε_0 is the electric permittivity of air. What is the potential in $\mathbf{R}^3_-$? Hint: Apply the method of images.

Solution

The potential φ satisfies Poisson's equation $-\varepsilon_0\Delta\varphi = Q\delta^3(\boldsymbol{r})$ satisfying the boundary condition at the interface:

$$\text{(i)} \qquad \lim_{z\to 0^-}\varphi(x,y,z) = \lim_{z\to 0^+}\varphi(x,y,z)$$

$$\text{(ii)} \qquad \lim_{z\to 0^-}\varepsilon\partial_z\varphi(x,y,z) = \lim_{z\to 0^+}\varepsilon_0\,\partial_z\varphi(x,y,z)$$

Applying the method of images, let q denote the image charge at $(0,0,-z_0) \in \mathbf{R}^3_-$. Then, the potential φ_+ in Ω is trivially given by

$$\varphi_+(x,y,z) = \frac{1}{4\pi\varepsilon_0}\left\{\frac{Q}{\sqrt{x^2+y^2+(z-z_0)^2}} + \frac{q}{\sqrt{x^2+y^2+(z+z_0)^2}}\right\}$$

On the other hand, the solution φ_- on $\mathbf{R}^3_-$ is obtained by replacing Q with some charge q' at $(0,0,z_0) \in (\mathbf{R}^3,\varepsilon)$ and replacing the entire space with electric permittivity ε. Then, the potential field defined on $\mathbf{R}^3_-$ is

$$\varphi_-(x,y,z) = \frac{1}{4\pi\varepsilon}\frac{q'}{\sqrt{x^2+y^2+(z-z_0)^2}}$$

Invoking the continuity condition (i) yields $\frac{1}{4\pi\varepsilon_0}(Q+q) = \frac{1}{4\pi\varepsilon}q'$, and applying condition (ii) yields $\frac{\varepsilon_0}{4\pi\varepsilon_0}(Q-q) = \frac{\varepsilon}{4\pi\varepsilon}q'$, whence

$$Q+q = \frac{\varepsilon_0}{\varepsilon}q' = \frac{\varepsilon_0}{\varepsilon}(Q-q) \Rightarrow q = -\frac{\varepsilon-\varepsilon_0}{\varepsilon+\varepsilon_0}Q \quad \text{and} \quad q' = \frac{2\varepsilon}{\varepsilon+\varepsilon_0}$$

Thus, the electric potential in $\mathbf{R}^3$ is given by

$$\varphi = \begin{cases} \dfrac{Q}{4\pi\varepsilon_0}\left\{\dfrac{1}{\sqrt{x^2+y^2+(z-z_0)^2}} - \dfrac{\varepsilon-\varepsilon_0}{\varepsilon+\varepsilon_0}\dfrac{1}{\sqrt{x^2+y^2+(z+z_0)^2}}\right\} & \text{on } \Omega \\[2ex] \dfrac{1}{4\pi\varepsilon_0}\dfrac{2\varepsilon}{\varepsilon+\varepsilon_0}\dfrac{Q}{\sqrt{x^2+y^2+(z-z_0)^2}} & \text{on } \mathbf{R}^3_- \end{cases}$$

References

1. Chang, D. 1992. *Fields and Wave Electromagnetics*. Reading, MA: Addison-Wesley.
2. Churchill, R. and Brown, J. 1990. *Complex Variables and Applications*. New York: McGraw-Hill.

3. Hsu, H. 1984. *Applied Vector Analysis*. New York: Harcourt Brace Jovanovich.
4. LePage, W. 1961. *Complex Variables and the Laplace Transform for Engineers*. New York: Dover.
5. Jackson, J. 1962. *Classical Electrodynamics*. New York: John Wiley & Sons Inc.
6. Rothwell, E. and Cloud, M.J. 2001. *Electromagnetics*. New York: CRC Press.
7. Smythe, W. 1950. *Static and Dynamic Electricity*. New York: McGraw-Hill.
8. Stratton, J. 1941. *Electromagnetic Theory*. New York: McGraw-Hill.
9. Wylie, C. Jr. 1960. *Advanced Engineering Mathematics*. New York: McGraw-Hill.
10. Zachmanoglou, E. and Thoe, D. 1976. *Introduction to Partial Differential Equations with Applications*. New York: Dover.

4

Transmission Line Theory

Transmission line theory is the study of electromagnetic waves propagating between two or more distinct conductors. The model is obtained by the judicious application of Maxwell's equations. It is typically derived via discrete circuit elements. The purpose here is to demonstrate how circuit theory is derived from field theory.

Indeed, the emphasis placed on the field-theoretic derivation is twofold: it has general applicability, and more important, it provides the basis for many engineering approximations and rules of thumb. This is particularly true in the chapter on antennae. Some useful references regarding the derivation of transmission lines from Maxwell's equations can be found in References [1,6,7], and for an informal and practical approach to the subject, refer to References [3–5].

4.1 Introduction

Recall from Theorem 1.4.1 and Corollary 1.4.2 that TEM cannot exist on a single conductor. Because, as shown below, waves defined by the transmission line equation are in TEM mode, it follows that transmission line structures comprise at least two conductors.

In practice, for very good conductors, an approximate TEM wave is sustained because conductivity $\sigma < \infty \Rightarrow E_{\|} \neq 0$ along the conductors, where $E_{\|}$ is the longitudinal (or axial) component of the electric field, (i.e., $E_{\|}$ is the field parallel to the direction of the TEM propagation). Physically, $E_{\|} \neq 0$ follows from the fact that $\sigma < \infty$ implies that the conductors have finite resistance (instead of zero resistance) and hence a driving voltage is needed to sustain the flow of charges, as energy is lost through the ohmic heating effect. Thus, the propagating wave is a quasi-TEM wave. It is not a perfect TEM wave because $E_{\|} \neq 0$. For good conductors, a TEM wave solution may be assumed for simplicity. Finally, a corollary of Section 1.4 is summarized below for future reference.

4.1.1. Proposition.

Let $M_{\pm} \subset \mathbf{R}^3$ be two disjoint conductors whose conductivity $\sigma_{\pm} >> 1$. Then, a TEM wave $(\boldsymbol{E}, \boldsymbol{B})$ propagating between $M_{\pm}$ induces a potential difference between M_- and M_+. In particular, the TEM wave induces a current density along $M_{\pm}$.

Proof

Without loss of generality, suppose $M_{\pm}$ is oriented such that the TEM wave is propagating along the z-axis. From Section 1.4, $\boldsymbol{E} = \boldsymbol{E}_{\perp}$ by definition, in particular, by Theorem 1.4.1, $\exists\phi$ such that $\boldsymbol{E}_{\perp} = -\nabla\varphi$. Because $M_{\pm}$ may be approximated as PECs, the boundary conditions on $M_{\pm}$ imply that $\varphi \mid M_{\pm} \equiv \varphi_{\pm}$ are constants, whence, $\delta\varphi = \varphi_+ - \varphi_-$ is the required potential difference, as claimed. To complete the proof, it suffices to observe that $B_z = 0$ implies that $\boldsymbol{n}_{\pm} \times \boldsymbol{B}_{\perp} = \mu \boldsymbol{J}_{\pm}$. The surface current is thus parallel to $\partial M_{\pm}$, as required. □

Thus, Proposition 4.1.1 establishes an equivalence between voltage–current and TEM waves. That is, a propagating TEM wave between two separate conductors will generate a potential difference between the two conductors and, equivalently, applying a time-harmonic voltage across two separate conductors will generate a TEM wave. The above theorem forms the basis for transmission line theory.

4.2 Transmission Line Equations

In this section, transmission line equations are derived from first principles via a field-theoretic method. It is then demonstrated that a pair of transmission lines can be approximated by a distributed line model commonly found in the literature. This thus justifies the use of a distributed line model in deriving the transmission line equation.

Transmission line equations are also known as the *telephone equations*. The following assumptions are made in the derivation.

- The conductors are imperfect; that is, $1 << \sigma < \infty$, where σ is the conductivity of the conductors, and without loss of generality, the conductors are assumed to have the same conductivity.
- The dielectric medium is imperfect; that is, $0 < \sigma_0 << 1$, where σ_0 is the conductivity of the surrounding homogeneous dielectric medium.
- The TEM solution holds between the pair of conductors, as $0 < \sigma_0 << 1 << \sigma$.
- The conductors are arbitrarily long with uniform cross-sections.

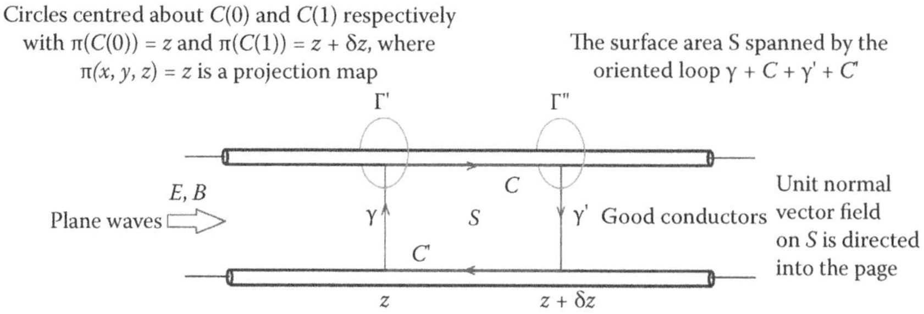

FIGURE 4.1
TEM waves propagating between two infinite transmission lines.

Figure 4.1 shows a TEM wave (or equivalently, some oscillating voltage source not shown) incident on an infinitely long pair of conductors. The closed path $\Gamma = \gamma \cup C \cup \gamma' \cup C'$ is oriented as shown in the figure and S is the rectangular surface that spans the oriented loop Γ. That is, $\Gamma(0) = \gamma(0)$ and $\Gamma(1) = \gamma(1)$. The unit vector $\boldsymbol{n}$ normal to S is directed into the page, consistent with the orientation of Γ (right-hand rule).

Assume for simplicity that γ is oriented in the $\mathbf{e}_y$ direction along the y-axis, C is oriented in the $\mathbf{e}_z$ direction along the z-axis and $\boldsymbol{n}$ is directed in the $+x$-axis direction, and suppose that the angular frequency of the incident TEM wave is ω. Suppose also that the conductors are very good conductors (viz., $\sigma >> \omega\varepsilon$ for the ω in question). Finally, recall that the following approximations are employed: (a) the general solution is a TEM solution and (b) a very small E_z-field on the surface of the conductors is assumed in order to overcome the ohmic loss due to finite conductivity of the conductors.

From Equation (1.15), invoking Stokes' theorem yields $\oiint_S \nabla \times \boldsymbol{E} \cdot \boldsymbol{n} d^2 r = \oint_\Gamma \boldsymbol{E} \cdot \boldsymbol{l} dr$ and hence,

$$-\frac{d}{dt}\oiint_S \boldsymbol{B} \cdot \boldsymbol{n} dS = \oint_\Gamma \boldsymbol{E} \cdot \boldsymbol{l} d\ell = \int_\gamma \boldsymbol{E} \cdot \boldsymbol{e}_y d\ell + \int_C \boldsymbol{E} \cdot \mathbf{e}_z d\ell + \int_{\gamma'} \boldsymbol{E} \cdot \mathbf{e}_y d\ell + \int_{C'} \boldsymbol{E} \cdot \mathbf{e}_z d\ell \tag{4.1}$$

where $\boldsymbol{n}$ is a unit normal vector field on S and $\boldsymbol{l}$ is the unit tangent vector field on Γ. Before proceeding further, recall that $\oiint_S \boldsymbol{B} \cdot \boldsymbol{n} dS = \Psi$ is just the magnetic flux that crosses the surface S, and *inductance* $L = L(S)$ is defined by $\Psi \equiv Li$, where i is the current flow around $\partial S = \gamma$.

Some comments are due. First, observe that the inductance $L = L(S)$ depends implicitly on the surface area S via the surface integral. Second, by definition, the potential difference along the path γ is given by $v(\gamma;t) = -\int_\gamma \boldsymbol{E} \cdot \mathbf{e}_y d\ell$, and likewise, the potential difference along the path γ' is given by $v(\gamma';t) = -\int_{\gamma'} \boldsymbol{E} \cdot \mathbf{e}_y d\ell = -v(-\gamma';t)$, where $(-\gamma')(t) \equiv \gamma'(1-t)\ \forall t \in [0,1]$ is the reverse orientation of γ'.

The potential difference $v(\gamma;t)$ and $v(\gamma';t)$ are well-defined because, by definition, TEM waves satisfy the Laplace equation and hence, the field is conservative. That is, the path integral is path-independent in a simply connected space as long as the endpoints are specified. By construction (see Figure 4.1) we may set $v(z,t) = v(\gamma;t)$ and $v(z+\delta z,t) = v(-\gamma';t)$.

The potential drop across C resulting from an imperfect conductor is a little bit more involved. For concreteness, let the loop Γ and the paths γ, γ', C, C' be parameterized by $0 \le s \le 1$. Fix some $s \in (0,1)$ such that Γ(s) = C(0) and a circle Γ′ centered at C(0) about the upper conductor along C, with π(C(0)) = z, where $\pi : \mathbf{R}^3 \to \mathbf{R}$ defined by $(x,y,z) \mapsto z$ is a projection map that projects a 3-vector onto its z-component; that is, $\pi(\boldsymbol{v}) = \boldsymbol{v} \cdot \mathbf{e}_z$ (see Figure 4.1). Then, $i_C(z,t) = \frac{1}{\mu}\oint_{\Gamma'} \boldsymbol{B} \cdot \mathbf{e}_\phi \, d\ell$ defines the flow of current along C, where $\mathbf{e}_\phi$ is a unit vector field tangent to Γ′. By Ohm's law,

$$v(C;t) = -\int_C E_z \, dz = \tfrac{1}{2} R \delta z i_C(z,t)$$

where $\frac{1}{2}R$ is the resistance per unit length of the respective conductors. Likewise, along C′, the voltage drop is given by

$$v(C';t) = -\int_{C'} E_z \, dz = -\tfrac{1}{2} R \delta z i_{C'}(z,t)$$

where, via Kirchhoff's current law, $i_{C'}(z,t) + i_C(z,t) = 0 \; \forall z,t$.

Indeed, this suggests that the current flowing from C to C′ along γ′ comprises essentially a displacement current; more on this point later. Thus, from Equation (4.1), it follows that

$$-\tfrac{d}{dt}\oiint_S \boldsymbol{B} \cdot \boldsymbol{n} d^2 \boldsymbol{x} = -v(z,t) + v(z+\delta z,t) + \tfrac{1}{2} R \delta z i_C(z,t) - \tfrac{1}{2} R \delta z i_{C'}(z,t) \tag{4.2}$$

and observe that $i_{C'}(z,t) = -i_{-C'}(z,t)$.

Now, recall from the preceding paragraph that the magnetic flux is defined by $\Psi = \int_S \boldsymbol{B} \cdot \boldsymbol{n} dS = Li$, where L is the inductance and i the current around Γ. Furthermore, observe trivially that $\lim_{\delta z \to 0} |S| \to 0$, where $|S|$ denotes the area of S and δz is the length of S. Hence, this motivates the following definition: $\tilde{\Psi} \equiv \lim_{\delta z \to 0} \frac{1}{\delta z} \int_S \boldsymbol{B} \cdot \boldsymbol{n} dS$, the magnetic flux per unit length. In particular, it follows that the inductance per unit length $\mathrm{L} = \frac{\tilde{\Psi}}{i}$ is also well-defined. Finally, from $-\frac{d}{dt}\int_S \boldsymbol{B} \cdot \boldsymbol{n} d^2 x = -L\partial_t i$ and hence, Equation (4.2) becomes

$$\partial_z v(z,t) = -\mathrm{L}\partial_t i(z,t) - \mathrm{R} i(z,t) \tag{4.3}$$

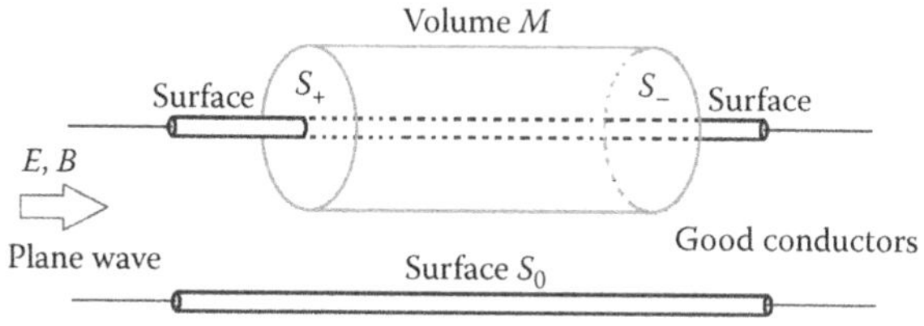

FIGURE 4.2
TEM waves propagating between two infinite transmission lines.

where

$$\partial_z v(z,t) = \lim_{\delta z \to 0} \tfrac{v(z+\delta z,t)-v(z,t)}{\delta z} \quad \text{and} \quad i(z,t) = \tfrac{1}{2} i_{C'}(z,t) + \tfrac{1}{2} i_{-C'}(z,t)$$

This is the first equation for the pair of transmission lines. To obtain the second transmission line equation, consider Figure 4.2.

The boundary of the small volume M is clearly seen to be $\partial M = S_+ \cup S_- \cup S_0$. Moreover, assume the same coordinate axes as in Figure 4.1. Namely, the z-direction is the direction of the wave propagation and the x–y plane defines the transverse (i.e., perpendicular) coordinates. Finally, suppose that the z-component of the surface S_+ is z and the z-component of the surface S_- is $z + \delta z$, where δz is the length of the cylindrical surface S_0.

From the equation of continuity (1.19), $-\frac{d}{dt} \oiiint_M \rho d^3x = \oiiint_M \nabla \cdot J d^3x$. Invoking the divergence theorem, $\oiiint_M \nabla \cdot J d^3x = \oiint_{\partial M} J \cdot d^2\boldsymbol{x}$, and Ohm's law, $\boldsymbol{J} = \sigma \boldsymbol{E}$, it follows that

$$\oiint_{\partial M} J \cdot \boldsymbol{n} d^2x = \iint_{S_+} J \cdot \boldsymbol{n}_+ d^2x + \iint_{S_-} J \cdot \boldsymbol{n}_- d^2x + \sigma \iint_{S_0} \boldsymbol{E} \cdot \boldsymbol{n} d^2x$$

where σ is the conductivity of the homogeneous medium wherein the two conductors are embedded. Set $i(z,t) = \iint_{S_+} \boldsymbol{J} \cdot \boldsymbol{n}_+ d^2x$ and $i(z+\delta z,t) = \iint_{S_-} \boldsymbol{J} \cdot \boldsymbol{n}_- d^2x$, where $\boldsymbol{n}_+ = -\boldsymbol{n}_-$ is the unit vector normal to $S_\pm$, respectively, to be the conduction current across the surfaces $S_\pm$. Hence, $\iint_{S_+} \boldsymbol{J} \cdot (-\boldsymbol{n}_-) d^2x + \iint_{S_-} \boldsymbol{J} \cdot \boldsymbol{n}_- d^2x = i(z+\delta z,t) - i(z,t)$.

The third term, $\sigma \iint_{S_0} \boldsymbol{E} \cdot \boldsymbol{n} d^2x$, denotes the transverse conduction contribution (it corresponds to the conductivity of a lossy dielectric medium) where $\boldsymbol{n}$ is the unit vector field on S_0. Moreover, recalling that $G(S_0) = \frac{\sigma}{v(z,t)} \iint_{S_0} \boldsymbol{E} \cdot \boldsymbol{n} d^2x$ defines the conductance, and noting trivially that $\lim_{\delta z \to 0} |S_0| = 0$ it follows that $G(S_0) \equiv \lim_{\delta z \to 0} \frac{1}{\delta z} \frac{\sigma}{v(z,t)} \iint_{S_0} \boldsymbol{E} \cdot \boldsymbol{n} d^2x$, is well-defined; this defines the conductance per unit length.

Finally, applying Gauss' law, $-\frac{d}{dt} \oiiint_M \rho d^3x = -\varepsilon \frac{d}{dt} \oiiint_M \nabla \cdot \boldsymbol{E} d^3x = -\varepsilon \frac{d}{dt} \iint_{\partial M} \boldsymbol{E} \cdot d^2\boldsymbol{x}$. Whence, from $Q = CV$, it follows that the capacitance

$C(\partial M) = \frac{\varepsilon}{v(z,t)} \frac{d}{dt} \int\int_{\partial M} \boldsymbol{E} \cdot d^2\boldsymbol{x}$ and in particular, the capacitance per unit length $C(S_0) = \lim_{\delta z \to 0} \frac{1}{\delta z} C(S_0)$ is well-defined, as the displacement current is zero across $S_+ \cup S_-$. Thus,

$$-C(S_0)\delta z\, \partial_t v(z,t) = i(z+\delta z,t) - i(z,t) + G(S_0)\delta z v(z,t) \tag{4.4}$$

and taking the limit as $\delta z \to 0$ yields the second transmission line equation:

$$\partial_z i(z,t) = -C(S_0)\partial_t v(z,t) - G(S_0)v(z,t) \tag{4.5}$$

As an aside, an alternative derivation of Equation (4.5) via (1.17) is given in Chapter 5. The above results can be formally epitomized as follows.

4.2.1 Theorem

Given a pair of semi-infinitely long conductors of conductivity $\sigma_0 >> 1$ that are parallel to the z-axis, suppose that the conductors have uniform cross-sections and are embedded in a homogeneous dielectric medium (ε, σ), where $0 < \sigma << 1 << \sigma_0$. Suppose a time-harmonic electromagnetic plane wave $(\boldsymbol{E}, \boldsymbol{B}, \omega)$ is incident on the pair of conductors at $z = 0$, where $\sigma_0 >> \varepsilon\omega$. Then, the induced voltage and current waves propagating along the conductor pair may be approximated by the TEM solution given by Equations (4.3) and (4.5). □

What is the explicit expression for the voltage v and current i from the coupled equations? To briefly see it, differentiate (4.3) with respect to z and replace $\partial_z i$ in the equation using (4.5):

$$\partial_z^2 v(z,t) = RGv(z,t) + (RC + LG)\partial_t v(z,t) + LC\partial_t^2 v(z,t) \tag{4.6}$$

where R,G, L, C are, respectively, resistance, conductance, inductance, and capacitance per unit length. Likewise, differentiating (4.5) with respect to z and replacing $\partial_z v$ in the equation with (4.3) yield:

$$\partial_z^2 i(z,t) = RGi(z,t) + (RC + LG)\partial_t i(z,t) + LC\partial_t^2 i(z,t) \tag{4.7}$$

Now, observe that Equations (4.6) and (4.7) have identical forms. Indeed, the equations are precisely the one-dimensional D'Alembert wave equation. Hence, the voltage and current that propagate along the conductors are precisely plane waves. By inspection, they propagate at speed $\frac{1}{\sqrt{LC}}$. In hindsight, it is almost clear that L, C, G are related to σ, μ, ε in some way.

4.2.2 Corollary

Given the conditions stated in Theorem 4.2.1, $LG = \mu\sigma$ and $LC = \mu\varepsilon$.

Proof

Suppose that the conductors are perfect conductors; that is, set R = 0. Then, Equations (4.6) and (4.7) reduce to

$$\partial_z^2 v(z,t) = \mathrm{LG}\,\partial_t v(z,t) + \mathrm{LC}\,\partial_t^2 v(z,t) \tag{4.8}$$

$$\partial_z^2 i(z,t) = \mathrm{LG}\,\partial_t i(z,t) + \mathrm{LC}\,\partial_t^2 i(z,t) \tag{4.9}$$

Now, recalling the wave Equation (1.28) and noting that $v(z,t) = -\int_\gamma \boldsymbol{E}\cdot\boldsymbol{l}\,d\ell$, (4.8) reduces to

$$-\int_\gamma \partial_z^2 \boldsymbol{E}\cdot\boldsymbol{l}\,d\ell = -\mathrm{LG}\int_\gamma \partial_t \boldsymbol{E}\cdot\boldsymbol{l}\,d\ell - \mathrm{LC}\int_\gamma \partial_t^2 \boldsymbol{E}\cdot\boldsymbol{l}\,d\ell$$

whence comparing the coefficients with Equation (1.28) yields

$$\partial_z^2 v(z,t) = \mu\varepsilon\,\partial_t^2 v(z,t) + \mu\sigma\,\partial_t v(z,t) \tag{4.10}$$

Likewise, recalling that $i(z,t) = \oint_C \boldsymbol{H}\cdot\boldsymbol{l}\,d\ell$, (4.9) has the same form as the wave Equation (1.29) for the magnetic flux density, yielding

$$\partial_z^2 i(z,t) = \mu\varepsilon\,\partial_t^2 i(z,t) + \mu\sigma\,\partial_t i(z,t) \tag{4.11}$$

whence, by inspection, it is clear that LG = μσ and LC = με, as required. □

Observe trivially from the above derivation that a transmission line can be represented by the *distributed parameter model* illustrated in Figure 4.3. Indeed, from the equivalence between (v,i) and $(\boldsymbol{E},\boldsymbol{B})$, it follows that voltage and current are subject to reflection at a boundary interface, just as electric and magnetic fields are. This topic is investigated in subsequent sections.

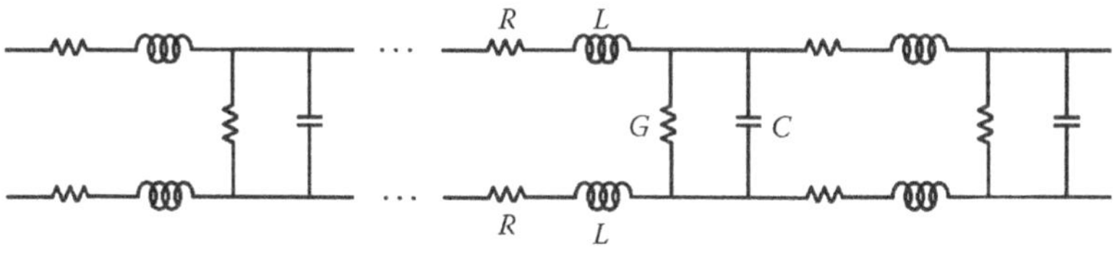

FIGURE 4.3
A distributed lumped parameter model representing a pair of transmission lines.

4.3 Characteristic Impedance and the Smith Chart

In this section, the concept of characteristic impedance is established. It has special significance in transmission line theory: to wit, the characteristic impedance is independent of the length of the conductors. This is established below. In Section 4.2, infinitely long conductors were studied. In contrast, finite conductors are investigated here. The motivation for this section runs as follows.

- Under what circumstances do finite conductors appear to be infinitely long conductors to a propagating TEM wave?
- What happens to an incident voltage and current at a boundary interface?
- What is the impedance along a transmission line from the perspective of an incident propagating wave?

For simplicity, consider a propagating time-harmonic TEM wave. Then, phasor transforming $(v,i) \to (V,I)$, Equations (4.3) and (4.4) can be rewritten as

$$\frac{dV(z)}{dz} = -RI(z) - i\omega LI(z) = -(R + i\omega L)I(z) \tag{4.12}$$

$$\frac{dI(z)}{dz} = -GV(z) - i\omega CV(z) = -(G + i\omega C)V(z) \tag{4.13}$$

The equations are now a pair of coupled first-order ordinary differential equations. The variable ω has been suppressed for convenience. So, differentiating Equation (4.12) with respect to z and substituting the value for $\frac{dI(z)}{dz}$ via (4.13):

$$\frac{d^2V(z)}{dz^2} = (R + i\omega L)(G + i\omega C)V(z) \equiv \gamma^2 V(z) \tag{4.14}$$

Likewise, differentiating (4.13) with respect to z yields:

$$\frac{d^2I(z)}{dz^2} = (R + i\omega L)(G + i\omega C)V(z) \equiv \gamma^2 I(z) \tag{4.15}$$

Observe the symmetry between the two equations. Before proceeding further, the *wave propagation constant* $\gamma \equiv \alpha + i\beta$ is evaluated explicitly.

First note that $(\alpha + i\beta)^2 = \gamma^2 = (R + i\omega L)(G + i\omega C)$. Hence, using the same technique as in Chapter 1, $\alpha^2 - \beta^2 + i2\alpha\beta = RG - \omega^2 LC + i\omega(RC + LG)$, leads to

$$\alpha^2 - \beta^2 = RG - \omega^2 LC \quad \text{and} \quad 2\alpha\beta = \omega(RC + LG)$$

Next, noting that $(\alpha^2+\beta^2)^2=(\alpha^2)^2+(\beta^2)^2+2\alpha^2\beta^2=(\alpha^2-\beta^2)^2+4\alpha^2\beta^2$, it follows that $\alpha^2+\beta^2=((RG-\omega^2LC)^2+\omega^2(RC+LG)^2)^{\frac{1}{2}}$. Hence, adding and subtracting the two equations, respectively, for $\alpha^2\pm\beta^2$ give

$$2\alpha^2=((RG-\omega^2LC)^2+\omega^2(RC+LG)^2)^{\frac{1}{2}}+RG-\omega^2LC$$

$$2\beta^2=((RG-\omega^2LC)^2+\omega^2(RC+LG)^2)^{\frac{1}{2}}-(RG-\omega^2LC)$$

In summary,

$$\alpha=\tfrac{1}{\sqrt{2}}\sqrt{\left\{(RG-\omega^2LC)^2+\omega^2(RC+LG)^2\right\}^{\frac{1}{2}}+RG-\omega^2LC} \tag{4.16a}$$

$$\beta=\tfrac{1}{\sqrt{2}}\sqrt{((RG-\omega^2LC)^2+\omega^2(RC+LG)^2)^{\frac{1}{2}}-(RG-\omega^2LC)} \tag{4.16b}$$

Now, observe trivially that $\sqrt{a^2+b^2}\geq\max\{a,b\}\ \forall a,b\in\mathbf{R}$. Thus, it follows that whenever

$$\omega<<\min\left\{\sqrt{\tfrac{RG}{LC}},\tfrac{1}{RC+LG}\right\}$$

by invoking the binomial expansion, Equation (4.16) reduces to

$$\alpha\approx\sqrt{RG}\left\{1+\tfrac{\omega^2}{8}\left(\tfrac{C}{G}+\tfrac{L}{R}\right)^2\right\}\quad\text{and}\quad\beta\approx\sqrt{RG}\,\tfrac{\omega}{2}\left(\tfrac{C}{G}+\tfrac{L}{R}\right)$$

To derive the expressions, it suffices to observe that $RG-\omega^2LC\approx RG$ and

$$\sqrt{1+\omega^2\left(\tfrac{RC+LG}{RG}\right)^2}\approx1+\tfrac{\omega^2}{2}\left(\tfrac{RC+LG}{RG}\right)^2=1+\tfrac{\omega^2}{2}\left(\tfrac{C}{G}+\tfrac{L}{R}\right)^2$$

In particular,

$$\lim_{\omega\to0}\alpha=\sqrt{RG}\quad\text{and}\quad\lim_{\omega\to0}\beta=0$$

and there is no wave propagation for

$$\omega<<\min\left\{\sqrt{\tfrac{RG}{LC}},\tfrac{1}{RC+LG}\right\},\text{ as }\gamma\approx\alpha$$

Next, consider the other extreme scenario wherein

$$\omega>>\max\left\{\tfrac{R}{L}+\tfrac{G}{C},\sqrt{\tfrac{RG}{LC}}\right\}$$

Then, by applying the binomial approximation once again, it follows that

$$\alpha \approx \tfrac{1}{2}\left\{R\sqrt{\tfrac{C}{L}} + G\sqrt{\tfrac{L}{C}}\right\} \quad \text{and} \quad \beta \approx \omega\sqrt{LC}\left\{1 + \tfrac{1}{8\omega^2}\left(\tfrac{R}{L} + \tfrac{G}{C}\right)^2\right\}$$

Once again, the derivation is similar to the former expressions: to wit, $RG - \omega^2 LC \approx -\omega^2 LC$ and

$$\sqrt{1 + \left(\tfrac{RC+LG}{\omega LC}\right)^2} \approx 1 + \tfrac{1}{2\omega^2}\left(\tfrac{RC+LG}{LC}\right)^2 = 1 + \tfrac{1}{2\omega^2}\left(\tfrac{R}{L} + \tfrac{G}{C}\right)^2$$

In particular,

$$\lim_{\omega\to\infty} \alpha \approx \tfrac{1}{2}\left\{R\sqrt{\tfrac{C}{L}} + G\sqrt{\tfrac{L}{C}}\right\} \equiv \tfrac{1}{2}\left\{\tfrac{R}{R_0} + GR_0\right\} \quad \text{and} \quad \lim_{\omega\to\infty} \beta \approx \omega\sqrt{LC}$$

where $R_0 = \sqrt{L/C}$ defines the *characteristic impedance* of a lossless transmission line.

Thus, it is clear from the above analysis that for very low frequencies, $\gamma \approx \alpha \approx \sqrt{RG}$, and the line appears to be purely resistive. On the other hand, for very high frequencies, $\Re e(\gamma)$ is independent of frequency. Finally, for a perfect conductor in a perfect dielectric, $\alpha = 0$ and $\beta = \omega\sqrt{LC} \Rightarrow \gamma = i\omega\sqrt{LC}\ \forall\omega$. These results are summarized formally below for future reference.

4.3.1 Proposition

Given a pair of infinitely long transmission lines $(C_{\pm}, L, C, R, G)$, for any positive number $\varepsilon > 0$, $\exists\omega_\varepsilon > 0$ such that $\forall\omega > \omega_\varepsilon$,

$$\text{a)} \quad \left|\alpha - \tfrac{1}{2}(RG_0 + GR_0)\right| < \varepsilon$$

$$\text{b)} \quad \left|\beta - \omega\sqrt{LC}\right| < \varepsilon$$

where $R_0 = \sqrt{\tfrac{L}{C}}$ is the characteristic impedance and $G_0 = \sqrt{\tfrac{C}{L}}$. In particular, up to first order in $\tfrac{1}{\omega}$,

$$\gamma \approx \tfrac{1}{2}(RG_0 + GR_0) + i\omega\sqrt{LC}\left\{1 + \tfrac{1}{8\omega^2}\left(\tfrac{R}{L} + \tfrac{G}{C}\right)^2\right\}$$

Proof

From the discussion above, given any $\varepsilon > 0$, provisionally choose

$$\omega >> \max\left\{\tfrac{R}{L} + \tfrac{G}{C}, \sqrt{\tfrac{RG}{LC}}\right\}$$

For concreteness, set

$$\omega > \omega_0 \equiv a \max\left\{\tfrac{R}{L} + \tfrac{G}{C}, \sqrt{\tfrac{RG}{LC}}\right\}$$

for any fixed constant $a >> 1$. Then, appealing to the binomial approximation, $(1+\delta)^r \approx 1 + r\delta + \frac{r(r-1)}{2!}\delta^2 + o(\delta^3)$, where $|\delta| << 1$ and $r \in \mathbf{R}$, on setting $\kappa = \frac{1}{2}(RG_0 + GR_0)$, it follows that up to $\frac{1}{\omega^2}$,

$$\begin{aligned}\alpha &\approx \tfrac{1}{\sqrt{2}}\omega\sqrt{LC}\left\{\tfrac{1}{2\omega^2}\left(\tfrac{R}{L}+\tfrac{G}{C}\right)^2 - \tfrac{1}{2\omega^2}\left(\tfrac{R}{L}+\tfrac{G}{C}\right)^2 \tfrac{1}{4\omega^2}\left(\tfrac{R}{L}+\tfrac{G}{C}\right)^2 + o\left(\tfrac{1}{\omega^6}\right)\right\}\\ &\approx \kappa\left\{1 - \tfrac{1}{4\omega^2}\left(\tfrac{R}{L}+\tfrac{G}{C}\right)^2\right\}^{\frac{1}{2}}\\ &\approx \kappa - \tfrac{\kappa}{4\omega^2}\left(\tfrac{R}{L}+\tfrac{G}{C}\right)^2\end{aligned}$$

whence, given $\varepsilon > 0$, choose ω such that

$$\tfrac{\kappa}{4\omega^2}\left(\tfrac{R}{L}+\tfrac{G}{C}\right)^2 < \varepsilon$$

In particular, set

$$\omega'_\varepsilon = \max\left\{\sqrt{\tfrac{\kappa}{\varepsilon}}\left(\tfrac{R}{L}+\tfrac{G}{C}\right), \omega_0\right\}$$

Then, $\forall \omega > \omega'_\varepsilon \Rightarrow \left|\alpha - \frac{1}{2}(RG_0 + GR_0)\right| < \varepsilon$.

Lastly, for case (b), via the binomial approximation, choose $\omega > \omega_0$, then

$$\beta \approx \tfrac{1}{\sqrt{2}}\omega\sqrt{LC}\sqrt{2 + \tfrac{1}{2\omega^2}\left(\tfrac{R}{L}+\tfrac{G}{C}\right)^2} \Rightarrow \beta - \omega\sqrt{LC} \approx 1 + \tfrac{\sqrt{LC}}{8\omega}\left(\tfrac{R}{L}+\tfrac{G}{C}\right)^2 + o\left(\tfrac{1}{\omega^3}\right)$$

whence choosing

$$\omega > \omega''_\varepsilon \equiv \max\left\{\tfrac{\sqrt{LC}}{8\varepsilon}\left(\tfrac{R}{L}+\tfrac{G}{C}\right)^2, \omega_0\right\} \Rightarrow \left|\beta - \omega\sqrt{LC}\right| < \varepsilon$$

So, it is evident that upon choosing $\omega_\varepsilon = \max\{\omega'_\varepsilon, \omega''_\varepsilon\}$, both (a) and (b) are satisfied whenever $\omega > \omega_\varepsilon$, as required. □

4.3.2 Definition

The wavelength λ of a wave propagating between a pair of transmission lines is defined by $\lambda = \frac{2\pi}{\beta}$, where $\beta = \Im m(\gamma)$.

For perfect conductors in a perfect dielectric, $\gamma = i\omega\sqrt{LC}$ and hence, $\lambda = \frac{2\pi}{\omega\sqrt{LC}}$. From this, it is clear that the *phase velocity* is $\upsilon = \frac{1}{\sqrt{LC}}$. For nonideal conductors and nonideal dielectrics, the phase velocity clearly becomes much more complicated.

The general solution to Equation (4.14) is

$$V(z) = V_+ e^{-\gamma z} + V_- e^{\gamma z} \tag{4.17}$$

The coefficient V_+ is the coefficient for the forward propagating (i.e., incident) wave. The coefficient V_- is the coefficient for the backward propagating (i.e., reflected) wave. In particular, $V_- \equiv 0$ if the transmission line is infinitely long.

As a corollary to Proposition 4.3.1, when the frequency of a propagating wave is sufficiently high, the attenuation of the wave falls off essentially as $e^{-\frac{1}{2}(RG_0+GR_0)z}$, irrespective of the frequency. So, what happens should the line be finite? To answer this question, recall from Section 4.2 regarding the equivalence between L, G, C and $(\mu, \sigma, \varepsilon)$:

$$LG = \mu\sigma \quad \text{and} \quad LC = \mu\varepsilon \tag{4.18}$$

Hence, changing the values of (μ, ε) will have an impact on the boundary between two media when a TEM wave is incident on the boundary. The equivalence established by Equation (4.16) implies that changing the values of the inductance and the capacitance of the line will affect how the voltage and current waves propagate. In particular, reflection will generally occur when the voltage wave is incident on the boundary between the conductors with different values of inductance and capacitance.

Reflection occurs in order to satisfy the boundary condition imposed by a change in L, C, G, or equivalently, in μ, σ, ε. For example, if a finite conductor is open with respect to ground, a physical boundary condition would be that the current at the end of the conductor be zero. Conversely, if a finite conductor were shorted to ground, a physical boundary condition would be that the voltage at the endpoint of the conductor be zero. A physical solution for the respective differential equations must then satisfy the given boundary conditions.

4.3.3 Example

Consider a pair of infinitely long perfect conductors such that

$$L = \begin{cases} L_1 & \text{for} \quad z < 0, \\ L_2 & \text{for} \quad z > 0, \end{cases} \quad \textit{and} \quad C = \begin{cases} C_1 & \text{for} \quad z < 0, \\ C_2 & \text{for} \quad z > 0. \end{cases}$$

That is, the conductor consists of a pair of semi-infinite conductors of different L, C. Recall that the propagating voltage wave must satisfy the wave equation

$$\partial_z^2 V_\alpha(z) = \gamma_\alpha^2 V_\alpha(z)$$

where $\gamma_\alpha^2 = -\omega^2 L_\alpha C_\alpha$, in conductor $C_\alpha, \alpha = 1,2$.

Let the solution in $z < 0$ be $V(z) = V_1^+ e^{\gamma_1 z} + V_1^- e^{-\gamma_1 z}$ and the solution in $z > 0$ be $V(z) = V_2^+ e^{-\gamma_2 z}$. Note that by assumption, $V_2^- = 0$ as there is no reflection from an infinite line (the absence of a boundary for all finite $z > 0$). Inasmuch as the potential must be continuous at the boundary $z = 0$, it follows that $V_1^+ + V_1^- = V_2^+$. Moreover, as $\frac{d^2 V(z)}{dz^2}$ is defined on the transmission line, it follows that $\frac{dV(z)}{dz}$ must also be continuous at $z = 0$. Hence,

$$-\gamma_1 V_1^+ + \gamma_1 V_1^- = -\gamma_2 V_2^+$$

The voltage $V_1^+ e^{\gamma_1 z}$ represents the incident wave, $V_1^- e^{-\gamma_1 z}$ corresponds to the reflected wave, and $V_2^+ e^{\gamma_2 z}$ the transmitted wave.

Now, observe that solving for $V_1^\pm$ as functions of V_2^+ at $z = 0$ yields

$$V_1^\pm = \tfrac{1}{2} V_2^+ \left(1 \pm \tfrac{\gamma_2}{\gamma_1}\right) \Rightarrow \frac{V_1^-}{V_1^+} = \frac{\gamma_1 - \gamma_2}{\gamma_1 + \gamma_2} \quad \text{and} \quad \frac{V_2^+}{V_1^+} = \frac{2\gamma_1}{\gamma_1 + \gamma_2}$$

which are constants. The former constant relates to the reflection of the wave at the boundary $z = 0$, whereas the latter relates to the transmission of the wave at the interface $z = 0$. The analysis for $(I_1^\pm, I_2^+)$ follows that of $(V_1^\pm, V_2^+)$ *mutatis mutandis*, leading to

$$\frac{I_1^-}{I_1^+} = \frac{\gamma_1 - \gamma_2}{\gamma_1 + \gamma_2} \quad \text{and} \quad \frac{V_2^+}{V_1^+} = \frac{2\gamma_1}{\gamma_1 + \gamma_2}$$

This analysis motivates the details sketched below. □

Having had a sneak preview of what's to follow from Example 4.3.3, consider for simplicity an infinitely long pair of transmission lines. Then, the general solutions to Equations (4.15) and (4.16) are, respectively, $V(z) = V_0^+ e^{-\gamma z} + V_0^- e^{\gamma z}$ and $I(z) = I_0^+ e^{-\gamma z} + I_0^- e^{\gamma z}$, where (V_0^+, I_0^+) are forward propagating waves and (V_0^-, I_0^-) are backward propagating (i.e., reflected) waves.

Next, noting that $\frac{d}{dz} V = -\gamma\{V_0^+ e^{-\gamma z} - V_0^- e^{\gamma z}\}$, it follows from Equation (4.12) that $\frac{d}{dz} V = -\gamma\{V_0^+ e^{-\gamma z} - V_0^- e^{\gamma z}\} = -(R + i\omega L) I$. That is, $V_0^+ e^{-\gamma z} - V_0^- e^{\gamma z} = \frac{R + i\omega L}{\gamma} \{I_0^+ e^{-\gamma z} + I_0^- e^{\gamma z}\}$ implies that

$$0 = \left\{V_0^+ - \tfrac{R + i\omega L}{\gamma} I_0^+\right\} e^{-\gamma z} - \left\{V_0^- + \tfrac{R + i\omega L}{\gamma} I_0^-\right\} e^{\gamma z}$$

Because this holds for all z, it follows immediately that $V_0^+ - \frac{R+i\omega L}{\gamma} I_0^+ = 0$ and $V_0^- + \frac{R+i\omega L}{\gamma} I_0^- = 0$; to wit,

$$\frac{V_0^+}{I_0^+} = \frac{R+i\omega L}{\gamma} = -\frac{V_0^-}{I_0^-}$$

As a side remark, the same result can be obtained via Equation (4.13) and $I(z) = I_0^+ e^{-\gamma z} + I_0^- e^{\gamma z}$:

$$I(z) = \frac{\gamma}{R+i\omega L}\left\{V_+ e^{-\gamma z} - V_- e^{\gamma z}\right\} = \sqrt{\frac{G+i\omega C}{R+i\omega L}}\left\{V_+ e^{-\gamma z} - V_- e^{\gamma z}\right\}$$

as expected. These results motivate the following definition.

4.3.4 Definition

Given a pair of parallel transmission lines $(C_\pm, \ell, R, L, C, G)$ of length $\ell \le \infty$, its characteristic impedance is given by $Z_0 = \sqrt{\frac{R+i\omega L}{G+i\omega C}}$.

The characteristic impedance is thus the impedance looking down a pair of infinite transmission lines. By definition, the characteristic impedance is, in general, frequency dependent. Clearly, for $\omega >> 1$ to be sufficiently large, that is, $R << \omega L$ and $G << \omega C$, then $Z_0 \approx R_0 \equiv \sqrt{\frac{L}{C}}$. That is, the characteristic impedance is approximately independent of the frequency. Explicitly, via the binomial expansion, up to first order in $1/\omega$,

$$Z_0 = \sqrt{\frac{R+i\omega L}{G+i\omega C}} \approx \sqrt{\frac{L}{C}}\left\{1 - \frac{i}{2\omega}\frac{R}{L}\right\}\left\{1 + \frac{i}{2\omega}\frac{C}{G}\right\} \approx R_0\left\{1 + \frac{i}{2\omega}\left(\frac{C}{G} - \frac{R}{L}\right)\right\}$$

4.3.5 Theorem

The characteristic impedance of a lossless pair of parallel transmission lines $(C_\pm, \ell, L, C)$ is independent of the angular frequency ω of a propagating time-harmonic wave.

Proof

Because the lines are lossless, $R = 0 = G$, and hence, from Definition 4.3.4, $Z_0 = \sqrt{L/C}$. □

From the previous discussion, a pair of transmission lines transmitting a very high frequency, monochromatic, time-harmonic TEM wave behave as if the lines were lossless. In particular, this holds for a square wave or trapezoidal wave provided the fundamental frequency is sufficiently large.

In what follows, only finite transmission lines are considered. For simplicity, let (γ_+, γ_-) denote the axes of a pair of parallel transmission lines (C_+, C_-) such that the lengths $|\gamma_+| = |\gamma_-|$. The pair of transmission lines (C_+, C_-)

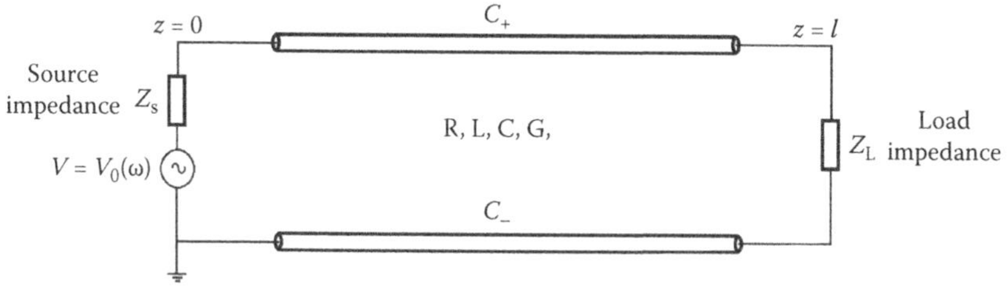

FIGURE 4.4
A pair of finite transmission lines.

is identified with (γ_+,γ_-), and where convenient, they are parameterized by $\gamma_\pm:[0,1]\to\mathbf{R}^3$. See Figure 4.4 for details. Finally, for notational convenience, let $(C_\pm,\ell,R,L,C,G)$ denote the above transmission line pair such that $|\gamma_+|=\ell=|\gamma_-|$. Note that although it is unfortunate that the axes of a pair of transmission lines (γ_+,γ_-) share the same symbol as the wave propagation constant γ, no confusion should arise on this account.

4.3.6 Proposition

Suppose a transmission line pair $(C_\pm,\ell,R,L,C,G)$ is connected to a source $(V_0(\omega),Z_S)$ and terminated by a load Z_L. Then, the input impedance $Z(z)$ toward the load at any point $z\in[0,\ell]$ on the transmission line is given by

$$Z(z)=Z_0\,\frac{Z_L+Z_0\tanh\gamma(\ell-z)}{Z_0+Z_L\tanh\gamma(\ell-z)} \tag{4.19}$$

Proof

The general solutions of Equations (4.15) and (4.16) at $z=\ell$ yield

$$V(\ell)=V_0^+e^{-\gamma\ell}+V_0^-e^{\gamma\ell} \quad \text{and} \quad I(\ell)=I_0^+e^{-\gamma\ell}+I_0^-e^{\gamma\ell}$$

Now, recalling that $Z_0=-\frac{V_0^-}{I_0^-}$, it follows at once that $I(\ell)=I_0^+e^{-\gamma\ell}-\frac{V_0^-}{Z_0}e^{\gamma\ell}$. Hence, solving for $V_0^\pm$ yield $V_0^\pm=\frac{1}{2}(V(\ell)\pm I(\ell)Z_0)e^{\pm\gamma\ell}$. Furthermore, by definition, $Z_L=\frac{V(\ell)}{I(\ell)}$ implies that

$$V_0^\pm=\tfrac{1}{2}I(\ell)(Z_L\pm Z_0)e^{\pm\gamma\ell}$$

and hence, substituting into the general solution for $V(z),I(z)$ leads directly to

$$V(z)=\tfrac{1}{2}I(\ell)\left\{(Z_L+Z_0)e^{\gamma(\ell-z)}+(Z_L-Z_0)e^{-\gamma(\ell-z)}\right\} \tag{4.20}$$

$$I(z)=\frac{1}{2}\frac{I(\ell)}{Z_0}\left\{(Z_L+Z_0)e^{\gamma(\ell-z)}-(Z_L-Z_0)e^{-\gamma(\ell-z)}\right\} \tag{4.21}$$

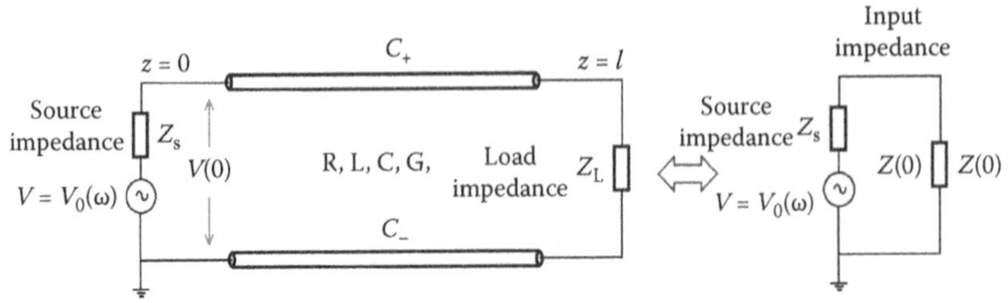

FIGURE 4.5
The equivalence between a transmission line and a lumped model.

Lastly, substituting $\cosh x = \frac{1}{2}\{e^x + e^{-x}\}$ and $\sinh x = \frac{1}{2}\{e^x - e^{-x}\}$ into the above two equations, it follows clearly that

$$Z(z) = \frac{V(z)}{I(z)} = Z_0 \frac{Z_L \cosh\gamma(\ell - z) + Z_0 \sinh\gamma(\ell - z)}{Z_0 \cosh\gamma(\ell - z) + Z_L \sinh\gamma(\ell - z)} = Z_0 \frac{Z_L + Z_0 \tanh\gamma(\ell - z)}{Z_0 + Z_L \tanh\gamma(\ell - z)} \qquad \square$$

In particular, it is obvious that the *input impedance* as seen by the source at $z = 0$ is

$$Z(0) = Z_0 \frac{Z_L + Z_0 \tanh\gamma\ell}{Z_0 + Z_L \tanh\gamma\ell}$$

Its physical significance is best illustrated in Figure 4.5.

4.3.7 Corollary

Suppose $(C_\pm^{(\infty)}, R, L, C, G)$ is a pair of infinitely long transmission lines and let $(C_\pm, \ell, R, L, C, G) \subset (C_\pm^{(\infty)}, R, L, C, G)$ be an arbitrary compact segment that is terminated by some load Z. That is, the pair $(\gamma_+(1), \gamma_-(1))$ is connected across Z. Then, $(C_\pm, \ell, R, L, C, G)$ is equivalent to $(C_\pm^{(\infty)}, R, L, C, G)$ if and only if $Z = Z_0$ for all $\ell > 0$.

Proof

From Proposition 4.3.6, $Z_L = Z_0$ if and only if $Z(z) = Z_0 \ \forall z \in [0, \ell]$, $\ell > 0$ arbitrary. $\square$

The above result thus illustrates an important feature of infinite transmission lines: they can be simulated by terminating a finite pair of transmission lines with their characteristic impedance. In this case, the finite transmission line is said to be *matched*.

Observe from Equation (4.20) and $V_0^\pm = \frac{1}{2} I(\ell)(Z_L \pm Z_0)$ that

$$V(z) = V_0^+ \left\{ e^{\gamma(\ell - z)} + \frac{Z_L - Z_0}{Z_L + Z_0} e^{-\gamma(\ell - z)} \right\} \tag{4.22}$$

for $0 \le z \le \ell$. In particular,

$$\frac{V_0^-}{V_0^+} = \frac{Z_L - Z_0}{Z_L + Z_0}$$

by definition, and

$$\frac{V_0^+ + V_0^-}{V_0^+} = \frac{2Z_L}{Z_L + Z_0}$$

Finally, recalling that V_0^- denotes the reflected wave whereas V_0^+ denotes the incident wave, the following definition ensues.

4.3.8 Definition

Given a pair of finite transmission lines $(C_\pm, \ell, R, L, C, G)$, the reflection coefficient at the load is $\Gamma(\ell) = \frac{Z_L - Z_0}{Z_L + Z_0}$ and the transmission coefficient at the load is $T(\ell) = \frac{2Z_L}{Z_L + Z_0}$.

Specifically, the reflected wave amplitude is $\Gamma(\ell)V_0^+$ and the transmitted wave amplitude is $T(\ell)V_0^+$. Furthermore, it is clear from Definition 4.3.8 that $1 + \Gamma(\ell) = T(\ell)$. That is, the sum of the incident wave and the reflected wave at the load is equal to the wave transmitted across the load. Lastly, observe that as impedance is, in general, complex, it is clear that $\Gamma(\ell) = |\Gamma(\ell)|e^{i\theta_\ell}$, where $\theta_\ell = \arg\Gamma(\ell)$.

4.3.9 Lemma

Given $(C_\pm, \ell, R, L, C, G)$, the local maxima of $|V|$ occur at $z = \ell - \frac{2n\pi + \theta_\ell}{2\beta}$ $\forall n = 0, 1, 2, \ldots$, and local minima of $|V|$ occur at $z = \ell - \frac{(2n-1)\pi + \theta_\ell}{2\beta}$ $\forall n = 0, 1, 2, \ldots$, where $\gamma = \alpha + i\beta$ is the wave propagation constant.

Proof

From Equation (4.22), $V(z) = V_0^+ e^{\alpha(\ell - z)} e^{i\beta(\ell - z)} \left\{1 + |\Gamma(\ell)| e^{i(\theta_\ell - 2\beta(\ell - z))} e^{-2\alpha(\ell - z)}\right\}$. That is,

$$|V(z)| = \left|V_0^+ e^{\alpha(\ell - z)}\right| \left|1 + |\Gamma(\ell)| e^{i(\theta_\ell - 2\beta(\ell - z))} e^{-2\beta(\ell - z)}\right|$$

Thus, it is quite evident that local maxima occur whenever $e^{i(\theta_\ell - 2\beta(\ell - z))} = 1$. Specifically,

$$e^{i(\theta_\ell - 2\beta(\ell - z))} = 1 \Rightarrow \theta_\ell - 2\beta(\ell - z) = -2n\pi \Rightarrow z = -\frac{2n\pi + \theta_\ell}{2\beta} + \ell \ \ \forall n = 0, 1, \ldots$$

as $z \ge 0$. Likewise, local minima occur whenever $e^{i(\theta_\ell - 2\beta(\ell - z))} = -1$. That is,

$$e^{i(\theta_\ell - 2\beta(\ell - z))} = -1 \Rightarrow \theta_\ell - 2\beta(\ell - z) = -(2n-1)\pi \Rightarrow z = -\frac{(2n-1)\pi + \theta_\ell}{2\beta} + \ell \ \ \forall n = 0, 1, \ldots$$

□

In view of Lemma 4.3.9, given $(C_{\pm},\ell,R,L,C,G)$, suppose $\ell>0$ is such that $\alpha\ell<<1$. Then, $e^{-\alpha\ell}\approx 1$ and hence,

$$\frac{\max|V|}{\min|V|}\approx\frac{1+|\Gamma(\ell)|}{1-|\Gamma(\ell)|}$$

which is a constant. This suggests the following definition. Suppose a lossless, finite transmission line, terminated at some load, is embedded in a lossless medium. Then, the *voltage standing wave ratio* is defined by

$$\mathcal{V}=\frac{\max|V|}{\min|V|}=\frac{1+|\Gamma(\ell)|}{1-|\Gamma(\ell)|}$$

Observe that for a low loss transmission line system $(C_{\pm},\ell,R,L,C,G)$, VSWR provides a convenient and well-defined way of quantifying standing waves that exist on the transmission line system. Indeed, when the load is matched with the characteristic impedance, $\Gamma(\ell)=0\Rightarrow\mathcal{V}=1$. On the other hand, when $\Gamma(\ell)=1$ (i.e., maximal reflection), then $\mathcal{V}=\infty$. Show, in Exercise 4.5.1, that if $\omega\geq 0$ is fixed, then $\alpha\ell<<1\Leftrightarrow\frac{\ell}{\lambda}<<1$, where λ is the wavelength of the propagating field.

4.3.10 Proposition

Given $(C_{\pm},\ell,R,L,C,G)$, let $\gamma=\alpha+i\beta$ be the wave propagation constant. Suppose that $\frac{\ell}{\lambda}<<1$, where λ is the wavelength of the propagating time-harmonic voltage wave. Then, the following two conditions hold.

a) Let $z=\underline{z}_n$ correspond to the minima of $V=V(z)$, and $z=\overline{z}_n$ correspond to the maxima of $V=V(z)$, where $\cdots<\underline{z}_n<\overline{z}_n<\underline{z}_{n+1}<\overline{z}_{n+1}<\cdots\ \forall n$. Then, $\overline{z}_n-\underline{z}_n\approx\frac{\lambda}{4}\ \forall n$.
b) Given VSWR $\mathcal{V}$, $|\Gamma(\ell)|=\frac{\mathcal{V}-1}{\mathcal{V}+1}$ and hence, for a known $\overline{z}_k$ for some k,

$$\theta_\ell=2\beta(\ell-\overline{z}_k)-2k\pi\Rightarrow\Gamma(\ell)=\frac{\mathcal{V}-1}{\mathcal{V}+1}e^{i2\beta(\ell-\overline{z}_k)-2k\pi}=\frac{\mathcal{V}-1}{\mathcal{V}+1}e^{i2\beta(\ell-\overline{z}_k)}$$

Similarly, $\Gamma(\ell)$ can be determined if $\mathcal{V}$ and any fixed $\underline{z}_k$ are known.

Proof

(a) From Lemma 4.3.9, $\overline{z}_n-\underline{z}_n=\frac{\pi}{2\beta}$. For $\alpha\ell<<1$, by Exercise 4.5.1, $\frac{\ell}{\lambda}<<1\Rightarrow\ \beta\approx\omega\sqrt{\mu\varepsilon}=\frac{2\pi}{\lambda}$, and the result thus follows.
(b) It is easy to see that

$$\mathcal{V}=\frac{1+|\Gamma(\ell)|}{1-|\Gamma(\ell)|}\Rightarrow\mathcal{V}\left(1-|\Gamma(\ell)|\right)=1+|\Gamma(\ell)|\Rightarrow|\Gamma|=\frac{\mathcal{V}-1}{\mathcal{V}+1}$$

Finally, $\Gamma=|\Gamma|e^{i\theta_\ell}$ and $\overline{z}_n=\ell-\frac{2n\pi+\theta_\ell}{2\beta}\ \forall n=0,1,2,\ldots$, from Lemma 4.3.9, yields the desired result. □

Because VSWR $\mathcal{V} = \frac{V_{max}}{V_{min}}$, it follows that should VSWR become very large, then V_{max} can become very large on parts of the transmission line. This voltage amplification phenomenon is known as the *Ferranti effect*. It is possible for breakdown voltage of the line to be attained via the Ferranti effect. This would clearly be detrimental in any digital circuit. However, this is generally more of a concern for power lines than for digital circuits.

4.3.11 Theorem

Given $(C_{\pm}, \ell, R, L, C, G)$, the condition $\frac{R}{L} = \frac{G}{C}$ constitutes a necessary and sufficient condition for the characteristic impedance of a transmission line to be independent of the angular frequency of propagation. That is, $Z_0 \equiv \sqrt{L/C}$ if and only if R/L=G/C.

Proof

Now,

$$Z_0 \equiv \sqrt{\frac{R + i\omega L}{G + i\omega C}}$$

can be rewritten as

$$\sqrt{\frac{R + i\omega L}{G + i\omega C}} = \sqrt{\frac{L}{C}}\sqrt{\frac{\frac{R}{L} + i\omega}{\frac{G}{C} + i\omega}}$$

Hence,

$$\frac{R}{L} = \frac{G}{C} \text{ if and only if } \sqrt{\frac{\frac{R}{L} + i\omega}{\frac{G}{C} + i\omega}} = 1$$

and thus yielding $Z_0 = \sqrt{L/C}$ for any frequency $\omega \geq 0$. □

A transmission line satisfying the condition stated in Theorem 4.3.11 is said to be *distortionless*. As a trivial corollary, note that for a perfect conductor (R = 0) in a lossless dielectric (G = 0), $Z_0 = \sqrt{L/C}$ automatically holds. However, this does not mean that if the conditions of Theorem 4.3.11 should be satisfied, there is no attenuation to the propagating TEM waves along (C_+, C_-). Indeed, attenuation will occur whenever $R \neq 0$: the energy needed to overcome the finite electrical conductivity manifests as heat, that is, as *Joule heating*. Theorem 4.3.11 merely states that the characteristic impedance will appear as if the transmission line system were lossless and hence no signal distortion for all angular frequencies of wave propagation.

4.3.12 Proposition

Given a pair of finite transmission lines $(C_{\pm}, \ell, R, L, C, G)$, suppose $\ell > 0$. If $\omega > 0$ satisfies $\beta\ell = \frac{1}{2}n\pi$ for any $n \in \mathbf{N}$, then $Z_L \in \mathbf{R} \Rightarrow Z(0) \in \mathbf{R}$ and $Z_L \in \mathbf{C} \Rightarrow Z(0) \in \mathbf{C}$.

Proof

The above assertions can be easily established by noting that

$$Z(0) = Z_0 \frac{Z_L + Z_0 \tanh\gamma\ell}{Z_0 + Z_L \tanh\gamma\ell} \quad \text{and} \quad \tanh(\alpha + i\beta)\ell = \frac{\tanh\alpha\ell + i\tan\beta\ell}{1 + i\tanh\alpha\ell\tan\beta\ell}$$

where $\gamma = \alpha + i\beta$. Then, for $\beta\ell = n\pi, n \in \mathbf{N}$, $\tan\beta\ell = 0$, and hence, $\tanh\gamma\ell = \tanh\alpha\ell \in \mathbf{R}$. Thus, $Z_L \in \mathbf{R} \Rightarrow Z(0) \in \mathbf{R}$ and $Z_L \in \mathbf{C} \Rightarrow Z(0) \in \mathbf{C}$. Next, if $\beta\ell = \frac{2n+1}{2}\pi, n \in \mathbf{N}$, then $\tanh\gamma\ell = \frac{1}{\tanh\alpha\ell} \in \mathbf{R}$ and hence, $Z_L \in \mathbf{R} \Rightarrow Z(0) \in \mathbf{R}$ and $Z_L \in \mathbf{C} \Rightarrow Z(0) \in \mathbf{C}$, once again, as claimed. □

The above proposition yields a simple condition under which the input impedance of a transmission line is resistive if the load is resistive. In general, it is clear from the proof that even if the load is resistive, the input impedance is complex. This section closes by demonstrating that there exists a conformal transformation mapping the input impedance (or the load impedance) onto the reflection coefficient. The graph of the mapping is called the *Smith Chart,* and it is often used to determine graphically the input impedance or load impedance, given that the reflection coefficient at the load is known.

4.3.13 Theorem

Given a pair of lossless transmission lines $(\gamma_{\pm}, \ell, L, C)$, set $\hat{Z}(z) = \frac{Z(z)}{Z_0}$ to be the normalized impedance along the transmission line. Then, there exists a conformal transformation ς mapping the ρ-space onto the $\hat{Z}$-space given by

$$\varsigma(\Gamma; z) = \frac{1+\rho(z)}{1-\rho(z)}$$

for any fixed $z \in [0, \ell]$, where $\Gamma = \Gamma(\ell)$ and $\rho = \Gamma e^{-i\beta(\ell - z)}$. In particular, on setting $\hat{Z} = \hat{R} + i\hat{X}$ and $\rho = \sigma + i\chi$, where $\hat{R}, \hat{X}, \sigma, \chi \in \mathbf{R}$, and the z-variable has been suppressed for simplicity,

$$\left(\sigma - \frac{\hat{R}}{1+\hat{R}}\right)^2 + \chi^2 = \frac{1}{(1+\hat{R})^2} \quad \text{and} \quad (\sigma - 1)^2 + \left(\chi - \frac{1}{\hat{X}}\right)^2 = \frac{1}{\chi^2}$$

Proof

From Equations (4.20) and (4.21), it is clear that

$$Z(z) = \frac{V(z)}{I(z)} = Z_0 \frac{e^{i\beta(\ell - z)} + \Gamma e^{-i\beta(\ell - z)}}{e^{i\beta(\ell - z)} - \Gamma e^{-i\beta(\ell - z)}} = Z_0 \frac{1 + \Gamma e^{-i2\beta(\ell - z)}}{1 - \Gamma e^{-i2\beta(\ell - z)}}$$

Hence,

$$\varsigma(\Gamma;z)=\hat{Z}(z)=\tfrac{1+\rho(z)}{1-\rho(z)}$$

is well-defined. Moreover as the mapping $\frac{az+b}{cz+d}$ $\forall z\in\mathbf{C}$ is a conformal transformation [2], it follows by inspection that ς is also conformal.

To complete the proof, observe by definition that $\hat{R}+i\hat{X}=\frac{1+\sigma+i\chi}{1-\sigma-i\chi}$. Hence, equating real and imaginary parts,

$$\tfrac{1+\sigma+i\chi}{1-\sigma-i\chi}=\tfrac{1-\sigma^2-\chi^2}{(1-\sigma)^2+\chi^2}+\tfrac{i2\chi}{(1-\sigma)^2+\chi^2}$$

yields

$$\hat{R}=\tfrac{1-\sigma^2-\chi^2}{(1-\sigma)^2+\chi^2}\quad\text{and}\quad\hat{X}=\tfrac{2\chi}{(1-\sigma)^2+\chi^2}$$

Rearranging, and noting that

$$\left(\sigma-\tfrac{\hat{R}}{1+\hat{R}}\right)^2-\left(\tfrac{\hat{R}}{1+\hat{R}}\right)^2=\sigma^2-\tfrac{2\sigma\hat{R}}{1+\hat{R}}$$

yields

$$\hat{R}=\tfrac{1-\sigma^2-\chi^2}{(1-\sigma)^2+\chi^2}\Leftrightarrow 0=\tfrac{\hat{R}}{1+\hat{R}}-\tfrac{2\sigma\hat{R}}{1+\hat{R}}+\sigma^2+\chi^2-\tfrac{1}{1+\hat{R}}\Leftrightarrow\left(\sigma-\tfrac{\hat{R}}{1+\hat{R}}\right)^2+\chi^2=\tfrac{1}{(1+\hat{R})^2}$$

Likewise, noting that $\left(\chi-\frac{1}{\hat{X}}\right)^2-\frac{1}{\hat{X}^2}=\chi^2-\frac{2\chi}{\hat{X}}$, it follows, from the above proof *mutatis mutandis*, that

$$\hat{X}=\tfrac{2\chi}{(1-\sigma)^2+\chi^2}\Leftrightarrow(\sigma-1)^2+\left(\chi-\tfrac{1}{\hat{X}}\right)^2=\tfrac{1}{\hat{X}^2}$$ □

It is clear that

$$\left(\sigma-\tfrac{\hat{R}}{1+\hat{R}}\right)^2+\chi^2=\tfrac{1}{(1+\hat{R})^2}$$

defines a circle of radius $\frac{1}{1+\hat{R}}$ centered at $\left(\frac{\hat{R}}{1+\hat{R}},0\right)$, and $(\sigma-1)^2+\left(\chi-\frac{1}{\hat{X}}\right)^2=\frac{1}{\chi^2}$ defines a circle of radius $\frac{1}{\chi}$ centered at $\left(1,\frac{1}{\hat{X}}\right)$. These circles are orthogonal to one another at the point wherein they intersect with one another. This forms the basis for the Smith Chart, and they define a coordinate system on the complex plane.

4.3.14 Remark

From the equivalence

$$\hat{R} = \frac{1-\sigma^2-\chi^2}{(1-\sigma)^2+\chi^2} \Leftrightarrow \left(\sigma - \frac{\hat{R}}{1+\hat{R}}\right)^2 + \chi^2 = \frac{1}{(1+\hat{R})^2}$$

the centers of the circles lie on the χ-axis. Moreover, $\hat{R} = 0 \Leftrightarrow \sigma^2 + \chi^2 = 1$; that is, $\hat{R} = 0 \Leftrightarrow |\rho| = 1$. Next, the equivalence

$$\hat{X} = \frac{2\chi}{(1-\sigma)^2+\chi^2} \Leftrightarrow (\sigma - 1)^2 + \left(\chi - \frac{1}{\hat{X}}\right)^2 = \frac{1}{\hat{X}^2}$$

evinces that the center of the circles lies on the line $\sigma = 1$. In particular, $\hat{X} > 0 \Leftrightarrow \chi > 0$ yields a condition for inductive reactance, and $\hat{X} < 0 \Leftrightarrow \chi < 0$ leads to a condition for capacitive reactance.

4.4 Impedance Matching and Standing Waves

A boundary point along a transmission line is typically the result of a change in impedance at a point by virtue of a load. The terms load and boundary are used interchangeably in this section when referring to a transmission line.

4.4.1 Example

Consider a transmission line pair $(C_{\pm}, \ell, R, L, C, G)$ connected to a source $(V_0(\omega), Z_S)$ and terminated by a load Z_L (see Figure 4.3). What is the input impedance if (a) the line is shorted and (b) the line is open. Finally, consider as a special case wherein $R = 0 = G$, that is, a perfect conductor embedded in a perfect dielectric medium.

(a) When the line is shorted to ground, the load is zero: $Z_L = 0$. Hence, by Equation (4.19),

$$Z(0) = Z_0 \frac{Z_0 \tanh \gamma\ell}{Z_0} = Z_0 \tanh \gamma\ell$$

(b) When the line is open, the load is infinite: $Z_L = \infty$, hence, by (4.19),

$$Z(0) = Z_0 \frac{Z_L}{Z_L \tanh \gamma\ell} = Z_0 \coth \gamma\ell$$

Now, set $Z_{(0)} = Z_0 \tanh \gamma\ell$ and $Z_{(\infty)} = Z_0 \coth \gamma\ell$. Then, $Z_{(0)} Z_{(\infty)} = Z_0^2 \Rightarrow Z_0 = \sqrt{Z_{(0)} Z_{(\infty)}}$, a rather startling result! That is, the characteristic impedance can be deduced immediately simply by knowing the input impedance when the line is shorted and when it is open.

Finally, to complete the example, consider the special case wherein R = 0 = G. Recalling that tanh iϕ = i tan ϕ, it follows at once that Equation (4.19) reduces to

$$Z(z) = Z_0 \frac{Z_L + iZ_0 \tan\beta(\ell - z)}{Z_0 + iZ_L \tan\beta(\ell - z)} \tag{4.23}$$

where $\beta = \Im m\gamma$ and in this instance, $\alpha = \Re e\gamma = 0$. Hence, for case (a), $Z_{(0)} = iZ_0 \tan\beta\ell$ and for (b), $Z_{(\infty)} = -iZ_0 \cot\beta\ell$. Thus, depending upon the length of the transmission line, the input impedance of a short or open circuit oscillates between being inductive and capacitive reactance; see the plot for illustration.

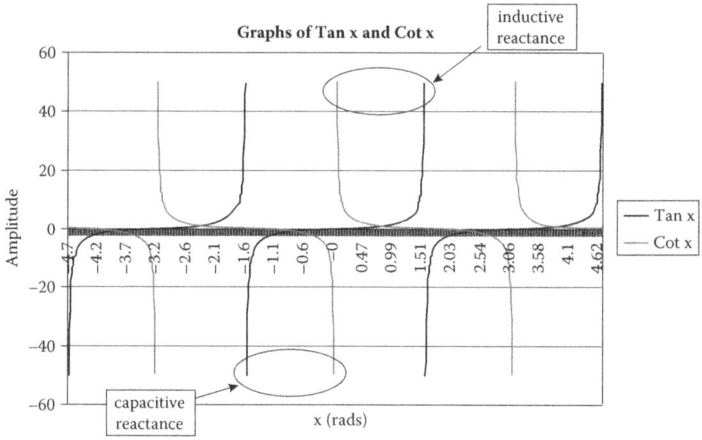

Explicitly, in view of the plot, for the case wherein R = 0 = G,

- $Z_{(0)}$ is inductive whenever $\ell \in \left[(n-1)\frac{\pi}{\beta}, (n-\frac{1}{2})\frac{\pi}{\beta}\right] \forall n \in \mathbf{N}$.
- $Z_{(0)}$ is capacitive whenever $\ell \in \left[(n-\frac{1}{2})\frac{\pi}{\beta}, n\frac{\pi}{\beta}\right] \forall n \in \mathbf{N}$.
- $Z_{(\infty)}$ is inductive whenever $\ell \in \left[(n-\frac{1}{2})\frac{\pi}{\beta}, n\frac{\pi}{\beta}\right] \forall n \in \mathbf{N}$.
- $Z_{(\infty)}$ is capacitive whenever $\ell \in \left[(n-1)\frac{\pi}{\beta}, (n-\frac{1}{2})\frac{\pi}{\beta}\right] \forall n \in \mathbf{N}$.

Unfortunately, for the general case wherein $\gamma \in \mathbf{C}$, no simple relationships similar to those given above exist. □

4.4.2 Proposition

Given a pair of finite transmission lines $(C_\pm, \ell, R, L, C, G)$ as shown in Figure 4.3, suppose $(V_0(\omega), Z_S)$ is the source at $z = 0$ and the transmission line terminates at some load Z_L. Then,

$$\forall z \in [0, \ell] \;\; V(z) = \frac{V_0(\omega) Z_0}{Z_0 + Z_S} \frac{e^{-\gamma z}}{1 - \Gamma(0)\Gamma(\ell) e^{-2\gamma\ell}} \left\{1 + \Gamma(\ell) e^{-2\gamma(\ell - z)}\right\}$$

Proof

Recall that the general solution of the voltage and current waves propagating along the transmission pair (4.20) and (4.21) can be rewritten as

$$V(z) = \tfrac{1}{2} I(\ell)(Z_L + Z_0)e^{\gamma\ell}\{e^{-\gamma z} + \Gamma(\ell)e^{-\gamma(2\ell - z)}\} \tag{4.24a}$$

$$I(z) = \tfrac{1}{2}\tfrac{I(\ell)}{Z_0}(Z_L + Z_0)e^{\gamma\ell}\{e^{-\gamma z} - \Gamma(\ell)e^{-\gamma(2\ell - z)}\} \tag{4.24b}$$

From Figure 4.4, it is clear from Ohm's law that $V_0(\omega) = I(0)Z_S + V(0)$, whence, appealing to Equation (4.24), $V_0(\omega) = Z_S \frac{I(\ell)}{2Z_0}(Z_L + Z_0)e^{\gamma\ell}\{1 - \Gamma(\ell)e^{-2\gamma\ell}\} + \frac{I(\ell)}{2}(Z_L + Z_0)e^{\gamma\ell}\{1 + \Gamma(\ell)e^{-2\gamma\ell}\}$. Thus,

$$\begin{aligned}
\tfrac{V_0(\omega)}{Z_L + Z_0}\tfrac{2}{I(\ell)}e^{-\gamma\ell} &= \tfrac{Z_S}{Z_0}\{1 - \Gamma(\ell)e^{-2\gamma\ell}\} + 1 + \Gamma(\ell)e^{-2\gamma\ell} \\
&= \tfrac{Z_S}{Z_0} + 1 + \Gamma(\ell)e^{-2\gamma\ell}\left\{1 - \tfrac{Z_S}{Z_0}\right\} \\
&= \tfrac{Z_S + Z_0}{Z_0}\{1 + \Gamma(0)\Gamma(\ell)e^{-2\gamma\ell}\}
\end{aligned}$$

where $\Gamma(0) = \frac{Z_S - Z_0}{Z_S + Z_0}$. In turn, this implies that

$$\tfrac{1}{2} I(\ell)(Z_L + Z_0)e^{\gamma\ell} = \tfrac{Z_0}{Z_S + Z_0}\tfrac{V_0(\omega)}{1 + \Gamma(0)\Gamma(\ell)e^{-2\gamma\ell}}$$

Substituting this back into Equation (4.24a) and noting that $e^{-\gamma z} + \Gamma(\ell)e^{-\gamma(2\ell - z)} = e^{-\gamma z}\{1 + \Gamma(\ell)e^{-2\gamma(\ell - z)}\}$, the assertion follows immediately. □

Provide an alternative proof for Proposition 4.4.2 (*cf.* Exercise 4.5.4) via the explicit reflection of waves occurring at each boundary interface, that is, at the source and at the load, respectively.

4.4.3 Corollary

Given the conditions of Proposition 4.4.2,

$$I(z) = \tfrac{V_0(\omega)}{Z_0 + Z_S}\tfrac{e^{-\gamma z}}{1 - \Gamma(0)\Gamma(\ell)e^{-2\gamma\ell}}\left\{1 - \Gamma(\ell)e^{-2\gamma(\ell - z)}\right\}, \forall z \in [0, \ell].$$

Proof

The proof follows that of Proposition 4.4.2 *mutatis mutandis*. □

In general, wave reflections are to be avoided as they have a tendency to form (partial) standing waves between the source and the load, leading to unwanted emissions and thereby rendering digital devices to be potentially noncompliant with regulatory agency requirements. Moreover, reflected waves may potentially cause electromagnetic interference by coupling onto adjacent lines, switching transistors on and off randomly. The following theorem also demonstrates why impedance matching is important; note, however, that for digital electronics, this aspect is usually immaterial.

4.4.4 Theorem

Given a finite transmission line system $(C_{\pm}, \ell, R, L, C, G)$, maximal power transfer occurs at the load if and only if $\Gamma(\ell) = 0$.

Proof

Let $p = p(z,t)$ denote the instantaneous power along $\gamma_{\pm}$. From Equation (4.22),

$$p(z) = V(z)I^*(z) = \frac{|V_0^+|^2}{Z_0}\{e^{2\alpha(\ell-z)} - |\Gamma(\ell)|e^{-i\theta_\ell}e^{i2\beta(\ell-z)} + |\Gamma(\ell)|e^{i\theta_\ell}e^{-i2\beta(\ell-z)} - |\Gamma(\ell)|^2 e^{-2\alpha(\ell-z)}\}$$

where $\gamma = \alpha + i\beta$. Thus, on setting $\langle p(z)\rangle = \frac{1}{2}\Re e\, p(z)$, it can be shown (see Exercise 4.5.2) that $P(\ell) = \frac{|V_0^+|^2}{2Z_0}\{1 - |\Gamma(\ell)|^2\}$. Whence, maximal power transfer $\bar{p}(\ell) = \max_{\Gamma}\langle p(\ell)\rangle \Leftrightarrow \Gamma(\ell) = 0$, as claimed. □

4.4.5 Corollary

Under the conditions of maximal power transfer along a pair of transmission lines toward a fixed load, the transmitted voltage across the load is $V(\ell) = \frac{Z_0}{Z_0 + Z_S} V_0(\omega) e^{-\gamma\ell}$.

Proof

By Theorem 4.4.4, maximal power transfer implies that $\Gamma(\ell) = 0$. The result thus follows at once from Proposition 4.4.2. □

Note that $\langle p(z)\rangle = \frac{1}{2}\Re e\, p(z)$ in the proof of Theorem 4.4.4 defines the *time-average power* (see the *time-average power density*) defined by $\langle S\rangle = \frac{1}{2}\Re e(\boldsymbol{E} \times \boldsymbol{H}^*)$, where S is the Poynting vector. From Theorem 4.4.4, it is clear that the absence of reflection results in maximal power transfer: all the transmitted power is absorbed by the load. Physically, this is obvious as reflection implies part of the incident energy is redirected away from the transmitted energy via $1 + \Gamma = T$ (transmission coefficient).

In spite of the less appealing nature of reflected waves, reflection has useful applications in the high-technology industry, for instance, in the memory architectures of personal computers. As a concrete example, suppose the input into an integrated circuit (IC) requires some fixed voltage V_0. Now, if the input impedance (into the IC) is extremely high relative to the characteristic line impedance, it is possible to utilize reflection at the input impedance to drive the line at $\frac{1}{2}V_0$ and thereby reduce power consumption. Explicitly, by designing the circuit such that $\Gamma \approx 1$ at the load, the resultant transmission coefficient at the input is $T = 1 + \Gamma \approx 2$: thus, $\frac{1}{2}V_0 \to V_0$, which is the required input voltage V_0, as claimed. For more details, see Exercise 4.5.3.

A transmission line is said to be *electrically long* if $|\gamma|\ell >> 1$, where γ is the wave propagation constant. Finally, define a transmission line to be *electrically short* if $|\gamma|\ell << 1$. It is clear from

$$Z(z) = Z_0 \frac{Z_L + Z_0 \tanh\gamma(\ell - z)}{Z_0 + Z_L \tanh\gamma(\ell - z)}$$

that $Z(0) \approx Z_L$, where Z_L is the load impedance at the end of the transmission line as $|\gamma|\ell << 1 \Rightarrow \tanh\gamma\ell \approx 0$. Thus, if a transmission line is electrically short, the voltage transmitted across the load is $V(\ell) = \frac{Z_L}{Z_0 + Z_L} V(0)$, where $V(0)$ is the output voltage from the source appearing at $z = 0$ of the transmission line. Thus, for an electrically short line, the load-line system forms a simple voltage divider.

4.4.6 Lemma

Given a pair of lossless transmission lines $(C_\pm, \ell, L, C)$ with some source (V_S, ω, Z_S) and load Z_L, there exists an infinite discrete set of angular frequencies $\{\omega_n\}$ such that the lines appear to be electrically short with respect to $\{\omega_n\}$.

Proof

For a lossless line, $\beta = \omega\sqrt{LC}$. From (4.23), set $\omega_n = \frac{n\pi}{\ell\sqrt{LC}}$ for $n = 1, 2, \ldots$. Then, $\tan\beta_n\ell = 0 \; \forall n \Rightarrow Z_{in} = Z(0) = Z_L$, as required. □

Observe from Proposition 4.4.2 that if $\Gamma(0), \Gamma(\ell) \neq 0$, waves will be reflected back and forth along a finite transmission line (see Exercise 4.5.4). Thus, impedance mismatch at the source and load leads to standing waves on the transmission line. An example below illustrates the technique of voltage reflection diagrams.

4.4.7 Example

Consider a finite transmission line system $(C_\pm, \ell, R, L, C, G)$ with reference to Figure 4.3 and the general equation $V(z) = V_0^+ e^{-\gamma z} + V_0^- e^{\gamma z}$. Suppose without loss of generality that $\Gamma(0) \neq 0 \neq \Gamma(\ell)$, and the lines are lossless for simplicity: $R = 0 = G$. Set $\upsilon = \frac{1}{\sqrt{LC}}$ to be the speed of wave propagation and $\tau = \frac{\ell}{\upsilon}$. Then, at $t = 0$, $V = V_0^+$. At $t = \tau$, the reflected wave is $V_0^- = \Gamma(\ell)V_0^+$. This wave travels back toward the source, whereupon it is reflected by the source impedance at $t = 2\tau$, and the reflected voltage is $V_1^+ = \Gamma(0)V_0^- = \Gamma(0)\Gamma(\ell)V_0^+$. Now, this reflected wave in turn propagates toward the load at $t = 3\tau$, yielding $V_1^- = \Gamma(\ell)V_1^+ = \Gamma(0)\Gamma^2(\ell)V_0^+$, and at $t = 4\tau$, the reflected wave is given by $V_2^+ = \Gamma(0)V_1^- = \Gamma^2(0)\Gamma^2(\ell)V_0^+$.

In general, it is easy to see inductively that $\forall n \geq 0$,

$$\begin{cases} V_n^+ = \Gamma^n(0)\Gamma^n(\ell)V_0^+ \text{ for } t = n\tau \\ V_n^- = \Gamma^{n+1}(0)\Gamma^n(\ell)V_0^+ \text{ for } t = (n+1)\tau \end{cases}$$

This leads to the *voltage reflection diagram*

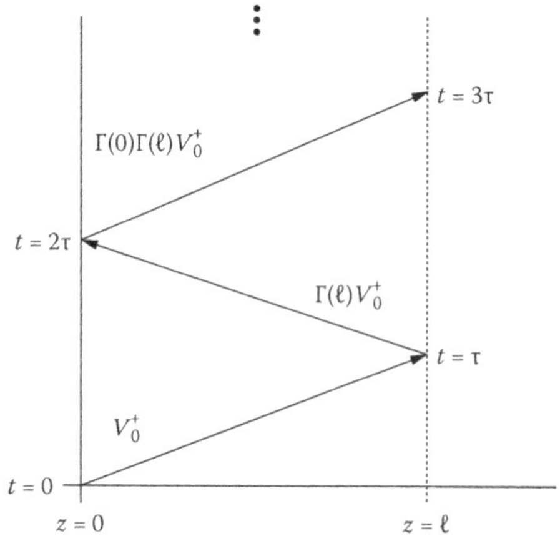

The gradient of the forward propagating wave is 1/v
The gradient of the backward propagating wave is −1/v

To complete the example, consider the plot of the transmitted voltage as a function of time for some fixed distance $0 < \hat{z} < \ell$ along the transmission line. For notational simplicity, set $\hat{\tau} = \frac{\hat{z}}{\upsilon}$. For $t \in [0, \hat{\tau})$, $V = 0$ at $z = \hat{z}$ as it takes finite time for the wave to propagate to $z = \hat{z}$. Namely, it takes $t = \hat{\tau}$ before the wave reaches $z = \hat{z}$.

For notational convenience, set $\tau_1^- = 0, \tau_1^+ = \hat{\tau}, \tau_2^{\pm} = 2\tau \pm \hat{\tau}, \ldots, \tau_n^{\pm} = n\tau \pm \hat{\tau}\ \forall n$. This yields the sequence

$$0 = \tau_1^- < \tau_1^+ < \tau < \tau_2^- < 2\tau < \tau_2^+ < \cdots < \tau_n^- < n\tau < \tau_n^+ < \cdots$$

Then, for $t \in [\tau_1^+, \tau_2^-)$, the transmitted wave is $V = V_0^+$. In particular, observe that at $t = \tau_1^+$, $\delta V(\hat{z}) = \lim_{z \to (\hat{z})^-} V(z) - \lim_{z \to (\hat{z})^+} V(\hat{z}) = V_0^+$ as on $z \in [0, \hat{z}]$, $V(z) = V_0^+$ whereas for $z \in (\hat{z}, \ell]$, $V(z) = 0$ as finite time is required for signal propagation (see the voltage reflection diagram).

Likewise, at $t = \tau_2^-$, it is clear from the voltage reflection diagram that on $z \in [0, \hat{z}]$, $V = V_0^+(1 + \Gamma(\ell))$, whereas for $z \in (\hat{z}, \ell]$, $V(z) = V_0^+$, whence, at $t = \tau_2^-$, the voltage at $z = \hat{z}$ is discontinuous by the amount $\delta V(\hat{z}) = \lim_{t \to (\hat{z})^-} V(\hat{z}) - \lim_{t \to (\hat{z})^+} V(\hat{z}) = \Gamma(\ell)V_0^+$.

Thus, by induction, it is clear that at each $t = n\hat{\tau}$, there exists a voltage discontinuity at $z = \hat{z}$ given by

$$\begin{cases} \delta V_{2n}(\hat{z}) = \Gamma^{n-1}(0)\Gamma^{n-1}(\ell)V_0^+ \\ \delta V_{2n+1}(\hat{z}) = \Gamma^{n-1}(0)\Gamma^{n}(\ell)V_0^+ \end{cases}$$

for all $n \geq 1$. This can be sketched as shown in Figure 4.6.

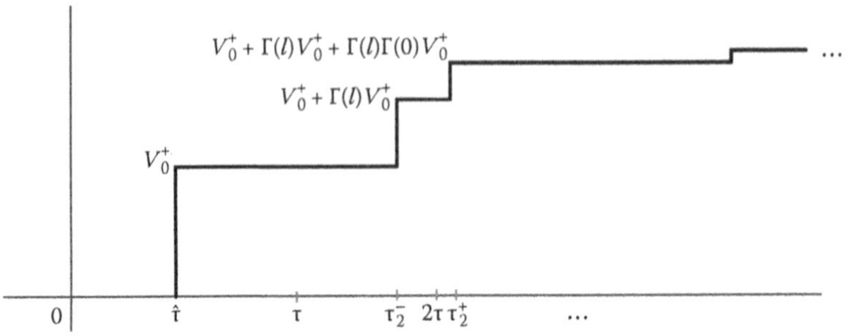

FIGURE 4.6
Successive reflection of voltage waves between source and load.

In general, the voltage at $\hat{z}$ for $t \in [\tau_n^-, \tau_n^+)$ is given by

$$V(\hat{z}) = \begin{cases} V_0^+(1+\Gamma(\ell)+\cdots+\Gamma^{\hat{n}-1}(0)\Gamma^{\hat{n}-1}(\ell)) & \text{if } n = 2\hat{n} \\ V_0^+(1+\Gamma(\ell)+\cdots+\Gamma^{\hat{n}-1}(0)\Gamma^{\hat{n}}(\ell)) & \text{if } n = 2\hat{n}+1 \end{cases}$$ □

4.4.8 Lemma

Given $(C_\pm, \ell, R, L, C, G)$, set $\gamma = \alpha + i\beta$ and suppose that $\beta\ell = n\pi$ for any fixed $n = 1, 2, \ldots$. Then, for arbitrary load Z_L,

$$\text{(a)} \quad \alpha\ell >> 1 \Rightarrow Z(0) \approx Z_0$$

$$\text{(b)} \quad \alpha\ell << 1 \Rightarrow Z(0) \approx Z_L$$

In particular, $Z(0)$ is purely resistive if Z_L is purely resistive.

Proof

From

$$\tanh(\alpha + i\beta)\ell = \frac{\tanh\alpha\ell + i\tan\beta\ell}{1 + i\tanh\alpha\ell\tan\beta\ell},$$

it is clear that $\tanh(\alpha + i\beta)\ell = \tanh\alpha\ell$ and the conclusion to (a) and (b) thus follows from

$$Z(z) = Z_0 \frac{Z_L + Z_0\tanh\gamma(\ell - z)}{Z_0 + Z_L\tanh\gamma(\ell - z)}.$$

Explicitly, it suffices to observe that $\alpha\ell >> 1 \Rightarrow \tanh\alpha\ell \approx 1$ whereas $\alpha\ell << 1 \Rightarrow \tanh\alpha\ell \approx 0$. □

4.4.9 Lemma

Given $(C_{\pm}, \ell, R, L, C, G)$, set $\gamma = \alpha + i\beta$ and suppose that $\beta\ell = \frac{2n-1}{2}\pi$ for any fixed $n = 1, 2, \ldots$. Then, for arbitrary load Z_L,

$$\text{a)} \quad \alpha\ell >> 1 \Rightarrow Z(0) \approx Z_0$$

$$\text{b)} \quad \alpha\ell << 1 \Rightarrow Z(0) \approx \frac{Z_0^2}{Z_L} = Z_L\left(\frac{Z_0}{Z_L}\right)^2$$

In particular, $Z(0)$ is purely resistive if Z_L is purely resistive.

Proof

The proof is similar to that of Lemma 4.4.8. □

Observe that the conclusions of Lemma 4.4.8 and Lemma 4.4.9 for the case wherein $\alpha\ell >> 1$ are identical. In particular, the input impedance is identical to the characteristic impedance of the transmission lines; that is, the input impedance is independent of the load. Hence, maximal power transfer occurs if the source $Z_S = Z_0$. More important, electromagnetic emissions are minimized under this condition. This is summarized below.

4.4.10 Corollary

Given $(C_{\pm}, \ell, R, L, C, G)$ with a source (V_S, ω, Z_S) and a fixed $\gamma = \alpha + i\beta$, if $\ell > 0$ satisfies $\beta\ell = \frac{n}{2}\pi$ for some $n \in \mathbf{N}$, and $\alpha\ell >> 0$, then for arbitrary load Z_L, maximal power transfer is achieved if $Z_S = Z_0$, and in particular, electromagnetic emissions are minimized.

Proof

By Lemma 4.4.8(a) and Lemma 4.4.9(a), the input impedance $Z(0) = Z_0$. Hence, setting $Z_S = Z_0$, yields (i) maximal power transfer and (ii) reflection suppression at the source and hence mitigating emissions for arbitrary loads. □

As a side remark, for a long trace on a printed circuit board (PCB) with a source (e.g., clocks) in the microwave regime, this corollary is still applicable provided that at the load the current is enhanced by a current source. For instance, the current source can be supplied via a parallel termination configuration, with a pull-up resistor to V_{cc} to supply the required current and a pull-down to complete the impedance match (see Figure 4.7).

On a side note regarding digital circuit design, is there an optimal placement along a trace for a series surface mount resistor to be placed in order to suppress emissions, assuming that its value is different from the characteristic impedance of the line for emission suppression? Clearly, placing the resistor R arbitrarily along the trace will result in unwanted reflections along the

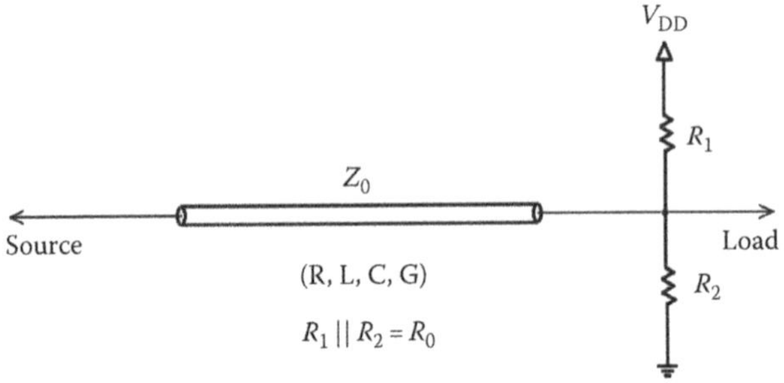

FIGURE 4.7
Active parallel termination scheme.

line. It is clear from transmission line theory that placing the resistor as close to the load as possible will optimize emissions suppression. This practice is commonly employed in the electronics industry.

4.4.11 Lemma

Given a pair of transmission lines $(C_{\pm}, \ell, R, L, C, G)$, the characteristic impedance Z_0 of the line is always inductive. That is, $\Im m(Z_0) \geq 0$.

Proof

Now,

$$Z_0 \equiv \sqrt{\frac{R+i\omega L}{G+i\omega C}} = R_0\left(\frac{\frac{R}{L}+i\omega}{\frac{G}{C}+i\omega}\right)^{\frac{1}{2}} = R_0\left(\frac{\frac{RG}{LC}+\omega^2+i\omega\left(\frac{G}{C}-\frac{R}{L}\right)}{\left(\frac{G}{C}\right)^2+\omega^2}\right)^{\frac{1}{2}}$$

where $R_0 = \sqrt{L/C}$. As before, setting

$$\left(\frac{\frac{RG}{LC}+\omega^2+i\omega\left(\frac{G}{C}-\frac{R}{L}\right)}{\left(\frac{G}{C}\right)^2+\omega^2}\right)^{\frac{1}{2}} = a+ib$$

gives

$$a^2-b^2 = \frac{\frac{RG}{LC}+\omega^2}{\left(\frac{G}{C}\right)^2+\omega^2} \quad \text{and} \quad 2ab = \frac{\omega\left(\frac{G}{C}-\frac{R}{L}\right)}{\left(\frac{G}{C}\right)^2+\omega^2}$$

whence, noting that

$$(a^2+b^2)^2 = (a^2-b^2)^2+(2ab)^2 \Rightarrow a^2+b^2 = \left\{\left(\frac{\frac{RG}{LC}+\omega^2}{\left(\frac{G}{C}\right)^2+\omega^2}\right)^2+\left(\frac{\omega\left(\frac{G}{C}-\frac{R}{L}\right)}{\left(\frac{G}{C}\right)^2+\omega^2}\right)^2\right\}^{\frac{1}{2}}$$

yields

$$a = \frac{1}{\sqrt{2}}\left\{\sqrt{\left(\frac{\frac{RG}{LC}+\omega^2}{\left(\frac{G}{C}\right)^2+\omega^2}\right)^2+\left(\frac{\omega\left(\frac{G}{C}-\frac{R}{L}\right)}{\left(\frac{G}{C}\right)^2+\omega^2}\right)^2}+\frac{\frac{RG}{LC}+\omega^2}{\left(\frac{G}{C}\right)^2+\omega^2}\right\}^{\frac{1}{2}}$$

$$b = \frac{1}{\sqrt{2}}\left\{\sqrt{\left(\frac{\frac{RG}{LC}+\omega^2}{\left(\frac{G}{C}\right)^2+\omega^2}\right)^2+\left(\frac{\omega\left(\frac{G}{C}-\frac{R}{L}\right)}{\left(\frac{G}{C}\right)^2+\omega^2}\right)^2}-\frac{\frac{RG}{LC}+\omega^2}{\left(\frac{G}{C}\right)^2+\omega^2}\right\}^{\frac{1}{2}}$$

Because trivially, $\sqrt{\alpha^2+\beta^2}-\alpha \geq \alpha-\alpha = 0$, it follows immediately that $b \geq 0$. That is, $\Im\mathrm{m}(Z_0) = R_0 b \geq 0$, as required. □

Thus, it is evident that the characteristic impedance of traces on a PCB cannot be capacitive regardless of its length. However, depending upon the load at the end of the transmission line and the length of the transmission line, the input impedance can clearly be made capacitive, inductive, or purely resistive.

4.4.12 Proposition

There exists at most a countably infinite set of angular frequencies $\{\omega_n\}$ such that the input impedance $Z(0)$ of a pair of lossless transmission lines $(C_\pm, \ell, \mathrm{L}, \mathrm{C})$ terminated by a purely resistive load R_L is real.

Proof

It suffices to note that

$$Z(0) = R_0 \frac{R_\mathrm{L} + iR_0 \tan\omega\sqrt{\mathrm{LC}}\ell}{R_0 + iR_\mathrm{L}\tan\omega\sqrt{\mathrm{LC}}\ell} = R_\mathrm{L}$$

if, $\tan\omega\sqrt{\mathrm{LC}}\ell = 0$, where $R_0 = \sqrt{\mathrm{L}/\mathrm{C}}$. This condition is satisfied if $\omega = \frac{n\pi}{\ell\sqrt{\mathrm{LC}}}$, for each $n = 1, 2, \ldots$. Likewise, if

$$\omega\ell\sqrt{\mathrm{LC}} = \frac{2n-1}{2}\pi,\ n \in \mathbf{N}$$

then $Z(0) = \frac{R_0^2}{R_\mathrm{L}} \in \mathbf{R}$. Thus, the desired countably infinite set of angular frequencies is precisely

$$\left\{\frac{n\pi}{\ell\sqrt{\mathrm{LC}}} : n \in \mathbf{N}\right\} \cup \left\{\frac{2n-1}{2}\frac{\pi}{\ell\sqrt{\mathrm{LC}}} : n \in \mathbf{N}\right\}$$

□

Now, consider the case wherein a lossless transmission line is terminated to ground: $Z_L = 0$. That is,

$$Z(z) = R_0 \frac{R_L + iR_0 \tan\beta(\ell - z)}{R_0 + iR_L \tan\beta(\ell - z)} = iR_0 \tan\beta(\ell - z)$$

Then, it is evident that along $\bar{z} \in [0,\ell]$ such that $\beta(\ell - \bar{z}) = n\pi$, $Z(\bar{z}) = 0$ and hence, those points may be shorted to ground without affecting the wave propagation. On the other hand, at points $\bar{z} \in [0,\ell]$ such that $\beta(\ell - \bar{z}) = \frac{2n-1}{2}\pi$, $Z(\bar{z}) = \pm i\infty$ and hence, those points may be cut without affecting the wave propagation along the line.

A similar analysis can be made for the case wherein a lossless transmission line is open. Indeed, in this instance,

$$R_L \to \infty \Rightarrow Z(z) = R_0 \frac{R_L + iR_0 \tan\beta(\ell - z)}{R_0 + iR_L \tan\beta(\ell - z)} = -iR_0 \cot\beta(\ell - z)$$

whence, along $\bar{z} \in [0,\ell]$ such that $\beta(\ell - \bar{z}) = n\pi$, $Z(\bar{z}) \to \pm i\infty$; thus, at these points, the line may be cut without affecting the wave propagation along the line. Likewise, along points $\bar{z} \in [0,\ell]$ such that $\beta(\ell - \bar{z}) = \frac{2n-1}{2}\pi$, $Z(\bar{z}) = 0$ and hence at these points may be grounded without affecting the wave propagation along the line.

4.4.13 Lemma

Given a pair of transmission lines $(C_\pm, \ell, R, G, L, C)$, given any $\varepsilon > 0$,

$$\forall\omega > \frac{1}{2}\frac{\left|\frac{G}{C} - \frac{R}{L}\right|}{(1+\varepsilon)^2 - 1} + \frac{1}{2}\sqrt{\left(\frac{\left|\frac{G}{C} - \frac{R}{L}\right|}{(1+\varepsilon)^2 - 1}\right)^2 - \frac{4\left((1+\varepsilon)^2\left(\frac{G}{C}\right)^2 - \frac{RG}{LC}\right)}{(1+\varepsilon)^2 - 1}} \Rightarrow \left|Z_0(\omega) - R_0\right| < R_0\left|1 + i\varepsilon\right|,$$

where $R_0 = \sqrt{L/C}$.

Proof

First, observe in general that $Z_0(\omega) = \sqrt{\frac{R + i\omega L}{G + i\omega C}}$ is a continuous function of ω. Hence, by continuity, for any given $\varepsilon > 0$, $\exists\omega_\varepsilon > 0$ satisfying the lemma. Thus, the proof is complete if ω_ε can be determined. Rearranging the expression for Z_0 yields:

$$\sqrt{\frac{R + i\omega L}{G + i\omega C}} = \sqrt{\frac{L}{C}}\left(\frac{1 - iR/\omega L}{1 - iG/\omega C}\right)^{\frac{1}{2}} = \sqrt{\frac{L}{C}}\left(\frac{(1 - iR/\omega L)(1 + iG/\omega C)}{1 + (G/\omega C)^2}\right)^{\frac{1}{2}} = \sqrt{\frac{L}{C}}\frac{1}{(1 + (G/\omega C)^2)^{1/2}}\left(1 + \frac{i}{\omega}\left(\frac{G}{C} - \frac{R}{L}\right) + \frac{RG}{\omega^2 LC}\right)^{\frac{1}{2}}$$

Hence, for any given $\varepsilon > 0$, it suffices to seek ω such that

$$\left|1 - \frac{1}{\sqrt{1 + (G/\omega C)^2}}\left\{1 + \frac{i}{\omega}\left(\frac{G}{C} - \frac{R}{L}\right) + \frac{RG}{\omega^2 LC}\right\}^{\frac{1}{2}}\right| < \varepsilon$$

because $|1+i\varepsilon| = \sqrt{1+\varepsilon^2} > \varepsilon$. So, from $|a|-|b| \le |a-b|$, it suffices for ω to satisfy the inequality:

$$\left| \frac{1}{\sqrt{1+(G/\omega C)^2}} \left(1+\frac{i}{\omega}\left(\frac{G}{C}-\frac{R}{L}\right)+\frac{RG}{\omega^2 LC}\right)^{\frac{1}{2}} \right| < 1+\varepsilon$$

This is equivalent to

$$\left|1+\frac{i}{\omega}\left(\frac{G}{C}-\frac{R}{L}\right)+\frac{RG}{\omega^2 LC}\right| < (1+\varepsilon)^2\left\{1+\left(\frac{G}{\omega C}\right)^2\right\}$$

However,

$$\left|1+\frac{i}{\omega}\left(\frac{G}{C}-\frac{R}{L}\right)+\frac{RG}{\omega^2 LC}\right| \le \left|1+\frac{RG}{\omega^2 LC}\right| + \left|\frac{i}{\omega}\left(\frac{G}{C}-\frac{R}{L}\right)\right|$$

Thus, it is enough to have

$$1+\frac{RG}{\omega^2 LC}+\frac{1}{\omega}\left|\frac{G}{C}-\frac{R}{L}\right| < (1+\varepsilon^2)\left\{1+\left(\frac{G}{\omega C}\right)^2\right\}$$

So, multiplying the inequality by ω^2 yields:

$$\omega^2((1+\varepsilon)^2-1)-\omega\left|\frac{G}{C}-\frac{R}{L}\right|+(1+\varepsilon)^2\left(\frac{G}{C}\right)^2-\frac{RG}{LC} > 0$$

The roots of the equation are:

$$\omega_{\pm}(\varepsilon) = \frac{1}{2}\frac{\left|\frac{G}{C}-\frac{R}{L}\right|}{(1+\varepsilon)^2-1} \pm \frac{1}{2}\sqrt{\left(\frac{\left|\frac{G}{C}-\frac{R}{L}\right|}{(1+\varepsilon)^2-1}\right)^2 - \frac{4\left((1+\varepsilon)^2\left(\frac{G}{C}\right)^2-\frac{RG}{LC}\right)}{(1+\varepsilon)^2-1}}$$

Therefore, set $\omega_\varepsilon = \omega_+(\varepsilon)$. Then, for any $\varepsilon > 0$, $\omega > \omega_\varepsilon$ implies that $\left|Z_0-\sqrt{L/C}\right| < \sqrt{L/C}\,|1+i\varepsilon|$ as required. □

The above lemma asserts trivially that for high enough frequencies, the characteristic impedance of the transmission line may be approximated with that of the lossless line up to first order in ε, where ε is some positive number.

Some comments regarding standing waves are in order. Consider Equation (4.22):

$$V(z) = V_0^+\left\{e^{\gamma(\ell-z)}+\Gamma(\ell)e^{-\gamma(\ell-z)}\right\}$$

Now, consider the case wherein $\Gamma(\ell)=1$. Then, by definition,

$$V(z) = V_0^+\left\{e^{\gamma(\ell-z)}+e^{-\gamma(\ell-z)}\right\} = 2V_0^+\cosh\gamma(\ell-z)$$

and hence, at the load, $V(\ell) = 2V_0^+$. That is, twice the incident voltage is transmitted to the load. On the other hand, for $\Gamma(\ell) = -1$,

$$V(z) = V_0^+ \left\{ e^{\gamma(\ell - z)} - e^{-\gamma(\ell - z)} \right\} = 2V_0^+ \sinh \gamma(\ell - z)$$

and hence, at the load, $V(\ell) = 0$ no voltage is transmitted to the load. Finally, define a standing wave along a transmission line to be a *partial standing wave* whenever $0 < |\Gamma(\ell)| < 1$. Here, the transmission line contains both traveling waves and (partial) standing waves.

Physically, standing waves are the result of energy stored as fields in a reactive line. Because reactive energy is not transmitted to the load, it is, in effect, reducing the total energy that the load can absorb from the incident wave. Indeed, this fact is expressed by the definition of the time-average power at any point along the transmission line:

$$\langle p(z) \rangle \equiv \tfrac{1}{T} \int_0^T p(t, z) dt = \tfrac{1}{2} \mathrm{Re}(P(z, \omega))$$

where T is a single period of the wave.

Mathematically, the time-average reactive power is always identical to zero. As mentioned above, this corresponds to the power that is unavailable to the load. To see this, consider the definition of time-average power, and noting that $c \in \mathbf{C} \Rightarrow \Re e\, c = \tfrac{1}{2}(c + c^*)$,

$$\begin{aligned}
p(z,t) &= \Re e\, v(z,t) \Re e\, i(z,t) \\
&= \tfrac{1}{2}\left\{ v(z,t) + v^*(z,t) \right\} \tfrac{1}{2}\left\{ i(z,t) + i^*(z,t) \right\} \\
&= \tfrac{1}{4}\left\{ v(z,t)i(z,t) + v^*(z,t)i^*(z,t) + v^*(z,t)i(z,t) + v(z,t)i^*(z,t) \right\} \\
&= \tfrac{1}{2}\left\{ \Re e(v(z,t)i(z,t)) + \Re e(v(z,t)i^*(z,t)) \right\} \\
&= \tfrac{1}{2}\left\{ \Re e(V(z)I(z)e^{i2\omega t}) + \Re e(V(z)I^*(z)) \right\}
\end{aligned}$$

The first term corresponds to the reactive term, and it is evident that the time average of that term vanishes. Hence, it does not contribute towards the load.

This chapter concludes with a brief analysis on time delay along a transmission line. This has strong implications in the design of high-speed digital circuits where timing is critical. Given $(C_\pm, \ell, R, L, C, G)$, set $\upsilon = \frac{1}{\sqrt{LC}}$. Then, the time it takes for a signal to propagate from the source to the load is $\tau_\ell = \ell\sqrt{LC}$.

There are a number of methods to delay a signal. The easiest is to increase the length of the trace. However, this will affect the input impedance,

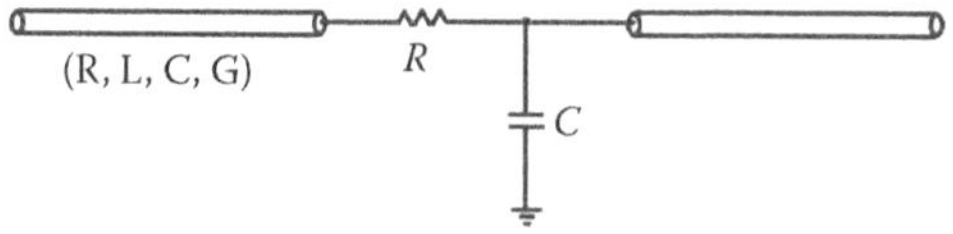

FIGURE 4.8
Implementing a simple RC-delay circuit.

assuming real estate on the PCB is available. A simple alternative is to implement an RC circuit as shown in Figure 4.8.

It is immediately clear that using an RC-delay circuit will affect the input resistance. Indeed, there will be a nonzero coefficient of reflection occurring on both sides of the transmission line where the RC-circuit is implemented. As this is undesirable, it is expedient to implement the circuit close to the load or source. Furthermore, to prevent reflection back from the source, an option is to add an impedance at the source such that the resultant time delay from the impedance and the RC-circuit meets the required specification.

4.4.14 Remark

Consider $(C_\pm, \ell, L, C)$ and suppose that the shortest data pulse propagating along the line is τ_d. If the load and source are not matched with the line, then clearly multiple reflections along the line will occur. Reflection occurring at the load will return to the load at $t = 2\tau$, where $\tau = \ell\sqrt{LC}$. Finally, suppose that the minimal pulse width $\delta\tau$ to which the load can respond is $\delta\tau \geq \tau_d$.

If $\ell > 0$ is such that $2\tau > \delta\tau$, then the reflected wave can potentially trigger the load, leading to multiple false data pulses. Hence, to avoid this problem, the easiest implementation is to require that $2\tau < \delta\tau$; then, multiple reflections will not be seen by the load as multiple data pulses. Suppose a design specification requires that $2\bar{n}\tau < \delta\tau$, for some $\bar{n} \in \mathbf{N}$. Then, it will suffice to choose $\ell > 0$ such that $2\bar{n}\ell\sqrt{LC} < \tau_d \Rightarrow \ell = \frac{\tau_d}{2\bar{n}\sqrt{LC}}$.

4.5 Worked Problems

4.5.1 Exercise

Given a finite transmission line system $(C_\pm, \ell, R, L, C, G)$, where the angular frequency ω of a time-harmonic wave is fixed, set $\alpha = \Re e\gamma$ and $\beta = \Im m\gamma$. Then, $\alpha\ell << 1 \Leftrightarrow \frac{\ell}{\lambda} << 1$.

Solution

Recalling that

$$\alpha = \omega\sqrt{\mu\varepsilon}\left\{\tfrac{1}{2}\left(\sqrt{1+\left(\tfrac{\sigma}{\omega\varepsilon}\right)^2}-1\right)\right\}^{\frac{1}{2}}$$

after some tedious manipulation, it is easy to show that

$$\alpha\ell << 1 \Leftrightarrow \tfrac{2}{(\ell\omega\sqrt{\mu\varepsilon})^2} >> \sqrt{1+\left(\tfrac{\sigma}{\omega\varepsilon}\right)^2}-1 \Leftrightarrow \left\{\tfrac{2}{(\ell\omega\sqrt{\mu\varepsilon})^2}+1\right\}^2 - 1 >> \left(\tfrac{\sigma}{\omega\varepsilon}\right)^2$$

Exploiting the identity $a^2 - b^2 = (a+b)(a-b)$, the above inequality is equivalent to

$$\tfrac{\sigma}{\omega\varepsilon} << \tfrac{2}{\ell\omega\sqrt{\mu\varepsilon}}\sqrt{1+\tfrac{1}{(\ell\omega\sqrt{\mu\varepsilon})^2}} \Leftrightarrow \sigma << \tfrac{2}{\ell}\sqrt{\tfrac{\varepsilon}{\mu}}\sqrt{1+\tfrac{1}{(\ell\omega\sqrt{\mu\varepsilon})^2}} = \tfrac{2}{\ell^2\omega\mu}\sqrt{1+(\ell\omega)^2\mu\varepsilon}$$

However, $\omega^2\mu\varepsilon = \frac{4\pi^2}{\lambda^2}$, whence

$$\sigma << \tfrac{2}{\ell^2\omega\mu}\sqrt{1+4\pi^2\left(\tfrac{\ell}{\lambda}\right)^2} \Rightarrow \sigma << \tfrac{1}{\ell\pi\upsilon\mu}\tfrac{\lambda}{\ell} \Leftrightarrow \tfrac{\ell}{\lambda} << 1$$

as claimed, where $\upsilon = f\lambda$. □

4.5.2 Exercise

Establish $P(\ell) = \frac{(V_0^+)^2}{2Z_0}\{1-|\Gamma(\ell)|^2\}$ in the proof of Theorem 4.4.2.

Solution

First, recall that $e^{i\theta} = \cos\theta + i\sin\theta$. Then, from

$$p(z) = V(z)I^*(z) = \tfrac{(V_0^+)^2}{Z_0}\{e^{2\alpha(\ell-z)} - |\Gamma(\ell)|e^{-i\theta_\ell}e^{i2\beta(\ell-z)} + |\Gamma(\ell)|e^{i\theta_\ell}e^{-i2\beta(\ell-z)} - |\Gamma(\ell)|^2 e^{-2\alpha(\ell-z)}\}$$

expanding the expression below and appealing to the identity $\sin(a+b) = \sin a \cos b + \cos a \sin b$,

$$\begin{aligned}&|\Gamma(\ell)|\{-e^{-i\theta_\ell}e^{i2\beta(\ell-z)} + e^{i\theta_\ell}e^{-i2\beta(\ell-z)}\}\\ &= |\Gamma(\ell)|\{\sin\theta_\ell\cos 2\beta(\ell-z) - \cos\theta_\ell\sin 2\beta(\ell-z)\}2i\\ &= i2|\Gamma(\ell)|\sin(\theta_\ell - 2\beta(\ell-z))\end{aligned}$$

and hence,

$$p(z) = V(z)I^*(z) = \frac{(V_0^+)^2}{Z_0}\left\{e^{2\alpha(\ell-z)} - |\Gamma(\ell)|^2 e^{-2\alpha(\ell-z)} + i2|\Gamma(\ell)|\sin(\theta_\ell - 2\beta(\ell - z))\right\}$$

Thus, $\langle p(z)\rangle = \frac{1}{2}\Re e\, p(z) = \frac{(V_0^+)^2}{2Z_0} e^{2\alpha(\ell-z)}\{1 - |\Gamma(\ell)|^2 e^{-4\alpha(\ell-z)}\}$
yielding

$$\langle p(\ell)\rangle = \frac{(V_0^+)^2}{2Z_0}\{1 - |\Gamma(\ell)|^2\} \Rightarrow \langle \bar{p}(\ell)\rangle = \max_{\Gamma}\langle p(\ell)\rangle \Leftrightarrow \Gamma(\ell) = 0$$

by inspection. □

4.5.3 Exercise

Consider the schematic diagram, where R = 0 = G.

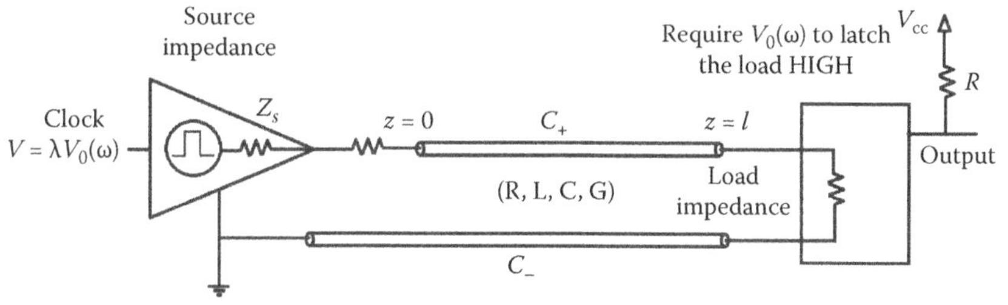

Suppose a clock can only supply $V = \lambda V_0$, for some $0 < \lambda < \frac{1}{2}$, and the required voltage by the load to signal High (or 1) is $V = V_0$. In the schematics shown, the input to the IC only depends upon the voltage level and not the current, the output is an open-drain, and the required voltage is supplied by V_{cc} via some pull-up resistor R. Furthermore, suppose $Z_S << Z_0 << Z_L$.

(a) Determine the length of the trace C_+ and series resistor r such that $V(\ell) = V_0(\omega)$ at the load impedance.

(b) Find the correct series resistor r such that no reflection will occur at the source.

(c) Are the stipulations of (a) and (b) compatible?

Solution

(a) By Proposition 4.4.2,

$$V(\ell) = \frac{\lambda V_0(\omega) Z_0}{Z_0 + Z_S^\ell} \frac{e^{-\gamma\ell}}{1 - \Gamma(0)\Gamma(\ell)e^{-2\gamma\ell}}\{1 + \Gamma(\ell)\}$$

where $Z_S' = Z_S + r$. By assumption, $Z_L >> Z_0 \Rightarrow \Gamma(\ell) \approx 1$. Hence,

$$V(\ell) = \frac{2\lambda V_0(\omega) R_0}{R_0 + Z_S'} \frac{e^{-i\beta\ell}}{1 - \Gamma(0)e^{-i2\beta\ell}}$$

Now, in order for $V(\ell) = V_0(\omega)$, it suffices that

$$1 = \frac{2\lambda R_0}{R_0 + Z_S'} \frac{e^{-i\beta\ell}}{1 - \Gamma(0)e^{-i2\beta\ell}}$$

be satisfied. Thus, $\Gamma(0) = \frac{Z_S' - R_0}{Z_S' + R_0}$ yields

$$\frac{2\lambda R_0}{R_0 + Z_S'} \frac{e^{-i\beta\ell}}{1 - \Gamma(0)e^{-i2\beta\ell}} = \frac{2\lambda R_0 e^{-i\beta\ell}}{Z_S' e^{-i\beta\ell}(e^{i\beta\ell} - e^{-i\beta\ell}) + R_0 e^{-i\beta\ell}(e^{i\beta\ell} + e^{-i\beta\ell})} = \frac{\lambda R_0}{Z_S' \sin\beta\ell + R_0 \cos\beta\ell}$$

Next, invoking the trigonometric identity $a \sin x + b \cos x = \sqrt{a^2 + b^2} \sin(x + y)$, where $y = \mathrm{sgn}(b) \arccos \frac{a}{\sqrt{a^2 + b^2}}$ and sgn(x) is the signum function defined by

$$\mathrm{sgn}(x) = \begin{cases} 1 & \text{if } x > 0 \\ -1 & \text{if } x < 0 \end{cases}$$

it follows at once that $\sin(\beta\ell + \phi_\lambda) = \frac{\lambda R_0}{\sqrt{R_0^2 + (Z_S')^2}}$ and hence,

$$\ell = \frac{1}{\beta}\left\{\arcsin \frac{\lambda R_0}{\sqrt{R_0^2 + (Z_S')^2}} - \varphi_\lambda\right\}$$

where $\varphi_\lambda = \arccos \frac{Z_S'}{\sqrt{R_0^2 + (Z_S')^2}}$, as desired.

(b) To prevent reflection from the source, $0 = \Gamma(0) = \frac{Z_S' - R_0}{Z_S' + R_0} \Rightarrow Z_S' = R_0 \Rightarrow r = R_0 - Z_S$.

(c) It is clear that (a) and (b) are compatible $\forall r \geq 0$. That is, it is possible to suppress reflection from the source and obtain the desired voltage transmission by appropriately selecting the length of the trace. □

4.5.4 Exercise

Prove Proposition 4.4.2 explicitly via reflection of waves.

Solution

Set $V_0^+ = \frac{V_S Z_0}{Z_0+Z_S} e^{-\gamma\ell}$ from the proof of Proposition 4.4.2. Then, via Example 4.4.7, $V(z) = V_0^+\{e^{\gamma(\ell-z)} + \Gamma(\ell)e^{-\gamma(\ell-z)}\}$ becomes

$$\begin{aligned} V(z) &= V_0^+\{e^{\gamma(\ell-z)} + \Gamma(\ell)e^{-\gamma(\ell-z)} + \Gamma(0)\Gamma(\ell)e^{\gamma(3\ell-z)} + \Gamma(0)\Gamma^2(\ell)e^{-\gamma(3\ell-z)} + \cdots \\ &\quad \Gamma^n(0)\Gamma^n(\ell)e^{\gamma((2n+1)\ell-z)} + \Gamma^n(0)\Gamma^{n+1}(\ell)e^{\gamma((2n+1)\ell-z)} + \cdots\} \\ &= V_0^+ e^{\gamma(\ell-z)}(1+\Gamma(\ell)e^{-2\gamma(\ell-z)})(1+\Gamma(0)\Gamma(\ell)e^{-2\gamma\ell} + \Gamma^2(0)\Gamma^2(\ell)e^{-4\gamma\ell} + \cdots) \\ &= V_0^+ e^{\gamma(\ell-z)}(1+\Gamma(\ell)e^{-2\gamma(\ell-z)})\{1+\Gamma(0)\Gamma(\ell)e^{-2\gamma\ell} + (\Gamma(0)\Gamma(\ell)e^{-2\gamma\ell})^2 + \cdots\} \\ &= V_0^+ e^{\gamma(\ell-z)}(1+\Gamma(\ell)e^{-2\gamma(\ell-z)})\frac{1}{1-\Gamma(0)\Gamma(\ell)e^{-2\gamma\ell}} \end{aligned}$$

yielding

$$V(z) = \frac{V_S Z_0}{Z_0+Z_S}\frac{e^{-\gamma z}}{1-\Gamma(0)\Gamma(\ell)e^{-2\gamma\ell}}\{1+\Gamma(\ell)e^{-2\gamma(\ell-z)}\}$$

as required. □

4.5.5 Exercise

Suppose a load on a lossless transmission line $(C_\pm, \ell, L, C)$ is capacitive: $Z_L = R_L - iX_L$, for some $X_C > 0$. Suppose further that the load is a digital logic input that responds to a minimal pulse width of τ_d. Find $(\tilde{Z}, \tilde{z})$ such that at

$$z = \tilde{z}, \quad \Gamma(\tilde{z}) = \frac{Z'(\tilde{z}) - R_0}{Z'(\tilde{z}) + R_0} = 0$$

where $Z'(\tilde{z}) = \tilde{Z}\,||\,Z(\tilde{z})$. Unwanted reflections are thus mitigated by the presence of the impedance $\tilde{Z}$ placed at $z = \tilde{z}$.

Solution

Consider the schematic diagram, and determine δZ such that the input impedanceat $z = l'$ is matched.

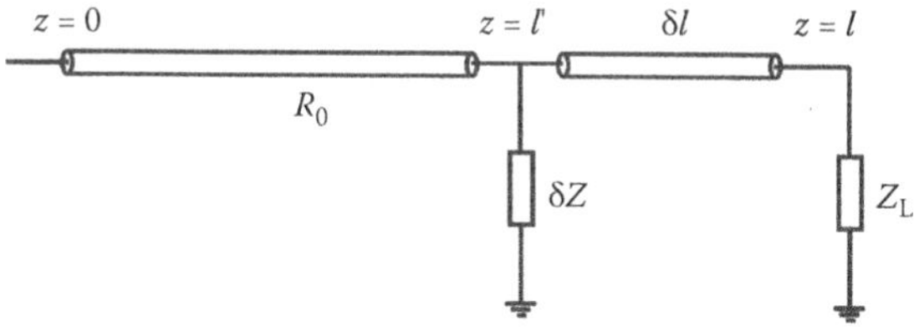

Now, set $Y' = \delta Y + \tilde{Y}_L$ to be the input admittance at $z = \ell'$, and let $\delta y = R_0 \delta Y$, $\tilde{y}_L = R_0 \tilde{Y}_L$ be the normalized admittance, where

$$\tilde{Y}_L = \frac{1}{R_0}\frac{R_0 + iZ_L \tan\beta\delta\ell}{Z_L + iR_0 \tan\beta\delta\ell}$$

Hence, the requirement that

$$\Gamma'(\ell') = \frac{Z'(\ell') - R_0}{Z'(\ell') + R_0} = 0$$

that is, that there be zero reflection at $z = \ell' \Rightarrow y' = 1$. In other words

$$1 = \delta y + \tilde{y}_L \Rightarrow \Re e(\delta y) = 1 - \Re e(\tilde{y}_L) \quad \text{and} \quad \Im m(\delta y) = -\Im m(\tilde{y}_L)$$

It is a somewhat tedious matter to show that

$$\tilde{y}_L = \frac{R_0 Z_L \sec^2\beta\delta\ell}{Z_L^2 + R_0^2 \tan^2\beta\delta\ell} + i\frac{(Z_L^2 - R_0^2)\tan^2\beta\delta\ell}{Z_L^2 + R_0^2 \tan^2\beta\delta\ell}$$

To complete the solution, observe from Remark 4.4.14 that $2\bar{n}\delta\ell\sqrt{LC} < \tau_d \Rightarrow \delta\ell < \frac{\tau_d}{2\bar{n}\sqrt{LC}}$. Hence, set $\delta\ell = \frac{\tau_d}{4\sqrt{LC}}$; then, the required matching impedance δZ is obtained via

$$\Re e(\delta y) = 1 - \frac{R_0 Z_L \sec^2\beta\delta\ell}{Z_L^2 + R_0^2 \tan^2\beta\delta\ell} \quad \text{and} \quad \Im m(\delta y) = -\frac{(Z_L^2 - R_0^2)\tan^2\beta\delta\ell}{Z_L^2 + R_0^2 \tan^2\beta\delta\ell}$$

Note: It is clear from the above analysis why the Smith Chart was developed to graphically determine the desired matching load. □

4.5.6 Exercise

Consider the schematic diagram for series termination versus parallel termination. Describe qualitatively the impact of the respective terminations on the rise time at the load. Suppose that $\ell >> \delta\ell$, where $\delta\ell$ is the distance between the resistor and the load.

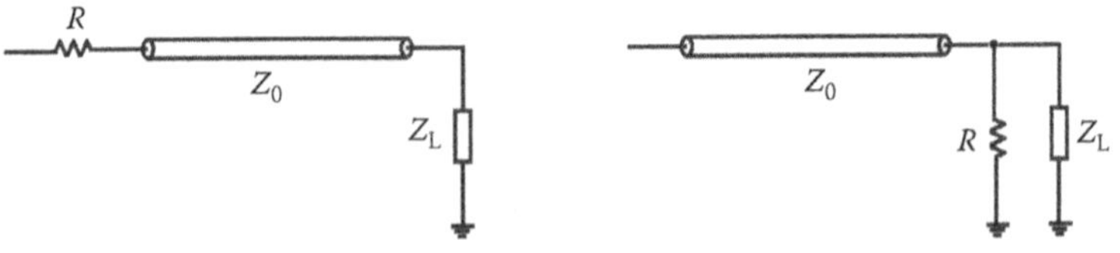

(a) Series termination (b) Parallel termination

Solution

(a) Series termination. Here, set $\sqrt{\frac{\tilde{L}}{\tilde{C}}} = R + \sqrt{\frac{L}{C}}$. Then, $\exists a > 1$ such that $\sqrt{\frac{\tilde{L}}{\tilde{C}}} = \sqrt{a\frac{L}{C}}$, whence, $\tilde{\tau} \approx \ell\sqrt{\alpha LC} > \ell\sqrt{LC} = \tau$, the propagation delay time. That is, the propagation delay of the input voltage wave is $\tilde{\tau}$ whereas the propagation delay of $V(0)$, just after the resistor R, is τ. From this, it is clear that the voltage reaches a peak at the load before it reaches a peak at the input side into the series resistor, that is, faster rise time at the load than at the input terminal. Thus, the far end switches faster with respect to the input terminal, as the series termination introduces a longer delay at the input voltage. Intuitively, recalling that a transmission line is inductive, introducing a series resistor leads to a delay induced by a series RL-time constant at the input terminal. Beyond the series termination, the propagating wave sees the characteristic impedance.

(b) Parallel termination. The analysis follows that of (a) *mutatis mutandis*. Here, the wave propagation does not see the load until it arrives at $z = \ell$. Hence, a propagation delay is introduced at the load and not at the input terminal. There is thus a delay in the rise time at the load with respect to the terminal. The scenario is thus the reverse of (a). In short, parallel termination is good for slowing down rise time for emissions reduction; however, series termination is used instead of parallel termination if rise time is critical. Intuitively, the wave propagating along the transmission line only sees the characteristic impedance until it arrives at the load; then, it sees a parallel RZ-impedance which, in the case of a capacitive load, will introduce an RC-time constant, delaying the rise time.

References

1. Cheng, D. 1989. *Field and Wave Electromagnetics*. Reading, MA: Addison-Wesley.
2. Churchill, R. and Brown, J. 1990. *Complex Variables and Applications*. New York: McGraw-Hill.
3. Johnson, H. and Graham, M. 1993. *High-Speed Digital Design*. Upper Saddle River, NJ: Prentice-Hall.
4. Kohonen, T. 1972. *Digital Circuits and Devices*. Englewood Cliffs, NJ: Prentice-Hall.
5. Krutz, R. 1988. *Interfacing Techniques in Digital Design*. New York: John Wiley & Sons.
6. Neff, Jr., H. 1981. *Basic Electromagnetic Fields*. New York: Harper & Row.
7. Plonsey, R. and Collin, R. 1961. *Principles and Applications of Electromagnetic Fields*. New York: McGraw-Hill.

5

Differential Transmission Lines

At the simplest level, a pair of differential lines comprises two transmission lines such that the current flowing along one line is equal to the current flowing in the opposite direction along the second line. Informally then, the resultant current flowing along the pair is zero. Equivalently, the transmitted voltage on one of the pair is 180 degrees out of phase with the voltage of the other line. Thus, each line acts as the ground for its counterpart, and in this sense, they have a common self-referencing ground.

From a theoretical perspective, differential lines are less sensitive to noise and have fewer emissions than nondifferential lines. Thus, they are often used in radio frequency (RF) design, especially for clock rates in the microwave regime. Notwithstanding, differential lines must still be matched or they can cause undesired emissions.

A knowledge of differential pairs is extremely important to EMC engineers, and an informal treatment can be found in References [8,10]. Differential transmission line theory is essentially a corollary of transmission line theory, and can thus be derived there from References [1,3–7]. This is the approach taken in this chapter, wherein the bulk of the derivation leverages the results from Chapter 4.

5.1 Differential Pair: Odd and Even Modes

The definition of an ideal differential pair is given below. The pair is assumed to be surrounded in a homogeneous dielectric medium. In practice, differential pairs are typically routed over a ground plane; more is said later. First, some notations are established below. In all that follows, where convenient and unless explicitly stated otherwise, let $\Omega \subset \mathbf{R}^3$ denote an open subset upon which circuits are modeled, and in addition, given a pair (C_+, C_-) of transmission lines of uniform cross-section $s_\pm$, let $\gamma_\pm : [0,1] \to \Omega$ be a path representing the axis of $C_\pm$. Finally, the Euclidean metric $\|\cdot\|$ in $\mathbf{R}^3$ is used whenever a distance function is invoked and it is also assumed that the cross-sectional area is much less than the length of the transmission line: $|s_\pm| << \|\gamma_\pm\|$.

5.1.1 Definition

Consider a pair of parallel transmission lines (C_+,C_-), and denote the lengths of the transmission lines by $|\gamma_+|,|\gamma_-|$, respectively. Suppose that the cross-section $|s_+(\gamma_+(s))| = |s_-(\gamma_-(s))|$ $\forall s \in [0,1]$, and (C_+,C_-) satisfies the following criteria:

$$\text{(a)} \quad |\gamma_+| = \ell = |\gamma_-|, \text{ for some } \ell > 0$$

$$\text{(b)} \quad d(\gamma_+(s),\gamma_-(s)) = \text{ constant } \forall s \in [0,1]$$

where $d(x,y) = \|x-y\|$ denotes the distance separating the points $x,y \in \Omega$,

$$\text{(c)} \quad I_+(s) + I_-(s) = 0 \ \forall s \in [0,1]$$

where $I_\pm(s) \equiv I(\gamma_\pm(s))$ denotes the current at the point $\gamma_\pm(s)$ along the lines $\gamma_\pm$, respectively,

$$(d) \quad R_0(C_+) = R_0(C_-)$$

where $R_0(C_\pm)$ denotes the characteristic impedance of $C_\pm$ with respect to a common ground.

Then, the pair (C_+,C_-) is said to be a *symmetric* (or *balanced*) *differential pair.* If condition (d) is violated, then the pair forms an *asymmetric* (or *unbalanced*) differential pair.

5.1.2 Remark

In the technology industry, a differential pair is often defined to be a pair (C_+,C_-) where criterion (b)—which essentially requires that the lines be parallel to each other—is relaxed due to physical constraints and design considerations.* Furthermore, it is trivially true that (d) holds if the pair is isolated in space, because then the characteristic impedance is defined by the pair. Where condition (d) comes into play is the following scenario: in a typical circuit layout, a differential pair often has a common ground plane. The characteristic impedance $R_0(C_\pm)$ is thus measured with respect to the ground plane in question. This is not to be confused with the differential impedance of the pair defined shortly.

* A serpentine differential pair is a case in point.

An idealized differential circuit is depicted in Figure 5.1.

- Z_{diff} denotes the characteristic differential impedance defined later.
- $R = Z_{\text{diff}}$ for a perfectly matched (symmetric) differential pair.
- The differential receiver is essentially a differential amplifier with the resultant output being the difference between the impressed voltages across R.

Let $V_{\pm}(R)$ denote the voltage developed across R by $V_{\pm}$ propagating along $C_{\pm}$. Then, the resultant output $V_{\text{out}} = V_{+}(R) - V_{-}(R)$ (modulo a scaling factor if the output is amplified by the differential amplifier). It is shown below that for a symmetrical differential line, $V_{\text{out}} = 2V_{+}(R)$.

5.1.3 Lemma

Let (C_{+}, C_{-}) be a symmetrical differential pair as illustrated in Figure 5.1. Suppose that the characteristic impedance of $C_{\pm}$ relative to a fixed ground plane is Z_0. Then, $V_{+}(s) = -V_{-}(s)$ for $0 \le s \le 1$, where $V_{\pm}(s) = V(\gamma_{\pm}(s))$ is the voltage at $\gamma_{\pm}(s)$ along $\gamma_{\pm}$.

Proof

First, recall from Chapter 4 that, heuristically, the voltage on one line can be induced along the other line via the distributed line model for a transmission line. This can be represented as

$$V_{+} = I_{+}Z_{++} + I_{-}Z_{+-} \tag{5.1}$$

where $Z_{++} = Z_0$ is the characteristic impedance of C_{+} relative to some fixed ground, and Z_{+-} is the transfer impedance that induces a voltage on C_{+} as a result of the current I_{-} flowing through C_{-}. By symmetry, the voltage propagating along C_{-} is given by

$$V_{-} = I_{+}Z_{-+} + I_{-}Z_{--} \tag{5.2}$$

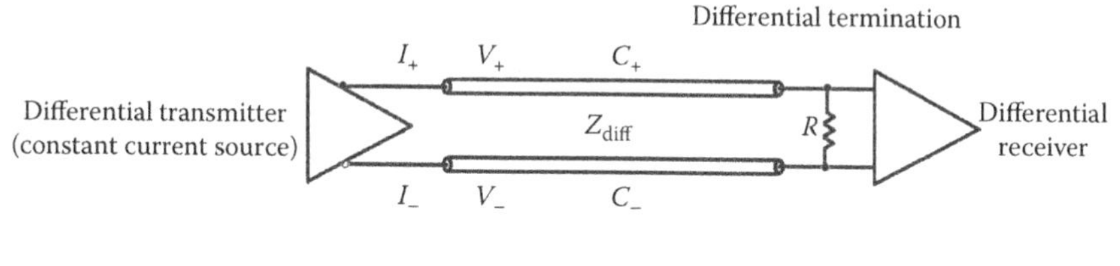

FIGURE 5.1
Schematic representation of an idealized symmetric differential pair.

where $Z_{--} = Z_0$ and Z_{-+} is the transfer impedance that induces a voltage on C_- as a result of the current I_+ flowing through C_+. Because the dielectric medium is linear and homogeneous with the magnetic permeability $\mu = \mu_0$, the transfer impedance must be symmetric: $Z_{+-} = \hat{Z} = Z_{-+}$, for some $\hat{Z}$, where $Z_{+-} \equiv \frac{V_+}{I_-}|_{I_+=0}$ and $Z_{-+} \equiv \frac{V_-}{I_+}|_{I_-=0}$. Adding Equations (5.1) and (5.2) and using Definition 5.1.1(c) yield the result: $V_+ + V_- = I_-\hat{Z} + I_+\hat{Z} = 0 \Rightarrow V_+ = -V_-$. □

5.1.4 Remark

In the proof of Lemma 5.1.3, let $\hat{Z} = \kappa Z_0$, where $\kappa > 0$ is some constant called the *differential coupling coefficient*. Then, Equation (5.1) becomes

$$V_+ = I_+ Z_{11} + I_- Z_{12} = I_+(1-\kappa)Z_0$$

The ratio $Z_{\text{odd}}^+ = \frac{V_+}{I_+} = (1-\kappa)Z_0$ is known as the *odd mode impedance* along C_+. By symmetry, $Z_{\text{odd}}^+ = Z_{\text{odd}} = Z_{\text{odd}}^-$. For this reason, it is known as the odd mode impedance of a differential pair, as $I_+ = -I_-$. Further details are outlined shortly.

5.1.5. Lemma.

Given a symmetric differential pair, the odd mode impedance is given by $Z_{\text{odd}} = Z_0 - \hat{Z}$.

Proof

By (5.1), $V_+ = I_+ Z_{11} + I_- Z_{12} = I_+(Z_0 - \hat{Z}) \Rightarrow Z_0 - \hat{Z} = \frac{V_+}{I_+} = Z_{\text{odd}}$, as required. □

The intent of the above brief account was to motivate an intuitive comprehension of a differential pair. A more rigorous approach via Maxwell's theory is sketched below. In particular, the odd and even mode impedances are derived. The analysis is a special case of multitransmission line theory covered in the following chapter. Indeed, much of the theoretical infrastructure has already been developed in Chapter 4.

Referring to Figure 5.2, consider initially a system of three conductors embedded in a homogeneous dielectric medium $(\Omega, \varepsilon, \mu, \sigma)$, where the middle conductor C_0 is grounded. Suppose $I_\pm = I_\pm(z,t)$ is propagating along conductor $C_\pm$. Now, observe that the results of transmission line theory outlined in Chapter 4 apply to each pair (C_+, C_0) and (C_-, C_0) by evaluating the field around loops γ_+ and γ_-, respectively, defined in Figure 5.2. However, cross-coupling also occurs between the pair (C_+, C_-), and the analysis carried out for $(C_\pm, C_0)$ also applies to (C_+, C_-).

Without loss of generality, consider the electric field $\mathbf{E}_+$ of an incident TEM wave on (C_+, γ_+). Appealing to Stokes' theorem, the integral version of Equation (1.15) is obtained:

$$\oint_{\gamma_+} \mathbf{E}_+ \cdot d\mathbf{l} = -\partial_t \oiint_{S_+} \mathbf{B}_+ \cdot d\mathbf{S} \tag{5.3}$$

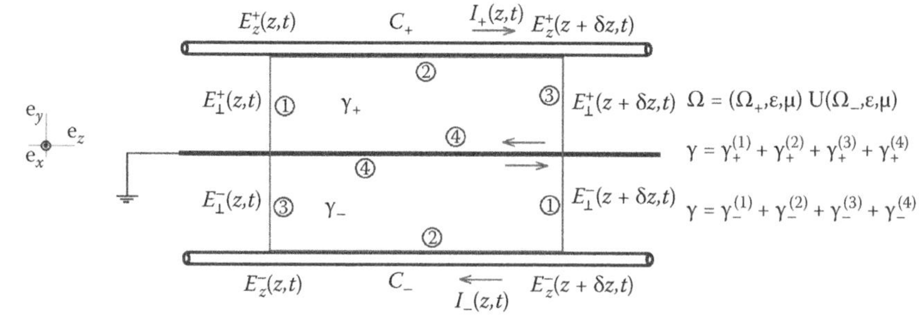

FIGURE 5.2
Differential transmission line pair supporting a TEM wave.

where $S_+ = S(\gamma_+) \subset \mathbf{R}^3$ is an arbitrary compact surface spanned by γ_+: $\partial S_+ = \gamma_+$, whence, following the argument in Chapter 4 *mutatis mutandis*, suppressing the (x,y) coordinates in the scalar and vector fields, yields

$$-\partial_t \oiint_{S_+} \boldsymbol{B}_+ \cdot d\boldsymbol{S} = -v_+(z,t;\gamma_+^{(1)}) + v_+(z+\delta z,t;\gamma_+^{(3)}) + \mathrm{R}_+\delta z i_+(z,t;\gamma_+^{(2)}) + \mathrm{R}_0 \delta z i_0(z,t;\gamma_+^{(4)}) \tag{5.4}$$

where $I_0 + I_+ + I_- = 0$ by Kirchhoff's current law, and R_0 is the resistance per unit length of the ground conductor. Similarly, $\oint_{\gamma_-} \boldsymbol{E}_- \cdot d\boldsymbol{l} = -\partial_t \oiint_{S_-} \boldsymbol{B}_- \cdot d\boldsymbol{S}$ yields

$$-\partial_t \oiint_{S_-} \boldsymbol{B}_- \cdot d\boldsymbol{S} = -v_-(z,t;\gamma_-^{(3)}) + v_-(z+\delta z,t;\gamma_-^{(1)}) + \mathrm{R}_-\delta z i_-(z,t;\gamma_-^{(2)}) + \mathrm{R}_0 \delta z i_0(z,t;\gamma_-^{(4)}) \tag{5.5}$$

where $S_- = S(\gamma_-) \subset \mathbf{R}^3$ is a compact surface spanned by γ_-.

Now, recall from Chapter 4 that the magnetic flux per unit length $\psi_\pm = \lim_{\delta z \to 0} \frac{1}{\delta z} \oiint_{S_\pm} \boldsymbol{B}_\pm \cdot d\boldsymbol{S}$, where $\delta z = \left|\gamma_\pm^{(4)}\right|$, is well-defined, and moreover, from the definition of magnetic flux $\Psi = Li$, it follows that

$$\psi_+ = \mathrm{L}_{++}\,\delta z I_+ + \mathrm{L}_{+-}\delta z I_- \tag{5.6}$$

where L_{++} is the *self-inductance* per unit length of conductor Γ_+, coupling the magnetic flux generated by current I_+ to S_+, and L_{+-} is the *mutual inductance* per unit length coupling the magnetic flux generated by current i_- to S_+.

Inasmuch as the limit $\lim_{\delta z \to 0} \frac{1}{\delta z} \oiint_{S_\pm} \boldsymbol{B}_\pm \cdot d\boldsymbol{S}$ is well-defined as $\lim_{\delta z \to 0} \frac{1}{\delta z} |S_\pm| < \infty$, it follows at once that $\lim_{\delta z \to 0} \frac{1}{\delta z} \oiint_{S_\pm} \partial_t \boldsymbol{B}_\pm \cdot d\boldsymbol{S} = \mathrm{L}_{++} \partial_t I_+(z,t) + \mathrm{L}_{+-} \partial_t I_-(z,t)$ together with Equation (4.4) yield

$$\partial_z v_+(z,t) = -(\mathrm{R}_+ + \mathrm{R}_0) I_+(z,t) - \mathrm{R}_0 I_-(z,t) - \mathrm{L}_{++} \partial_t I_+(z,t) - \mathrm{L}_{+-} \partial_t I_-(z,t). \quad (5.7)$$

By symmetry,

$$\partial_z v_-(z,t) = -(\mathrm{R}_- + \mathrm{R}_0) I_-(z,t) - \mathrm{R}_0 I_+(z,t) - \mathrm{L}_{--} \partial_t I_-(z,t) - \mathrm{L}_{-+} \partial_t I_+(z,t). \quad (5.8)$$

5.1.6 Lemma

Given a system of three parallel conductors (C_+, C_-, C_0) as depicted in Figure 5.2, if $i_-(z,t) + i_+(z,t) = 0 \; \forall z, t \geq 0$, then the triple (C_+, C_-, C_0) is equivalent to the pair (C_+, C_-) where C_0 is removed from the system of conductors.

Proof

From Kirchhoff's current law, $0 = I_0 + I_+ + I_-$, $I_- + I_+ = 0 \Rightarrow I_0 \equiv 0 \; \forall z, t \geq 0$ and similarly, the potential is also identically zero, whence, C_0 may be removed from the system without affecting the field distribution from $C_\pm$, as required. □

The above result is the reason why a differential pair is said to have a *self-referencing ground*; that is, it possesses a "virtual" ground whereby the physical ground conductor may be removed without affecting the electromagnetic field distribution of the system. Thus, in view of Remark 5.1.7, the ground conductor may be removed by setting $\mathrm{R}_0 = 0$ when considering a differential pair; that is, $I_-(z,t) = -I_+(z,t) \; \forall z, t \geq 0$. This simplifies Equations (5.7) and (5.8) to

$$\partial_z v_+(z,t) = -\mathrm{R}_+ I_+(z,t) - (\mathrm{L}_{++} - \mathrm{L}_{+-}) \partial_t I_+(z,t) \quad (5.9)$$

$$\partial_z v_-(z,t) = \mathrm{R}_- I_+(z,t) + (\mathrm{L}_{--} - \mathrm{L}_{-+}) \partial_t I_+(z,t) \quad (5.10)$$

5.1.7 Remark

Note in passing from Equation (5.7) that for a balanced differential pair where $I_-(z,t) = -I_+(z,t) \; \forall z, t \geq 0$, (5.7) reduces automatically to (5.9) without the need to set $\mathrm{R}_0 = 0$. This implies that R_0 is arbitrary and hence, we may choose $\mathrm{R}_0 = 0$ without any loss of generality.

Indeed, Equations (5.9) and (5.10) can be rewritten more compactly as

$$\partial_z \begin{pmatrix} v_+ \\ v_- \end{pmatrix} = -\begin{pmatrix} \mathrm{R}_+ & \mathrm{L}_{++} - \mathrm{L}_{+-} \\ -\mathrm{R}_- & \mathrm{L}_{-+} - \mathrm{L}_{--} \end{pmatrix} \begin{pmatrix} I_+ \\ \partial_t I_+ \end{pmatrix}$$

This furnishes a field-theoretic proof for Lemma 5.1.3 given above.

5.1.8 Lemma

Suppose (C_+, C_-) is a balanced differential pair (*viz.*, $\mathrm{R}_+ = \mathrm{R} = \mathrm{R}_-$), for some fixed impedance per unit length R. Then, $v_+(z,t) = -v_-(z,t)\ \forall z, t \geq 0$.

Proof

Adding Equations (5.9) and (5.10) yields

$$\partial_z(v_+(z,t) + v_-(z,t)) = (-\mathrm{R}_+ + \mathrm{R}_-)I_+(z,t) - (\mathrm{L}_{++} - \mathrm{L}_{--})\partial_t I_+(z,t).$$

Because the differential lines are balanced, $\mathrm{RL} = \sigma\mu =$ constant $\Rightarrow \mathrm{L}_{++} = \mathrm{L}_{--}$ and hence, $\partial_t(v_+ + v_-) = 0$ implies that $v_+ = -v_- + k$ for some constant k, where σ, μ are, respectively, the conductivity and magnetic permeability of the medium surrounding (C_+, C_-). Furthermore, "balanced" implies the pair (C_+, C_-) has identical characteristic impedance with respect to a common ground and hence, $k \equiv 0$, as required. □

To complete the analysis, the current variation along the conductors is developed below. Consider the cylinder $C_+ = S_+(z) \times [z, z + \delta z]$ shown in Figure 5.3, where

$$S_+(z) = \left\{(x,y) \in \mathbf{R}^2 : (x - x_+)^2 + (y - y_+)^2 \leq r_+^2\right\}$$

$(x_+, y_+, z) \in C_+$ is a point along the axis of C_+. Let $\gamma_z^+ = \partial S_+(z)$ and $\gamma_{z+\delta z}^+ = \partial S_+(z + \delta z)$. Finally, set $C_+^\circ = S_+(z) \times (z, z + \delta z)$.

By appealing to Maxwell's second equation $\nabla \times \boldsymbol{B} = \mu\varepsilon\partial_t \boldsymbol{E} + \mu\sigma\boldsymbol{E}$ and invoking Stokes' theorem, it follows from Figure 5.3 that

$$\oiint_{\partial C_+^\circ} \nabla \times \boldsymbol{B} \cdot \mathrm{d}\boldsymbol{S} = \oint_{\partial^2 C_+^\circ} \boldsymbol{B} \cdot \mathrm{d}\boldsymbol{l} \equiv 0$$

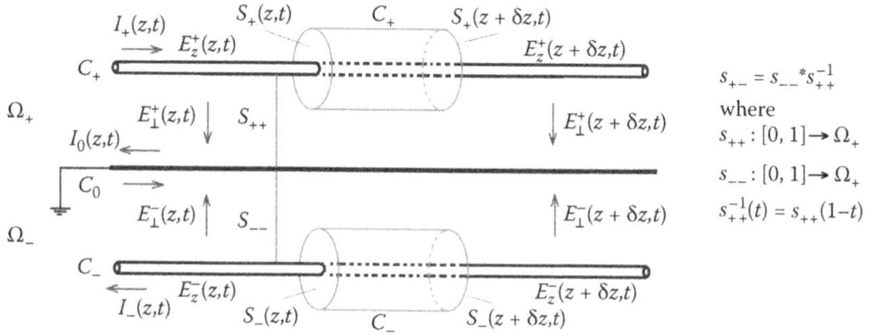

FIGURE 5.3
Current variation of TEM wave along multiple conductors.

where $\partial^2 C_+^\circ = \partial(\partial C_+^\circ)$ is the boundary of the boundary, which is precisely the empty set.* Hence the integral vanishes.

In Figure 5.3, $s_{+-} = s_{--} * s_{++}^{-1}$, where $s_{++}^{-1}(t) = s_{++}(1-t)$ and

$$s_{--} * s_{++}^{-1}(t) \equiv \begin{cases} s_{--}(2t) & \text{for } 0 \le t \le \frac{1}{2} \\ s_{++}^{-1}(2t-1) & \text{for } \frac{1}{2} \le t \le 1 \end{cases}$$

Moreover, observe that as the field is assumed to be quasi-static, the potential difference between two points is essentially independent of the path chosen, and hence the choice of paths depicted in Figure 5.3. The paths $s_{\alpha\beta}$, for $\alpha,\beta \in \{+,-\}$, are chosen such that they are normal to both $\partial C_\pm^\circ$ for simplicity.

For convenience, denote $\boldsymbol{E} = \boldsymbol{E}_+ + \boldsymbol{E}_-$ and $\boldsymbol{B} = \boldsymbol{B}_+ + \boldsymbol{B}_-$, where $\boldsymbol{E}_\pm(\boldsymbol{B}_\pm)$ defines the electric (magnetic) field generated, respectively, by conductors $C_\pm$. Furthermore, let $i_+(z,t)$ denote the current propagating along C_+ entering $S_+(z)$ and $i_+(z+\delta z,t)$ the current exiting $S_+(z+\delta z)$. Physically, it can be seen that for $0 < \sigma < \infty$, some of the current propagating along C_+ will also be conducted via the medium to C_0 and C_-. Hence, it is intuitively clear that $i_+(z+\delta z,t) \ne i_+(z,t)$ when the medium is lossy. Indeed, even if the medium were lossless, it is clear via the distributed line model that the existence of a distributed capacitance renders the inequality to hold in general.

Now, by Figure 5.3, $\partial C_+^\circ = \partial C_+ \cup S_+(z) \cup S_+(z+\delta z)$ is the boundary of the differential cylinder of length δz along C_+, whence Maxwell's equation yields

$$0 = \mu \int_{\partial C_+^\circ} \boldsymbol{J} \cdot \boldsymbol{n} d^2x + \mu\varepsilon \tfrac{d}{dt} \int_{\partial C_+^\circ} \boldsymbol{E} \cdot \boldsymbol{n} d^2x = \mu \int_{\partial C_+^\circ} \boldsymbol{J}_\perp \cdot \boldsymbol{n} d^2x + \mu\varepsilon \tfrac{d}{dt} \int_{\partial C_+^\circ} \boldsymbol{E}_\perp \cdot \boldsymbol{n} d^2x \tag{5.11}$$

under the assumption of quasi-TEM propagation: $\boldsymbol{E}_\parallel \cdot \boldsymbol{n} = 0$, where $|\boldsymbol{E}_\parallel| << |\boldsymbol{E}_\perp|$ and $\boldsymbol{E} = \boldsymbol{E}_\perp + \boldsymbol{E}_\parallel$, the sum of the transverse and longitudinal components, respectively.

Next, because the medium surrounding the conductors is an imperfect dielectric of conductivity $0 < |\sigma_0| < \infty$, $\oint\!\!\oint_{\partial C_+^\circ} \boldsymbol{J}_+ \cdot d\boldsymbol{S} \ne 0$. Hence, $\oint\!\!\oint_{\partial C_+^\circ} \boldsymbol{J}_\perp \cdot \boldsymbol{n} d^2x = \sigma_0 \oint\!\!\oint_{\partial C_+^\circ} \boldsymbol{E}_\perp \cdot \boldsymbol{n} d^2x$ defines the conduction current through Ω_+ from C_+ to C_0 and C_- across ∂C_+°. Thus, by construction, invoking the

* Intuitively, this can be seen by considering a 2-sphere: the boundary of a three-dimensional ball. It is obvious that a 2-sphere has no boundary as every two-dimensional neighborhood about any point on the sphere is completely contained in the sphere. Thus, the boundary of a 2-sphere is empty.

definition of resistance $R = \left\{ \int_{\gamma} \boldsymbol{E}_{\perp} \cdot \mathrm{d}\boldsymbol{l} \right\} \left\{ \sigma \oiint_{S} \boldsymbol{E}_{\perp} \cdot \boldsymbol{n} \mathrm{d}^2 x \right\}^{-1}$, it follows at once that via Kirchhoff's current law,

$$\sigma_0 \oiint_{\partial C_+^\circ} \boldsymbol{E}_{\perp} \cdot \mathrm{d}\boldsymbol{S} = \tfrac{\delta z}{R_{++}} \int_{s_+} \boldsymbol{E}_{\perp} \cdot \mathrm{d}\boldsymbol{l} + \tfrac{\delta z}{R_{+-}} \int_{s_{+-}} \boldsymbol{E}_{\perp} \cdot \mathrm{d}\boldsymbol{l} \equiv G_{++}\delta z \int_{s_+} \boldsymbol{E}_{\perp} \cdot \mathrm{d}\boldsymbol{l}$$
$$+ G_{+-}\delta z \int_{s_{+-}} \boldsymbol{E}_{\perp} \cdot \mathrm{d}\boldsymbol{l} \tag{5.12}$$

where $s_{++}(s_{+-})$ is a path from $C_+ \cap C_+$ to $C_0(C_-)$, and $R_{++}(R_{+-})$ is the resistance per unit length along the segment $[z, z + \delta z]$.

From Equation (5.11), evaluating term by term yields:

$$\int_{S_+(z)} \boldsymbol{J}_{\perp} \cdot \boldsymbol{n}_+ \, \mathrm{d}S = -I_+(z)$$

$$\int_{S_+(z+\delta z)} \boldsymbol{J}_{\perp} \cdot \boldsymbol{n}_+ \, \mathrm{d}S = I_+(z + \delta z)$$

$$\int_{C_+} \boldsymbol{J}_{\perp} \cdot \boldsymbol{n}_+ \mathrm{d}^2 x = \sigma \int_{C_+} \boldsymbol{E}_{\perp} \cdot \boldsymbol{n}_+ \mathrm{d}^2 x$$

However, noting that $\int_{C_+} \boldsymbol{J}_{\perp} \cdot \boldsymbol{n}_+ \mathrm{d}^2 x$ is the conduction current flowing across C_+, from Ohm's law, $I = GV$, where G is the conductance, it follows immediately that $\sigma \int_{C_+} \boldsymbol{E}_{\perp}^{+} \cdot \boldsymbol{n}_+ \mathrm{d}^2 x = G_{++}\delta z v_+$ with G_{++} denoting the conductance per unit length and δz is the length of the cylinder C_+. Likewise, $\sigma \int_{C_+} \boldsymbol{E}_{\perp}^{-} \cdot \boldsymbol{n}_+ \mathrm{d}^2 x = G_{+-}\delta z (v_+ - v_-)$. Hence,

$$\sigma \int_{C_+} \boldsymbol{E}_{\perp} \cdot \boldsymbol{n}_+ \, \mathrm{d}S = G_{+-}\delta z (v_+ - v_-) + G_{++}\delta z v_+ \tag{5.13a}$$

Next, via Figure 5.3, the displacement current across the segment C_+ is determined by $\varepsilon \partial_t \int_{C_+} \boldsymbol{E}_{\perp}^{+} \cdot \boldsymbol{n}_+ \mathrm{d}S$. Indeed, $Q = CV \Rightarrow i = \partial_t Q = C \partial_t V$. Hence, on setting C_{+-} to be the capacitance per unit length between the pair (C_+, C_-) and C_{++} the pair (C_+, C_0), it follows at once that

$$\varepsilon \partial_t \int_{C_+} \boldsymbol{E}_{\perp}^{+} \cdot \boldsymbol{n}_+ \, \mathrm{d}S = C_{+-}\delta z \, \partial_t (v_+ - v_-) + C_{++}\delta z \, \partial_t v_+ \tag{5.13b}$$

whence, applying Equations (5.13a) and (5.13b) to $0 = \mu \int_{\bar{C}_+} \boldsymbol{J}_{\perp} \cdot \boldsymbol{n} \mathrm{d}^2 x + \mu\varepsilon \tfrac{\mathrm{d}}{dt} \int_{\bar{C}_+} \boldsymbol{E}_{\perp} \cdot \boldsymbol{n} \mathrm{d}^2 x$ and taking the limit $\delta z \to 0$ yields

$$\partial_z I_+(z,t) = -(G_{+-} + G_{++}) v_+ + G_{+-} v_- - (C_{+-} + C_{++}) \partial_t v_+ + C_{+-} \partial_t v_- \tag{5.14}$$

By symmetry,

$$\partial_z I_-(z,t) = -(G_{-+} + G_{--})v_- + G_{-+}v_+ - (C_{-+} + C_{--})\partial_t v_- + C_{-+}\partial_t v_+ \quad (5.15)$$

Indeed, in the case of a balanced differential pair, where $v_+ = -v_-$, Equations (5.14) and (5.15) reduce to

$$\partial_t \begin{pmatrix} I_+ \\ I_- \end{pmatrix} = \begin{pmatrix} -(G_{++} + 2G_{+-}) & -(C_{++} + 2C_{+-}) \\ G_{--} + 2G_{+-} & C_{--} + 2C_{+-} \end{pmatrix} \begin{pmatrix} v_+ \\ \partial_t v_+ \end{pmatrix}$$

Now, consider a balanced differential pair (C_+, C_-) satisfying the respective conditions:

$$\text{(a)} \quad v_+ = -v_-$$

$$\text{(b)} \quad v_+ = v_-$$

where the incident field is assumed to be time harmonic; that is, $\mathbf{E}_\pm(z,t) = \mathbf{E}_\pm(z)e^{i\omega t}$ and the same symbol is used for convenience should no confusion arise. Recall that this means the replacement $\partial_t \to i\omega$ may be freely made.

5.1.9 Lemma

Given a symmetric differential pair $(C_\pm, C_0)$, where C_0 is the ground conductor*, $v_+(z,t) = v_-(z,t) \Rightarrow I_+(z,t) = I_-(z,t)\ \forall z, t \geq 0$.

Proof

For a symmetric pair, $L_{++} = L_{--}$ and $R_+ = R_-$. The result thus follows trivially from Equations (5.7) and (5.8). □

5.1.10 Remark

It is clear from the proof of Lemma 5.1.9 that as R_0 does not cancel out, or equivalently, as $I_0 = -2I_+ \neq 0$, R_0 cannot be set arbitrarily, as in the case for $v_+ = -v_-$.

For case (a): This case was examined above. Whence, taking the ratio $\frac{\partial_z v_+}{\partial_z I_+}$ yields

$$Z'^2 \equiv \frac{v_+}{I_+}\frac{\partial_z v_+}{\partial_z I_+} = \frac{R_+ + i\omega(L_{++} - L_{+-})}{G_{++} + 2G_{+-} + i\omega(C_{++} + 2C_{+-})}$$

For case (b): From Remark 5.1.10, taking the ratio $\frac{\partial_z v_+}{\partial_z I_+}$ yields

$$Z''^2 \equiv \frac{v_+}{I_+}\frac{\partial_z v_+}{\partial_z I_+} = \frac{R_+ + 2R_0 + i\omega(L_{++} + L_{+-})}{G_{++} + 2G_{+-} + i\omega C_{++}}$$

These two cases lead to the following respective definitions.

* In the case of a printed circuit board, this represents the ground plane.

5.1.11 Definition

Given a symmetric differential pair $(C_{\pm}, C_0)$, the *odd mode* is defined by the condition $v_+(z,t) = -v_-(z,t)\ \forall z, t \geq 0$, and the *odd mode impedance* Z_{odd} is defined by

$$Z_{odd} = \sqrt{\frac{R_+ + i\omega(L_{++} - L_{+-})}{G_{++} + 2G_{+-} + i\omega(C_{++} + 2C_{+-})}}$$

In particular, for a lossless system,

$$Z_{odd} = \sqrt{\frac{L_{++} - L_{+-}}{C_{++} + 2C_{+-}}}$$

5.1.12 Definition

Given a symmetric differential pair $(C_{\pm}, C_0)$, the *even mode* is defined by the condition $v_+(z,t) = v_-(z,t)\ \forall z, t \geq 0$, and the *even mode impedance* Z_{even} is defined by

$$Z_{even} = \sqrt{\frac{R_+ + 2R_0 + i\omega(L_{++} + L_{+-})}{G_{++} + 2G_{+-} + i\omega C_{++}}}$$

In particular, for a lossless system,

$$Z_{even} = \sqrt{\frac{L_{++} + L_{+-}}{C_{++}}}$$

5.1.13 Proposition

Given a symmetric differential pair $(C_{\pm}, C_0)$, under the condition of even mode propagation, there is no capacitive coupling between (C_+, C_-).

Proof

From Equation (5.14), on setting $v_+ = v_-$, the two terms involving C_{+-} cancel out.

□

Lastly, given a symmetric differential pair $(C_{\pm}, C_0)$, the impedance seen by the wave propagating between the pair (C_+, C_-) for $v_+ = -v_-$ is called the *differential impedance*, where C_- is considered as a reference. As the potential difference $v_{diff} = v_+ - v_- = 2v_+$; whereas the current along C_+ is $I_+ = -I_-$, it follows at once that the differential impedance is defined by $Z_{diff} = \frac{v_{diff}}{I_+} = \frac{2v_+}{I_+} = 2Z_{odd}$. Explicitly,

$$Z_{diff} = 2\sqrt{\frac{R_+ + i\omega(L_{++} - L_{+-})}{G_{++} + 2G_{+-} + i\omega(C_{++} + 2C_{+-})}} \tag{5.16}$$

and for a lossless system,

$$Z_{\text{diff}} = 2\sqrt{\frac{L_{++}-L_{+-}}{C_{++}+2C_{+-}}}$$

By symmetry, the impedance seen by a wave propagating between the pair (C_+,C_-) for $v_+ = v_-$, along a common ground Γ_0 is called the *common mode impedance*. Here, the pair (C_+,C_-) is treated as a single conductor and the analysis reduces to the pair $(\tilde{C},C_0)$, where $\tilde{C} = (C_+,C_-)$. Although it is possible to go through the entire field argument carried out above, it is easier to observe that (a) the common mode potential is $v_{cm} = v_+$, and (b) the common mode current $I_{cm} = I_+ + I_- = 2I_+$, whence, the common mode impedance is

$$Z_{cm} = \frac{v_{cm}}{I_{cm}} = \frac{v_+}{2I_+} = \frac{1}{2}Z_{\text{even}}$$

That is,

$$Z_{cm} = \frac{1}{2}\sqrt{\frac{R_+ +2R_0+i\omega(L_{++}+L_{+-})}{G_{++}+2G_{+-}+i\omega C_{++}}} \tag{5.17}$$

and for a lossless system,

$$Z_{cm} = \frac{1}{2}\sqrt{\frac{L_{++}+L_{+-}}{C_{++}}}$$

The derivation of differential impedance and common mode impedance via distributed lumped model is left as an exercise for the attentive reader; see Exercises 5.4.1 and 5.4.2.

To complete the analysis of odd and even modes along a symmetric differential pair, consider qualitatively, the fields generated by the respective modes. First, consider the odd mode wave propagation. Here, recall that the propagating current and hence voltage field along C_+ are 180 degrees out of phase with respect to the current and voltage fields propagating along C_-. By the right-hand rule, the magnetic field density is as illustrated, where the odd mode current along C_+ is directed into the page, and the current along C_- is directed off the page. Suppose $\gamma_\pm = (x_\pm,0)$, for some $x_- < 0 < x_+$ with the z-component suppressed for simplicity. Then, along $J_0 = \{(x,0): x \in (x_- + a, x_+ - a)\}$, where $a > 0$ is the radius of the cable, the magnetic field density crowds along J_0, whereas the magnetic field density is much weaker on $\mathbf{R}^2 - J_0$, that is, away from J_0. Physically, because the currents are oppositely directed—that is, $I_+ = -I_-$—by Definition 5.1.11, the odd mode inductance is reduced by the mutual inductance and hence the magnetic flux, for a fixed current, must trivially decrease correspondingly via $\Psi_B = LI$.

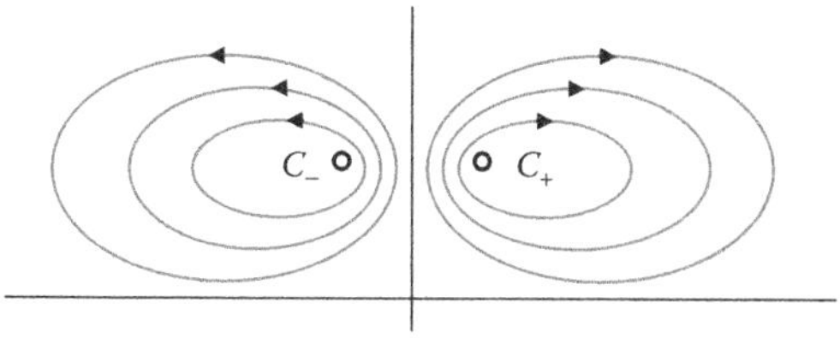

Odd mode current propagating along a differential pair.

By contrast, observe that the electric field has the profile shown in the figure, where odd mode voltage, $V_- = -V_+$, via Definition 5.1.11, leads to the enhancement of the resultant capacitance via the mutual capacitance. Physically, the surface charge density crowds around the boundary $\partial C_\pm$ of the lines about a small neighborhood of J_0, and away from J_0, the charge density on $\partial C_\pm$ decreases (see the electric field lines in the figure), whence, for a fixed potential difference between $C_\pm$, $Q = CV$ implies that the resultant surface charge density on $\partial C_\pm$ is enhanced by the increase by the mutual capacitance.

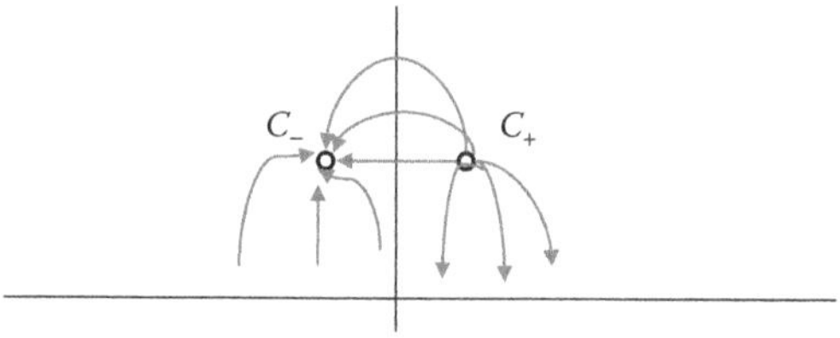

Odd mode voltage propagating along a differential pair.

Finally, to complete the picture, it is clear from Definition 5.1.12, that the scenario for even mode propagation is different from odd mode propagation. First of all, in the case of the even mode current propagation, as $I_- = I_+$, it is clear via the right-hand rule that the magnetic field density on J_0 cancels, leading to the diagram. In a sense, the magnetic field is enhanced for even mode propagation. Physically, via Definition 5.1.12, the even mode inductance is increased by the mutual inductance. Hence, the magnetic flux, for a fixed current, must clearly increase from $\Psi_B = LI$.

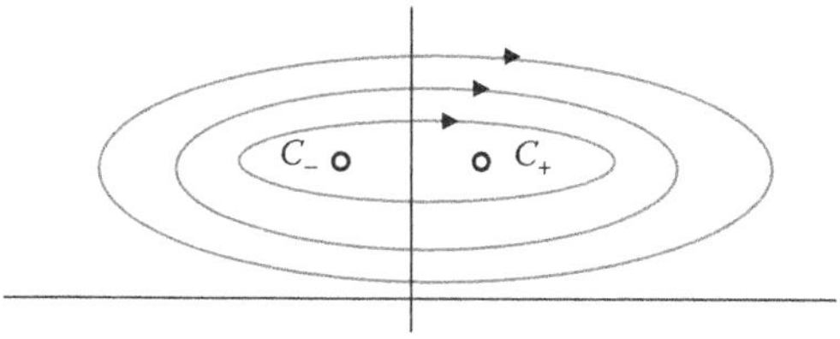

Even mode current propagating along a differential pair.

The scenario with the voltage profile for even mode propagation can similarly be sketched. Thus, via Definition 5.1.12 and $Q = CV$, it is clear that the even mode capacitance is decreased by the mutual capacitance and hence, the resultant electric field decreases correspondingly (see the odd mode propagation electric field). The electric flux is low on J_0 due to mutual repulsion of the electric field, as illustrated in the diagram (see the mutual repulsion of the magnetic flux density for the odd mode propagation depicted above).

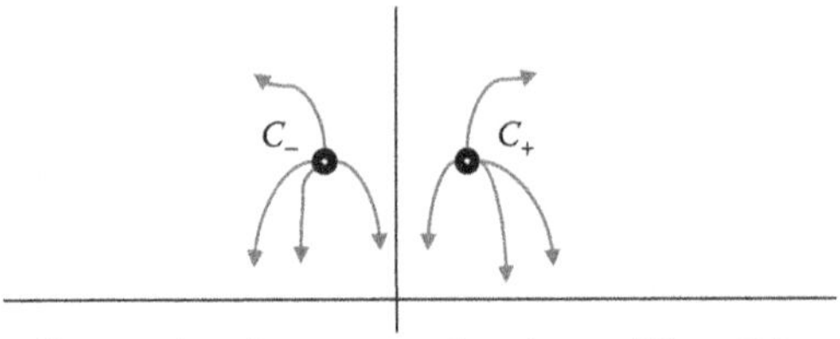

Even mode voltage propagating along a differential.

5.1.14 Remark

Given a lossless differential pair $(C_\pm, C_0, \mu, \varepsilon)$ defined by the following distributed parameter model, where the propagation waves are assumed without loss of generality to be time harmonic:

$$\partial_z\begin{pmatrix} v_+ \\ v_- \end{pmatrix} = -i\omega\begin{pmatrix} L_{++} & L_{+-} \\ L_{-+} & L_{--} \end{pmatrix}\begin{pmatrix} i_+ \\ i_- \end{pmatrix}$$

$$\partial_z\begin{pmatrix} i_+ \\ i_- \end{pmatrix} = -i\omega\begin{pmatrix} C_{++} & -C_{+-} \\ -C_{-+} & C_{--} \end{pmatrix}\begin{pmatrix} v_+ \\ v_- \end{pmatrix}$$

Set $L_{+-} = \kappa'\sqrt{L_{++}L_{--}}$ and $C_{+-} = \kappa''\sqrt{C_{++}C_{--}}$, for some coupling constants $\kappa', \kappa'' \in [0,1]$. Then, $\sqrt{\frac{L_{++}}{C_{++}}} = \sqrt{Z_+Z_-}$ and $\kappa' = \kappa''$, where $Z_\pm$ is the characteristic impedance of $C_\pm$ with respect to C_0. To see this, it suffices to observe that

$$\sqrt{\frac{L_{++}}{C_{++}}} = \sqrt{\frac{\mu}{\varepsilon}} = \sqrt{\frac{L_{--}}{C_{--}}}$$

Next, by definition, $\mathbf{LC} = \mu\varepsilon\mathbf{I} \Rightarrow L_{++}C_{+-} - L_{+-}C_{--} = 0$ and $L_{+-}C_{++} - L_{--}C_{+-} = 0$, where

$$\mathbf{I} = \begin{pmatrix} 1 & 0 \\ 0 & 1 \end{pmatrix}$$

whence

$$\left(\frac{L_{+-}}{C_{+-}}\right)^2 = \frac{L_{++}}{C_{++}}\frac{L_{--}}{C_{--}} \Rightarrow \frac{L_{+-}}{C_{+-}} = \sqrt{Z_+Z_-} = \frac{\mu}{\varepsilon}$$

Moreover,

$$\frac{L_{+-}}{C_{+-}} = \frac{\kappa'}{\kappa''}\frac{\mu}{\varepsilon} \Rightarrow \kappa' = \kappa''$$

as claimed. □

5.2 Impedance Matching Along a Differential Pair

In the previous section, the odd and even mode impedance along a differential pair were defined. In this section, matching the impedance of a differential pair is crucial when designing a matched differential line to prevent unwanted reflections. In particular, as $Z_{\text{odd}} \neq Z_{\text{even}}$, it follows that informally, different prescriptions may be required to prevent reflection from odd/even mode propagation. This is clearly important when designing for signal integrity and electromagnetic interference.

Indeed, in view of Remark 5.1.4, the coupling constant κ allows the pair (C_+, C_-) to be analyzed as a single transmission pair wherein C_- is taken to be the reference ground, and the resultant current flowing along C_+ is $I_+ + \kappa I_- = (1-\kappa)I_+$. Intuitively, the cross-coupling is the result of mutual inductance and mutual capacitance between the differential pair; this should be clear from Chapter 4. This is indeed obvious by expressing $\hat{Z}$ in terms of Z_0 via the coupling constant: $\hat{Z} = \kappa Z_0$.

The transfer impedance $\hat{Z}$ is a quantitative characterization of the induced voltage appearing on C_+ as a result of current flowing in C_-. For instance, switching occurring on an adjacent trace, called the *aggressor*, can induce a voltage (noise) on the trace lying next to it; this trace is called a *victim* trace.

5.2.1 Lemma

Given a symmetric differential pair (C_+, C_-), the differential coupling coefficient κ satisfies $0 \le \kappa \le 1$.

Proof

The pair of Equations (5.1) and (5.2) can be written in matrix form as $\boldsymbol{V} = \boldsymbol{ZI}$. Explicitly,

$$\begin{pmatrix} V_+ \\ V_- \end{pmatrix} = \begin{pmatrix} Z_0 & \hat{Z} \\ \hat{Z} & Z_0 \end{pmatrix} \begin{pmatrix} I_+ \\ I_- \end{pmatrix}$$

The time-average power delivered to the load is $\langle P\rangle = \frac{1}{2}\mathrm{Re}(\boldsymbol{V}^* \cdot \boldsymbol{I})$, whence from

$$\begin{pmatrix} V_+ \\ V_- \end{pmatrix} = \begin{pmatrix} Z_0 & \hat{Z} \\ \hat{Z} & Z_0 \end{pmatrix}\begin{pmatrix} I_+ \\ I_- \end{pmatrix} = \begin{pmatrix} Z_0 I_+ + \hat{Z} I_- \\ \hat{Z} I_+ + Z_0 I_- \end{pmatrix}$$

$$\boldsymbol{V}^* \cdot \boldsymbol{I} = \begin{pmatrix} Z_0 I_+^* + \hat{Z} I_-^* \\ \hat{Z} I_+^* + Z_0 I_-^* \end{pmatrix}^t \begin{pmatrix} I_+ \\ I_- \end{pmatrix} = (Z_0 I_+^* + \hat{Z} I_-^*) I_+ + (\hat{Z} I_+^* + Z_0 I_-^*) I_- = 2Z_0 |I_+|^2 (1-\kappa)$$

Thus, $\langle P\rangle = Z_0 |I_+|^2 (1-\kappa)$. However, $\langle P\rangle \geq 0 \Rightarrow \kappa \leq 1$. Clearly, $\kappa \geq 0$ because the power delivered to the load cannot be greater than that of the power generated by the source. Hence, $0 \leq \kappa \leq 1$, as claimed. □

From the proof of Lemma 5.2.1, define an alternative expression for the time-average power in terms of forward and backward propagating waves (*cf.* Exercise 5.4.4).

The odd mode impedance is the effective characteristic impedance of C_+ (respectively, C_-) of a differential pair (C_+, C_-) if C_- (respectively, C_+) were treated as the reference ground. In particular, because the solution to Maxwell's equations is linear, a differential pair (C_+, C_-) can be separately analyzed by investigating C_+ and C_- separately and then summing the two solutions. The summation of the two solutions is again a solution because of linearity (i.e., the principle of superposition).

Now, recalling that given a symmetric differential pair (C_+, C_-), the characteristic differential impedance $Z_{\mathrm{diff}} = Z_{\mathrm{odd}} + Z_{\mathrm{odd}} = 2(1-\kappa)Z_0$, it follows that the differential lines of Figure 5.1 are matched if $R = 2(1-\kappa)Z_0$. Suppose a common mode noise of voltage δV is (simultaneously) propagating along the pair (C_+, C_-) depicted in Figure 5.1. That is, $V_\pm \to V'_\pm = V_\pm + \delta V$. Then, the resultant output signal at the receiver is $V_{\mathrm{out}} = V'_+ - V'_- = V_+ - V_-$. Consequently, a symmetrical differential pair is immune to common mode noise.

5.2.2 Remark

For completeness, consider Equations (5.1) and (5.2) again. Given a symmetric differential pair (C_+, C_-), suppose that the currents (I_+, I_-) along the lines satisfy $I_+(t) = I_-(t)\ \forall t \in [0,1]$. Then, the signal (I_+, I_-) is called *common mode*. It is trivial to establish that $V_+(z) = V_-(z)\ \forall z \in [0, \ell]$ for common mode via (5.1) and (5.2).

5.2.3 Lemma

Given a symmetric differential pair, the even mode impedance is given by $Z_{\mathrm{even}} = Z_0 + \hat{Z}$.

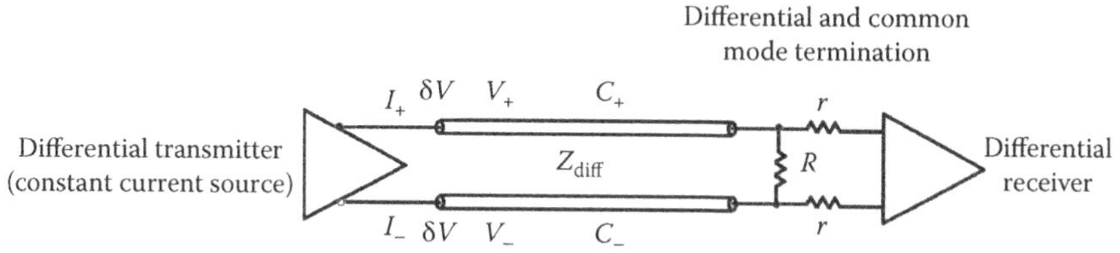

FIGURE 5.4
Differential termination scheme with common mode noise.

Proof

By (5.1), $V_+ = I_+Z_{++} + I_-Z_{+-} = I_+(Z_0 + \hat{Z}) \Rightarrow Z_0 + \hat{Z} = \frac{V_+}{I_+} = Z_{even}$, as required. □

5.2.4 Proposition

Suppose a common mode noise δV couples onto a symmetric differential pair (C_+, C_-). In order to prevent δV from reflecting back to the source along (C_+, C_-), a proper termination scheme for Figure 5.1 is shown in Figure 5.4. The *π-termination* scheme comprises a series resistor placed very close to the parallel differential termination resistor along each line. The proper values for (R,r) are:

$$r = Z_{even} \tag{5.18}$$

$$R = \frac{Z_{diff} Z_{even}}{Z_{even} - Z_{odd}} \tag{5.19}$$

Proof

First, observe by definition that (C_+, C_-) is at equipotential with respect to δV. In particular, there is zero (noise) current flowing through R as it is at the same potential as the pair (C_+, C_-). Thus, the proper termination for the noise is $r = Z_{even}$ by Remark 4.2.3.

To complete the proof, recall that the impedance seen by odd mode propagation is Z_{odd}. Hence, in order to match the impedance, $Z_{odd} = \left\{\frac{2}{R} + \frac{1}{r}\right\}^{-1} \Rightarrow 2Z_{odd}r = R(r - Z_{odd})$. Rearranging yields $R = \frac{Z_{diff} Z_{even}}{Z_{even} - Z_{odd}}$, where $Z_{diff} = 2Z_{odd}$ via (5.16) and $r = Z_{even}$ from (a) evaluated above. □

5.2.5 Remark

Observe that the termination scheme shown in Figure 5.4 for Proposition 5.2.4 depends on the placement of R. Explicitly, the results of Proposition 5.2.4 do not hold for the layout shown in Figure 5.5.

To see this, note the following. First, by Remark 4.2.3, $r = Z_{even}$. Second, by Definition 4.2.2, $r + \frac{1}{2}R = Z_{odd}$. And because $Z_{even} = Z_0 + \hat{Z} > Z_0 - \hat{Z} = Z_{odd}$, it thus follows that $r + \frac{1}{2}R = Z_{even} + \frac{1}{2}R > Z_{odd}$ $\forall R \geq 0$. Hence, the placement of

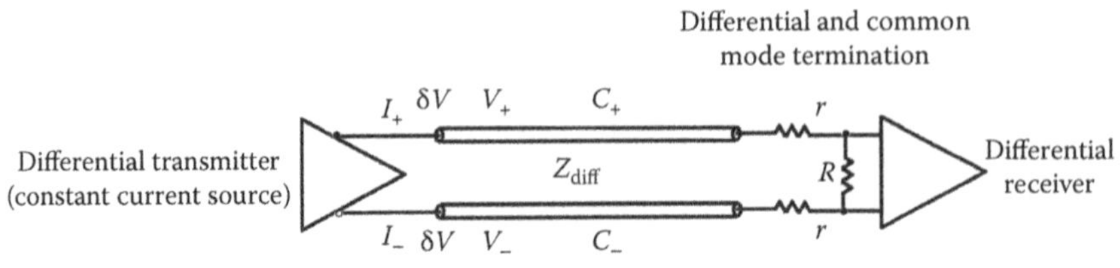

FIGURE 5.5
Improper termination scheme for even and odd mode signals.

(R,r) is critical in the proper termination of a differential pair (C_+,C_-). In particular, the termination scheme of Proposition 5.2.4 will correctly terminate both odd and even mode propagations, preventing unwanted reflections back along the differential pair. Clearly (R,r) must be placed as close to the differential receiver as possible to mitigate false triggers arising from reflections between (R,r) and the load (*cf.* Remark 4.4.14).

5.2.6 Remark

In the technology industry, the scheme shown in Figure 5.6 is often used for differential matching: termination via capacitor to ground as depicted in the schematic diagram, where the presence of the capacitor is to filter out high frequency noise. In Exercise 5.4.3, the reader is ask to comment on the validity of the scheme. In particular, is the capacitor necessary to filter high-frequency noise for emissions suppression? If so, under what conditions?

5.3 Field Propagation Along a Differential Pair

For simplicity, consider an infinitely long differential pair over an infinite ground plane, as portrayed in Figure 5.7(a), where the transmission lines are of radius $a > 0$ and the centers of the transmission lines are a distance $h > a$ above the ground plane. Remember from Chapter 4 that a TEM wave propagating

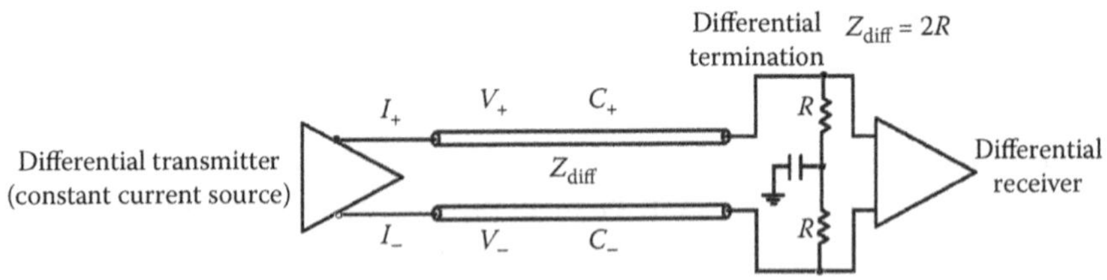

FIGURE 5.6
Differential termination with a bypassing capacitor.

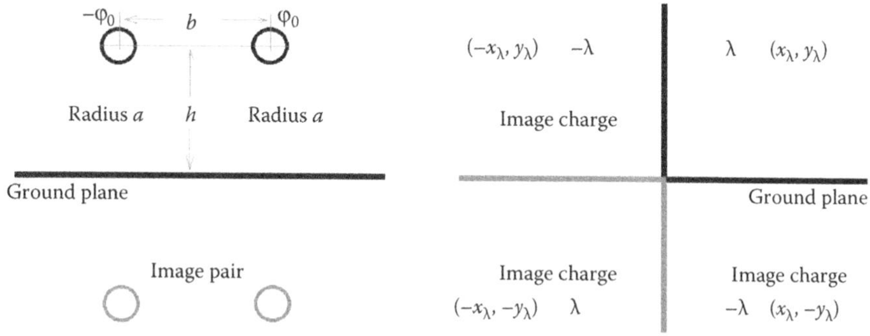

(a) Differential pair of fixed radii

(b) Equivalent line charge representation

FIGURE 5.7
Potential of a differential pair via the method of images.

along a transmission line satisfies the Laplace equation on the plane normal to the direction of propagation. In addition, recall from Chapter 3 that a charged, infinitely long cylinder over an infinite ground plane is equivalent to a suitably placed infinite line charge. This suggests that the scalar potential can be obtained via Figure 5.7(b), with the boundary of the differential lines suitably placed. That is, via the method of images, Figure 5.7(a) is equivalent to Figure 5.7(b).

For simplicity, let λ denote the line charge as depicted in Figure 5.7(b). For notational convenience, set $\boldsymbol{r}_\lambda^{(1)} = (x_\lambda, y_\lambda)$, $\boldsymbol{r}_\lambda^{(2)} = (-x_\lambda, y_\lambda)$, $\boldsymbol{r}_\lambda^{(3)} = (-x_\lambda, -y_\lambda)$, and $\boldsymbol{r}_\lambda^{(4)} = (x_\lambda, -y_\lambda)$, and also set $\boldsymbol{r}^{(i)} = \boldsymbol{r} - \boldsymbol{r}_\lambda^{(i)}\ \forall i$, where $\boldsymbol{r} = (x, y)$. Then, the potential at an arbitrary point $\boldsymbol{r} \neq \boldsymbol{r}_\lambda^{(i)}\ \forall i$ is given by

$$\varphi(x,y) = -\frac{\lambda}{2\pi\varepsilon}\left\{\ln\frac{1}{r^{(1)}} - \ln\frac{1}{r^{(2)}} + \ln\frac{1}{r^{(3)}} - \ln\frac{1}{r^{(4)}}\right\} = \frac{\lambda}{2\pi\varepsilon}\left\{\ln\frac{r^{(1)}}{r^{(2)}} + \ln\frac{r^{(3)}}{r^{(4)}}\right\} \tag{5.20}$$

Referring to Figure 5.7(b), an equipotential loop about $\boldsymbol{r}_\lambda^{(1)}$ satisfies

$$K = \frac{r^{(1)}}{r^{(2)}}\frac{r^{(3)}}{r^{(4)}}$$

for some constant $K > 0$. Without loss of generality, set $K = K_1 K_2$, where $K_1 = \frac{r^{(1)}}{r^{(2)}}$, $K_2 = \frac{r^{(3)}}{r^{(4)}}$. Then, evaluating $K_1 = \frac{r^{(1)}}{r^{(2)}}$ and $K_2 = \frac{r^{(3)}}{r^{(4)}}$ lead to

$$\left(x - \frac{K_1+1}{K_1-1}x_\lambda\right)^2 + (y - y_\lambda)^2 = \left(\frac{2\sqrt{K_1}}{K_1-1}x_\lambda\right)^2 \tag{5.21}$$

$$\left(x + \frac{K_2+1}{K_2-1}x_\lambda\right)^2 + (y + y_\lambda)^2 = \left(\frac{2\sqrt{K_2}}{K_2-1}x_\lambda\right)^2 \tag{5.22}$$

Equations (5.21) and (5.22) describe the equipotential loops about $r_\lambda^{(1)}$ and $r_\lambda^{(3)}$, respectively. By inspection, the loops are circles with their respective centers and radii given by

$$\left\{\left(\tfrac{K_1+1}{K_1-1}x_\lambda, y_\lambda\right), \tfrac{2\sqrt{K_1}}{K_1-1}x_\lambda\right\} \text{ and } \left\{\left(-\tfrac{K_2+1}{K_2-1}x_\lambda, -y_\lambda\right), \tfrac{2\sqrt{K_2}}{K_2-1}x_\lambda\right\}$$

By the method of images, the images must have the same radii as the original loop about $r_\lambda^{(1)}$. Hence, $K_1 = \sqrt{K} = K_2$, and in particular,

$$x_0 = \tfrac{K_1+1}{K_1-1}x_\lambda \text{ and } a = \tfrac{2\sqrt{K}}{K-1}x_\lambda \Rightarrow 0 = K - 1 - \tfrac{2x_0}{a}\sqrt{K} \Rightarrow \sqrt{K} = \frac{x_0+\sqrt{x_0^2-a^2}}{a}$$

Also, $x_0^2 = \left(\tfrac{K_1+1}{K_1-1}\right)^2 x_\lambda^2$ and $a^2 = \tfrac{4K}{(K-1)^2}x_\lambda^2 \Rightarrow x_0^2 - a^2 = x_\lambda^2\left\{\left(\tfrac{K+1}{K-1}\right)^2 - \tfrac{4K}{(K-1)^2}\right\}$

However, it is clear that

$$\left(\tfrac{K+1}{K-1}\right)^2 - \tfrac{4K}{(K-1)^2} = \tfrac{K^2+1-2K}{(K-1)^2} = 1$$

Hence, $x_\lambda = \sqrt{x_0^2 - a^2}$. Thus, given the center of the cable $(x_0, y_0) \equiv (x_0, y_\lambda)$, the location of the equivalent line charge (x_λ, y_λ) can be determined. Indeed, suppose the transmission line is at some fixed potential φ_0. Then,

$$\varphi_0 = \tfrac{\lambda}{2\pi\varepsilon}\ln\sqrt{K} \Rightarrow \lambda = 2\pi\varepsilon\varphi_0\left\{\ln\frac{x_0+\sqrt{x_0^2-a^2}}{a}\right\}^{-1}$$

yielding the line charge as a function of potential to which the transmission line is charged. The above analysis is summarized in the following lemma.

5.3.1 Lemma

Given a lossless, infinitely long differential pair $(C_\pm, a, b, h)$, where $a > 0$ is the radii of the transmission lines, $b > 0$ the distance of separation between the centers of (C_+, C_-), and $h > 0$ the distance between the respective centers of $C_\pm$ and the ground plane [see Figure 5.7(a)] the differential pair and their images are given, respectively, by $C_\pm = \bar{B}(r_0^\pm, a) \times \mathbf{R}$ and $C'_\pm = \bar{B}(r_0'^\pm, a) \times \mathbf{R}$, where

$$\bar{B}(r_0^\pm, a) = \{(x,y) \in \mathbf{R}^2 : (x \mp x_0)^2 + (y - y_0)^2 \le a^2\}$$

$$\bar{B}'(r_0'^\pm, a) = \{(x,y) \in \mathbf{R}^2 : (x \mp x_0)^2 + (y + y_0)^2 \le a^2\}$$

are represented by infinite line charges $(\gamma_\pm, \pm\lambda)$ and their images $(\gamma'_\pm, \pm\lambda)$ defined by

$$\gamma_\pm = \{(\pm x_\lambda, y_\lambda, z) : -\infty < z < \infty\} \quad \text{and} \quad \gamma'_\pm = \{(\pm x_\lambda, -y_\lambda, z) : -\infty < z < \infty\}$$

with $x_\lambda = \sqrt{x_0^2 - a^2}$ and $y_\lambda = y_0$ □

5.3.2 Proposition

Given a pair of infinitely long, lossless, differential transmission lines $C_\pm$ over an infinite ground plane, suppose that $C_\pm$ are charged, respectively, to $V(\omega) = \pm\varphi_0(\omega)$. Set $\Omega = \mathbf{R}^3 - (C_+ \cup C_-)$, where $C_\pm = \bar{B}(r_0^\pm, a) \times \mathbf{R}$ are the differential lines. Then, the potential at $\boldsymbol{r} = (x, y, z) \in \Omega$ is given by

$$\varphi(x,y) = -\frac{\varphi_0}{\ln\left\{x_0 + \sqrt{x_0^2 - a^2}\right\} - \ln a} \ln \sqrt{\frac{\{(x+x_\lambda)^2 + (y-y_\lambda)^2\}\{(x-x_\lambda)^2 + (y+y_\lambda)^2\}}{\{(x-x_\lambda)^2 + (y-y_\lambda)^2\}\{(x+x_\lambda)^2 + (y+y_\lambda)^2\}}}$$

In particular, for

$$\frac{b+h}{r} << \frac{1}{\sqrt{2}}, \quad |\varphi(x,y)| < \left| \frac{\varphi_0}{\ln\left\{x_0 + \sqrt{x_0^2 - a^2}\right\} - \ln a} \right| \delta^2$$

where $\delta = 2\sqrt{2}\,\frac{r_\lambda}{r}$.

Proof

The first assertion follows trivially from Equation (5.20). Thus, it remains to establish the last assertion. Now,

$$\frac{\left\{(x+x_\lambda)^2 + (y-y_\lambda)^2\right\}\left\{(x-x_\lambda)^2 + (y+y_\lambda)^2\right\}}{\left\{(x-x_\lambda)^2 + (y-y_\lambda)^2\right\}\left\{(x+x_\lambda)^2 + (y+y_\lambda)^2\right\}} \approx \frac{1 + 2(x_\lambda x - y_\lambda y)/r^2}{1 - 2(x_\lambda x + y_\lambda y)/r^2} \frac{1 - 2(x_\lambda x - y_\lambda y)/r^2}{1 + 2(x_\lambda x + y_\lambda y)/r^2} = \frac{1 - 4(x_\lambda x - y_\lambda y)^2/r^4}{1 - 4(\boldsymbol{r} \cdot \boldsymbol{r}_\lambda)^2/r^4}$$

Moreover, noting that $r_\lambda = \frac{1}{2}\sqrt{b^2 + h^2} \le \frac{1}{2}(b+h) << \frac{1}{2\sqrt{2}} r < r$, it follows via the binomial approximation that

$$\frac{\left\{(x+x_\lambda)^2 + (y-y_\lambda)^2\right\}\left\{(x-x_\lambda)^2 + (y+y_\lambda)^2\right\}}{\left\{(x-x_\lambda)^2 + (y-y_\lambda)^2\right\}\left\{(x+x_\lambda)^2 + (y+y_\lambda)^2\right\}} \approx \frac{1 - 4(x_\lambda x - y_\lambda y)^2/r^4}{1 - 4(\boldsymbol{r} \cdot \boldsymbol{r}_\lambda)^2/r^4} \approx \left\{1 - \frac{4(x_\lambda x - y_\lambda y)^2}{r^4}\right\}\left\{1 + \frac{4(\boldsymbol{r} \cdot \boldsymbol{r}_\lambda)^2}{r^4}\right\}$$

$$= 1 + 4\frac{(\boldsymbol{r} \cdot \boldsymbol{r}_\lambda)^2 - (x_\lambda x - y_\lambda y)^2}{r^4}$$

Now, noting that $\ln(1+\delta) \approx \delta$ for $|\delta| << 1$, it follows at once that

$$\ln \sqrt{\frac{\left\{(x+x_\lambda)^2 + (y-y_\lambda)^2\right\}\left\{(x-x_\lambda)^2 + (y+y_\lambda)^2\right\}}{\left\{(x-x_\lambda)^2 + (y-y_\lambda)^2\right\}\left\{(x+x_\lambda)^2 + (y+y_\lambda)^2\right\}}} \approx 2\frac{(\boldsymbol{r} \cdot \boldsymbol{r}_\lambda)^2 - (x_\lambda x - y_\lambda y)^2}{r^4} = 8\frac{y_\lambda}{r}\frac{y}{r}\frac{x_\lambda}{r}\frac{x}{r} < 8\frac{y_\lambda}{r}\frac{x_\lambda}{r} < 8\left(\frac{r_\lambda}{r}\right)^2$$

Finally, set $\delta = 2\sqrt{2}\frac{r_\lambda}{r}$. Then, $\frac{r_\lambda}{r} < \frac{1}{2\sqrt{2}} \Rightarrow \delta << 1$, and

$$\ln\sqrt{\frac{\{(x+x_\lambda)^2+(y-y_\lambda)^2\}\{(x-x_\lambda)^2+(y+y_\lambda)^2\}}{\{(x-x_\lambda)^2+(y-y_\lambda)^2\}\{(x+x_\lambda)^2+(y+y_\lambda)^2\}}} < 8\frac{y_\lambda}{r}\frac{x_\lambda}{r} \le \delta^2$$

whence

$$|\varphi(x,y)| < \left|\frac{\varphi_0}{\ln\left\{x_0+\sqrt{x_0^2-a^2}\right\}-\ln a}\right|\delta^2$$

as asserted. □

5.3.3 Remark

The above proposition is the reason why EMC engineers often state that fields cancel out for differential lines; that is, away from a differential pair, the electromagnetic field cancels. Clearly, that is an oversimplification: the fields do not cancel out identically. However, at a sufficiently large distance away from the pair, the potential field falls off approximately as $\left(\frac{r_\lambda}{r}\right)^2$. Thus, a differential pair does mitigate emissions, although it is clear from the proof that emissions do not vanish, a common incorrect assumption made by EMC engineers. Indeed, even at an intuitive level, it is clear that fields do not identically cancel out: a differential pair can be looked upon as a dipole string (as opposed to point charge dipole). Thus, the fields do not cancel out, albeit they fall off very much faster than the field of a single line charge.

5.3.4 Corollary

Given a differential pair stated in Proposition 5.3.2, suppose $\|r - r_\lambda\| = \delta > a$. Then,

$$\left|\frac{2\varphi_0}{\ln\left\{x_0+\sqrt{x_0^2-a^2}\right\}-\ln a}\ln\frac{\delta^2+r_\lambda\cdot r}{\delta^2+4r_\lambda\cdot r}\right| < |\varphi(x,y)| < \left|\frac{2\varphi_0}{\ln\left\{x_0+\sqrt{x_0^2-a^2}\right\}-\ln a}\ln\left\{1+\frac{4r_\lambda\cdot r}{\delta^2}\right\}\right|$$

Proof

Now,

$$\frac{\{(x+x_\lambda)^2+(y-y_\lambda)^2\}\{(x-x_\lambda)^2+(y+y_\lambda)^2\}}{\{(x-x_\lambda)^2+(y-y_\lambda)^2\}\{(x+x_\lambda)^2+(y+y_\lambda)^2\}} = \frac{\delta^2+4x_\lambda x}{\delta^2}\frac{\delta^2+4y_\lambda y}{\delta^2+4r_\lambda\cdot r}$$

Thus, noting the following inequality

$$\frac{\delta^2+r_\lambda\cdot r}{\delta^2}\,\frac{\delta^2}{\delta^2+4r_\lambda\cdot r}<\frac{\delta^2+4x_\lambda x}{\delta^2}\,\frac{\delta^2+4y_\lambda y}{\delta^2+4r_\lambda\cdot r}<\frac{\delta^2+4r_\lambda\cdot r}{\delta^2}$$

yields the desired result at once. □

In particular, the above corollary provides a bound for the scalar potential near one of the differential lines.

5.3.5 Remark

It is clear from the above discussion that the potential generated by the differential pair at an arbitrary point $\boldsymbol{r}\in\mathbf{R}^3_+-(C_+\cup C_-)$ satisfies $V(\boldsymbol{r})=\varphi(x,y)e^{-i\beta z}$, and hence, the electric field is given by $\boldsymbol{E}(\boldsymbol{r}) = -\nabla V(\boldsymbol{r})$.

Now, observe that the capacitance per unit length C_{++} of C_+ with respect to the ground plane is found as follows.

By definition, the capacitance per unit length C of two infinitely long cables of uniform radius a, their axes separated by a distance d, is given by

$$C=\pi\varepsilon\left\{\ln\frac{d+\sqrt{d^2+4a^2}}{2a}\right\}^{-1}$$

whence it follows at once that

$$C_{++}=\frac{\lambda}{\varphi_0+\varphi_0}=\pi\varepsilon\left\{\ln\frac{h+\sqrt{h^2-a^2}}{a}\right\}^{-1}$$

Likewise, the mutual capacitance C_{+-} is, by definition,

$$C_{+-}=\pi\varepsilon\left\{\ln\frac{b+\sqrt{b^2-4a^2}}{2a}\right\}^{-1}$$

In particular, the total capacitance per unit length

$$C_+=C_{++}+C_{+-}=\pi\varepsilon\left\{\left(\ln\frac{b+\sqrt{b^2-4a^2}}{2a}\right)^{-1}+\left(\ln\frac{h+\sqrt{h^2-a^2}}{a}\right)^{-1}\right\}$$

For odd mode propagation, the mutual inductance can be found via Equation (5.16):

$$L_{+-}=L_{++}-\tfrac{1}{4}Z^2_{\text{diff}}(C_++C_{+-})=L_+-\tfrac{1}{4}Z^2_{\text{diff}}\pi\varepsilon\left\{\left(\ln\frac{b+\sqrt{b^2-4a^2}}{2a}\right)^{-1}+2\left(\ln\frac{h+\sqrt{h^2-a^2}}{a}\right)^{-1}\right.$$

Hence, any current noise pulse δi propagating along one of the differential pair will induce a voltage spike on the corresponding line via $\delta v=-L_{+-}\frac{d}{dt}\delta i$.

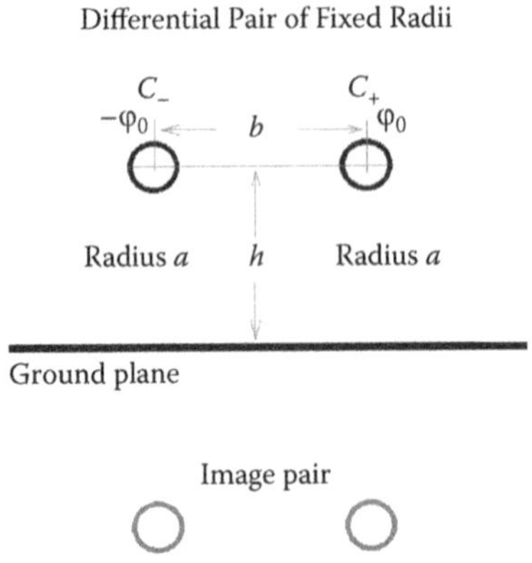

FIGURE 5.8
Potential of a differential pair via the method of images.

Finally, observe that for the (lossless, symmetric) differential pair illustrated in Figure 5.8, the "return" current via the ground plane is nonzero; this fact is occasionally falsely assumed to be zero by less careful EMC engineers.

5.3.6 Proposition

Given the lossless, symmetric differential pair illustrated in Figure 5.8, the ratio of the image current on the ground plane of $C_\pm$ with respect to the current propagating along $C_\pm$ satisfies

$$\frac{I_{+-}}{I_{++}} = 2\ln\frac{h+\sqrt{h^2-a^2}}{a}\left\{\ln\frac{b+\sqrt{b^2-4a^2}}{2a}\right\}^{-1} \tag{5.23}$$

where I_{+-} is the return current between (C_+, C_-) and I_{++} is the return current between C_+ and the ground plane.

Proof

From $Q = CV$, it is clear that

$$\frac{I_{+-}}{I_{++}} = \frac{dQ_{+-}}{dt}\left(\frac{dQ_{++}}{dt}\right)^{-1} = \frac{Q_{+-}}{Q_{++}}$$

From the preceding discussion,

$$C_{++} = \frac{\lambda}{\varphi_0+\varphi_0} = \pi\varepsilon\left\{\ln\frac{h+\sqrt{h^2-a^2}}{a}\right\}^{-1} \quad \text{and} \quad C_{+-} = \pi\varepsilon\left\{\ln\frac{b+\sqrt{b^2-4a^2}}{2a}\right\}^{-1}$$

whence, noting that $\delta V_{+-} = \varphi_+ + \varphi_+ = 2\varphi_+$ is the potential difference between (C_+, C_-), and $\delta V_{++} = \varphi_+$ is the potential difference between C_+ and the ground plane, yields the result at once. □

It is clear from Proposition 5.3.5 that in a typical scenario, such as for traces buried within a PCB—known as a *stripline* as opposed to a *microstrip* wherein the traces are on top of a PCB—the following usually holds: $\frac{1}{2}b >> h$. Hence, in this instance, $I_{+-} < I_{++}$ and most of the return current occurs along the ground plane.

5.4 Worked Problems

5.4.1 Exercise

Derive the odd and even mode impedance via the distributed parameter model for a lossless differential pair. Comment on the time delay for the odd mode versus even mode propagation along a differential pair.

Solution

Consider the following pair of transmission lines depicted in Figure 5.9. The capacitor coupling the two lines (i.e., mutual capacitance $C_{12} = C_{21}$) and the inductor coupling the two lines (i.e., mutual inductance $L_{12} = L_{21}$) are also illustrated once for simplicity.

Now, observe that the current (I_1, γ_1) induces, via L_{12}, a voltage on γ_2: $V_{21} = -L_{12}\frac{dI_1}{dt}$. By symmetry, (I_2, γ_2) induces a voltage on γ_1: $V_{12} = -L_{12}\frac{dI_2}{dt}$, where for simplicity, (γ_1, γ_2) represents the respective axes of a pair (C_1, C_2) of thin transmission lines. Similarly, the voltage (V_2, γ_2) induces a current, via the mutual capacitance C_{12}, on γ_1 as follows. The instantaneous charge per unit length induced on γ_1 is $\delta Q_1 = C_{12}(V_1 - V_2)$, whence, $Q_1 = C_{10}V_1 + \delta Q_1 = C_{11}V_1 - C_{12}V_2$. Likewise, $Q_2 = C_{20}V_2 + \delta Q_2 = C_{22}V_2 - C_{12}V_1$.

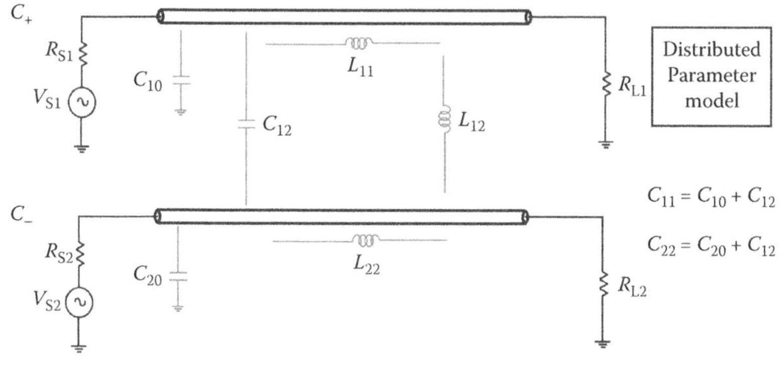

FIGURE 5.9
Mutual inductance and mutual capacitance coupling the two lines.

Thus, appealing to transmission line theory,

$$\partial_z V_1 = -L_{11}\,\partial_t I_1 - L_{12}\,\partial_t I_2,\ \ \partial_z I_1 = -C_{11}\,\partial_t V_1 - C_{12}\,\partial_t (V_1 - V_2)$$

$$\partial_z V_2 = -L_{12}\,\partial_t I_1 - L_{22}\,\partial_t I_2,\ \ \partial_z I_2 = -C_{12}\,\partial_t (V_2 - V_1) - C_{22}\,\partial_t V_2$$

That is, $\partial_z \boldsymbol{V} = -\boldsymbol{L}\partial_t \boldsymbol{I}$ and $\partial_z \boldsymbol{I} = -\mathbf{C}\partial_t \boldsymbol{V}$, where

$$\boldsymbol{V} = \begin{pmatrix} V_1 \\ V_2 \end{pmatrix}, \boldsymbol{I} = \begin{pmatrix} I_1 \\ I_2 \end{pmatrix}, \mathbf{C} = \begin{pmatrix} C_{11} + C_{12} & -C_{12} \\ -C_{12} & C_{22} + C_{12} \end{pmatrix} \text{ and } \mathbf{L} = \begin{pmatrix} L_{11} & L_{12} \\ L_{12} & L_{22} \end{pmatrix}$$

Now, for an odd mode (i.e., differential) current $I_1 = -I_2 \Rightarrow \partial_z V_k = -(L_{kk} - L_{12})\partial_t I_k$ for $k = 1,2$. For a symmetric differential pair, $L_1 = L_2 \Rightarrow L' = L_{kk} - L_{12}\ \forall k$. Hence, $\partial_z V_k = -L'\partial_t I_k$ for a differential signal. Thus, define $L_{odd} \equiv L'$ to be the *odd mode inductance* for a differential pair. Likewise, for odd mode current, $V_1 = -V_2 \Rightarrow \partial_z I_k = -(C_{kk} + 2C_{12})\partial_t V_k$, and for a symmetric differential pair, $C_1 = C_2 \Rightarrow C' = C_{kk} + 2C_{12}\ \forall k$. Thus, define $C_{odd} \equiv C'$ to be the *odd mode capacitance* for a differential pair.

In summary, for a differential pair, the transmission line equations for odd mode become

$$\begin{cases} \partial_z V_k = -L_{odd}\,\partial_t I_k \\ \partial_z I_k = -C_{odd}\,\partial_t V_k \end{cases}$$

for $k = 1,2$. In particular, by definition, the odd mode impedance is

$$Z_{odd} = \sqrt{\frac{L_{odd}}{C_{odd}}} = \sqrt{\frac{L_{kk} - L_{12}}{C_{kk} + 2C_{12}}}$$

for the transmission line γ_k, as required. As an aside, observe that $V_{diff} = V_1 - V_2 = 2V_1 \Rightarrow Z_{odd} = 2Z_{odd}$, consistent with the derivation of Section 5.1. For symmetric lines, $L_{kk} = L_0$ and $C_{kk} = C_0$ for all k = 1,2, where $L_0(C_0)$ is the inductance (capacitance) per unit length of a single line defining its characteristic impedance $R_0 = \sqrt{\frac{L_0}{C_0}}$.

To complete this problem, consider even mode propagation. Here, $I_1 = I_2 \Rightarrow \partial_z V_k = -(L_{kk} + L_{12})\partial_t I_k$, and $V_1 = V_2 \Rightarrow \partial_z I_k = -C_{kk}\,\partial_t V_k\ \forall k = 1,2$. That is, for even mode,

$$\begin{cases} \partial_z V_k = -L_{even}\,\partial_t I_k \equiv -(L_{kk} + L_{12})\partial_t I_k \\ \partial_z I_k = -C_{even}\,\partial_t V_k \equiv -C_{kk}\,\partial_t V_k \end{cases}$$

Hence, the even mode impedance is

$$Z_{\text{even}} = \sqrt{\frac{L_{\text{even}}}{C_{\text{even}}}} = \sqrt{\frac{L_{kk}+L_{12}}{C_{kk}}} \equiv \sqrt{\frac{L_0+L_{12}}{C_0}}$$

As a passing remark, it is clear from the above analysis that the time delay in odd mode versus even mode propagation is, in general, different:

$$\tau_{\text{odd}} = \ell\sqrt{L_{\text{odd}}C_{\text{odd}}} \quad \text{and} \quad \tau_{\text{even}} = \ell\sqrt{L_{\text{even}}C_{\text{even}}}$$

where ℓ is the length of the differential pair. Finally, observe from the above derivation that

$$L_{\text{odd}} < L_0 < L_{\text{even}} \quad \text{and} \quad C_{\text{even}} < C_0 < C_{\text{odd}}$$

5.4.2 Exercise

Provide an alternative derivation for the relationship between common mode and even mode impedance.

Solution

For notational simplicity, the transmission line pair (C_+, C_-) is identified with its respective axes (γ_+, γ_-). This convention is adopted in all subsequent exercises. First, recall from (4.1) that along γ_+, $V_+ = I_+Z_0 + I_-\hat{Z} = I_+Z_0 + I_+\kappa Z_0 = I_+(1+\kappa)Z_0$. Thus, the effective characteristic impedance for common mode waves propagating along γ_+ is $Z_+ = \frac{V_+}{I_-} = (1+\kappa)Z_0$. That is, the current I_+ flowing along γ_+ sees an effective impedance of $Z_{\text{even}} = Z_+$. Because $I_- = I_+$ along γ_-, it follows that the total current conducting along (γ_+, γ_-) is $I = I_+ + I_- = 2I_+$. Hence, the pair (γ_+, γ_-) appears to be a common mode signal as parallel lines connected to some virtual ground. That is, the resultant characteristic impedance of (γ_+, γ_-) is $\frac{1}{Z_{\text{cm}}} = \frac{1}{Z_{\text{eve}}} + \frac{1}{Z_{\text{even}}} \Rightarrow Z_{\text{cm}} = \frac{1}{2}Z_{\text{even}}$. The resultant characteristic impedance $Z_{\text{cm}} = \frac{1}{2}(1+\kappa)Z_0$ for a common mode signal is called the *common mode impedance* of (γ_+, γ_-). □

Exercise 5.4.3

Refer to the question posed in Remark 5.2.6, where the differential pair is assumed to be symmetric.

Solution

If a high-frequency noise is a common mode noise, then, by assumption, it will perfectly cancel out at the load, and hence, it is not necessary to add the capacitor to filter out the noise. On the other hand, if the noise only appears on either γ_+ or γ_-, then the only way the high-frequency noise can be by-passed without affecting the input impedance is to implement it as illustrated in Remark 5.2.6 for impedance-matching purposes. To see this, it suffices to note that by construction, the capacitor is at zero potential with respect to $\gamma_\pm$, and hence, its presence will not disturb the differential impedance match afforded by $2R$.

In practice, this implementation depends upon the real estate of the printed circuit board, and for very high frequency noise, the added surface mount and vias will potentially increase the inductance in the matching circuit, and may perturb the effectiveness of the impedance matching scheme. □

Exercise 5.4.4

Given a symmetric differential pair* (γ_+,γ_-) denote $\gamma_+ \cup \gamma_-$ to be the pair that incorporates mutual interaction between the pair of lines: $\gamma_+ \cup \gamma_- \equiv \gamma_+ \oplus \gamma_- + \gamma_+ \otimes \gamma_-$, where $\gamma_+ \oplus \gamma_-$ defines the pair (γ_+,γ_-) in the absence of mutual interaction; that is, γ_+,γ_- are over a common ground plane such that they are infinitely far away from each other, and $\gamma_+ \otimes \gamma_-$ denotes the mutual interaction between the pair defined by the mutual inductance and mutual capacitance. For simplicity, assume that the differential pair is lossless in all that follows, and in particular, for (b) and (c) below, assume that $Z_{S\pm} = Z_0 = Z_{L\pm}$.

(a) Show that the time-average power can be expressed as $\langle P\rangle = \frac{1}{2}(\boldsymbol{a}^\dagger\boldsymbol{a} - \boldsymbol{b}^\dagger\boldsymbol{b})$, where $\boldsymbol{a} = (a_+,a_-)$, $\boldsymbol{a}^\dagger = (a_+^*,a_-^*)$, $\boldsymbol{b} = (b_+,b_-)$ and $\boldsymbol{b}^\dagger = (b_+^*,b_-^*)$,

$$a_\pm = \frac{V_\pm + Z_0 I_\pm}{2\sqrt{Z_0}} \quad \text{and} \quad b_\pm = \frac{V_\pm - Z_0 I_\pm}{2\sqrt{Z_0}}$$

(b) Define the *far-end cross-talk* induced on γ_- by γ_+ by $\delta V_{-,+} = k_+ \frac{d}{dt} V_+(t-\tau)$, where k_+ is the far end cross-talk *coupling coefficient* [5,6]. By definition, $\delta V_{-,+}$ is the induced voltage near the source of γ_- (adjacent to the source of γ_+). Show that

$$k_+ = -\frac{1}{2}\tau\left(\frac{L_{+-}}{L_0} - \frac{C_{+-}}{C_0}\right)$$

where $\tau = \ell\sqrt{\mu\varepsilon}$.

(c) Define the *near-end cross-talk* induced on γ_- by γ_+ by $\delta V_{-,+} = k_-(V_+(t) - V_+(t-2\tau))$, where k_- is the near-end cross-talk *coupling coefficient*. By definition, $\delta V_{-,+}$ is the induced voltage near the source of γ_- (adjacent to the source of γ_+). Show that

$$k_- = \frac{1}{4}\left(\frac{L_{+-}}{L_0} + \frac{C_{+-}}{C_0}\right)$$

* Recall that transmission lines and their respective axes are identified for simplicity.

Solution

(a) By definition, $V_+ = V_1^+ e^{-i\beta z} + V_1^- e^{i\beta z}$ and $I_+ = \frac{V_1^+}{Z_0} e^{-i\beta z} - \frac{V_1^-}{Z_0} e^{i\beta z}$. So, set

$$a_+(z) = \frac{V_1^+}{\sqrt{Z_0}} e^{-i\beta z} \quad \text{and} \quad a_+(z) = \frac{V_1^+}{\sqrt{Z_0}} e^{-i\beta z}$$

Then, substituting back into the original equations yields

$$\begin{pmatrix} V_+ \\ I_+ \end{pmatrix} = \begin{pmatrix} \sqrt{Z_0} & \sqrt{Z_0} \\ \frac{1}{\sqrt{Z_0}} & -\frac{1}{\sqrt{Z_0}} \end{pmatrix} \begin{pmatrix} a_+ \\ b_+ \end{pmatrix}$$

and hence, solving for (a_+, b_+) via inversion yields

$$\begin{pmatrix} a_+ \\ b_+ \end{pmatrix} = \frac{1}{2} \begin{pmatrix} \frac{1}{\sqrt{Z_0}} & \sqrt{Z_0} \\ \frac{1}{\sqrt{Z_0}} & -\sqrt{Z_0} \end{pmatrix} \begin{pmatrix} V_+ \\ I_+ \end{pmatrix}$$

By symmetry, $V_- = V_2^+ e^{-i\beta z} + V_2^- e^{i\beta z}$ and $I_- = \frac{V_2^+}{Z_0} e^{-i\beta z} - \frac{V_2^-}{Z_0} e^{i\beta z}$, where $V_2^\pm = -V_1^\pm$ from the assumption of a symmetric differential pair, yield

$$\begin{pmatrix} a_- \\ b_- \end{pmatrix} = \frac{1}{2} \begin{pmatrix} \frac{1}{\sqrt{Z_0}} & \sqrt{Z_0} \\ \sqrt{Z_0} & -\sqrt{Z_0} \end{pmatrix} \begin{pmatrix} V_- \\ I_- \end{pmatrix}$$

That is,

$$a_\pm = \frac{V_\pm + Z_0 I_\pm}{2\sqrt{Z_0}} \quad \text{and} \quad b_\pm = \frac{V_\pm - Z_0 I_\pm}{2\sqrt{Z_0}}$$

Note that by construction, $\boldsymbol{a}, \boldsymbol{b}$ represents forward propagating and backward propagating waves, respectively.

Next, appealing to the definition of time-average power $\langle P \rangle = \frac{1}{2}\Re\mathrm{e}(\boldsymbol{V}^\dagger \boldsymbol{I})$ for the differential pair, it follows at once that $\langle P \rangle = \frac{1}{2}\Re\mathrm{e}(V_+^* I_+) + \frac{1}{2}\Re\mathrm{e}(V_-^* I_-)$, where $\boldsymbol{V} = (V_+, V_-)$ and $\boldsymbol{I} = (I_+, I_-)$, whence substituting $V_\pm = \sqrt{Z_0}(a_\pm + b_\pm)$ and $I_\pm = \frac{1}{\sqrt{Z_0}}(a_\pm - b_\pm)$ into $\boldsymbol{V}^\dagger \boldsymbol{I}$ yields

$$\begin{aligned} \boldsymbol{V}^\dagger \boldsymbol{I} = V_+^* I_+ + V_-^* I_- &= a_+^* a_+ + a_-^* a_- - b_+^* b_+ - b_-^* b_- + \\ &\quad - a_+^* b_+ + b_+^* a_+ - a_-^* b_- + b_-^* a_- \\ &= |a_+|^2 + |a_-|^2 - |b_+|^2 - |b_-|^2 + i2\left\{\Im\mathrm{m}(b_+^\dagger a_+) + \Im\mathrm{m}(b_-^\dagger a_-)\right\} \end{aligned}$$

and hence, $\langle P\rangle = \frac{1}{2}\Re e(\mathbf{V}^{+}\mathbf{I}) = \frac{1}{2}\left\{|a_{+}|^{2} + |a_{-}|^{2} - |b_{+}|^{2} - |b_{-}|^{2}\right\} = \frac{1}{2}(\mathbf{a}^{+}\mathbf{a} - \mathbf{b}^{+}\mathbf{b})$, where for any complex $z = x + iy \Rightarrow z - z^{*} = i2y$ was invoked. The time-average power can thus be expressed as the difference between forward and backward travelling waves.

(b) Referring to Example 4.4.7, the forward propagating wave in time-domain can be expressed as

$$V^{+}\left(t - \tfrac{z}{v}\right) = V_0\left(t - \tfrac{z}{v}\right) + \Gamma(0)\Gamma(\ell)V_0\left(t - 2\tau - \tfrac{z}{v}\right) + \cdots$$

with

$$V_0\left(t - \tfrac{z}{v}\right) = \tfrac{Z_0}{Z_0 + Z_S} V_S(t)e^{-\gamma z}$$

and the backward propagating wave in time-domain can be expressed as

$$V^{-}\left(t + \tfrac{z}{v}\right) = \Gamma(\ell)V_0\left(t - 2\tau + \tfrac{z}{v}\right) + \Gamma(0)\Gamma(\ell)^{2}V_0\left(t - 4\tau + \tfrac{z}{v}\right) + \cdots$$

where $v = \frac{1}{\sqrt{\mu\varepsilon}}$ is the wave propagation speed and $\tau = \frac{\ell}{v}$ is the propagation delay time, thus defining $V\left(t - \frac{z}{v}\right) = V^{+}\left(t - \frac{z}{v}\right) + V^{-}\left(t - \frac{z}{v}\right) \Rightarrow V(t) = V^{+}(t) + \Gamma(\ell)V^{+}(t - 2\tau)$, where $\tau = \frac{\ell}{v}$ is the propagation delay time. In particular, at the source $z = 0$,

$$\begin{aligned} V(t) &= V^{+}(t) + \Gamma(\ell)V^{+}(t - 2\tau) \\ &= V_0(t) + \Gamma(0)\Gamma(\ell)V_0(t - 2\tau) + \cdots + \Gamma(\ell)\{V_0(t - 2\tau) + \Gamma(0)\Gamma(\ell)V_0(t - 4\tau) + \cdots\} \\ &= V_0(t) + \left\{1 + \tfrac{1}{\Gamma(0)}\right\}\{\Gamma(0)\Gamma(\ell)V_0(t - 2\tau) + \Gamma^{2}(0)\Gamma^{2}(\ell)V_0(t - 4\tau) + \cdots\} \end{aligned}$$

and at the load $z = \ell$, noting that $V^{-}\left(t - \frac{\ell}{v}\right) = \Gamma(\ell)V^{+}\left(t + \frac{\ell}{v} - 2\tau\right) = \Gamma(\ell)V^{+}(t - \tau)$

$$\begin{aligned} V\left(t - \tfrac{\ell}{v}\right) &= V^{+}\left(t - \tfrac{\ell}{v}\right) + \Gamma(\ell)V^{+}\left(t - \tfrac{\ell}{v}\right) \\ &= (1 + \Gamma(\ell))\{V_0(t - \tau) + \Gamma(0)\Gamma(\ell)V_0(t - 3\tau) + \Gamma^{2}(0)\Gamma^{2}(\ell)V_0(t - 5\tau)\cdots\} \end{aligned}$$

From (a),

$$a_{+} = \tfrac{V_{+} + Z_0 I_{+}}{2\sqrt{Z_0}} \Rightarrow \partial_z a_{+} = \tfrac{1}{2\sqrt{Z_0}}\left\{\partial_z V_{+} + Z_0\,\partial_z I_{+}\right\}$$

For simplicity, suppose the waves are time harmonic. Then, from Remark 5.1.14,

$$\partial_z V_{+} = -i\omega\left\{L_{++}I_{+} + L_{+-}I_{-}\right\} \quad \text{and} \quad \partial_z I_{+} = -i\omega\left\{C_{++}V_{+} - C_{+-}V_{-}\right\}$$

Thus, substituting the above relations yields

$$\begin{aligned}\partial_z a_+ &= -\tfrac{i\omega}{2\sqrt{Z_0}}\{L_{++}I_+ + Z_0C_{++}V_+ + L_{+-}I_- - Z_0C_{+-}V_2\}\\ &= -i\omega Z_0C_{++}a_+ + i\omega\tfrac{Z_0C_{+-}V_- - L_{+-}I_-}{2\sqrt{Z_0}}\end{aligned}$$

Now, noting that $\frac{L_{+-}}{C_{+-}} = Z_{++}Z_{--} = Z_0^2$ by assumption, it follows immediately that

$$i\omega\tfrac{Z_0C_{+-}V_- - L_{+-}I_-}{2\sqrt{Z_0}} = i\omega Z_0C_{+-}\tfrac{V_- - \frac{L_{+-}}{Z_0C_{+-}}I_-}{2\sqrt{Z_0}} = i\omega Z_0C_{+-}\tfrac{V_- - Z_0I_-}{2\sqrt{Z_0}} = i\omega Z_0C_{+-}b_-$$

Hence, $\partial_z a_+ = -i\omega Z_0C_{++}a_+ + i\omega Z_0C_{+-}b_-$. The evaluation of $\partial_z a_-$ follows that of $\partial_z a_+$ *mutatis mutandis*. In particular, $\partial_z a_- = i\omega Z_0C_{+-}b_+ - i\omega Z_0C_{--}a_-$. Similarly, evaluating $\partial_z b_\pm$ yields $\partial_z b_+ = i\omega Z_0C_{++}b_+ - i\omega Z_0C_{+-}a_-$ and $\partial_z b_- = -i\omega Z_0C_{+-}a_+ + i\omega Z_0C_{--}b_-$, whence, observing trivially that $\omega Z_0C_{\pm\pm} = \omega\sqrt{L_{\pm\pm}C_{\pm\pm}} \equiv \frac{\omega}{\upsilon} \equiv \beta$, with $\upsilon = \frac{1}{\sqrt{L_{\pm\pm}C_{\pm\pm}}}$ being the speed of a wave propagating in a homogeneous medium—that is, for the case wherein $L_\pm = L_0$ and $C_\pm = C_0$—it follows clearly that

$$\partial_z\begin{pmatrix} a_+ \\ a_- \end{pmatrix} = -i\begin{pmatrix} \beta & 0 \\ 0 & \beta \end{pmatrix}\begin{pmatrix} a_+ \\ a_- \end{pmatrix} + i\begin{pmatrix} 0 & \omega Z_0C_{+-} \\ \omega Z_0C_{+-} & 0 \end{pmatrix}\begin{pmatrix} b_+ \\ b_- \end{pmatrix}$$

$$\partial_z\begin{pmatrix} b_+ \\ b_- \end{pmatrix} = -i\begin{pmatrix} 0 & \omega Z_0C_{+-} \\ \omega Z_0C_{+-} & 0 \end{pmatrix}\begin{pmatrix} a_+ \\ a_- \end{pmatrix} + i\begin{pmatrix} \beta & 0 \\ 0 & \beta \end{pmatrix}\begin{pmatrix} b_+ \\ b_- \end{pmatrix}$$

Next, note that $\omega Z_0C_{+-}b_- = \frac{1}{2\sqrt{Z_0}}\omega Z_0C_{+-}(V_- - Z_0I_-)$. Specifically, motivated by the fact that the mutual capacitive and mutual inductive coupling induce the following waves propagating towards the load:

$$V_{+-} = -i\omega L_{+-}I_+ + i\omega Z_0C_{+-}V_+ = -i\omega\left\{\tfrac{L_{+-}}{Z_0} - Z_0C_{+-}\right\}V_+ = i\omega Z_0C_{+-}(V_+ - Z_0I_-)$$

this suggests the decomposition:

$$\omega Z_0C_{+-}(V_- - Z_0I_-) = -\tfrac{1}{2}\omega\left\{\tfrac{L_{+-}}{Z_0} - Z_0C_{+-}\right\}(V_- + Z_0I_-) + \tfrac{1}{2}\omega\left\{\tfrac{L_{+-}}{Z_0} + Z_0C_{+-}\right\}(V_- - Z_0I_-)$$

That is,

$$-\omega Z_0C_{+-}b_- = \tfrac{1}{2}\omega\left\{\tfrac{L_{+-}}{Z_0} - Z_0C_{+-}\right\}a_- - \tfrac{1}{2}\omega\left\{\tfrac{L_{+-}}{Z_0} + Z_0C_{+-}\right\}b_-$$

Likewise, it can be seen by symmetry that

$$-\omega Z_0 C_{+-} b_+ = \tfrac{1}{2}\omega\left\{\tfrac{L_{+-}}{Z_0} - Z_0 C_{+-}\right\}a_+ - \tfrac{1}{2}\omega\left\{\tfrac{L_{+-}}{Z_0} + Z_0 C_{+-}\right\}b_+$$

Thus,

$$\partial_z\begin{pmatrix} a_+ \\ a_- \end{pmatrix} = -i\begin{pmatrix} \beta & 0 \\ 0 & \beta \end{pmatrix}\begin{pmatrix} a_+ \\ a_- \end{pmatrix} - i\begin{pmatrix} 0 & \kappa_- \\ \kappa_- & 0 \end{pmatrix}\begin{pmatrix} a_+ \\ a_- \end{pmatrix} + i\begin{pmatrix} 0 & \kappa_+ \\ \kappa_+ & 0 \end{pmatrix}\begin{pmatrix} b_+ \\ b_- \end{pmatrix}$$

$$= -i\begin{pmatrix} \beta & \kappa_- \\ \kappa_- & \beta \end{pmatrix}\begin{pmatrix} a_+ \\ a_- \end{pmatrix} + i\begin{pmatrix} 0 & \kappa_+ \\ \kappa_+ & 0 \end{pmatrix}\begin{pmatrix} b_+ \\ b_- \end{pmatrix}$$

where $\kappa_\pm = \tfrac{1}{2}\omega\left\{\tfrac{L_{+-}}{Z_0} \pm Z_0 C_{+-}\right\}$.

On setting

$$\mathbf{U} = \begin{pmatrix} \beta & \kappa_- \\ \kappa_- & \beta \end{pmatrix} \quad \text{and} \quad \mathbf{V} = \begin{pmatrix} 0 & \kappa_+ \\ \kappa_+ & 0 \end{pmatrix}$$

the above partial differential equation leads to the pair of transmission line equations:

$$\partial_z \boldsymbol{V} = -i(\mathbf{U}+\mathbf{V})Z_0\boldsymbol{I} \quad \text{and} \quad \partial_z \boldsymbol{I} = -i(\mathbf{U}-\mathbf{V})\tfrac{1}{Z_0}\boldsymbol{V}$$

Thus, from $\partial_z \boldsymbol{V} = -i\omega\mathbf{L}\boldsymbol{I}$ and $\partial_z \boldsymbol{I} = -i\omega\mathbf{C}\boldsymbol{V}$, where

$$\mathbf{L} = \begin{pmatrix} L_{++} & L_{+-} \\ L_{+-} & L_{--} \end{pmatrix} \quad \text{and} \quad \mathbf{C} = \begin{pmatrix} C_{++} & -C_{+-} \\ -C_{+-} & C_{--} \end{pmatrix}$$

it is clear by definition that the impedance matrix is given by

$$\mathbf{Z}^2 \equiv \mathbf{L}\mathbf{C}^{-1} = Z_0^2(\mathbf{U}+\mathbf{V})(\mathbf{U}-\mathbf{V}) \Rightarrow \mathbf{Z} = Z_0\mathbf{H},$$

where $\mathbf{H} = \sqrt{(\mathbf{U}+\mathbf{V})(\mathbf{U}-\mathbf{V})}$, and the reflection coefficient matrix by $\Gamma = (\mathbf{Z}-Z_0\mathbf{I})(\mathbf{Z}+Z_0\mathbf{I})^{-1}$, with $\mathbf{I}$ being the identity 2×2 matrix. Finally, substituting the results from (a) into $\partial_z^2 \boldsymbol{a}$ and $\partial_z^2 \boldsymbol{b}$ yields

$$\partial_z^2 \boldsymbol{a} = -(\mathbf{U}+\mathbf{V})(\mathbf{U}-\mathbf{V})\boldsymbol{a} \quad \text{and} \quad \partial_z^2 \boldsymbol{b} = -(\mathbf{U}+\mathbf{V})(\mathbf{U}-\mathbf{V})\boldsymbol{b}$$

Now, note first that $\mathbf{U}, \mathbf{V}$ are diagonalizable and hence, so is $\mathbf{H}$. Furthermore, observe that for a matrix

$$\mathbf{M} = \begin{pmatrix} a & b \\ b & a \end{pmatrix}$$

its eigenvalues are $m_{\pm} = a \pm b$, and the diagonalization of $\mathbf{M}$ is $\text{diag}(m_+, m_-)$. Next, in all that follows, by an abuse of notations, denote $\mathbf{H}$ by its diagonalization for convenience, because the original solution can easily be obtained from the inverse similarity transformation diagonalizing M; see Chapter 6 for more details. Thus,

$$\mathbf{H} \rightarrow \mathbf{H} = \begin{pmatrix} H_+ & 0 \\ 0 & H_- \end{pmatrix}$$

where $H_{\pm}$ are the two eigenvalues of $\mathbf{H}$, with $H_{\pm}^2 = \beta^2 + \kappa_-^2 - \kappa_+^2 \pm 2\beta\kappa_- = (\beta \pm \kappa_-)^2 - \kappa_+^2$.

Specifically,

$$H_+^2 = (\beta + \kappa_- + \kappa_+)(\beta + \kappa_- - \kappa_+) \Rightarrow H_+ = \omega\sqrt{(\mathrm{L}_0 + \mathrm{L}_{+-})(\mathrm{C}_0 - \mathrm{C}_{+-})}$$

and

$$H_-^2 = (\beta - \kappa_- + \kappa_+)(\beta - \kappa_- - \kappa_+) \Rightarrow H_- = \omega\sqrt{(\mathrm{L}_0 - \mathrm{L}_{+-})(\mathrm{C}_0 + \mathrm{C}_{+-})}$$

are the explicit eigenvalues. Moreover, as $\boldsymbol{a}(\boldsymbol{b})$ is the forward (backward) propagating wave, the results of Chapter 4 (*cf.* Proposition 4.4.2) generalize immediately to the coupled system of transmission line equations:

$$V(z) = \mathbf{V}_\mathrm{S}\mathbf{Z}(\mathbf{Z} + \mathbf{Z}_\mathrm{S})^{-1}\mathrm{e}^{-\mathrm{i}\mathbf{H}z}\{1 - \Gamma(0)\Gamma(\ell)\mathrm{e}^{-\mathrm{i}2\mathbf{H}\ell}\}^{-1}\{\mathbf{I} + \Gamma(\ell)\mathrm{e}^{-\mathrm{i}2\mathbf{H}(\ell - z)}\} \quad (5.24)$$

where the source impedance

$$\mathbf{Z}_\mathrm{S} = \begin{pmatrix} Z_{\mathrm{S}+} & 0 \\ 0 & Z_{\mathrm{S}-} \end{pmatrix} \quad \text{and} \quad \mathbf{V}_\mathrm{S} = \begin{pmatrix} V_{\mathrm{S}+} \\ V_{\mathrm{S}-} \end{pmatrix}$$

is the source of $\gamma_{\pm}$.

Following [5], let $\mathbf{e}_{\pm}$ denote the normalized eigenvectors of $\mathbf{H}$ associated with the eigenvalues $H_{\pm}$, where $\mathbf{e}_{\pm} = \frac{1}{\sqrt{2}}(1, \pm 1)$. It is easy to see that $\text{diag}(m_+, m_-) = \mathbf{PHP}^{\dagger}$, where $\dagger$ denotes the conjugate transpose, and $\mathbf{P} = (\mathbf{e}_+, \mathbf{e}_-)$, for all $a, b \neq 0$. Intuitively, $\mathbf{e}_+$ corresponds to the waves propagating along

$\gamma_\pm$ in the parallel direction $(\rightarrow,\rightarrow)$, $\mathbf{e}_-$ corresponds to waves along $\gamma_\pm$ in the antiparallel direction $(\rightarrow,\leftarrow)$. That is,

$$\mathbf{e}_+ \leftrightarrow \begin{pmatrix} \xrightarrow{\gamma_+ \text{propagation}} \\ \xrightarrow{\gamma_- \text{ propagation}} \end{pmatrix} \quad \text{and} \quad \mathbf{e}_- \leftrightarrow \begin{pmatrix} \xrightarrow{\gamma_+ \text{ propagation}} \\ \xleftarrow{\gamma_- \text{ propagation}} \end{pmatrix}$$

Hence, $\mathbf{e}_+$ defines the even mode and $\mathbf{e}_-$ defines the odd mode.

Now $\mathcal{B} = \{\mathbf{e}_+, \mathbf{e}_-\}$ defines a basis as $\mathbf{e}_+ \cdot \mathbf{e}_- = 0$, and it decouples $\gamma_+ \cup \gamma_- \rightarrow \gamma_+ \oplus \gamma_-$ in the following fashion. By definition, set $\boldsymbol{V} = V^{(+)}\mathbf{e}_+ + V^{(-)}\mathbf{e}_-$, where $V^{(\pm)} = \frac{1}{\sqrt{2}}(V_+ \pm V_-)$. Then, noting that

$$\exp\begin{pmatrix} a & 0 \\ 0 & b \end{pmatrix} = \begin{pmatrix} 1 & 0 \\ 0 & 1 \end{pmatrix} + \begin{pmatrix} a & 0 \\ 0 & b \end{pmatrix} + \frac{1}{2!}\begin{pmatrix} a & 0 \\ 0 & b \end{pmatrix}^2$$

$$+ \cdots = \begin{pmatrix} 1 + a + \frac{1}{2!}a^2 + \cdots & 0 \\ 0 & 1 + a + \frac{1}{2!}a^2 + \cdots \end{pmatrix}$$

that is,

$$\exp\begin{pmatrix} a & 0 \\ 0 & b \end{pmatrix} = \begin{pmatrix} e^a & 0 \\ 0 & e^b \end{pmatrix}$$

it follows upon expanding Equation (5.24), and defining

$$Z_\pm = \sqrt{\frac{L_0 \pm L_{+-}}{C_0 \mp C_{+-}}} \quad \text{and} \quad \Gamma_\pm = \frac{Z - Z_\pm}{Z + Z_\pm}$$

where $Z_\pm, \Gamma_\pm$ correspond, respectively, to the characteristic impedance and coefficient of reflection of the even(odd) modes, that

$$\mathbf{Z}(\mathbf{Z} + \mathbf{Z}_S)^{-1} = \begin{pmatrix} H_+ & 0 \\ 0 & H_- \end{pmatrix}\begin{pmatrix} H_+ + Z_{S+}^{(0)} & 0 \\ 0 & H_- + Z_{S-}^{(0)} \end{pmatrix}^{-1} = \begin{pmatrix} \frac{H_+}{H_+ + Z_{S+}^{(0)}} & 0 \\ 0 & \frac{H_-}{H_- + Z_{S-}^{(0)}} \end{pmatrix}$$

where $Z_{S\pm}^{(0)} = \frac{Z_{S\pm}}{Z_0}$, and likewise, on setting $Z_{L\pm}^{(0)} = \frac{Z_{L\pm}}{Z_0}$,

$$\mathbf{I} - \Gamma(0)\Gamma(\ell)e^{-i2\mathbf{H}\ell} = \begin{pmatrix} 1 & 0 \\ 0 & 1 \end{pmatrix} - \begin{pmatrix} \frac{H_+ - Z_{S+}^{(0)}}{H_+ + Z_{S+}^{(0)}} & 0 \\ 0 & \frac{H_- - Z_{S-}^{(0)}}{H_- + Z_{S-}^{(0)}} \end{pmatrix}\begin{pmatrix} \frac{H_+ - Z_{L+}^{(0)}}{H_+ + Z_{L+}^{(0)}} & 0 \\ 0 & \frac{H_- - Z_{L-}^{(0)}}{H_- + Z_{L-}^{(0)}} \end{pmatrix}\begin{pmatrix} e^{-i2H_+\ell} & 0 \\ 0 & e^{-i2H_+\ell} \end{pmatrix}$$

and hence, the system is decoupled via the diagonalization scheme, and further simplification thus yields

$$V^{(\pm)} = V_S^{(\pm)} \tfrac{Z_\pm}{Z_\pm + Z_{S\pm}} e^{-iH_\pm z} \left\{1 - \Gamma_\pm(0)\Gamma_\pm(\ell) e^{-i2H_\pm \ell}\right\}^{-1} \left\{1 + \Gamma_\pm(\ell) e^{-i2H_\pm(\ell - z)}\right\} \quad (5.25)$$

Note, as a side comment, that the speed of wave propagation for the even/odd mode is given, respectively, by

$$v_\pm = \tfrac{1}{\sqrt{(L_0 \pm L_{+-})(C_0 \mp C_{+-})}}$$

and hence $\beta_\pm = \omega v_\pm$ is the respective even/odd mode wave number.

By construction, $V = V_+ \mathbf{e}_1 + V_- \mathbf{e}_2$, where $\mathbf{e}_1 = (1,0), \mathbf{e}_2 = (0,1)$. The linear operator mapping the basis $\mathcal{B}$ onto the standard basis $\mathcal{B}_0 = \{\mathbf{e}_1, \mathbf{e}_2\}$ can be shown, after some algebraic manipulation, to be

$$\mathbf{T} = \tfrac{1}{\sqrt{2}} \begin{pmatrix} 1 & 1 \\ 1 & -1 \end{pmatrix}$$

Thus, $\mathbf{T}^{-1} = \mathbf{T} \Rightarrow (V_+, V_-) = V^{(+)}\mathbf{T}(\mathbf{e}_+) + V^{(-)}\mathbf{T}(\mathbf{e}_-)$. Explicitly, $V_\pm = \tfrac{1}{\sqrt{2}}\{V^{(+)} \pm V^{(-)}\}$ and similarly, $V_S^{(\pm)} = \tfrac{1}{\sqrt{2}}\{V_{S+} \pm V_{S-}\}$.

To determine the cross-talk coefficient on γ_-, it suffices to assume without loss of generality that $V_{S-} = 0$. Then, the far end cross-talk is determined by setting $z = \ell$. That is,

$$V_-(\ell) = \tfrac{1}{\sqrt{2}}\{V^{(+)}(\ell) - V^{(-)}(\ell)\}$$

Furthermore, observe that

$$\tfrac{Z_\pm}{Z_\pm + Z_{S\pm}} = \tfrac{1}{2} \tfrac{Z_\pm + Z_{S\pm} - Z_{S\pm} + Z_\pm}{Z_\pm + Z_{S\pm}} = \tfrac{1}{2}(1 - \Gamma_\pm(0))$$

Hence,

$$V_-(\ell) = \tfrac{1}{2}\left\{\tfrac{1}{2}(1 - \Gamma_+(0))(V_{S+} + V_{S-})(1 - \Gamma_+(0)\Gamma_+(\ell)e^{-i2H_\pm \ell})^{-1} e^{-iH_+\ell}(1 + \Gamma_+(\ell))\right\} -$$

$$\tfrac{1}{2}\left\{\tfrac{1}{2}(1 - \Gamma_-(0))(V_{S+} - V_{S-})(1 - \Gamma_-(0)\Gamma_-(\ell)e^{-i2H_\pm \ell})^{-1} e^{-iH_+\ell}(1 + \Gamma_-(\ell))\right\}$$

Finally, observe that under conditions of *weak coupling*, $|L_{+-}| << |L_{\pm\pm}|$ and $|C_{+-}| << |C_{\pm\pm}|$. Remember moreover, it was assumed herein that $L_\pm = L_0$ and $C_\pm = C_0$. Thus,

$$H_\pm = \omega\sqrt{(L_0 \pm L_{+-})(C_0 \mp C_{+-})} \approx \omega\sqrt{L_0 C_0}\sqrt{1 \pm \tfrac{L_{+-}}{L_0} \mp \tfrac{C_{+-}}{C_0}}$$

In particular, we may define, up to first order in

$$\frac{L_{+-}}{L_0}, \frac{C_{+-}}{C_0}$$

(i.e., the conditions of weak coupling) $H_{\pm} = \beta \pm \delta H$, where

$$\beta = \omega\sqrt{L_0 C_0} \quad \text{and} \quad \delta H = \tfrac{1}{2}\beta\left(\tfrac{L_{+-}}{L_0} - \tfrac{C_{+-}}{C_0}\right)$$

Likewise,

$$Z_{\pm} = \sqrt{\tfrac{L_0 \pm L_{+-}}{C_0 \mp C_{+-}}} = Z_0\sqrt{\left(1 \pm \tfrac{L_{+-}}{L_0}\right)\left(1 \pm \tfrac{C_{+-}}{C_0}\right)^{-1}} \approx Z_0\left\{1 \pm \tfrac{1}{2}\left(\tfrac{L_{+-}}{L_0} + \tfrac{C_{+-}}{C_0}\right)\right\} = Z_0 \pm \tfrac{1}{2} Z_0\left(\tfrac{L_{+-}}{L_0} + \tfrac{C_{+-}}{C_0}\right)$$

under conditions of weak coupling.

Moreover, on defining the differential pair reflection coefficients

$$\Gamma_{\pm}(z) = \tfrac{Z_{\pm}(z) - Z_{\pm}}{Z_{\pm}(z) + Z_{\pm}}$$

if the source and load impedances are matched to Z_0, then the individual uncoupled reflection coefficients under the conditions of weak coupling along $\gamma_{\pm}$ are:

$$\Gamma_0^{(\pm)} = \tfrac{Z_{\pm} - Z_0}{Z_{\pm} + Z_0} = \pm\tfrac{1}{2}\left(\tfrac{L_{+-}}{L_0} + \tfrac{C_{+-}}{C_0}\right)\left\{2 \pm \tfrac{1}{2}\left(\tfrac{L_{+-}}{L_0} + \tfrac{C_{+-}}{C_0}\right)\right\}^{-1} \approx \pm\tfrac{1}{4}\left(\tfrac{L_{+-}}{L_0} + \tfrac{C_{+-}}{C_0}\right)$$

Lastly, set $\tau_{\pm} = \ell\frac{H_{\pm}}{\omega}$ to be the propagation delay time. Then, under weak coupling, $\tau_{\pm} \approx \tau \pm \delta\tau$, where $\delta\tau = \frac{\ell}{\omega}\delta H$ and $\tau = \frac{\ell}{\omega}\beta$.

In terms of $\kappa_{\pm}$, noting that

$$\kappa_{\pm} = \tfrac{1}{2}\beta\left(\tfrac{L_{+-}}{L_0} \pm \tfrac{C_{+-}}{C_0}\right)$$

it is clear that

$$\delta H = \tfrac{1}{2}\beta\left(\tfrac{L_{+-}}{L_0} - \tfrac{C_{+-}}{C_0}\right) = \kappa_{-}, \quad \Gamma_0^{(\pm)} \approx \pm\tfrac{1}{2}\tfrac{\kappa_{+}}{\beta}, \quad \delta\tau = \tfrac{\ell}{\omega}\kappa_{-} = \tfrac{\tau}{\beta}\kappa_{-}$$

In time-domain, from the expansion of $V\left(t - \frac{\ell}{v}\right)$ given at the beginning of this problem outlined above leads trivially to

$$V_{-}(\ell) = \tfrac{1}{2}(1 - \Gamma_{+}(0))(1 + \Gamma_{+}(\ell))\{V_0(t - \tau_{+}) + \Gamma_{+}(0)\Gamma_{+}(\ell)V_0(t - 3\tau_{+}) + \cdots\}$$
$$- \tfrac{1}{2}(1 - \Gamma_{-}(0))(1 + \Gamma_{-}(\ell))\{V_0(t - \tau_{-}) + \Gamma_{-}(0)\Gamma_{-}(\ell)V_0(t - 3\tau_{-}) + \cdots\}$$

whence, under the assumptions of weak coupling, set $\delta\Gamma = \frac{1}{2}\frac{\kappa_{+}}{\beta}$, which by definition, is first order in $\frac{L_{+-}}{L_0}, \frac{C_{+-}}{C_0}$. Hence, define $\delta\Gamma(0) = \delta\Gamma + \varepsilon'$ and

$\delta\Gamma(\ell) = \delta\Gamma + \varepsilon''$, for some $\varepsilon', \varepsilon'' = o(\delta\Gamma^2)$. Then, $\varepsilon' - \varepsilon'' \sim o(\delta\Gamma^2)$, and via the assumption $Z_{S\pm} = Z_0 = Z_{L\pm}$,

$$(1 - \Gamma_{\pm}(0))(1 + \Gamma_{\pm}(\ell)) = 1 - \Gamma_{\pm}(0)\Gamma_{\pm}(\ell) \approx 1$$

Thus, noting via Taylor's expansion (to first order) that $V_0(t - \tau_{\pm}) \approx V_0(t - \tau) \mp \partial_t V_0(t - \tau)\delta\tau$, it is clear that

$$V_-(\ell) \approx \tfrac{1}{2}(V_0(t - \tau_+) - V_0(t - \tau_-)) = \tfrac{1}{2}\{V_0(t - \tau) - \partial_t V_0(t - \tau)\delta\tau - V_0(t - \tau) - \partial_t V_0(t - \tau)\delta\tau\},$$

that is, $V_-(\ell) \approx -\partial_t V_0(t - \tau)\delta\tau$. Thus,

$$k_+ = -\delta\tau = -\tfrac{\ell}{\omega}\kappa_- = -\tfrac{1}{2}\tfrac{\ell}{\omega}\beta\left(\tfrac{L_{+-}}{L_0} - \tfrac{C_{+-}}{C_0}\right) = -\tfrac{1}{2}\tau\left(\tfrac{L_{+-}}{L_0} - \tfrac{C_{+-}}{C_0}\right)$$

as required.

(c) The near-end cross-talk is determined as (b) above by setting $z = 0$. Under the conditions of weak coupling, via $V_-(0) = \tfrac{1}{\sqrt{2}}\{V^{(+)}(0) - V^{(-)}(0)\}$, following the derivation above for $V_-(\ell)$,

$$V_-(\ell) = \tfrac{1}{2}\left\{\tfrac{1}{2}(1 - \Gamma_+(0))(V_{S+} + V_{S-})(1 - \Gamma_+(0)\Gamma_+(\ell)e^{-i2H_{\pm}\ell})^{-1}(1 + \Gamma_+(\ell)e^{iH_+\ell})\right\}$$
$$- \tfrac{1}{2}\left\{\tfrac{1}{2}(1 - \Gamma_-(0))(V_{S+} - V_{S-})(1 - \Gamma_-(0)\Gamma_-(\ell)e^{-i2H_{\pm}\ell})^{-1}(1 + \Gamma_-(\ell)e^{iH_+\ell})\right\}$$

Noting that $V_0(t) = \tfrac{1}{2}V_{S+}(t)$ when $Z_{S\pm} = Z_0 = Z_{L\pm}$, and recalling that

$$V_0(t) + \Gamma_+(\ell)\sum_{m>0}\Gamma_+^{m-1}(0)\Gamma_+^{m-1}(\ell)V_0(t - 2m\tau) = V_0(t)\{1 + \Gamma_+(\ell)e^{iH_+\ell}\}\{1 - \Gamma_+(0)\Gamma_+(\ell)e^{-iH_+\ell}\}^{-1}$$

weak coupling yields $\Gamma_+(0) - \Gamma_-(0), \Gamma_+(\ell) - \Gamma_-(\ell) \sim \delta\Gamma_+ - \delta\Gamma_- = \tfrac{\kappa_+}{\beta}$ and hence

$$\begin{aligned}
V_-(0) &\approx \tfrac{1}{2}(1 - \Gamma_+(0))\left\{V_0(t) + \left(1 + \tfrac{1}{\Gamma_+(0)}\right)\Gamma_+(0)\Gamma_+(\ell)V_0(t - 2\tau)\right\} \\
&\quad - \tfrac{1}{2}(1 - \Gamma_-(0))\left\{V_0(t) + \left(1 + \tfrac{1}{\Gamma_-(0)}\right)\Gamma_-(0)\Gamma_-(\ell)V_0(t - 2\tau)\right\} \\
&= \tfrac{1}{2}(1 + \Gamma_0^{(+)})\left\{V_0(t) + \left(1 - \tfrac{1}{\Gamma_0^{(+)}}\right)\Gamma_0^{(+)}\Gamma_0^{(+)}V_0(t - 2\tau)\right\} \\
&\quad - \tfrac{1}{2}(1 + \Gamma_0^{(-)})\left\{V_0(t) + \left(1 - \tfrac{1}{\Gamma_0^{(-)}}\right)\Gamma_0^{(-)}\Gamma_0^{(-)}V_0(t - 2\tau)\right\} \\
&= \tfrac{1}{2}(1 + \Gamma_0^{(+)})\left\{V_0(t) - \left(1 - \Gamma_0^{(+)}\right)\Gamma_0^{(+)}V_0(t - 2\tau)\right\} \\
&\quad - \tfrac{1}{2}(1 + \Gamma_0^{(-)})\left\{V_0(t) - \left(1 - \Gamma_0^{(-)}\right)\Gamma_0^{(-)}V_0(t - 2\tau)\right\}
\end{aligned}$$

Finally, noting that $\Gamma_0^{(+)} - \Gamma_0^{(-)} = \frac{\kappa_+}{\beta}$, it is clear that

$$V_-(0) \approx \tfrac{1}{2}(\Gamma_0^{(+)} - \Gamma_0^{(-)})V_0(t) - \tfrac{1}{2}(\Gamma_0^{(+)} - \Gamma_0^{(-)})V_0(t-2\tau)$$
$$\approx \tfrac{1}{2}\tfrac{\kappa_+}{\beta}(V_0(t) - V_0(t-2\tau))$$

and hence, by definition,

$$k_- = \tfrac{1}{2}\tfrac{\kappa_+}{\beta} = \tfrac{1}{4}\left(\tfrac{L_{+-}}{L_0} + \tfrac{C_{+-}}{C_0}\right)$$

as desired. □

Exercise 5.4.5

The electrical implementation of a *transition minimized differential signaling* (TMDS) signaling protocol is given below [2]: this is a low-cost implementation for transmitting high-frequency digital data packages that minimizes electromagnetic interference. It is typically employed in the design of graphics interfaces for computers, high-definition televisions, and graphic cards.

Suppose that the transmission lines are lossless, and the differential impedance Z_2 of the cable differs from that of the differential pair: $Z_1 \neq Z_2$. Suppose further that $R_s \neq \frac{1}{2}Z_1$ is the source impedance. What is the best way to match the impedance between the transmitter and the TMDS cable in order to minimize undue reflections? Refer to method (b) versus method (c) of Figure 5.10.

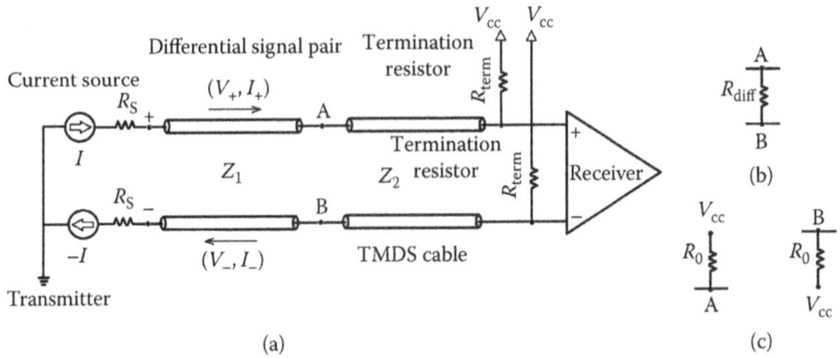

FIGURE 5.10
A simple model for a TMDS data pair.

Solution

The only place on the schematics to implement an impedance match is between A, B. First, observe that if scheme (b) were implemented, where $R_{diff} = Z_1$, the resultant potential difference along the TMDS cable will be halved. This can be easily seen (via the superposition principle) schematically as follows.

The discrete lumped equivalent circuit representation for γ_+ is Z_1 in series with $R = \frac{1}{2}R_{diff}$, whence, the voltage output is that of the potential divider: $V_A = V_+ \frac{R}{R+Z_1} = \frac{1}{2}V_+$, where V_+ is the incident voltage.

By symmetry, on γ_-, $V_B = V_- \frac{R}{R+Z_1} = \frac{1}{2}V_-$, whence $|V_{diff}| = |V_A - V_B| = |V_+|$, which is half the amplitude at the differential load.

Finally, consider (c). Here, on γ_+, set $R_0 = Z_1$ in order to match the impedance along γ_+. Set $V_{cc} = V_+$. Then, $V_A = V_{cc} - V_+ \frac{R}{R+Z_1} = \frac{1}{2}V_+$ and $V_B = V_- \frac{R}{R+Z_1} - V_{cc}$. That is, $V_B = -V_+ \frac{R}{R+Z_1} - V_{cc} = \frac{3}{2}V_+$ and hence, $|V_{diff}| = |V_A - V_B| = 2|V_+|$, as required. □

Exercise 5.4.6

Describe an alternative method to terminate a differential line such that both odd and even modes are properly terminated.

Solution

T-termination of differential lines is an alternative termination scheme; see the schematics shown in Figure 5.11.

Now, for odd mode, by definition, the potential drop across R is zero. Hence, $2R_1 = Z_{diff}$; that is, $R_1 = Z_{odd}$. Next, consider an even mode noise propagating along the differential pair toward the receiver. First, observe that via the principle of superposition, in the absence of any coupling, the differential pair $\gamma_+ \cup \gamma_-$ is equivalent to $\tilde{\gamma}_+ \oplus \tilde{\gamma}_-$, where the direct sum $\oplus$ denotes the

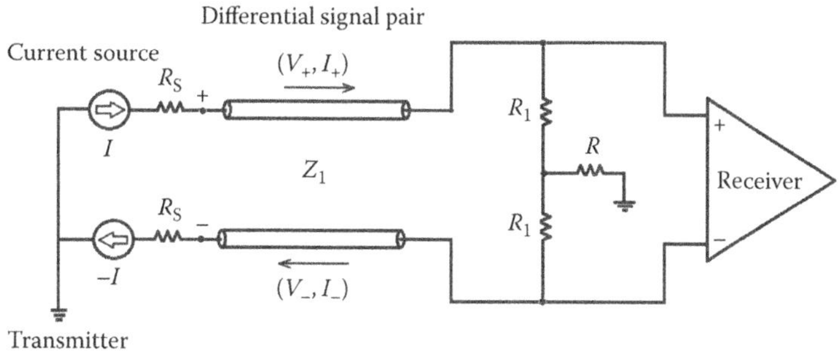

FIGURE 5.11
T-termination scheme for a differential pair.

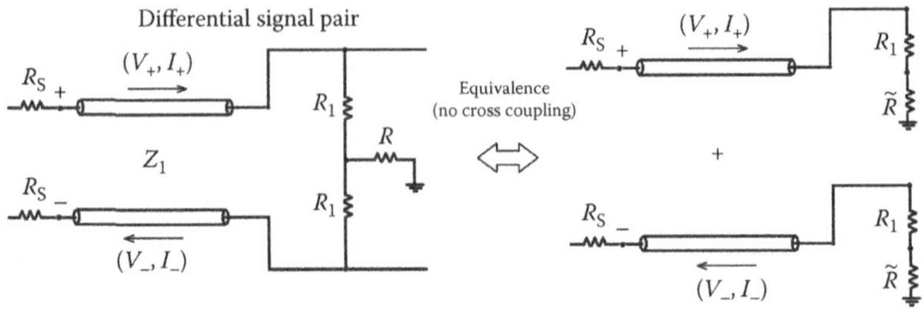

FIGURE 5.12
Equivalent *T-termination* scheme in the absence of cross-coupling.

absence of cross-coupling (the disjoint union). Here, by definition, the load of $\gamma_\pm$ is $\tilde{Z} = R_1 + \tilde{R}$, where $R = \tilde{R} \| \tilde{R} \Rightarrow \tilde{R} = \frac{1}{2}R$ (cf. Figure 5.12).

Hence, by construction, $Z_{\text{even}} = R_1 + \tilde{R} \Rightarrow R = 2(Z_{\text{even}} - Z_{\text{odd}})$, as required. □

References

1. Chang, D. 1992. *Field and Wave Electromagnetics*. Reading, MA: Addison-Wesley.
2. Digital Visual Interface. 1999. *Digital Display Working Group*, Rev. 1.0.
3. Knockaert, J., Peuteman, J., Catrysse, J., and Belmans, R. 2009. General equations for the characteristic impedance matrix and termination network of multiconductor transmission lines. In *Proceedings of the IEEE Int. Conf. Industrial Technology*, pp. 595–600.
4. Neff, Jr., H. 1981. *Basic Electromagnetic Fields*. New York: Harper & Row.
5. Orfanidis, S. 2002. *Electromagnetic Waves and Antenna*. Rutgers University, ECE Dept., http://www.ece.rutgers.edu/~orfanidi/ewa/.
6. Paul, C. 2002: Solution of the transmission-line equations under the weak-coupling assumption. *IEEE Trans. Electromagn. Compat.* 44(3), 413–423.
7. Paul, C. 1994. *Analysis of Multiconductor Transmission Lines*. New York, John Wiley & Sons.
8. Paul, C. 2006. *Introduction to Electromagnetic Compatibility*. Hoboken, NJ: John Wiley & Sons.
9. Plonsey, R. and Collin, R. 1961. *Principles and Applications of Electromagnetic Fields*. New York: McGraw-Hill.
10. Sengupta, D. and Liepa, V. 2006. *Applied Electromagnetics and Electromagnetic Compatibility*. Hoboken, NJ: John Wiley & Sons.

6

Cross-Talk in Multiconductor Transmission Lines

In this section, the phenomenon of cross-talk from transmission lines coupling electromagnetically is derived from Maxwell's theory. To facilitate the discussion of cross-talk, the concepts of mutual capacitance, mutual impedance, and mutual inductance are introduced; see, for example, References [4,5,7,9]. The concept of mutual coupling leads naturally to the study of multiconductor transmission lines. Indeed, a glimpse into the concept of cross-talk was introduced in Chapter 5; in particular, see Exercise 5.4.4.

The last two sections comprise elements of multiconductor transmission line theory, followed by the concept of scattering parameters. Scattering parameters have great applications in microwave engineering, the details of which can be found in References [5,6,8]. In particular, for readers wishing to pursue these topics in greater depth, Reference [6] is a classic exposition on multiconductor transmission line theory and that of microwave engineering is given in [8].

6.1 Reciprocity Theorem and Mutual Capacitance

An important result employed in this section is called *Green's reciprocity theorem*; see, for example, References [2,3,9,10] for various equivalent reciprocity theorems.

6.1.1 Theorem (Green's Reciprocity)

Given a system $\{(C_i,\rho_i)\}_{i=1}^{n}$ of charged conductors, where ρ_i is the charge density on conductor C_i, and $C_i \cap C_j = \varnothing\ \forall i \neq j$, let V_i be a fixed potential on $C_i\ \forall i = 1,\dots,n$. Then, given another system $\{(C_i',\rho_i')\}_{i=1}^{n}$ of charged conductors, with $C_i' \cap C_j' = \varnothing$ and $C_i \cap C_j' = \varnothing\ \forall i \neq j$,

$$\sum_i \int_\Omega \rho_i \varphi_i' d^3x = \sum_i \int_\Omega \rho_i' \varphi_i d^3x$$

where $\Omega = \mathbf{R}^3 - \bigcup_i (C_i \cup C_i')$, and the potential φ_i satisfies Poisson's equation $-\Delta\varphi_i = \frac{1}{\varepsilon}\rho_i$ subject to the Dirichlet boundary condition $\varphi_i \mid \partial C_i = V_i$, and the same also applies to the triple $(\varphi_i', V_i', C_i')\ \forall i$.

Proof

The proof is an easy matter via mathematical induction. First, consider the case where $n = 1$. Then, via the second Green's identity, $\int_\Omega \varphi_1 \Delta \varphi_1' - \varphi_1' \Delta \varphi_1 = \int_{\partial\Omega} \varphi_1 \partial_n \varphi_1' - \varphi_1' \partial_n \varphi_1$, substituting Poisson's equation yields

$$\tfrac{1}{\varepsilon}\int_\Omega (-\varphi_1 \rho_i' + \varphi_1' \rho_1) = \int_{\partial\Omega} \varphi_1 \partial_n \varphi_1' - \varphi_1' \partial_n \varphi_1 = 0$$

as $\varphi_1 \mid \partial\Omega, \varphi_1' \mid \partial\Omega$ are constants implies that $\partial_n \varphi_1 \mid \partial\Omega = 0 = \partial_n \varphi_1' \mid \partial\Omega$, and hence, $\int_\Omega \varphi_1 \rho_1' = \int_\Omega \varphi_1' \rho_1$. So, suppose that for some fixed $k > 1$, the following holds:

$$\sum_{i=1}^{k} \int_\Omega \rho_i \varphi_i' d^3 x = \sum_{i=1}^{k} \int_\Omega \rho_i \varphi_i' d^3 x$$

Consider $n = k + 1$.

Then

$$\sum_{i=1}^{n} \int_\Omega \varphi_i \Delta \varphi_i' - \varphi_i' \Delta \varphi_i = \sum_{i=1}^{n} \int_{\partial\Omega} \varphi_i \partial_n \varphi_i' - \varphi_i' \partial_n \varphi_i$$

from Green's identity. However, noting that

$$\sum_{i=1}^{n} \int_\Omega \varphi_i \Delta \varphi_i' - \varphi_i' \Delta \varphi_i = \sum_{i=1}^{k} \int_\Omega \varphi_i \Delta \varphi_i' - \varphi_i' \Delta \varphi_i + \int_\Omega \varphi_{k+1} \Delta \varphi_{k+1}' - \varphi_{k+1}' \Delta \varphi_{k+1},$$

the inductive assumption implies immediately that the first $k = n - 1$ summation vanishes, leaving

$$\int_\Omega \varphi_{k+1} \Delta \varphi_{k+1}' - \varphi_{k+1}' \Delta \varphi_{k+1} = \int_{\partial\Omega} \varphi_{k+1} \partial_n \varphi_{k+1}' - \varphi_{k+1}' \partial_n \varphi_{k+1}.$$

From which, Gauss' law and $\partial_n \varphi_{k+1} \mid \partial\Omega = 0 = \partial_n \varphi_{k+1}' \mid \partial\Omega$ together yield $\int_\Omega \varphi_{k+1} \rho_{k+1}' = \int_\Omega \varphi_{k+1}' \rho_{k+1}$, hence, $\sum_{i=1}^{n} \int_\Omega \rho_i \varphi_i' d^3 x = \sum_{i=1}^{n} \int_\Omega \rho_i' \varphi_i d^3 x$, and induction thus follows, as desired. □

Indeed, observe trivially as an immediate corollary that for the pair of systems $\{(\varphi_i, C_i)\}_{i=1}^n$ and $\{(\varphi_i', C_i')\}_{i=1}^n$, the restriction of $\varphi_i(\varphi_i')$ to $C_i(C_i')$ leads immediately to the relation

$$\sum_{i=1}^{n} Q_i V_i' = \sum_{i=1}^{n} Q_i' V_i \tag{6.1}$$

where $Q_i = \int_{C_i} \rho_i \, d^3 x$ and $Q_i' = \int_{C_i'} \rho_i' d^3 x$ for all $i = 1, \ldots, n$. Establish this in Exercise 6.5.1.

6.1.2 Example

As an application of Theorem 6.1.1, consider a system $\{C_i\}$ of disjoint grounded n conductors and suppose that conductor C_0 has a total charge Q_0. Determine the induced charge Q_k on C_k, for some k, as a function of Q_0.

Let V_0 be the equipotential on C_0 as a result of some free charge Q_0 placed on the conductor. Let $Q_i' \ \forall i=1,\ldots,n,$ denote the induced charges on $\{C_i\}_{i=1}^n$, if $\{C_i\}_{i=1}^n$ were raised to potential $\{V_i'\}$, and likewise, let V_0' denote the potential on C_0 if $Q_0 = 0$ and Q_i are free charges placed on $C_i \ \forall i=1,\ldots,n$. Then, from the above corollary,

$$\sum_{i=0}^{n} Q_i V_i' = \sum_{i=0}^{n} Q_i' V_i \Rightarrow Q_0 V_0' + Q_1 V_1' + \cdots + Q_n V_n' = Q_0 V_0 + Q_1 V_1 + \cdots + Q_n V_n = 0$$

as $V_i = 0 \ \forall i = 1,\ldots, n$ by assumption (the conductors are grounded), and $Q_0 = 0$. Whence, noting from the conservation of charge that $Q_0 + Q_1 + \cdots + Q_n = 0$, it follows that

$$Q_i = -\left\{V_i' - \sum_{k \neq i} V_k'\right\}^{-1} \left\{Q_0 \sum_i V_i' - \sum_{j \neq i} Q_j \sum_{k \neq j}^{(j)} V_k'\right\}$$

where $\sum_k^{(i)} \alpha_k = \alpha_1 + \cdots + \alpha_{i-1} + \alpha_{i+1} + \cdots$. To see this, it suffices to expand and regroup the expression:

$$\begin{aligned} 0 &= Q_0 V_0' + Q_1 V_1' + \cdots + Q_n V_n' \\ &= Q_0 V_0' + Q_1 V_1' + \{Q_0 - Q_1 - Q_3 - \cdots - Q_n\} V_2' + \cdots + \{Q_0 - Q_1 - Q_3 - \cdots - Q_{n-1}\} V_n' \end{aligned}$$

□

The Lorentz reciprocity theorem stated in Chapter 3 is given in a slightly different form below with a shorter proof for instructive purposes, and to preserve the continuity of the exposition.

6.1.3 Theorem (Lorentz Reciprocity)

Given $(\Omega, \varepsilon, \mu)$, where $\Omega = \mathbf{R}^3 - (C_1 \cup C_2)$ and (C_i, J_i), for $i = 1,2$, are compact sources with time harmonic current densities J_i defined on C_i. Then, on Ω, $\nabla \cdot (E \times B' - E' \times B) = 0$. In particular, $\int_{C_1 \cup C_2} E \cdot J' \, d^3x = \int_{C_1 \cup C_2} E' \cdot J d^3x$.

Proof

First, recall the vector identity $\nabla \cdot A \times B = B \cdot \nabla \times A - A \cdot \nabla \times B$. Then, clearly,

$$\begin{aligned} \nabla \cdot (E \times B' - E' \times B) &= B' \cdot \nabla \times E - E \cdot \nabla \times B' - E' \cdot \nabla \times B + B \cdot \nabla \times E' \\ &= -i\omega B' \cdot B - i\omega\mu\varepsilon E \cdot E' - i\omega\mu J' - \\ &\quad\ i\omega E' \cdot E + i\omega\mu\varepsilon B \cdot B' + i\omega\mu J \\ &= i\omega\mu J - i\omega\mu J' \end{aligned}$$

whence, on Ω, $J, J' = 0 \Rightarrow \nabla \cdot (E \times B' - E' \times B) = 0$. To establish the second result, it suffices to apply the divergence theorem:

$$\int_{\Omega} \nabla \cdot (E \times B' - E' \times B) d^3x = \int_{\partial\Omega} (E \times B' - E' \times B) \cdot n d^2x$$

$$= i\omega\mu \int_{\Omega} (E' \cdot J - E \cdot J') d^3x = 0$$

as $J|_{\Omega}, J'|_{\Omega} \equiv 0$ completing the proof. □

6.1.4 Remark

From Theorem 6.1.3, and noting that $V \propto E, I \propto J$, it follows that given two thin conductors (C_i, V_i, I_i), at potential V_i and current I_i, for $i = 1,2$, with respect to some fixed ground, then $V_1 I_2 = V_2 I_1$; see Exercise 6.5.2 for a rigorous proof. More generally, for a system of thin conductors $\{(C_i, V_i, I_i)\}$ and $\{(C_i', V_i', I_i')\}$, for $i = 1,\ldots, n$, $\sum_{i=1}^{n} V_i I_i' = \sum_{i=1}^{n} V_i' I_i$. Establish this in Exercise 6.5.2. Finally, it ought to be pointed out that the full statement of Lorentz reciprocity involves introducing a fictitious magnetic charge density $\tilde{J}$; see, for example Reference [1]. This was deliberately left out in the proof of Theorem 6.1.3. To wit, the statement of the theorem in generality is:

$$\int_{\partial\Omega} (E_1 \times H_2 - E_2 \times H_1) \cdot d^2x = \int_{\Omega} (E_2 \cdot J_1 + H_1 \cdot \tilde{J}_2 - E_1 \cdot J_2 - H_2 \cdot \tilde{J}_1) d^3x$$

where $\tilde{J}_i$ is the (fictitious) magnetic current density on C_i, and $B_i = \mu H_i$.

Consider a system $\{C_i\}$ of $N + 1$ conductors as illustrated in Figure 6.1. Suppose that the conductors are of arbitrary cross-sections, and their lengths run parallel to one another. For simplicity, assume that the conductors are of identical lengths ℓ. Fix one conductor as the ground conductor and denote this by the 0th conductor C^0. All potential references are made with respect to the conductor C_0.

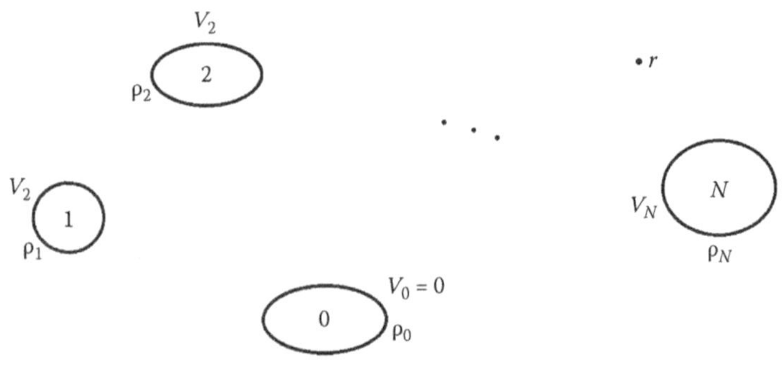

FIGURE 6.1
Cross-section of conductors in a homogeneous dielectric medium.

Suppose that the conductor C_i is charged to a uniform charge density of ρ_i for each $i = 1,\ldots, N$. Under electrostatic conditions, the resultant charge on the system comprising the $N+1$ conductors must sum to zero. Whence, the charge density on the ground conductor C_0 (here, denoted by conductor 0) must satisfy

$$\rho_0 = -(\rho_1 + \cdots + \rho_N) \tag{6.2}$$

Set $\Omega = \mathbf{R}^3 - \cup_i C_i$ for simplicity. Then the electric field at an arbitrary point $\boldsymbol{r} \in \Omega$ is the sum of the electric field contribution from each conductor: $\boldsymbol{E}(\boldsymbol{r}) = \sum_{i=0}^{N} \boldsymbol{E}_i(\boldsymbol{r})$. From Chapter 3, it is clear that the electric field $E(\boldsymbol{r}) = -\nabla\varphi$ is defined by the Laplace equation:

$$\begin{cases} \Delta\varphi = 0 \quad \text{on } \Omega, \\ -\varepsilon\nabla\varphi \cdot \boldsymbol{n}_i = \rho_i \quad \forall i = 0,\ldots,N, \end{cases} \tag{6.3}$$

where $\boldsymbol{n}_i$ is the unit normal vector field on ∂C_i.

By definition, $V_i = -\int_{\gamma_i} \boldsymbol{E} \cdot d\boldsymbol{l}$, where $\gamma_i \subset \Omega$ is a path satisfying $\gamma_i(0) \in C_0$ and $\gamma_i(1) \in C_i$. Furthermore, from $Q = CV$, it is clear by invoking Gauss' law and the divergence theorem, that $V_i = \sum_{j=0}^{N} p_{ij} Q_j$, where $Q_i = \varepsilon \int_{C_i} \boldsymbol{E} \cdot d^3\boldsymbol{x} = \varepsilon \int_{\partial C_i} \boldsymbol{E} \cdot \boldsymbol{n}_i d^2\boldsymbol{x}$, whence,

$$V_i = -\int_{\gamma_i} \boldsymbol{E} \cdot d\boldsymbol{l} = -\sum_{j=0}^{N} \int_{\gamma_i} \boldsymbol{E}_j \cdot d\boldsymbol{l} = \sum_{j=0}^{N} p_{ij} Q_j \Rightarrow p_{ij} = -\tfrac{1}{Q_j} \int_{\gamma_i} \boldsymbol{E}_j \cdot d\boldsymbol{l}$$

This leads to the following result.

6.1.5 Lemma

Given a system of charged, thinly insulated conductors illustrated in Figure 6.1, the charge Q_i on the conductor C_i is related to the collection $\{(V_k, C_k)\}$ via

$$\begin{pmatrix} V_1 \\ \vdots \\ V_N \end{pmatrix} = \begin{pmatrix} P_{11} & \cdots & P_{1N} \\ \vdots & \ddots & \vdots \\ P_{N1} & \cdots & P_{NN} \end{pmatrix} \begin{pmatrix} Q_1 \\ \vdots \\ Q_N \end{pmatrix} \tag{6.4}$$

where $P_{ij} = p_{ij} - p_{i0} \ \forall i, j$.

Proof

From (6.1), $Q_0 + Q_1 + \cdots + Q_N = 0$ (charge conservation), whence,

$$V_i = \sum_{j=0}^{N} p_{ij} Q_j = \sum_{j=1}^{N} p_{ij} Q_j - p_{i0}(Q_1 + \cdots + Q_N) = \sum_{j=0}^{N} (p_{ij} - p_{i0}) Q_j,$$

as required. □

6.1.6 Lemma

The matrix $\mathbf{P} = (P_{ij})$ of (6.4) is symmetric: $P_{ij} = P_{ji}\ \forall i, j$.

Proof

Now, by (6.1), $\Sigma_{i=1}^{n} Q_i V_i' = \Sigma_{i=1}^{n} Q_i' V_i$; that is, $\boldsymbol{Q}' \cdot \boldsymbol{V} = \boldsymbol{Q} \cdot \boldsymbol{V}'$, where

$$\boldsymbol{V} = (V_1, \ldots, V_N), \boldsymbol{V}' = (V_1', \ldots, V_N'), \boldsymbol{Q} = (Q_1, \ldots, Q_N), \boldsymbol{Q}' = (Q_1', \ldots, Q_N')$$

Thus, by definition,

$$\boldsymbol{Q}' \cdot \boldsymbol{V} = \sum_i \sum_j Q_i' P_{ij} Q_j = \sum_i Q_i' P_{ii} Q_i + \sum_{i<j} \sum_j Q_i' P_{ij} Q_j + \sum_{i>j} \sum_j Q_i' P_{ij} Q_j$$

and

$$\boldsymbol{Q} \cdot \boldsymbol{V}' = \sum_i \sum_j Q_i P_{ij} Q_j' = \sum_i Q_i P_{ii} Q_i' + \sum_{i<j} \sum_j Q_i P_{ij} Q_j' + \sum_{i>j} \sum_j Q_i P_{ij} Q_j'$$

Hence, from $\boldsymbol{Q}' \cdot \boldsymbol{V} = \boldsymbol{Q} \cdot \boldsymbol{V}'$, the above expansions imply immediately that

$$\sum_{i<j} \sum_j Q_i P_{ij} Q_j' = \sum_{i>j} \sum_j Q_j P_{ji} Q_i' = \sum_{i<j} \sum_j Q_i P_{ji} Q_j'$$

$$\sum_{i>j} \sum_j Q_i P_{ij} Q_j' = \sum_{i<j} \sum_j Q_i' P_{ij} Q_j = \sum_{i>j} \sum_j Q_i P_{ji} Q_j'$$

via relabeling the dummy indices, and hence, $P_{ij} = P_{ji}\ \forall i, j$, as claimed. □

Indeed, it is easy to see from Equation (6.4) that as P is a constant, it clearly follows from

$$V_i = P_{i1} Q_1 + P_{i2} Q_2 + \cdots + P_{iN} Q_N$$

that on setting $Q_k = 0\ \forall k \neq i,\ V_i = P_{ik} Q_k \Rightarrow P_{ik} = \frac{V_i}{Q_k}$; that is, when all conductors C_k for $k \neq i$ are discharged except for C_i. Furthermore, recalling that $Q = CV$, it is evident that this can be generalized to a multiconductor system $\{C_i\}$ via $\boldsymbol{Q} = \mathbf{C}\boldsymbol{V}$. Indeed, based on physical considerations, for passive devices—that is, devices that are not anisotropic (such as ferromagnetic materials, plasma) or active generators—it is clear that $\mathbf{P}$ is invertible and in particular, $\mathbf{P}^{-1} = \mathbf{C}$. Moreover, by Lemma 6.1.6, $\mathbf{C}$ is symmetric as it is the inverse of a symmetric matrix. These observations justify the following definition.

6.1.7 Definition

Given the pair V,Q such that $V = \mathbf{P}Q$, the coefficients P_{ij} are called the *coefficients of inductance*. Moreover,

$$P_{ij} = \frac{V_i}{Q_j}\bigg|_{Q_k=0\forall k\neq j}$$

Likewise, the coefficients C_{ij} of $Q = \mathbf{C}V$, are called the *coefficients of capacitance*, and

$$C_{ij} = \frac{Q_i}{V_j}\bigg|_{V_k=0\forall k\neq j}$$

when all conductors C_k for $k \neq i$ are grounded.

6.1.8 Remark

Physically, the coefficient of capacitance C_{ij}, for $i \neq j$, expresses the mutual capacitance between C_i, C_j. That is, it represents the capacitive coupling between C_i, C_j. On the other hand, C_{ii} represents the self-capacitance: it is the capacitance between $C_i, \bigcup_{j\neq i} C_j \cup C_0$, when $\bigcup_{j\neq i} C_j$ is grounded. Explicitly, consider $Q = \mathbf{C}V$:

$$Q_i = C_{i1}V_1 + C_{i2}V_2 + \cdots + C_{in}V_n \equiv c_{i1}(V_i - V_1) + \cdots + c_{ii}V_i + \cdots + c_{in}(V_i - V_n)$$

where C_{ij} denotes the mutual capacitance between C_i, C_j for $i \neq j$, and C_{ii} denotes the capacitance between C_i,C_0. Then, by definition, equating the coefficients of V_i,

$$C_{ij} = \begin{cases} \sum_{k\geq 1} c_{ik} & \text{for } i = j, \\ c_{ij} & \text{for } i \neq j. \end{cases}$$

That is, C_{ii} is the sum of the mutual capacitance between C_i,C_j for each $j \neq i$.

6.1.9 Lemma

Given the pair $Q = \mathbf{C}V$, $C_{ii} > 0$ and $C_{ij} < 0$ $\forall i, j$.

Proof

By definition,

$$C_{ij} = \frac{Q_i}{V_j}\bigg|_{V_k=0\forall k\neq j}$$

Hence, suppose without loss of generality that $V_j > 0$ as $V_j < 0$ follows *mutatis mutandis*. Then, $Q_j > 0 \Rightarrow Q_i < 0$, and $C_{ii} = \frac{Q_i}{V_i}$ implies that $V_i > 0 \Rightarrow Q_i > 0$. □

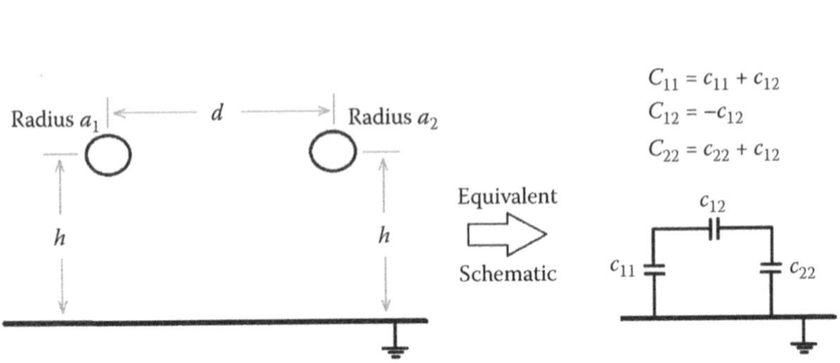

FIGURE 6.2
Capacitive coupling among conductors and the ground plane.

6.1.10 Example

The above theory is illustrated by applying it to a system of two conductors illustrated in Figure 6.2, where the conductors

$$C_i = \left\{(x,y) : (x + \tfrac{\chi_i}{2}d)^2 + (y - h_i)^2 \leq a_i^2\right\} \times \mathbf{R}$$

for $i = 1,2$, and

$$\chi_i = \begin{cases} 1 & \text{if } i = 1, \\ -1 & \text{if } i = 2. \end{cases}$$

Finally, set $C_i(0) = \{(x,y) : (x + \frac{\chi_i}{2}d)^2 + (y - h_i)^2 \leq a_i^2\}$.

By the method of images, the field resulting from $C_i(0)$ (see Chapter 3) can be determined as follows. Via the principle of superposition, there are two cases to be considered. First, consider the case wherein C_2 is absent. Let ρ_1 denote the charge per unit length. Then, the equivalent line charge in C_1 is displaced from the center $O_1 = (-\frac{1}{2}d, h)$ to $O_1' = (-\frac{1}{2}d, h - d_1)$, where $d_1 = \frac{a_1^2}{2h}$. By symmetry, the equivalent line charge ρ_2 in C_2, in the absence of C_1, is displaced from the center $O_2 = (\frac{1}{2}d, h)$ to $O_2' = (\frac{1}{2}d, h - d_2)$, where $d_2 = \frac{a_2^2}{2h}$.

Next, consider the pair C_1, C_2 in the absence of the ground plane. The locations of the equivalent line charges are determined as follows. First, without loss of generality, relative to a new origin $O_{\hat{\delta}} = (\hat{\delta}, 0)$, for some $\hat{\delta}$ such that the equivalent line charges are located, respectively, at $\hat{O}_1' = (-\hat{x}, h)$ and $\hat{O}_2' = (\hat{x}, h)$. Recalling from Example 3.2.4, it is clear that the respective x-coordinates of C_1, C_2 are

$$\hat{x}_i^2 = \hat{x}^2 - a_i^2 \tag{6.5}$$

for $i = 1,2$. Hence, $\hat{x}_1^2 + a_1^2 = \hat{x}_2^2 + a_2^2 \Rightarrow \hat{x}_1^2 - \hat{x}_2^2 = a_2^2 - a_1^2$. That is,

$$(\hat{x}_1 - \hat{x}_2)(\hat{x}_1 + \hat{x}_2) = a_2^2 - a_1^2$$

Furthermore, observe by definition that $d = \hat{x}_1 + \hat{x}_2$. Whence, $\hat{x}_1 - \hat{x}_2 = \frac{a_2^2 - a_1^2}{d}$ and thus

$$\hat{x}_1 = \frac{d^2 + a_1^2 - a_2^2}{2d} \tag{6.6}$$

$$\hat{x}_2 = \frac{d^2 + a_2^2 - a_1^2}{2d} \tag{6.7}$$

Without loss of generality, suppose that $a_1 < a_2$. Then, by definition, $\hat{\delta} < 0$ with respect to $x = 0$, and in particular, $\hat{x}_1 + |\hat{\delta}| = \frac{1}{2}d \Rightarrow |\hat{\delta}| = \frac{a_2^2 - a_1^2}{2d}$. That is, $\hat{\delta} = \frac{a_1^2 - a_2^2}{2d}$. Thus, in the absence of the ground plane, the equivalent charges are located, with respect to the origin $O = (0,0)$, at $O_1'' = (-\hat{x} + \hat{\delta}, h)$ and $O_2'' = (\hat{x} + \hat{\delta}, h)$, respectively.

From the above analysis, it is clear that the new locations of the equivalent line charges are, respectively, $\bar{O}_1 = (-\hat{x} + \hat{\delta}, h - d_1)$ and $\bar{O}_2 = (\hat{x} + \hat{\delta}, h - d_2)$. To see this, it suffices to note that in the former case (see Figure 6.3) the presence of the ground plane exerts a force f_1' on the equivalent line charge located at the axis of C_1 such that it is translated to $O_1' = (-\frac{1}{2}d, h - d_1)$, and in the latter scenario, the presence of the grounded conductor C_2 exerts a force f_1'' on the equivalent line charge located at the axis of C_1 such that it is translated to $O_1'' = (-\hat{x} + \hat{\delta}, h)$. Hence, the resultant location in the presence of the ground plane and C_1 leads to the translation of the line charge initially at $O_1 = (-\frac{1}{2}d, h)$ to $\bar{O}_1 = (-\hat{x} + \hat{\delta}, h - d_1)$ under the force $f_1 = f_1' + f_1''$. The new location $\bar{O}_2$ of the line charge in C_2 follows that of $\bar{O}_1$ *mutatis mutandis*.

Consequently, the resultant equivalent line charges and their respective image charges are:

$$\bar{O}_1 = (-\hat{x} + \hat{\delta}, h - d_1),\ \bar{O}_2 = (\hat{x} + \hat{\delta}, h - d_2),\ \bar{O}_1^* = (-\hat{x} + \hat{\delta}, -h + d_1),$$
$$\bar{O}_2^* = (\hat{x} + \hat{\delta}, -h + d_2)$$

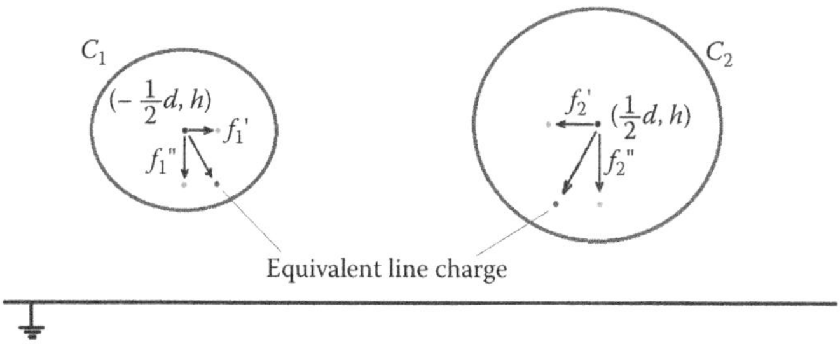

FIGURE 6.3
Equivalent line charges for the two respective charged conductors.

Next, recall from Example 3.2.4 that the potential of an infinite line charge λ is $\varphi = -\frac{\lambda}{2\pi\varepsilon_0}\ln\frac{r}{r_\infty}$, for some reference point r_∞; it follows at once that the potential above the ground plane is

$$\Phi = -\frac{\lambda}{2\pi\varepsilon_0}\ln\sqrt{\frac{(x+\hat{x}-\hat{\delta})^2+(y-h+d_1)^2}{(x+\hat{x}-\hat{\delta})^2+(y+h-d_1)^2}} + \frac{\lambda}{2\pi\varepsilon_0}\ln\sqrt{\frac{(x-\hat{x}-\hat{\delta})^2+(y-h+d_2)^2}{(x-\hat{x}-\hat{\delta})^2+(y+h-d_2)^2}},\ y>0 \tag{6.8}$$

where, without loss of generality, λ is any fixed line charge on C_1 and $-\lambda$ the charge on C_2. Then, by definition, the capacitance per unit length is between C_1, C_2 is $C_{12} = \frac{\lambda}{\Phi}$. That is,

$$C_{12} = 2\pi\varepsilon_0\left\{\ln\sqrt{\frac{(x+\hat{x}-\hat{\delta})^2+(y-h+d_1)^2}{(x+\hat{x}-\hat{\delta})^2+(y+h-d_1)^2}\,\frac{(x-\hat{x}-\hat{\delta})^2+(y+h-d_2)^2}{(x-\hat{x}-\hat{\delta})^2+(y-h+d_2)^2}}\right\}^{-1} \tag{6.9}$$

Similarly, the coefficient of capacitance per unit length between C_1 and ground is

$$c_{11} = 2\pi\varepsilon_0\left\{\ln\sqrt{\frac{(x+\hat{x}-\hat{\delta})^2+(y-h+d_1)^2}{(x+\hat{x}-\hat{\delta})^2+(y+h-d_1)^2}}\right\}^{-1} \tag{6.10}$$

and that between C_2 and ground is

$$c_{22} = 2\pi\varepsilon_0\left\{\ln\sqrt{\frac{(x-\hat{x}-\hat{\delta})^2+(y-h+d_2)^2}{(x-\hat{x}-\hat{\delta})^2+(y+h-d_2)^2}}\right\}^{-1}, \tag{6.11}$$

as required. □

6.2 Mutual Inductance and Mutual Impedance

From Chapter 5, it is clear that cross-talk is also contributed by mutual inductance. Thus, along the vein of Section 6.1, consider a system of $N + 1$ conductors running parallel to one another, each of length ℓ. Let the 0th conductor be a grounded conductor and the remaining N conductors be measured relative to the grounded conductor.

Let the ith conductor C_i carry a current I_i and $\gamma_i = \partial C_i(0)$ be a circular path encircling the boundary of a cross-section of the conductor. Then, invoking Kirchhoff's current law, $I_1 + \cdots + I_N = -I_0$. That is, the ground conductor provides the return current pathway for the circuit.

At any point (x,y,z), the total magnetic field intensity is given by the sum of magnetic field contributions from the $N + 1$ conductors:

$$B(x,y,z) = \sum_{i=0}^{N} B_i(x,y,z) \tag{6.12}$$

where, without loss of generality, set $z = 0$ in all that follows. Furthermore, for simplicity, assume that the conductor cross-sectional areas S_i' are constants. Then,

$$\boldsymbol{B}_i(x,y,z) = \tfrac{\mu_0}{4\pi} \iint_{S_i'} d^2S' J_i(x_i',y_i',z')\left\{(x-x_i')^2+(y-y_i')^2+(z-z')^2\right\}^{-\frac{3}{2}} \begin{pmatrix} y_i'-y \\ x-x_i' \end{pmatrix} \tag{6.13}$$

with J_i being the surface current density flowing along conductor C_i, $x_i' = x' + x_i$, $y_i' = y' + y_i$ and (x_i, y_i, z) is the axis of C_i, for some fixed pair (x_i, y_i).

Next, recall that the magnetic flux Ψ_i per unit length intersecting the area $S(\gamma_i)$ spanned by a loop γ_i around $C_i(0)$ is given by $\Psi_i = \lim_{\delta z \to 0} \tfrac{1}{\delta z} \iint_{\partial S(\gamma_i)\times[0,\delta z]} \boldsymbol{B}\cdot \boldsymbol{n}_i d^2 x$, where $\boldsymbol{n}_i$ is the unit vector normal to $\partial S(\gamma_i)\times[0,\delta z]$; see Chapter 4 on transmission line theory. Hence, by Equation (6.12),

$$\Psi_i = \lim_{\delta z\to 0} \sum_k \tfrac{1}{\delta z} \iint_{\partial S(\gamma_k)\times[0,\delta z]} \boldsymbol{B}_k \cdot \boldsymbol{n}_k d^2 x \equiv \sum_k \mathrm{L}_{ik} I_k \tag{6.14}$$

where L_{ik} is the mutual inductance per unit length between C_i, C_k. See Figure 6.4. In particular, by Equation (6.13),

$$\mathrm{L}_{ij} = \tfrac{\Psi_i}{I_j}\Big|_{I_k=0,k\neq j} \tag{6.15}$$

That is, the mutual inductance between conductors i and j is defined by making conductors $k \neq j$ open circuits. Indeed, Equation (6.15) can be defined by the matrix:

$$\boldsymbol{\Psi} = \mathbf{L}\boldsymbol{I} \tag{6.16}$$

where

$$\boldsymbol{\Psi} = \begin{pmatrix} \Psi_1 \\ \vdots \\ \Psi_N \end{pmatrix}, \mathbf{L} = \begin{pmatrix} L_{11} & \cdots & L_{1N} \\ \vdots & \ddots & \vdots \\ L_{N1} & \cdots & L_{NN} \end{pmatrix}, \boldsymbol{I} = \begin{pmatrix} I_1 \\ \vdots \\ I_N \end{pmatrix}$$

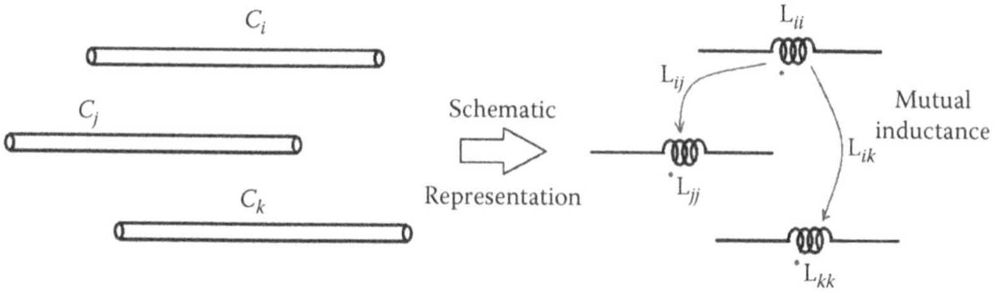

FIGURE 6.4
Voltage coupling via mutual inductance between two conductors.

6.2.1 Lemma

Given Equation (6.16), the matrix **L** is symmetric: $L_{ij} = L_{ji}\ \forall i, j$.

Proof

Just as with C_{ij}, L_{ij} depends only on geometry, and in particular, the coefficients are constants. Now, consider a loop γ_i bounded by (C_i, C_0) where C_0 is the ground conductor: $\gamma_i \cap \partial C_i \neq \varnothing$ and $\gamma_i \cap \partial C_0 \neq \varnothing$. Let $S(\gamma_i)$ denote the surface spanned by γ_i and $\boldsymbol{n}_i$ be the unit normal on $S(\gamma_i)$. Then, the magnetic flux Ψ_{ij} linking C_i with respect to γ_i due to current flowing in C_j is, by definition,

$$\Psi_{ij} = \int_{S(\gamma_i)} \boldsymbol{B}_j \cdot \boldsymbol{n}_i d^2x = \tfrac{\mu_0 I_j}{4\pi} \int_{S(\gamma_i)} \int_{\gamma_j} \nabla \times \tfrac{dl_j}{R} \cdot \boldsymbol{n}_i d^2x$$

By Stokes' theorem,

$$\int_{S(\gamma_j)} \nabla \times \tfrac{dl_j}{R} \cdot \boldsymbol{n}_i d^2x = \int_{\gamma_j} \tfrac{dl_j}{R} \cdot \tfrac{dl_i}{R} \tag{6.17}$$

whence,

$$\Psi_{ij} = \tfrac{\mu_0 I_j}{4\pi} \int_{S(\gamma_i)} \int_{\gamma_j} \nabla \times \tfrac{dl_j}{R} \cdot \boldsymbol{n}_i d^2x = \tfrac{\mu_0 I_j}{4\pi} \int_{\gamma_i} \int_{\gamma_j} \frac{dl_j \cdot dl_i}{R}$$

By definition,

$$L_{ij} \equiv \tfrac{\Psi_{ij}}{I_j} = \frac{\mu_0}{4\pi} \int_{\gamma_i} \int_{\gamma_j} \tfrac{dl_j}{R} \cdot \tfrac{dl_i}{R} = \tfrac{\mu_0}{4\pi} \int_{\gamma_j} \int_{\gamma_i} \frac{dl_i \cdot dl_j}{R} \equiv L_{ji}$$

and hence, the arbitrariness of γ_i, γ_j concludes the proof. □

6.2.2 Example

The above theory is illustrated by applying it to a system of two parallel conductors of radii $a > 0$ (*cf.* Figure 6.5), where the lines are assumed to be infinitely long, separated by a distance $d > a$, and their respective axes are a distance $h > 0$ above the ground plane.

Determine the resultant voltage at the loads of C_1, C_2 respectively.

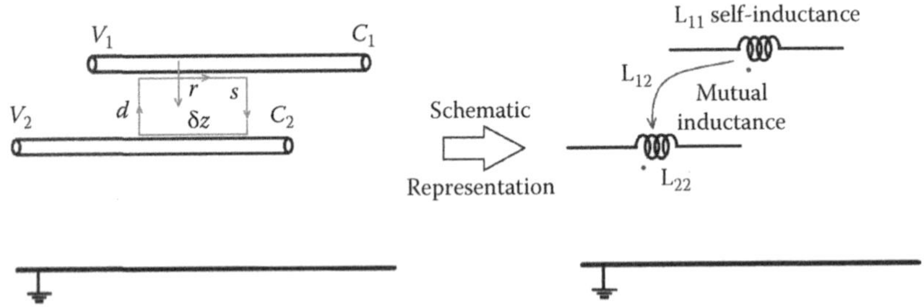

FIGURE 6.5
Voltage coupling via mutual inductance.

Now, via superposition principle, it suffices to consider $V_2 = 0$ and determine the induced voltage δV_{21} along C_2. Then, the resultant voltage on C_2 is thus $V_2 + \delta V_{21}$ to first order, because δV_{21} will in turn induce a voltage δv_{21} on C_1, and this voltage will in turn induce a voltage $\delta^2 V_{21} \equiv \delta(\delta V_{21})$ on C_2, ad infinitum: $V_2 \to V_2 + \delta V_{21} + \delta^2 V_{21} + \delta^3 V_{21} + \cdots$. By symmetry, on setting $V_1 = 0$, the resultant voltage is, to first order, $V_1 + \delta V_{21}$.

Therefore, suppose that $V_2 = 0$ and $V_1 = V_1(t)$ is some time harmonic voltage propagating along C_1. Then, $\delta V_{21} = -\mathrm{L}_{21}\frac{dI_1}{dt}$. From Chapter 5, recall that $\mathrm{L}_{21} = \frac{\Psi_2}{I_1}\Big|_{I_2=0}$, where $\Psi_2|_{I_2=0}\,\delta z = \int_S \boldsymbol{B}|_{I_2=0}\cdot \boldsymbol{n}d^2x$ with $\boldsymbol{n}$ being the unit normal vector field on S and $\boldsymbol{B}|_{I_2=0}$ is determined by the field along C_1 and the return current I_0'' along C_2, where $I_0 + I_1 = 0$ and $I_0 = I_0' + I_0''$, with I_0' being the return current on the ground plane.

By inspecting Figure 6.5 (*cf.* Example 6.1.8) it is clear that the boundary charge density induced on $C_2(0)$ is given by $\rho_{s,2} = -\varepsilon_0 \nabla\Phi\cdot\boldsymbol{n}_2$, where Φ is defined by Equation (6.8) and $\boldsymbol{n}_2$ is the unit normal vector field on $\partial C_2(0)$. Thus, the equivalent return current density along C_2 is

$$J_0'' = \frac{d}{dt}\int_{\partial C_2(0)} \rho_{s,2}(t)\,dl = \frac{d\lambda(t)}{dt}\frac{1}{2\pi}\int_{\partial C_2(0)} \nabla \ln\sqrt{\frac{(x+\hat{x}-\hat{\delta})^2+(y+h-d_1)^2}{(x+\hat{x}-\hat{\delta})^2+(y-h+d_1)^2}\,\frac{(x-\hat{x}-\hat{\delta})^2+(y-h+d_2)^2}{(x-\hat{x}-\hat{\delta})^2+(y+h-d_2)^2}}\cdot\boldsymbol{n}_2\,dl \tag{6.18}$$

Likewise, the current density along C_1 is given by

$$J_1 = \frac{d}{dt}\int_{\partial C_1(0)} \rho_{s,1}(t)\,dl = \frac{d\lambda(t)}{dt}\frac{1}{2\pi}\int_{\partial C_1(0)} \nabla \ln\sqrt{\frac{(x+\hat{x}-\hat{\delta})^2+(y+h-d_1)^2}{(x+\hat{x}-\hat{\delta})^2+(y-h+d_1)^2}\,\frac{(x-\hat{x}-\hat{\delta})^2+(y-h+d_2)^2}{(x-\hat{x}-\hat{\delta})^2+(y+h-d_2)^2}}\cdot\boldsymbol{n}_1\,dl \tag{6.19}$$

From this, the magnetic density at an arbitrary point $(x,y)\in\mathbf{R}^2_+ - (C_1(0)\cup C_2(0))$ resulting from (C_1, C_2) is

$$\boldsymbol{B}_{12}|_{I_2=0} = \frac{\mu_0}{2\pi}\frac{d\lambda(t)}{dt}\pi a^2\left\{\frac{\kappa_1}{\sqrt{(x+\hat{x}-\hat{\delta})^2+(y-h+d_1)^2}} + \frac{\kappa_2''}{\sqrt{(x-\hat{x}-\hat{\delta})^2+(y-h+d_2)^2}}\right\}\mathbf{e}_\phi$$

where κ_1, κ_2'' are defined by Equations (6.18) and (6.19), respectively:

$$\kappa_1 = \frac{1}{2\pi}\int_{\partial C_1(0)} \nabla \ln\sqrt{\frac{(x+\hat{x}-\hat{\delta})^2+(y+h-d_1)^2}{(x+\hat{x}-\hat{\delta})^2+(y-h+d_1)^2}\,\frac{(x-\hat{x}-\hat{\delta})^2+(y-h+d_2)^2}{(x-\hat{x}-\hat{\delta})^2+(y+h-d_2)^2}}\cdot\boldsymbol{n}_1\,dl$$

$$\kappa_2'' = \frac{1}{2\pi}\int_{\partial C_2(0)} \nabla \ln\sqrt{\frac{(x+\hat{x}-\hat{\delta})^2+(y+h-d_1)^2}{(x+\hat{x}-\hat{\delta})^2+(y-h+d_1)^2}\,\frac{(x-\hat{x}-\hat{\delta})^2+(y-h+d_2)^2}{(x-\hat{x}-\hat{\delta})^2+(y+h-d_2)^2}}\cdot\boldsymbol{n}_2\,dl$$

By definition, the magnetic flux density per unit length across the rectangular loop $S = [a,d - a] \times [0,\delta z]$ at $y = h$ is:

$$\lim_{\delta z \to 0} \Psi_2 \big|_{I_2=0} = \lim_{\delta z \to 0} \tfrac{1}{\delta z} \tfrac{\mu_0}{2\pi} \tfrac{d\lambda}{dt} \int_0^{\delta z} \int_a^{d-a} \left\{ \frac{\kappa_1}{\sqrt{(x+\hat{x}-\hat{\delta})^2+d_1^2}} + \frac{\kappa_2''}{\sqrt{(x-\hat{x}-\hat{\delta})^2+d_2^2}} \right\} dx\, dz$$

$$= \tfrac{\mu_0}{2\pi} \tfrac{d\lambda}{dt} \left\{ \kappa_1 \ln \frac{\sqrt{(d-a+\hat{x}-\hat{d})^2+d_1^2}+d-a+\hat{x}-\hat{\delta}}{\sqrt{(a+\hat{x}-\hat{d})^2+d_1^2}+a+\hat{x}-\hat{\delta}} + \kappa_2'' \ln \frac{\sqrt{(d-a-\hat{x}-\hat{d})^2+d_2^2}+d-a-\hat{x}-\hat{\delta}}{\sqrt{(a-\hat{x}-\hat{d})^2+d_2^2}+a-\hat{x}-\hat{\delta}} \right\}$$

and hence,

$$L_{21} = \frac{\Psi_2}{I_1}\bigg|_{I_2=0} = \frac{\mu_0}{2\pi} \left\{ \kappa_1 \ln \frac{\sqrt{(d-a+\hat{x}-\hat{d})^2+d_1^2}+d-a+\hat{x}-\hat{\delta}}{\sqrt{(a+\hat{x}-\hat{d})^2+d_1^2}+a+\hat{x}-\hat{\delta}} + \kappa_2'' \ln \frac{\sqrt{(d-a-\hat{x}-\hat{d})^2+d_2^2}+d-a-\hat{x}-\hat{\delta}}{\sqrt{(a-\hat{x}-\hat{d})^2+d_2^2}+a-\hat{x}-\hat{\delta}} \right\}$$

That is, the resultant voltage along C_2 is thus

$$V_2 - \frac{\mu_0}{2\pi} \left\{ \kappa_1 \ln \frac{\sqrt{(d-a+\hat{x}-\hat{d})^2+d_1^2}+d-a+\hat{x}-\hat{\delta}}{\sqrt{(a+\hat{x}-\hat{d})^2+d_1^2}+a+\hat{x}-\hat{\delta}} + \kappa_2'' \ln \frac{\sqrt{(d-a-\hat{x}-\hat{d})^2+d_2^2}+d-a-\hat{x}-\hat{\delta}}{\sqrt{(a-\hat{x}-\hat{d})^2+d_2^2}+a-\hat{x}-\hat{\delta}} \right\} \frac{dI_1}{dt}$$

By symmetry, the resultant voltage along C_1 is

$$V_1 - \frac{\mu_0}{2\pi} \left\{ \kappa_1 \ln \frac{\sqrt{(d-a+\hat{x}-\hat{d})^2+d_1^2}+d-a+\hat{x}-\hat{\delta}}{\sqrt{(a+\hat{x}-\hat{d})^2+d_1^2}+a+\hat{x}-\hat{\delta}} + \kappa_2'' \ln \frac{\sqrt{(d-a-\hat{x}-\hat{d})^2+d_2^2}+d-a-\hat{x}-\hat{\delta}}{\sqrt{(a-\hat{x}-\hat{d})^2+d_2^2}+a-\hat{x}-\hat{\delta}} \right\} \frac{dI_2}{dt}$$ □

Having established the mutual inductance and capacitance for a system of conductors, the mutual impedance can now be defined. In view of Equations (6.4) and (6.16), define

$$\begin{aligned} V_1 &= Z_{11}I_1 + \cdots + Z_{1n}I_n \\ &\ddots \\ V_n &= Z_{n1}I_1 + \cdots + Z_{n1}I_n \end{aligned} \tag{6.20}$$

for n-conductors; that is, $\mathbf{V} = \mathbf{ZI}$. Then, it is clear by definition that for each fixed i, $V_i = Z_{ij}I_j$ if $I_k = 0\ \forall k \neq j$. That is, $Z_{ij} \equiv \frac{V_i}{I_j}\big|_{I_k=0\ \forall k\neq j}$ is well-defined as Z_{ij} are constants for all $i,j = 1,\ldots, n$. Thus, the *mutual impedance* Z_{ij} is obtained by keeping each conductor $C_k\ \forall k \neq j$ an open circuit. The matrix **Z** is also called the *transfer impedance matrix* or simply the *Z-matrix*.

6.2.3 Example

Consider the schematic diagram shown in Figure 6.6. It illustrates qualitatively the impact of a nonzero, finite, mutual impedance between two conductors C_1, C_2. In the illustrated scenario, C_1 is assumed to be an open circuit

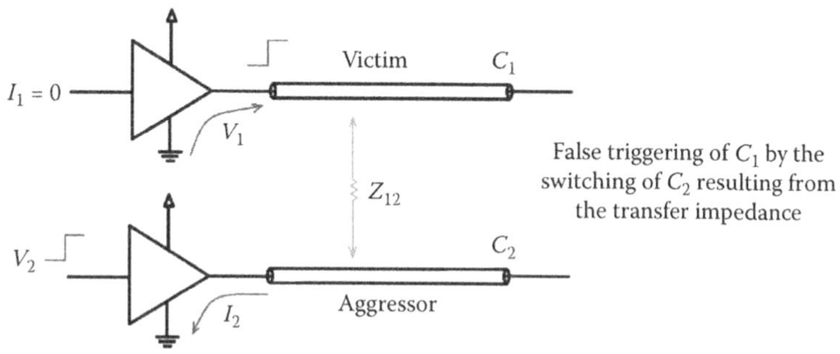

FIGURE 6.6
Induced voltage along victim line via transfer impedance.

at the load (not illustrated): $I_1 = 0$. When C_2 is switching and hence $I_2 \neq 0$, the transfer impedance induces a potential difference along C_1 according to: $V_1 = Z_{12}I_2$. From an application perspective, if V_1 is sufficiently large—for instance, large enough to switch on a transistor—then this will cause a false bit to be transmitted, causing signal integrity issues. In this particular scenario, the effect is called a *ground bounce* wherein the sudden current draw by C_2 induces a voltage spike along C_1. □

Returning to Equation (6.20), it is clear that the transfer matrix **Z** for passive physical systems (i.e., networks) encountered is invertible: if this were false, then a fixed set of currents $\{I_i\}$ would induce multivalued voltages $\{V_i^{(k)}\}$, which is clearly not observed in passive physical networks: that is, the voltages are not uniquely defined. Hence, for most practicable applications,

$$I = \mathbf{Y}V = \mathbf{YZ}I \equiv I \Leftrightarrow \mathbf{YZ} = \mathbf{I}$$

where **I** is the identity matrix. That is, $\mathbf{Z}^{-1} = \mathbf{Y}$ exists, assuming a linear passive network, and there is thus a one-to-one correspondence between voltage and current for ohmic systems, as required. For obvious reasons, **Y** is called the *mutual admittance matrix.*

6.2.4 Lemma

Given a linear passive network defined by Equation (6.20), the mutual impedance matrix **Z** is symmetric.

Proof

Now, for any fixed i, $I_k = 0\ \forall k \neq j \Rightarrow V_i = Z_{ij}I_j \Leftrightarrow V_iI_j = Z_{ij}I_j^2$. Likewise, for any fixed j, $I_k = 0\ \forall k \neq i \Rightarrow V_j = Z_{ji}I_i \Leftrightarrow V_jI_i = Z_{ji}I_i^2$. Hence, by the reciprocity theorem via Remark 6.1.4, $V_iI_j = V_jI_i \Rightarrow Z_{ij} = Z_{ji}$ on setting $I_i = I_j$ (as I_i, I_j are arbitrary and Z_{ij} are constants). □

Indeed, it is clear by definition that $Z_{mn} \equiv -\frac{1}{I_n}\int_{\gamma_{mn}}(\mathbf{E}_m - \mathbf{E}_n)\cdot d\boldsymbol{l}$, where $I_k = 0\ \forall k \neq n$ and $\mathbf{E}_k$ is the electric field resulting from conductor C_k. That is, the mutual impedance measures the potential difference between C_m and C_n when current I_n is conducting along C_n, with the remaining conductors kept open.

6.2.5 Definition

Given a linear system defined by $\mathbf{V} = \mathbf{Z}\boldsymbol{I}$, the system is said to be lossless if $\Re\mathrm{e}(Z_{ij}) = 0\ \forall i,j$. In particular, $\mathbf{Z} = \mathrm{i}\hat{\mathbf{Z}}$, where $\hat{Z}_{ij} = \Im\mathrm{m}(Z_{ij})$ for all i,j. Finally, on setting $\mathbf{Y} = \mathbf{Z}^{-1}$, $\boldsymbol{I} = \mathbf{Y}\mathbf{V}$, where $Y_{ij} = \frac{I_i}{V_j}|_{V_k=0\ \forall k\neq j}$; that is, C_k is grounded $\forall k \neq j$.

6.2.6 Remark

For a lossless linear network, the time-average power $\langle P\rangle = \frac{1}{2}\Re\mathrm{e}(\boldsymbol{V}\cdot\boldsymbol{I}^*)$ implies that $\langle P\rangle = \frac{1}{2}\Re\mathrm{e}\langle \boldsymbol{ZI},\boldsymbol{I}^*\rangle = \frac{1}{2}\Re\mathrm{e}\sum_{i,j}\mathrm{i}\hat{Z}_{ij}I_iI_j^* = 0$. That is, the net power delivered to the network is zero. Thus, the energy is stored reactively.

6.2.7 Example

Consider a 2-port network illustrated in Figure 6.7. Is it possible to express the general 2-port network as Figure 6.7(a) or (b)? Because **Z** is symmetric, it depends only on three parameters instead of four: $\{Z_{11}, Z_{12}, Z_{22}\}$. However, Figure 6.7(a) or (b) depends only on three parameters $\{R_1, R_2, R_3\}$ or $\{r_1, r_2, r_3\}$, respectively. Hence, the 2-port network can be represented by (a) or (b).

Now, consider the T-network depicted in Figure 6.7(a). Express Z_{ij} in terms of R_k for $k = 1,2,3$. By definition, $Z_{11} = \frac{V_1}{I_1}$ with $I_2 \equiv 0$. Then, inspecting Figure 6.7(a), $Z_{11} = R_1 + R_3$, as $I_2 = 0 \Rightarrow R_2$-branch is an open circuit. Likewise,

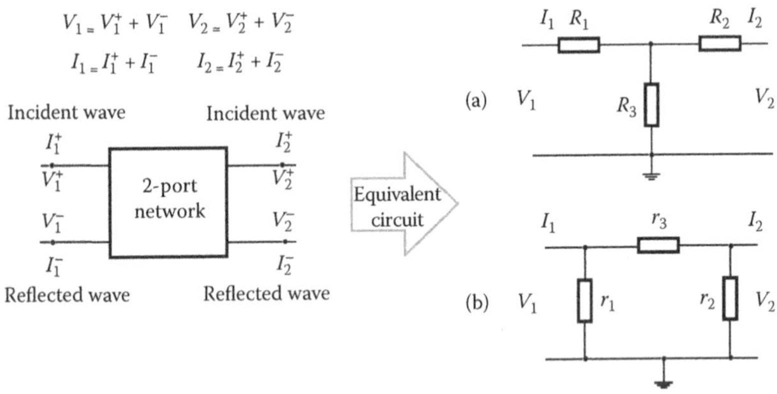

FIGURE 6.7
A general 2-port network and its equivalent circuit representation.

$Z_{22} = \frac{V_2}{I_2}$ with $I_1 = 0$ leads to the R_1-branch of the circuit being open. Hence, $Z_{22} = R_2 + R_3$. Finally, for $Z_{12} = \frac{V_1}{I_2}|_{I_1=0}$, $I_1 = 0$ implies that the potential drop across R_3 is V_1. Thus, $V_1 = R_3 I_2 \Rightarrow Z_{12} = R_3$. Indeed, it is easy to see that $Z_{21} = Z_{12}$: $I_2 = 0 \Rightarrow V_2 = R_3 I_1 \Rightarrow Z_{21} = R_3 = Z_{12}$. Equivalently, expressing R_i in terms of Z_{ij} yields: $R_1 = Z_{11} - Z_{12}$, $R_2 = Z_{22} - Z_{12}$, $R_3 = Z_{12}$.

Next, consider Figure 6.7(b): a Π-network. Set $Y_i = r_i \; \forall i = 1,2,3$. Then, in terms of admittance, via Definition 6.2.5, on setting $V_2 = 0$ and $V_1, I_1 \neq 0$, $Y_{11} = \frac{I_1}{V_1} = Y_1 + Y_3$. Thus,

$$Z_{11} = \frac{1}{Y_1+Y_3} \frac{r_1 r_3}{r_1+r_3}$$

Likewise, on setting $V_1 = 0$ and $V_2, I_2 \neq 0$, $Y_{22} = \frac{I_2}{V_2} = Y_2 + Y_3$, and hence, $Z_{22} = \frac{1}{Y_2+Y_3} = \frac{r_2 r_3}{r_2+r_3}$. Finally, setting $V_1 = 0$ and $V_2, I_1 \neq 0$, it follows from Kirchhoff's current law that $I_1 + Y_3 V_2 = 0 \Rightarrow Y_3 = -\frac{I_1}{V_2} \equiv Y_{12}$ by definition. Moreover, it is easy to see that for $V_2 = 0$ and $V_1, I_2 \neq 0$, $I_2 + Y_3 V_1 = 0 \Rightarrow Y_3 = -\frac{I_2}{V_1} \equiv Y_{21} = Y_{12}$, as expected. Finally, for completeness, expressing Y_i in terms of Y_{ij} yields:

$$Y_1 = Y_{11} + Y_{12}, \; Y_2 = Y_{22} + Y_{12} \quad \text{and} \quad Y_3 = -Y_{12}$$

In particular,

$$r_1 = \frac{1}{Y_{11}+Y_{12}}, \; r_2 = \frac{1}{Y_{22}+Y_{12}} \quad \text{and} \quad r_3 = -\frac{1}{Y_{12}}$$ □

6.3 Multiconductor Transmission Lines and Cross-Talk

In this section, a pair of transmission lines is generalized to $n > 2$ transmission lines. For simplicity, assume that the conductors $\{C_i\}$ are parallel to one another, embedded in a homogeneous medium $(\mathbf{R}^3, \mu, \varepsilon)$, where the conductor C_0 is taken to be the ground reference. Indeed, the analysis for multitransmission lines parallels that of the differential pair expounded in Section 5.1.

6.3.1 Theorem

Given a system of multitransmission lines $\{C_i\}$ in $\Omega = (\mathbf{R}^3, \mu, \varepsilon) - \cup_{i=0}^{n} C_i$, where the cross-sections of C_i are constants, and the conductor losses are sufficiently small so that the TEM approximation may be employed, the system is then defined by

$$\partial_z \boldsymbol{V} = -\mathbf{R}\boldsymbol{I} - \mathbf{L}\partial_t \boldsymbol{I} \tag{6.21}$$

$$\partial_z \boldsymbol{I} = -\mathbf{G}\boldsymbol{V} - \mathbf{C}\partial_t \boldsymbol{V} \tag{6.22}$$

where $\boldsymbol{V} = (V_1, \ldots, V_n)$, $\boldsymbol{I} = (I_1, \ldots, I_n)$

$$\mathbf{R} = \begin{bmatrix} R_1 + R_0 & R_0 & \cdots & R_0 \\ R_0 & R_2 + R_0 & & \vdots \\ \vdots & & \ddots & R_0 \\ R_0 & \cdots & R_0 & R_n + R_0 \end{bmatrix},$$

$$\mathbf{L} = \begin{bmatrix} L_{11} & L_{12} & \cdots & L_{1n} \\ L_{21} & L_{22} & \cdots & L_{2n} \\ \vdots & & \ddots & \vdots \\ L_{n1} & L_{n2} & \cdots & L_{nn} \end{bmatrix}$$

$$\mathbf{G} = \begin{bmatrix} \sum_{i=1}^{n} G_{1i} & -G_{12} & \cdots & -G_{1n} \\ -G_{21} & \sum_{i=1}^{n} G_{2i} & \cdots & -G_{2n} \\ \vdots & \vdots & \ddots & \vdots \\ -G_{n1} & -G_{n2} & \cdots & \sum_{i=1}^{n} G_{ni} \end{bmatrix},$$

$$\mathbf{C} = \begin{bmatrix} \sum_{i=1}^{n} C_{1i} & -C_{12} & \cdots & -C_{1n} \\ -C_{21} & \sum_{i=1}^{n} C_{2i} & \cdots & -C_{2n} \\ \vdots & \vdots & \ddots & \vdots \\ -C_{n1} & -C_{n2} & \cdots & \sum_{i=1}^{n} C_{ni} \end{bmatrix}$$

R_i is the resistance per unit length along C_i, and L_{ij}, G_{ij}, C_{ij} are, respectively, the mutual inductance, mutual admittance, and mutual capacitance per unit length between C_i and C_j.

Proof

Essentially, the argument follows that of Section 5.1 *mutatis mutandis*. First, recall that $I_0 + \sum_{i=1}^{n} I_i = 0$, where I_0 is the return current along the ground conductor C_0. Then, the voltage variation along C_i is determined as follows. Consider a differential rectangular loop between C_i and C_0; see Figure 5.2, where C_+ corresponds to C_i and C_- corresponds to C_0. Let $\boldsymbol{E}_\perp$ denote the electric field from $\{C_i\}$ normal to $\{C_i\}$, and $\boldsymbol{E}_{i,\parallel}$ the electric field on ∂C_i and parallel to C_i, where, by assumption, $\|\boldsymbol{E}_{i,\parallel}\| << \|\boldsymbol{E}_\perp\|$. Likewise, let $\boldsymbol{B}_\perp$ denote the magnetic field density from $\{C_i\}$ normal to $\{C_i\}$.

Next, invoke Maxwell's integral equation: $\int_{\partial S_i} \boldsymbol{E}_i \cdot \mathrm{d}\boldsymbol{l} = -\frac{\mathrm{d}}{\mathrm{d}t}\int_{S_i} \boldsymbol{B}_\perp \cdot \boldsymbol{n}\mathrm{d}^2x$, where $\boldsymbol{E}_i = \boldsymbol{E}_\perp + \boldsymbol{E}_{i,\|}$. Then, along the loop $\partial S_i = \gamma_\uparrow \cup \gamma_\rightarrow \cup \gamma_\downarrow \cup \gamma_\leftarrow$, where $a \xrightarrow{\gamma_\uparrow} b$, $b \xrightarrow{\gamma_\rightarrow} c$, $c \xrightarrow{\gamma_\downarrow} d$, and lastly, $d \xrightarrow{\gamma_\leftarrow} a$; then by definition, $\int_{\partial S_i} \boldsymbol{E}_\perp \cdot \mathrm{d}\boldsymbol{l}$ yields:

$$V_i(z) = -\int_a^b \boldsymbol{E}_\perp \cdot \mathrm{d}\boldsymbol{l}$$

$$\delta V_i = -\int_c^b \boldsymbol{E}_{i,\|} \cdot \mathrm{d}\boldsymbol{l} \equiv \mathrm{R}_i \delta z I_i$$

$$V_i(z+\delta z) = -\int_c^b \boldsymbol{E}_\perp \cdot \mathrm{d}\boldsymbol{l} = \int_b^c \boldsymbol{E}_\perp \cdot \mathrm{d}\boldsymbol{l}$$

$$\delta V_{i,0} = -\int_c^b \boldsymbol{E}_{i,\|} \cdot \mathrm{d}\boldsymbol{l} \equiv \mathrm{R}_0 \delta z I_0 = -\mathrm{R}_0 \delta z \sum_{k=1}^n I_k$$

whence $\int_{\partial S_i} \boldsymbol{E}_\perp \cdot \mathrm{d}\boldsymbol{l} = -V_i(z) + \mathrm{R}_i \delta z I_i + V_i(z+\delta z) + \mathrm{R}_0 \delta z \sum_{k=1}^n I_k$, where the time variable is suppressed for notational convenience, and $|\gamma_\rightarrow| = \delta z = |\gamma_\leftarrow|$.

Similarly, considering the surface area S_i spanned by the loop ∂S_i, the magnetic flux across S_i is $\Psi_i = \int_{S_i} \boldsymbol{B}_\perp \cdot \boldsymbol{n}\mathrm{d}^2x$. Because $\boldsymbol{B}_\perp$ is the contribution from each C_k, it thus follows by definition $\Psi = LI$ that $\psi_i \equiv \lim_{\delta z \to 0} \frac{1}{\delta z}\Psi_i = \mathrm{L}_{i1}I_1 + \cdots + \mathrm{L}_{in}I_n$ defines the magnetic flux per unit length cutting the surface area S_i. Thus,

$$\partial_t \psi_i \equiv \lim_{\delta z \to 0} \frac{1}{\delta z}\partial_t \Psi_i = \lim_{\delta z \to 0} \frac{1}{\delta z}\frac{\mathrm{d}}{\mathrm{d}t}\int_{S_i} \boldsymbol{B}_\perp \cdot \boldsymbol{n}\mathrm{d}^2x = \mathrm{L}_{i1}\,\partial_t I_1 + \cdots + \mathrm{L}_{in}\,\partial_t I_n$$

where we recall by definition that

$$\mathrm{L}_{ij} = \left.\frac{\psi_i}{I_j}\right|_{I_k = 0\ \forall k \neq j}$$

yielding

$$\begin{aligned} \partial_z V_i &= -\mathrm{R}_i I_i + \mathrm{R}_0 I_0 - \mathrm{L}_{i1}\,\partial_t I_1 - \cdots - \mathrm{L}_{in}\,\partial_t I_n \\ &= -\mathrm{R}_0 I_1 - \cdots - (\mathrm{R}_0 + \mathrm{R}_i)I_i - \cdots - \mathrm{R}_0 I_n - \mathrm{L}_{i1}\,\partial_t I_1 - \cdots - \mathrm{L}_{in}\,\partial_t I_n \end{aligned}$$

Likewise, the current variation along C_i is determined by inspecting Figure 5.3, and invoking Maxwell's equation in integral form $\int_S \nabla \times \boldsymbol{B} \cdot \boldsymbol{n}\mathrm{d}^2x = \mu\int_S \boldsymbol{J} \cdot \boldsymbol{n}\mathrm{d}^2x + \mu\frac{\mathrm{d}}{\mathrm{d}t}\int_S \boldsymbol{E} \cdot \boldsymbol{n}\mathrm{d}^2x$, for any compact surface S. Indeed, it is instructive to summarize the argument carried out in Section 5.1 again.

First, consider some differential cylinder c_i of length δz around c_i, let $C_i^-(z), C_i^+(z+\delta z) \subset C_i$ denote the end caps of the cylinder, and set $\partial C_i = \partial C_i^\circ \cup C_i^-(z) \cup C_i^+(z+\delta z)$. By construction, if ∂C_i denotes the boundary of the cylinder C_i, then $\partial(\partial C_i) \equiv \partial^2 C_i \equiv \varnothing$ (as the boundary of a boundary is the empty set). Hence, by Stokes' theorem, $\int_{\partial C_i} \nabla \times \boldsymbol{B} \cdot \boldsymbol{n} d^2x = \int_{\partial^2 C_i} \boldsymbol{B} \cdot d\boldsymbol{l} = 0$. Thus,

$$0 = \mu \int_{\partial C_i} \boldsymbol{J} \cdot \boldsymbol{n} d^2x + \mu\varepsilon \tfrac{d}{dt} \int_{\partial C_i} \boldsymbol{E} \cdot \boldsymbol{n} d^2x$$

Now,

$$\int_{\partial C_i} \boldsymbol{J} \cdot \boldsymbol{n} d^2x = \int_{\partial C_i^\circ} \boldsymbol{J} \cdot \boldsymbol{n}_\perp d^2x - \int_{C_i^-(z)} \boldsymbol{J} \cdot \boldsymbol{n}_\| d^2x + \int_{C_i^+(z+\delta z)} \boldsymbol{J} \cdot \boldsymbol{n}_\| \, d^2x$$

where $\boldsymbol{n}_\perp (\boldsymbol{n}_\|)$ is a unit normal vector field on $\partial C_i^\circ (C_i^\pm)$. By definition, $\int_{C_i^-(z)} \boldsymbol{J} \cdot \boldsymbol{n}_\| \, d^2x = -I_i(z)$ and $\int_{C_i^+(z+\delta z)} \boldsymbol{J} \cdot \boldsymbol{n}_\| \, d^2x = I_i(z+\delta z)$. Furthermore, $\int_{\partial C_i^\circ} \boldsymbol{J} \cdot \boldsymbol{n}_\perp d^2x \approx \sigma \int_{\partial C_i^\circ} \boldsymbol{E}_\perp \cdot \boldsymbol{n}_\perp d^2x$, where the quasi-TEM approximation is invoked: $\boldsymbol{E}_\| \approx 0$. And from $V = IR$, and noting that $\int_{\partial C_i^\circ} \boldsymbol{J} \cdot \boldsymbol{n}_\perp d^2x$ defines the total conduction current across ∂C_i° between C_i, C_j along the segment $[z, z + \delta z]$ of C_i, for each $j = 1, \ldots, n$,

$$\sigma \int_{\partial C_i^\circ} \boldsymbol{E}_\perp \cdot \boldsymbol{n}_\perp d^2x \equiv \sum_{j \neq i} Y_{ij} \delta z (V_i - V_j) + Y_{ii} \delta z (V_i - 0)$$

where V_k is the potential of C_k with respect to C_0, and Y_{ij} is the admittance per unit length between C_k, C_j, and finally, Y_{ii} is the admittance per unit length between C_i, C_0.

Next, $\mu\varepsilon \tfrac{d}{dt} \int_{\partial C_i} \boldsymbol{E} \cdot \boldsymbol{n} d^2x = \mu\varepsilon \tfrac{d}{dt} \int_{\partial C_i^\circ} \boldsymbol{E}_\perp \cdot \boldsymbol{n}_\perp d^2x$. Thus, via Gauss' theorem and $Q = CV$,

$$\varepsilon \int_{\partial C_i^\circ} \boldsymbol{E}_\perp \cdot \boldsymbol{n}_\perp d^2x = C_{ii} \delta z V_i + \sum_{j \neq i} C_{ij} \delta z (V_i - V_j)$$

where C_{ij} is the mutual capacitance per unit length between C_i, C_j, and C_{ij} is the capacitance between C_i, C_0. whence, in the limit as $\delta z \to 0$,

$$\partial_z I_i = -\left(-G_{i1} V_i - \cdots + \sum_{j=1}^{n} G_{ij} V_j - \cdots - G_{in} V_n \right)$$

$$-(-C_{i1} \partial_t V_1 - \cdots + \sum_{j=1}^{n} C_{ij} \partial_t V_j - \cdots - C_{in} \partial_t V_n)$$

and the theorem is thus established. □

6.3.2 Corollary

Given the conditions of Theorem 6.3.1, the following wave equations are satisfied:

$$\partial_z^2 V = \mathbf{RG}V + (\mathbf{RC} + \mathbf{LG})\partial_t V + \mathbf{LC}\partial_t^2 V \tag{6.23}$$

$$\partial_z^2 \boldsymbol{I} = \mathbf{GR}\boldsymbol{I} + (\mathbf{CR} + \mathbf{GL})\partial_t \boldsymbol{I} + \mathbf{CL}\partial_t^2 \boldsymbol{I} \tag{6.24}$$

In particular, $\mathbf{LC} = \mu\varepsilon\mathbf{I} = \mathbf{CL}$ and $\mathbf{LG} = \mu\sigma\mathbf{I} = \mathbf{GL}$, where $\boldsymbol{I}$ is the identity matrix, and hence, the waves propagate along C_i at $v = \frac{1}{\sqrt{\mu\varepsilon}}$.

Proof

The proof is similar to that in Chapter 4; the proof is left as a warm-up exercise (see Exercise 6.5.3). □

6.3.3 Remark

From Corollary 6.3.2, it follows immediately that in a homogeneous medium, $\mathbf{L} = \mu\varepsilon\mathbf{C}^{-1}$ and $\mathbf{G} = (\sigma/\varepsilon)\mathbf{C}^{-1}$; see Chapter 4 for a transmission line pair.

Now, the phenomenon of cross-talk outlined in Chapter 5 can be extended once again to multiconductor transmission lines. In particular, the analysis follows that of Chapter 5 along a similar vein. To facilitate the discussion, the following definition is made for notational convenience.

6.3.4 Definition

Given a system of n transmission lines $\{C_i\}$, define $\bigcup_{i=1}^{n} C_i = \oplus_{i=1}^{n} C_i + \otimes_{i=1}^{n} C_i$, where the first term represents the isolated coupling of the pair $\{C_i C_0\}$—that is, where the cross-talk between $\{C_i, C_j\}\ \forall j \neq 0, i$ is absent—and the second term represents the cross-talk between $\{C_i, C_j\}\ \forall j \neq 0, i$. Call $\oplus$ the *direct sum* and $\otimes$ the *direct product*.

The generalization to n transmission lines is carried out briefly below. First, from Corollary 6.3.2, the generalized wave equation can be easily obtained. Then, referring to Exercise 5.4.4, the technique to diagonalize the coefficient matrices can be employed to decouple the multiconductor lines: $\gamma_1 \cup \cdots \cup \gamma_n \rightarrow \gamma_1 \oplus \cdots \oplus \gamma_n$. First, observe that Corollary 6.3.2 can be expressed as follows.

6.3.5 Corollary

Under the conditions of Theorem 6.3.1, where the propagating waves are assumed to be time harmonic, there exists a coordinate transformation $\mathbf{P}$ such that Equations (6.23) and (6.24) can be diagonalized. Explicitly, rewriting (6.21) and (6.22) as

$$\partial_z^2 V = \mathbf{UW}V \quad \text{and} \quad \partial_z^2 \boldsymbol{I} = \mathbf{WU}\boldsymbol{I}$$

where $\mathbf{U} = \mathbf{R} + i\omega\mathbf{L}$ and $\mathbf{W} = \mathbf{G} + i\omega\mathbf{C}$, defining $\tilde{V} = \mathbf{P}V$ and $\tilde{I} = \mathbf{P}^{t}I$ lead to the decoupled system of wave equations:

$$\tilde{V}_k = \tilde{V}_k^{+} e^{-\sqrt{\lambda_k}(\ell - z)} + \tilde{V}_k^{-} e^{\sqrt{\lambda_k}(\ell - z)} \quad \text{and} \quad \tilde{I}_k = \tilde{I}_k^{+} e^{-\sqrt{\lambda_k}(\ell - z)} - \tilde{I}_k^{-} e^{\sqrt{\lambda_k}(\ell - z)}$$

for $i = 1, \ldots, n$, where $\{\lambda_1, \ldots, \lambda_n\}$ is the set of eigenvalues of **UW**.

Proof

From Proposition 6.3.1, inasmuch as it is not clear in general that **R** and **G** commute, it follows that **U** and **W** might not commute. However, because **L**, **C**, **R**, **G** are diagonalizable, it follows immediately that **U** and **W** are diagonalizable. So, let $\{\mathbf{P}_1, \mathbf{P}_2\}$ be the respective coordinate transformations (i.e., similarity transformations) that diagonalize **U** and **W**:

$$\mathbf{P}_1^{-1}\mathbf{U}\mathbf{P}_1 = \mathrm{diag}(\xi_1, \ldots, \xi_n) \quad \text{and} \quad \mathbf{P}_2^{-1}\mathbf{W}\mathbf{P}_2 = \mathrm{diag}(\zeta_1, \ldots, \zeta_n)$$

where $\xi_i, \zeta_i \ \forall i = 1, \ldots, n$ are defined later.

Then clearly, $(\mathbf{P}_1^{-1}\mathbf{U}\mathbf{P}_2)(\mathbf{P}_2^{-1}\mathbf{W}\mathbf{P}_1) = (\mathbf{P}_2^{-1}\mathbf{W}\mathbf{P}_1)(\mathbf{P}_1^{-1}\mathbf{U}\mathbf{P}_2)$ as diagonal matrices commute. Furthermore, noting that

$$(\mathbf{P}_1^{-1}\mathbf{U}\mathbf{P}_2\mathbf{P}_2^{-1}\mathbf{W}\mathbf{P}_1)^{t} = \mathbf{P}_1^{t}\mathbf{W}^{t}(\mathbf{P}_2^{-1})^{t}\mathbf{P}_2^{t}\mathbf{U}^{t}(\mathbf{P}_1^{-1})^{t} = (\mathbf{P}_2^{-1}\mathbf{W}\mathbf{P}_1)^{t}(\mathbf{P}_1^{-1}\mathbf{U}\mathbf{P}_2)^{t},$$

as $\mathbf{L}, \mathbf{C}, \mathbf{R}, \mathbf{G}$ are symmetric imply that $\mathbf{U}, \mathbf{V}$ are symmetric. Hence, the fact that diagonal matrices commute implies at once that $(\mathbf{P}_2^{-1}\mathbf{W}\mathbf{P}_1)^{t}(\mathbf{P}_1^{-1}\mathbf{U}\mathbf{P}_2)^{t} = (\mathbf{P}_1^{-1}\mathbf{U}\mathbf{P}_2)^{t}(\mathbf{P}_2^{-1}\mathbf{W}\mathbf{P}_1)^{t}$. That is,

$$\mathbf{P}_1^{t}\mathbf{W}(\mathbf{P}_2^{-1})^{t}\mathbf{P}_2^{t}\mathbf{U}(\mathbf{P}_1^{-1})^{t} = (\mathbf{P}_2^{-1}\mathbf{W}\mathbf{P}_1)(\mathbf{P}_1^{-1}\mathbf{U}\mathbf{P}_2) \Rightarrow \mathbf{P}_1^{t} = \mathbf{P}_2^{-1}.$$

Equivalently, $\mathbf{P}_2 = (\mathbf{P}_1^{-1})^{t} \Rightarrow \mathbf{P}_2^{t} = \mathbf{P}_1^{-1}$. Hence, setting $\mathbf{P} = \mathbf{P}_1^{-1} \Rightarrow \mathbf{P}_2 = \mathbf{P}^{t}$.

In the light of the above analysis, and noting that $\mathbf{U}, \mathbf{W}$ are symmetric, it follows at once that

$$\mathbf{P}^{t}\mathbf{U}\mathbf{P} = \mathrm{diag}(\xi_1, \ldots, \xi_n) \quad \text{and} \quad (\mathbf{P}^{-1})^{t}\mathbf{W}\mathbf{P}^{-1} = \mathrm{diag}(\zeta_1, \ldots, \zeta_n),$$

where $\{(w_{1,i}, \xi_i)\}$ denotes the eigenvectors and eigenvalues of $\mathbf{U}$, and $\{(w_{2,i}, \zeta_i)\}$ the eigenvectors and eigenvalues of $\mathbf{W}$. That is,

$$\mathbf{U}w_{1,i} = \xi_i w_{1,i} \quad \text{and} \quad \mathbf{W}w_{2,i} = \zeta_i w_{2,i} \ \forall i = 1, \ldots, n.$$

Then, $\mathbf{P} = [w_{1,1}, \ldots, w_{1,n}]$ and is the required matrix to diagonalise $\mathbf{U}, \mathbf{W}$ respectively, with $w_{2,i}$ related to $w_{1,i}$ via $\mathbf{P}^{-1} = [w_{2,1}, \ldots, w_{2,n}]$.

For notational simplicity, given a diagonalisable matrix $\mathbf{D}$, set $\mathbf{D}_\Delta = \text{diag}(\lambda_1,\ldots,\lambda_n)$ to be the diagonalisation of $\mathbf{D}$, where $\{\lambda_i\}$ are eigenvalues of $\mathbf{D}$. Then, on setting $\mathbf{D} = \mathbf{UW}$,

$$\partial_z^2 \boldsymbol{V} = \mathbf{UW}\boldsymbol{V} \Leftrightarrow \partial_z^2 \tilde{\boldsymbol{V}} = \mathbf{D}_\Delta \tilde{\boldsymbol{V}} \Rightarrow \tilde{\boldsymbol{V}} = e^{-\sqrt{\Lambda}(\ell - z)}\tilde{\boldsymbol{V}}_+ + e^{\sqrt{\Lambda}(\ell - z)}\tilde{\boldsymbol{V}}_-$$

is the general solution of the decoupled system, for some $\tilde{V}_\pm$. Explicitly,

$$\partial_z^2 \begin{pmatrix} \tilde{V}_1 \\ \tilde{V}_2 \\ \vdots \\ \tilde{V}_n \end{pmatrix} = \begin{pmatrix} \lambda_1 & 0 & \cdots & 0 \\ 0 & \lambda_2 & \cdots & 0 \\ \vdots & \vdots & \ddots & 0 \\ 0 & \cdots & 0 & \lambda_n \end{pmatrix} \begin{pmatrix} \tilde{V}_1 \\ \tilde{V}_2 \\ \vdots \\ \tilde{V}_n \end{pmatrix}$$

$$= \begin{pmatrix} \lambda_1 \tilde{V}_1 \\ \lambda_2 \tilde{V}_2 \\ \vdots \\ \lambda_n \tilde{V}_n \end{pmatrix} \Rightarrow \tilde{V}_i = \tilde{V}_i^+ e^{-\sqrt{\lambda_i}(\ell - z)} + \tilde{V}_i^- e^{\sqrt{\lambda_i}(\ell - z)} \ \forall i.$$

Similarly, the current propagation is: $\partial_z^2 \boldsymbol{I} = \mathbf{WU}\boldsymbol{I} \Rightarrow \tilde{\boldsymbol{I}} = e^{-\sqrt{\Lambda}(\ell - z)}\tilde{\boldsymbol{I}}_+ - e^{\sqrt{\Lambda}(\ell - z)}\tilde{\boldsymbol{I}}_-$, for some $\tilde{\boldsymbol{I}}_\pm$. That is,

$$\partial_z^2 \begin{pmatrix} \tilde{I}_1 \\ \tilde{I}_2 \\ \vdots \\ \tilde{I}_n \end{pmatrix} = \begin{pmatrix} \lambda_1 & 0 & \cdots & 0 \\ 0 & \lambda_2 & \cdots & 0 \\ \vdots & \vdots & \ddots & 0 \\ 0 & \cdots & 0 & \lambda_n \end{pmatrix} \begin{pmatrix} \tilde{I}_1 \\ \tilde{I}_2 \\ \vdots \\ \tilde{I}_n \end{pmatrix}$$

$$= \begin{pmatrix} \lambda_1 \tilde{I}_1 \\ \lambda_2 \tilde{I}_2 \\ \vdots \\ \lambda_n \tilde{I}_n \end{pmatrix} \Rightarrow \tilde{I}_i = \tilde{I}_i^+ e^{-\sqrt{\lambda_i}(\ell - z)} + \tilde{I}_i^- e^{\sqrt{\lambda_i}(\ell - z)} \ \forall i.$$

Whence, the desired solutions for the voltage and current propagations are obtained via the linear transformation $\gamma_1 \cup \cdots \cup \gamma_n \xrightarrow{\mathbf{P}^t} \gamma_1 \oplus \cdots \oplus \gamma_n$: $\boldsymbol{V} = \mathbf{P}^{-1}\tilde{\boldsymbol{V}}$ and $\boldsymbol{I} = \mathbf{P}^t \tilde{\boldsymbol{I}}$, as required. □

6.3.6 Proposition

The characteristic impedance matrix for the multiconductor transmission line system is given by $\mathbf{Z}_0 = \mathbf{W}^{-1}\mathbf{P}^t e^{\sqrt{\Lambda}}\mathbf{P}$.

Proof

From $\partial_z \boldsymbol{I} = -\mathbf{W}\boldsymbol{V}$, via Corollary 6.3.5,

$$\boldsymbol{V} = -\mathbf{W}^{-1}\mathbf{P}^{t}\sqrt{\Lambda}\left\{-e^{-\sqrt{\Lambda}(\ell-z)}\tilde{\boldsymbol{I}}^{+} - e^{\sqrt{\Lambda}(\ell-z)}\tilde{\boldsymbol{I}}^{-}\right\} = \mathbf{W}^{-1}\mathbf{P}^{t}\sqrt{\Lambda}\mathbf{P}\mathbf{P}^{t}\left\{e^{-\sqrt{\Lambda}(\ell-z)}\tilde{\boldsymbol{I}}^{+} + e^{\sqrt{\Lambda}(\ell-z)}\tilde{\boldsymbol{I}}^{-}\right\}$$

whence, motivated by the fact that if $\mathbf{M}$ were any symmetric matrix, then $\mathbf{M}e^{\pm\sqrt{\Lambda}(\ell-z)} = e^{\pm\sqrt{\Lambda}(\ell-z)}\mathbf{M}$, it follows clearly that

$$\boldsymbol{V} = \mathbf{Z}_0\mathbf{P}^{t}\left\{e^{-\sqrt{\Lambda}(\ell-z)}\tilde{\boldsymbol{I}}^{+} + e^{\sqrt{\Lambda}(\ell-z)}\tilde{\boldsymbol{I}}^{-}\right\}$$

where $\mathbf{Z}_0 \equiv \mathbf{W}^{-1}\mathbf{P}^{t}\sqrt{\Lambda}\mathbf{P}$, as asserted. □

Note in passing that at the source and load, respectively, $\boldsymbol{V}(0) = \boldsymbol{V}_S - \mathbf{Z}_S\boldsymbol{I}(0)$ and $\boldsymbol{V}(\ell) = \boldsymbol{V}_L + \mathbf{Z}(\ell)\boldsymbol{I}(\ell)$, via Kirchhoff's voltage law, where $\mathbf{Z}_S = \mathrm{diag}(Z_{S,1},\ldots,Z_{S,n})$ is the source impedance matrix, and $\mathbf{Z}(\ell) = \mathrm{diag}(Z_1(\ell),\ldots,Z_n(\ell))$ is the load impedance matrix, $\boldsymbol{V}_s$ is the voltage source at $z = 0$ and lastly, $\boldsymbol{V}_L$ is the voltage source at $z = \ell$. Then, the near-end and far-end voltages can be determined by the following standard technique (e.g., see Reference [6]).

First, consider the coupled system expressed in terms of $\boldsymbol{V}(0),\boldsymbol{I}(0)$:

$$\begin{pmatrix} \boldsymbol{V}(z) \\ \boldsymbol{I}(z) \end{pmatrix} = \begin{pmatrix} \mathbf{M}_{11}(z) & \mathbf{M}_{12}(z) \\ \mathbf{M}_{21}(z) & \mathbf{M}_{22}(z) \end{pmatrix}\begin{pmatrix} \boldsymbol{V}(0) \\ \boldsymbol{I}(0) \end{pmatrix} \tag{6.25}$$

for some yet to be determined matrix coefficients $\mathbf{M}_{ij}$ $\forall i, j = 1,2$. Then, for $z = \ell$, Equation (6.25) implies at once that

$$\boldsymbol{V}_L + \mathbf{Z}(\ell)\boldsymbol{I}(\ell) = \mathbf{M}_{11}(\ell)(\boldsymbol{V}_S - \mathbf{Z}_S\boldsymbol{I}(0)) + \mathbf{M}_{12}(\ell)\boldsymbol{I}(0) \tag{6.26}$$

$$\boldsymbol{I}(\ell) = \mathbf{M}_{21}(\ell)(\boldsymbol{V}_S - \mathbf{Z}_S\boldsymbol{I}(0)) + \mathbf{M}_{22}(\ell)\boldsymbol{I}(0) \tag{6.27}$$

6.3.7 Lemma

Given (6.25), at $z = \ell$, the matrix coefficients are defined by

$$\mathbf{M}_{11}(\ell) = \mathbf{Z}_0\mathbf{P}^{t}\cosh\left(\sqrt{\Lambda}\ell\right)(\mathbf{P}^{t})^{-1}\mathbf{Z}_0^{-1}$$

$$\mathbf{M}_{12}(\ell) = \mathbf{Z}_0\mathbf{P}^{t}\sinh\left(\sqrt{\Lambda}\ell\right)(\mathbf{P}^{t})^{-1}$$

$$\mathbf{M}_{21}(\ell) = \mathbf{P}^{t}\sinh\left(\sqrt{\Lambda}\ell\right)(\mathbf{P}^{t})^{-1}\mathbf{Z}_0^{-1}$$

$$\mathbf{M}_{22}(\ell) = \mathbf{P}^{t}\cosh\left(\sqrt{\Lambda}\ell\right)(\mathbf{P}^{t})^{-1}$$

Proof

From the proof of Proposition 6.3.6,

$$V(0) = \mathbf{Z}_0\mathbf{P}^t\left\{e^{-\sqrt{\Lambda}\ell}\tilde{I}^+ + e^{\sqrt{\Lambda}\ell}\tilde{I}^-\right\} \quad \text{and} \quad I(0) = \mathbf{P}^t\left\{e^{-\sqrt{\Lambda}\ell}\tilde{I}^+ - e^{\sqrt{\Lambda}\ell}\tilde{I}^-\right\}$$

Hence, rearranging the two equations to express $\tilde{I}^{\pm}$ in terms of $V(0)$, $I(0)$ yields

$$\tilde{I}^{\pm} = \tfrac{1}{2}e^{\pm\sqrt{\Lambda}\ell}(\mathbf{P}^t)^{-1}\left\{\mathbf{Z}_0^{-1}V(0) \pm I(0)\right\}$$

Then, substituting into Equation (6.25) leads to

$$\begin{aligned}V(\ell) &= \mathbf{Z}_0\mathbf{P}^t\left\{\tfrac{1}{2}\left(e^{\sqrt{\Lambda}\ell} + e^{-\sqrt{\Lambda}\ell}\right)(\mathbf{P}^t)^{-1}\mathbf{Z}_0^{-1}V(0) + \tfrac{1}{2}\left(e^{\sqrt{\Lambda}\ell} - e^{-\sqrt{\Lambda}\ell}\right)(\mathbf{P}^t)^{-1}I(0)\right\}\\ &= \mathbf{Z}_0\mathbf{P}^t\cosh\left(\sqrt{\Lambda}\ell\right)(\mathbf{P}^t)^{-1}\mathbf{Z}_0^{-1}V(0) + \mathbf{Z}_0\mathbf{P}^t\sinh\left(\sqrt{\Lambda}\ell\right)(\mathbf{P}^t)^{-1}I(0),\end{aligned}$$

and

$$\begin{aligned}I(\ell) &= \mathbf{P}^t\left\{\tfrac{1}{2}\left(e^{\sqrt{\Lambda}\ell} - e^{-\sqrt{\Lambda}\ell}\right)(\mathbf{P}^t)^{-1}\mathbf{Z}_0^{-1}V(0) + \tfrac{1}{2}\left(e^{\sqrt{\Lambda}\ell} + e^{-\sqrt{\Lambda}\ell}\right)(\mathbf{P}^t)^{-1}I(0)\right\}\\ &= \mathbf{P}^t\sinh\left(\sqrt{\Lambda}\ell\right)(\mathbf{P}^t)^{-1}\mathbf{Z}_0^{-1}V(0) + \mathbf{P}^t\cosh\left(\sqrt{\Lambda}\ell\right)(\mathbf{P}^t)^{-1}I(0)\end{aligned}$$

So, comparing the two expressions with (6.25) and equating coefficients,

$$\mathbf{M}_{11}(\ell) = \mathbf{Z}_0\mathbf{P}^t\cosh\left(\sqrt{\Lambda}\ell\right)(\mathbf{P}^t)^{-1}\mathbf{Z}_0^{-1}$$

$$\mathbf{M}_{12}(\ell) = \mathbf{Z}_0\mathbf{P}^t\sinh\left(\sqrt{\Lambda}\ell\right)(\mathbf{P}^t)^{-1}$$

$$\mathbf{M}_{21}(\ell) = \mathbf{P}^t\sinh\left(\sqrt{\Lambda}\ell\right)(\mathbf{P}^t)^{-1}\mathbf{Z}_0^{-1}$$

$$\mathbf{M}_{22}(\ell) = \mathbf{P}^t\cosh\left(\sqrt{\Lambda}\ell\right)(\mathbf{P}^t)^{-1}$$

□

6.3.8 Proposition

Given (V_S, V_L), the near-end and far-end currents and voltages of $\{C_i\}$ satisfy

$$I(0) = \mathbf{N}(\ell)^{-1}\left\{V_L + (\mathbf{Z}(\ell)\mathbf{M}_{21}(\ell) - \mathbf{M}_{11}(\ell))V_S\right\}$$

$$I(\ell) = \mathbf{M}_{21}(\ell)\left(\mathbf{I} + \mathbf{Z}_S\mathbf{N}(\ell)^{-1}\mathbf{M}_{11}(\ell)\right)V_S + \mathbf{M}_{22}(\ell)\mathbf{N}(\ell)^{-1}V_L$$

$$V(0) = \left\{\mathbf{I} - \mathbf{Z}_S\mathbf{N}(\ell)^{-1}(\mathbf{Z}(\ell)\mathbf{M}_{21}(\ell) - \mathbf{M}_{11}(\ell))\right\}V_S - \mathbf{Z}_S\mathbf{N}(\ell)^{-1}V_L$$

$$\begin{aligned}V(\ell) = &\left\{\mathbf{M}_{11}(\ell) + (\mathbf{M}_{12}(\ell) - \mathbf{M}_{11}(\ell)\mathbf{Z}_S)\mathbf{N}(\ell)^{-1}(\mathbf{Z}(\ell)\mathbf{M}_{21}(\ell) - \mathbf{M}_{11}(\ell))\right\}V_S\\ &+ (\mathbf{M}_{12}(\ell) - \mathbf{M}_{11}(\ell)\mathbf{Z}_S)\mathbf{N}(\ell)^{-1}V_L,\end{aligned}$$

where $\mathbf{N}(\ell) = \mathbf{M}_{12}(\ell) - \mathbf{M}_{11}(\ell)\mathbf{Z}_S - \mathbf{Z}(\ell)\left(\mathbf{M}_{22}(\ell) - \mathbf{M}_{21}(\ell)\mathbf{Z}_S\right)$.

Proof

Substituting Equation (6.25) into (6.26),

$$V_L + \mathbf{Z}(\ell)\boldsymbol{I}(\ell) = \mathbf{M}_{11}(\ell)V_S - \mathbf{M}_{11}(\ell)\mathbf{Z}_S\boldsymbol{I}(0) + \mathbf{M}_{12}(\ell)\boldsymbol{I}(0)$$

$$= \mathbf{M}_{11}(\ell)V_S + \{\mathbf{M}_{12}(\ell) - \mathbf{M}_{11}(\ell)\mathbf{Z}_S\}\boldsymbol{I}(0)$$

and from (6.25),

$$\boldsymbol{I}(\ell) = \mathbf{M}_{21}(\ell)V_S + \{\mathbf{M}_{22}(\ell) - \mathbf{M}_{21}(\ell)\mathbf{Z}_S\}\boldsymbol{I}(0)$$

whence,

$$V_L + \mathbf{Z}(\ell)\mathbf{M}_{21}(\ell)V_S + \{\mathbf{Z}_S\mathbf{M}_{22}(\ell) - \mathbf{M}_{21}(\ell)\mathbf{Z}_S\}\boldsymbol{I}(0) + \mathbf{M}_{12}(\ell)\boldsymbol{I}(0)$$

$$= \mathbf{M}_{11}(\ell)V_S + \{\mathbf{M}_{12}(\ell) - \mathbf{M}_{11}(\ell)\mathbf{Z}_S\}\boldsymbol{I}(0)$$

$\Rightarrow$

$$\boldsymbol{I}(0) = \mathbf{N}(\ell)^{-1}\{V_L + (\mathbf{Z}(\ell)\mathbf{M}_{21}(\ell) - \mathbf{M}_{11}(\ell))V_S\}$$

where $\mathbf{N}(\ell) = \mathbf{M}_{12}(\ell) - \mathbf{M}_{11}(\ell)\mathbf{Z}_S - \mathbf{Z}(\ell)\{\mathbf{M}_{22}(\ell) - \mathbf{M}_{21}(\ell)\mathbf{Z}_S\}$. Substituting $\boldsymbol{I}(0)$ into the above expression for $\boldsymbol{I}(\ell)$ yields the result stated in the proposition. Finally, substituting $\boldsymbol{I}(0)$ into $V(0) = V_S - \mathbf{Z}_S\boldsymbol{I}(0)$ and $V(\ell) = \mathbf{M}_{11}V_S + \{\mathbf{M}_{12}(\ell) - \mathbf{M}_{11}(\ell)\mathbf{Z}_S\}\boldsymbol{I}(0)$ concludes the proof. □

In many cases, $\mathbf{V}_L \equiv 0$, and Proposition 6.3.6 thus yields the near-end and far-end cross-talk for the multiconductor transmission line $\{C_i\}$.

6.3.9 Corollary

Given a system of multiconductor transmission lines $\{C_i\}$ depicted in Figure 6.8, the induced near-end cross-talk at $z = 0$ and far-end cross-talk at $z = \ell$ on the system are given by:

$$\boldsymbol{I}(0) = \mathbf{N}(\ell)^{-1}\{\mathbf{Z}(\ell)\mathbf{M}_{21}(\ell) - \mathbf{M}_{11}(\ell)\}V_S$$

$$\boldsymbol{I}(\ell) = \mathbf{M}_{21}(\ell)\{\mathbf{I} + \mathbf{Z}_S\mathbf{N}(\ell)^{-1}\mathbf{M}_{11}(\ell)\}V_S$$

$$V(0) = \{\mathbf{I} - \mathbf{Z}_S\mathbf{N}(\ell)^{-1}(\mathbf{Z}(\ell)\mathbf{M}_{21}(\ell) - \mathbf{M}_{11}(\ell))\}V_S$$

$$V(\ell) = \{\mathbf{M}_{11}(\ell) + (\mathbf{M}_{12}(\ell) - \mathbf{M}_{11}(\ell)\mathbf{Z}_S)\mathbf{N}(\ell)^{-1}(\mathbf{Z}(\ell)\mathbf{M}_{21}(\ell) - \mathbf{M}_{11}(\ell))\}V_S \qquad \square$$

6.3.10 Lemma

Given a matrix $\mathbf{A} = (A_{ij})$, define $\|\mathbf{A}\| = \max_{i,j}|A_{ij}|$. Suppose that $\{C_i\}$ is embedded in a homogeneous medium such that the transmission lines satisfy $\ell \ll \lambda$, for some time harmonic voltage sources, where $\lambda = \min_k \lambda_k$, and λ_k is

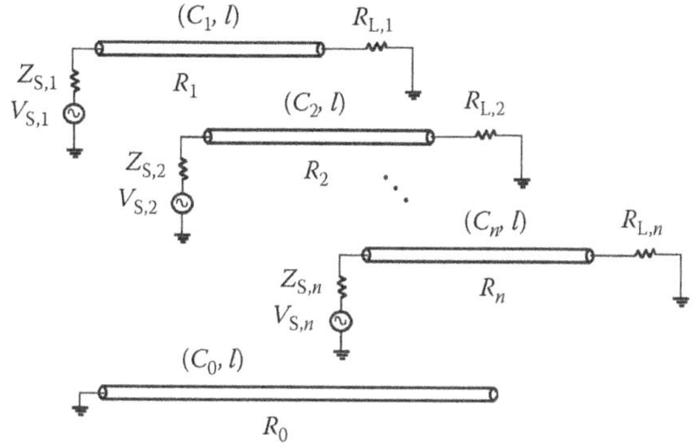

FIGURE 6.8
Cross-talk induced on a system of multiconductor transmission lines.

the wavelength of the wave propagation along C_k, with $\ell = |C_k|\ \forall k$. If $\|\mathbf{Z}_0^{-1}\mathbf{Z}_S - \mathbf{Z}^{-1}(\ell)(\mathbf{Z}_S - \mathbf{Z}_0)\| \leq 1$ and $\{C_i\}$ is lossless, then to first order in $\frac{\ell}{\lambda}$, $\mathbf{N}^{-1}(\ell) \approx -(\mathbf{Z}_S + \mathbf{Z}(\ell))^{-1}\{\mathbf{I} + i\beta\ell(\mathbf{Z}_S + \mathbf{Z}(\ell))^{-1}(\mathbf{Z}_0 + \mathbf{Z}(\ell)\mathbf{Z}_0^{-1}\mathbf{Z}_S)\}$.

Proof

The proof is left as an exercise; see Exercise 6.5.4. □

6.3.11 Proposition

Suppose $\{C_i\}$ is embedded in a homogeneous medium such that the transmission lines satisfy $\ell \ll \lambda$, for some time harmonic voltage sources, where $\lambda = \min_k \lambda_k$, and λ_k is the wavelength of the wave propagation along C_k, with $\ell = |C_k|\ \forall k$. Suppose that $\|\mathbf{Z}_0^{-1}\mathbf{Z}_S - \mathbf{Z}^{-1}(\ell)(\mathbf{Z}_S - \mathbf{Z}_0)\| \leq 1$ and $\{C_i\}$ is lossless. Then, for $V_S \neq 0, V_L = 0$, the cross-talk noise can be approximated to first order in $\frac{\ell}{\lambda}$ by

$$I(0) \approx (\mathbf{Z}_S + \mathbf{Z}(\ell))^{-1}\left\{\mathbf{I} + i\beta\ell\left((\mathbf{Z}_S + \mathbf{Z}(\ell))^{-1}\left(\mathbf{Z}_0 + \mathbf{Z}(\ell)\mathbf{Z}_0^{-1}\mathbf{Z}_S\right) - \mathbf{Z}(\ell)\mathbf{Z}_0^{-1}\right)\right\}V_S$$

$$I(\ell) \approx i\beta\ell\mathbf{Z}_0^{-1}\left\{\mathbf{I} - \mathbf{Z}_S(\mathbf{Z}_S + \mathbf{Z}(\ell))^{-1}\right\}V_S$$

$$V(0) \approx \left\{\mathbf{I} - \mathbf{Z}_S(\mathbf{Z}_S + \mathbf{Z}(\ell))^{-1}\left(\mathbf{I} + i\beta\ell\left(\Xi(\ell) - \mathbf{Z}(\ell)\mathbf{Z}_0^{-1}\right)\right)\right\}V_S$$

$$V(\ell) \approx \left\{\mathbf{I} - \mathbf{Z}_S(\mathbf{Z}_S + \mathbf{Z}(\ell))^{-1} + i\beta\ell\left(\mathbf{Z}_0(\mathbf{Z}_S + \mathbf{Z}(\ell))^{-1} - \mathbf{Z}_S(\mathbf{Z}_S + \mathbf{Z}(\ell))^{-1}(\Xi - \mathbf{Z}(\ell)\mathbf{Z}_0^{-1})\right)\right\}V_S$$

where $\Xi(\ell) = i\beta\ell(\mathbf{Z}_S + \mathbf{Z}(\ell))^{-1}(\mathbf{Z}_0 + \mathbf{Z}(\ell)\mathbf{Z}_0^{-1}\mathbf{Z}_S)$.

Proof

This is an immediate consequence of Corollary 6.3.7 and Lemma 6.3.8. See Exercise 6.5.5. □

6.4 S-Parameters: Scattering Parameters

No exposition on electromagnetic theory for EMC engineers is complete without incorporating a cursory review on S-parameter theory. By way of introduction, "S" is short for *scattering*, and the origin of the term is best illustrated in Figure 6.9. The motivation is to determine the *i*-load impedance without knowing the makeup of the black box by observing the scattering (i.e., reflected) waves traveling back toward the source. Note that this applies to a linear network system under various steady-state conditions. The special case is first formulated, and then followed by the general theory. An example illustrating a 2-port network is also presented in some depth.

In formulating the special case, the characteristic impedance across each port is assumed to be the same; that is, the characteristic impedance of the *i*-port is identical with that of the *j*-port $\forall i,j$. In the general case, this stipulation is dropped. So, referring to Figure 6.9, let T_k denote the *k*-terminal at along the *k*-transmission pair $C_k^{\pm}$ such that the voltage and current phases are 0, for simplicity. That is, along the pair $\{C_k^+, C_k^-\}$ terminating at the kth port of the black box, let T_k denote the terminal along $\{C_k^+, C_k^-\}$, near the port, such that the phase is zero. This can be easily accomplished via a coordinate transformation.

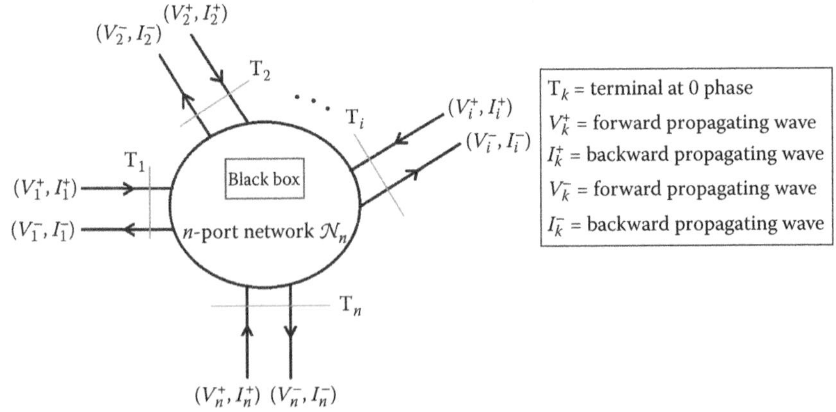

FIGURE 6.9
Propagating waves along *n*-pairs of transmission lines entering some network.

6.4.1 Lemma

Given $\mathbf{X} = \mathbf{Q}\mathbf{Y}$, where $\mathbf{X} = (a_1 X_1, \ldots, a_n X_n)$, $\mathbf{Y} = (b_1 Y_1, \ldots, b_n Y_n)$, and $\mathbf{Q} = (Q_{ij})$, for some constants $a_i, b_i, i = 1, \ldots, n$, there exists a transformation $(\mathbf{X}, \mathbf{Y}) \mapsto (\tilde{\mathbf{X}}, \tilde{\mathbf{Y}})$ such that $\tilde{\mathbf{X}} = (X_1, \ldots, X_n)$ and $\tilde{\mathbf{Y}} = (Y_1, \ldots, Y_n)$.

Proof

The proof is trivial. First, observe that

$$\begin{pmatrix} a_1 X_1 \\ \vdots \\ a_n X_n \end{pmatrix} = \mathrm{diag}(a_1, \ldots, a_n) \begin{pmatrix} X_1 \\ \vdots \\ X_n \end{pmatrix} \quad \text{and}$$

$$\begin{pmatrix} b_1 Y_1 \\ \vdots \\ b_n Y_n \end{pmatrix} = \mathrm{diag}(b_1, \ldots, b_n) \begin{pmatrix} Y_1 \\ \vdots \\ Y_n \end{pmatrix}$$

Hence, on setting $\mathbf{A} = \mathrm{diag}(a_1, \ldots, a_n)$ and $\mathbf{B} = \mathrm{diag}(b_1, \ldots, b_n)$, it follows at once that $\tilde{\mathbf{X}} = \tilde{\mathbf{Q}}\tilde{\mathbf{Y}}$, where $\tilde{\mathbf{Q}} = \mathbf{A}^{-1}\mathbf{Q}\mathbf{B}$. □

To continue with the above exposition, suppose in general that the n-port network is connected to $(C_k^{\pm}, Z_{0,k}, \ell_i, V_{S,k}, Z_{S,k})$, where ℓ_k is the length of the transmission lines $C_k^{\pm}$, $Z_{0,k} \leftrightarrow \gamma_k$ is the characteristic impedance of (C_k^+, C_k^-), and $(V_{S,k}, Z_{S,k})$ is the source at $z_k = 0$, where γ_k is the wave propagation constant. By assumption, $Z_{0,i} \equiv Z_0$, for some $Z_0 \neq 0$. Note further that by construction, T_k is located at $z_k = \ell_k$ and that Lemma 6.4.1 is indeed equivalent to translating the origin along $C_k^{\pm}$ to T_k: $z_k \mapsto z_k - \ell_k$. Thus, recalling that

$$V_k(z) = V_k^+ e^{-\gamma(\ell - z)} + V_k^- e^{\gamma(\ell - z)} \Rightarrow V_k(0) = V_k^+ e^{-\gamma_k \ell_k} + V_k^- e^{\gamma_k \ell_k}$$

$$I_k(z) = I_k^+ e^{-\gamma(\ell - z)} - I_k^- e^{\gamma(\ell - z)} \Rightarrow I_k(0) = I_k^+ e^{-\gamma_k \ell_k} - I_k^- e^{\gamma_k \ell_k}$$

by Lemma 6.4.1, $(V_k(0), I_k(0))$ can be transformed into

$$(V_k(0), I_k(0)) \mapsto (\tilde{V}_k(0), \tilde{I}_k(0)) = (V_k^+ + V_k^-, I_k^+ + I_k^-)$$

That is, via Lemma 6.4.1, T_k can be transformed into a zero-phase terminal for the n-port network. In particular, without loss of generality, we define T_k to be the zero-phase terminal in all that follows.

Then, for the n-port network illustrated in Figure 6.9, the *S-parameter* formalism of the network is defined by expressing the reflected waves as a function of the incident waves:

$$\begin{pmatrix} V_1^- \\ \vdots \\ V_n^- \end{pmatrix} = \begin{pmatrix} S_{11} & \cdots & S_{1n} \\ \vdots & \ddots & \vdots \\ S_{n1} & \cdots & S_{nn} \end{pmatrix} \begin{pmatrix} V_1^+ \\ \vdots \\ V_n^+ \end{pmatrix} \tag{6.28}$$

where $\mathbf{S} = (S_{ij})$ is the *S-matrix* (or *scattering matrix*) defining the n-port network. Because the S-matrix is a constant, it follows at once by construction that

$$S_{ij} = \left.\frac{V_i^-}{V_j^+}\right|_{V_k^+=0\ \forall k\neq j} \tag{6.29}$$

Verify this assertion in Exercise 6.5.6. Some comments are due. First, $V_k^+ = 0$ is obtained by (i) setting the source at $z_k = -\ell_k$ along $C_k^\pm$ to be zero, and (ii) matching the source impedance to the characteristic impedance of $C_k^\pm$. Note that (i) is clearly necessary, and (ii) is also necessary to ensure that any waves traveling from the k-port toward the source of $C_k^\pm$ are completely absorbed by the source (technically, "load", in this instance). These two conditions are clearly sufficient to ensure that $V_k^+ = 0$.

In short, each $C_k^\pm$, for $k \neq j$, is connected to a matched load (with the source removed). Second, S_{ij} for $i \neq j$, can be intuitively viewed as the transmission coefficients, and S_{ij} can be viewed as the reflection coefficients; these two observations are true when all other ports are matched with their characteristic impedance. For instance, when some of the ports are not matched, then S_{ij} is not necessarily equal to the reflection coefficient Γ_i at T_i, that is, waves traveling out of the i-port will be the superposition of reflected waves from the nonmatched loads terminating the k-port Similar comments hold for S_{ij}, $i \neq j$, by interchanging reflection with transmission.

6.4.2 Definition

Given a linear time-invariant network defined by $V = \mathbf{Z}I$, the network is said to be *reciprocal* if $\mathbf{Z}$ is symmetric: $Z_{ij} = Z_{ji}\ \forall i, j$. And the network is said to be *lossless* if $\Re\mathrm{e}(Z_{ij}) = 0\ \forall i, j$.

Clearly, as the inverse of a symmetric matrix is symmetric, it follows at once that $\mathbf{Z} = \mathbf{Z}^t \Rightarrow \mathbf{Y} = \mathbf{Y}^t$, where $\mathbf{Y} = \mathbf{Z}^{-1}$.

6.4.3 Proposition

The S-matrix $\mathbf{S}$ is symmetric; that is, $S_{ij} = S_{ji}\ \forall i, j$, whenever the n-port network is reciprocal.

Proof

By definition, $V^{\pm} = \frac{1}{2}(V \pm \mathbf{Z}_0 I)$. This follows from the fact that $V = V^+ + V^-$ and $I = I^+ - I^-$, where $I^{\pm} = \mathbf{Z}_0^{-1}V^{\pm}$, whence, from $V = \mathbf{Z}I$,

$$V^{\pm} = \tfrac{1}{2}(\mathbf{Z} \pm \mathbf{Z}_0)I \Rightarrow (\mathbf{Z} - \mathbf{Z}_0)I = \tfrac{1}{2}\mathbf{S}(\mathbf{Z} + \mathbf{Z}_0)I$$

That is, $\mathbf{S} = (\mathbf{Z} - \mathbf{Z}_0)(\mathbf{Z} + \mathbf{Z}_0)^{-1}$, and thus, $\mathbf{Z}, \mathbf{Z}_0$ are symmetric imply that $\mathbf{S}$ is also symmetric, as asserted. □

In light of the above proof, the following lemma is evident. For notational convenience, let $\mathcal{N}_n$ denote the n-port network illustrated in Figure 6.9.

6.4.4 Lemma

Let $\mathbf{S}$ be the S-matrix associated with $\mathcal{N}_n$. Then, $S_{kk} = 0\ \forall k$ if each k-port of $\mathcal{N}_n$ is matched.

Proof

By definition, $V^- = \mathbf{S}V^+$. Hence, by assumption, $V_k^- = S_{kk}V_k^+ = 0 \Rightarrow S_{kk} \equiv 0\ \forall k$. □

6.4.5 Theorem

The network $\mathcal{N}_n$ is lossless if and only if $\mathbf{S}^{\dagger}\mathbf{S} = \mathbf{I}$; that is, $\mathbf{S}$ is unitary.

Proof

First, recall that the time-average power is defined by

$$\langle P_k^{\pm} \rangle = \tfrac{1}{2}\Re\mathrm{e}\{(V_k^{\pm})^* I_k^{\pm}\} = \tfrac{1}{2}\frac{|V_k^{\pm}|^2}{Z_0}$$

where $\mathbf{Z}_0$ is the characteristic impedance of (C_k^+, C_k^-) connected to the k-port of $\mathcal{N}_n$. Here, $\langle P_k^+ \rangle (\langle P_k^- \rangle)$ denote the incident (reflected) power at the k-port. That is, $\langle P_k^+ \rangle$ represents the time-average power entering $\mathcal{N}_n$ whereas $\langle P_k^- \rangle$ represents the power exiting $\mathcal{N}_n$. Thus, $\mathcal{N}_n$ is lossless if and only if $\langle P^+ \rangle = \langle P^- \rangle$, where $\langle P^{\pm} \rangle = \Sigma_k \langle P_k^{\pm} \rangle$. Physically, this means that the power absorbed by $\mathcal{N}_n$ must be zero (i.e., no loss!).

Now, the resultant power absorbed by $\mathcal{N}_n$ is defined by $\langle \delta P \rangle = \langle P^+ \rangle - \langle P^- \rangle$. On setting $\mathbf{Z}_0 = \mathrm{diag}(Z_0, \ldots, Z_0)$ and noting that

$$\langle P^+ \rangle = \sum_k \langle P_k^+ \rangle = \tfrac{1}{2}\mathbf{Z}_0^{-1}(\mathbf{V}^+)^{\dagger}\mathbf{V}^+$$

$$\langle P^- \rangle = \sum_k \langle P_k^- \rangle = \tfrac{1}{2}\mathbf{Z}_0^{-1}(\mathbf{S}\mathbf{V}^+)^{\dagger}\mathbf{S}\mathbf{V}^+ = \tfrac{1}{2}\mathbf{Z}_0^{-1}(\mathbf{V}^+)^{\dagger}\mathbf{S}^{\dagger}\mathbf{S}\mathbf{V}^+$$

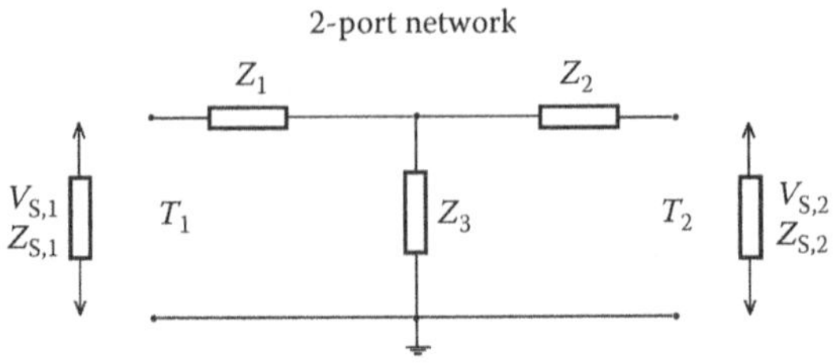

FIGURE 6.10
Determination of the S-matrix for a T-network.

it follows that lossless condition is equivalent to

$$0 = \langle \delta P \rangle = \tfrac{1}{2}\mathrm{diag}^{-1}(Z_0,\ldots,Z_0)\left\{(\mathbf{V}^+)^\dagger \mathbf{V}^+ - (\mathbf{V}^+)^\dagger \mathbf{S}^\dagger \mathbf{S}\mathbf{V}^+\right\}$$

$$= \tfrac{1}{2}\mathrm{diag}^{-1}(Z_0,\ldots,Z_0)(\mathbf{V}^+)^\dagger\left\{\mathbf{I} - \mathbf{S}^\dagger \mathbf{S}\right\}\mathbf{V}^+$$

and hence, $\mathbf{V}^+$ is arbitrary implies at once that $\mathbf{I} - \mathbf{S}^\dagger \mathbf{S} \equiv 0$, as required. □

6.4.6 Example

Consider the 2-port network depicted in Figure 6.10. Determine the S-parameters for the T-filter given passive elements $\{Z_1, Z_2, Z_3\}$ given that T_1, T_2 are matched. Note that in Figure 6.10, only the source impedance is shown at T_1; the source is left out for simplicity. At T_2, only a load is present. Matching Z_L at T_2 to the T-filter can be accomplished as follows. Set

$$Z_{\mathrm{in}}^{(1)} = Z_1 + \left\{\tfrac{1}{Z_3} + \tfrac{1}{Z_2 + Z_L}\right\}^{-1}$$

This is the impedance looking into T_1. Then, impedance matching port 2 at T_2 yields:

$$Z_{\mathrm{in}}^{(1)} = Z_1 + \left\{\tfrac{1}{Z_3} + \tfrac{1}{Z_2 + Z_L}\right\}^{-1} = Z_L \Rightarrow$$

$$Z_L = \tfrac{1}{2}\left\{Z_1 - Z_2 + \sqrt{(Z_1 - Z_2)^2 + 4(Z_1 Z_2 + Z_1 Z_3 + Z_2 Z_3)}\right\}$$

where the positive solution was chosen for physical reasons. Thus, terminating T_2 with Z_L leads to

$$S_{11} = \left.\tfrac{V_1^-}{V_1^+}\right|_{V_2^+ = 0} = \Gamma_1 = \tfrac{Z_{\mathrm{in}}^{(1)} - Z_L}{Z_{\mathrm{in}}^{(1)} + Z_L} = 0$$

For S_{22}, interchanging $Z_1 \leftrightarrow Z_2 \Rightarrow Z_S = \frac{1}{2}\left\{Z_2 - Z_1 + \sqrt{(Z_2 - Z_1)^2 + 4(Z_1Z_2 + Z_2Z_3 + Z_1Z_3)}\right\}$, and hence, by assumption, $Z_{in}^{(2)} = Z_{S,1}$ yields

$$S_{22} = \left.\frac{V_2^-}{V_2^+}\right|_{V_1^+ = 0} = \Gamma_2 = \frac{Z_{in}^{(2)} - Z_S}{Z_{in}^{(2)} + Z_S} = 0$$

Finally, regarding S_{21}, observing that T_2 is matched implies that $V_1^- \equiv 0$, that is, no reflection, and similarly, $V_2^+ \equiv 0$ as there can be no reflection from a matched load. Hence, $V_1 = V_1^+ + V_1^- \Rightarrow V_1^+ = V_1$ and $V_L = V_2^+ + V_2^- \Rightarrow V_2^- = V_L$, where $V_1 = V_S \frac{Z}{Z+Z_S}$ and Z is the resultant impedance defined by

$$Z = Z_S + Z_1 + \left\{\frac{1}{Z_3} + \frac{1}{Z_2 + Z_L}\right\}^{-1}.$$

Let I denote the current conducting through the source Z_S. Then, by construction,

$$I = \frac{V_S}{Z} = \frac{V_1}{Z - Z_S}$$

So, on setting $I = I' + I''$, where I' is the current through Z_L and I'' the current through Z_3, it follows that

$$I' = I\frac{Z_3}{Z_3 + Z_2 + Z_L}$$

and hence,

$$V_L = Z_L I' = \frac{Z_L}{Z - Z_S}\frac{Z_3}{Z_2 + Z_3 + Z_L}V_1$$

Thus,

$$S_{21} = \left.\frac{V_2^-}{V_1^+}\right|_{V_2^+ = 0} = \frac{Z_L}{Z - Z_S}\frac{Z_3}{Z_2 + Z_3 + Z_L} = S_{12}$$

by appealing to the symmetry of **S**. □

The general S-parameter theory is considered below. First, recall the essence of the S-matrix formulation. It is based on measuring the incident power and reflected power occurring at the ports of a linear network. In the present formulation, each transmission line pair (C_k^+, C_k^-) has characteristic impedance $Z_{0,k}$, not necessarily identical for each k. Second, recall that the time-average power transmitted is defined by $P = \frac{1}{2}\Re e(\boldsymbol{V} \cdot \boldsymbol{I}^*)$. Hence, along (C_k^+, C_k^-), the time-average power is $\langle P_k^\pm \rangle = \frac{1}{2}\frac{|V_k^\pm|^2}{Z_{0,k}}$, motivating the definition:

$$a_k^\pm = \frac{V_k^\pm}{\sqrt{Z_{0,k}}} \propto \sqrt{\langle P_k^\pm \rangle}$$

In what follows, assume for simplicity that (C_k^+, C_k^-) is lossless and hence, $Z_{0,k} \in \mathbf{R}$ and $V_k^\pm(z) = V_{0,k}^\pm e^{\pm i\beta_k \ell_k}$ $\forall k$.

6.4.7 Remark

Observe that by definition, $a_k^{\pm} = \frac{1}{2\sqrt{Z_{0,k}}}(V_k \pm Z_{0,k}I_k)$. To see this, it suffices to recall that $V_k(z) = V_k^+(z) + V_k^-(z)$ and $I_k(z) = I_k^+(z) - I_k^-(z)$, where $V_k^{\pm} = Z_{0,k}I_k^{\pm}$. From this, it follows that $Z_{0,k}I_k(z) = V_k^+(z) - V_k^-(z) \Rightarrow a_k^+(z) + a_k^-(z) = \frac{1}{\sqrt{Z_{0,k}}}(V_k^+ + V_k^-) = \frac{1}{\sqrt{Z_{0,k}}}V_k$ and $a_k^+(z) - a_k^-(z) = \frac{1}{\sqrt{Z_{0,k}}}(V_k^+ - V_k^-) = \sqrt{Z_{0,k}}\,I_k(z)$ whence, adding and subtracting the expressions for $a_k^{\pm}$ yields $a_k^{\pm} = \frac{1}{2\sqrt{Z_{0,k}}}(V_k \pm Z_{0,k}I_k)$, as claimed.

6.4.8 Lemma

The pair (a_k^+, a_k^-) defines the transmitted and reflected power along (C_k^+, C_k^-) by $\langle P_k^{\pm}\rangle = \frac{1}{2}|a_k^{\pm}|^2$.

Proof

By definition,

$$\langle P_k^{\pm}\rangle = \frac{1}{2}\frac{|V_k^{\pm}|^2}{Z_{0,k}} = \frac{1}{2}\frac{V_k^{\pm}}{\sqrt{Z_{0,k}}}\left(\frac{V_k^{\pm}}{\sqrt{Z_{0,k}}}\right)^* = \frac{1}{2}a_k^{\pm}\left(a_k^{\pm}\right)^* = \frac{1}{2}\left|a_k^{\pm}\right|^2$$

as required. □

6.4.9 Corollary

The time-average power across T_k satisfies $\langle P_k\rangle = \langle P_k^+\rangle - \langle P_k^-\rangle$.

Proof

By definition, $\langle P_k\rangle = \frac{1}{2}\Re\mathrm{e}(V_k I_k^*) = \frac{1}{2}\Re\mathrm{e}\left\{\frac{V_k}{\sqrt{Z_{0,k}}}\sqrt{Z_{0,k}}\,I_k^*\right\}$. From Remark 6.4.6,

$$\begin{aligned}\langle P_k\rangle &= \frac{1}{2}\Re\mathrm{e}\left\{\left(a_k^+ + a_k^-\right)\left(a_k^{+*} - a_k^{-*}\right)\right\}\\ &= \frac{1}{2}\Re\mathrm{e}\left\{\left|a_k^+\right|^2 - \left|a_k^-\right|^2 + a_k^- a_k^{+*} - a_k^+ a_k^{-*}\right\}.\end{aligned}$$

By assumption, (C_k^+, C_k^-) is lossless implies that $a_k^{\pm} = a_{0,k}^{\pm}e^{\pm i\beta_k z_k}$, where $a_{0,k}^{\pm} \in \mathbf{R}$. Hence,

$$a_k^- a_k^{+*} - a_k^+ a_k^{-*} = a_{0,k}^+ a_{0,k}^-\left(e^{-i2\beta_k z_z} - e^{i2\beta_k z_z}\right) = -i2a_{0,k}^+ a_{0,k}^- \sin 2\beta_k z_z$$

which is purely imaginary. Thus,

$$\langle P_k\rangle = \frac{1}{2}\left\{\left|a_k^+\right|^2 - \left|a_k^-\right|^2\right\} = \langle P_k^+\rangle - \langle P_k^-\rangle$$

by Lemma 6.4.8. □

6.4.10 Definition

Given a linear network $\{\mathcal{N}_n, C_k^{\pm}, Z_{0,k}\}$, where the transmission lines are lossless, the *S-parameters* are defined by

$$\begin{pmatrix} a_1^- \\ \vdots \\ a_n^- \end{pmatrix} = \begin{pmatrix} S_{11} & \cdots & S_{1n} \\ \vdots & \ddots & \vdots \\ S_{n1} & \cdots & S_{nn} \end{pmatrix} \begin{pmatrix} a_1^+ \\ \vdots \\ a_n^+ \end{pmatrix} \tag{6.30}$$

where

$$S_{ij} = \left. \frac{a_i^-}{a_j^+} \right|_{a_k^+ = 0 \ \forall k \neq j}$$

with each k-port matched for all $k \neq j$.

6.4.11 Remark

Provisionally, set $\boldsymbol{a}^- = \hat{\mathbf{S}}\boldsymbol{a}^+$ and $\boldsymbol{V}^- = \mathbf{S}\boldsymbol{V}^+$. Then, by definition,

$$\hat{S}_{ij} = \frac{a_i^-}{a_j^+} = \frac{V_i^-}{\sqrt{Z_{0,i}}} \frac{\sqrt{Z_{0,j}}}{V_j^+} = \frac{V_i^-}{V_j^+} \sqrt{\frac{Z_{0,j}}{Z_{0,i}}} = S_{ij} \sqrt{\frac{Z_{0,j}}{Z_{0,i}}}$$

Hence, $Z_{0,k} = Z_0 \ \forall k \Rightarrow \hat{S}_{ij} = S_{ij} \ \forall i, j$. Thus, this justifies denoting the S-matrix in Equation (6.30) by $\mathbf{S}$ instead of $\hat{\mathbf{S}}$. In particular, it is clear that $\mathbf{S}$ is symmetric does not imply that $\hat{\mathbf{S}}$ is also symmetric. This is because unless $Z_{0,i} = Z_{0,j}$, $S_{ij} = S_{ji} \nRightarrow \hat{S}_{ij} = \hat{S}_{ji}$, in general. This immediately yields the following theorem.

6.4.12 Theorem

Suppose the network $\{\mathcal{N}_n, C_k^{\pm}, Z_{0,k}\}$ is reciprocal. Then, $\mathbf{S}$ is symmetric if and only if $Z_{0,k} = Z_0 \ \forall k$, for some fixed Z_0. □

On the other hand, by inspecting the proof of Theorem 6.4.4, replacing $\text{diag}(Z_0, \ldots, Z_0)$ with $\text{diag}(Z_{0,1}, \ldots, Z_{0,n})$ yields the same result. This is summarized below for convenience.

6.4.13 Corollary

The network $\{\mathcal{N}_n, C_k^{\pm}, Z_{0,k}\}$ is lossless if and only if $\mathbf{S}$ is unitary. □

This section concludes by showing how $\mathbf{Z}$ and $\mathbf{S}$ are related. Note in passing that the Z-matrix is determined by either opening or shorting out the circuit in the case of the Y-matrix. In contrast, the S-matrix is determined by impedance matching at the ports. From this, it is clear that the S-matrix has a greater utility over the Y-matrix as shorting out terminals has the potential to cause irreversible damage to circuit components.

6.5.14. Theorem.

Given a network $\{\mathcal{N}_n, C_k^{\pm}, Z_{0,k}\}$, **Z** and **S** are related by

$$\mathbf{Z} = \left\{\mathbf{I} - \mathbf{Z}_0\mathbf{S}\mathbf{Z}_0^{-1}\right\}\mathbf{Z}_0\left\{\mathbf{I} - \mathbf{S}\right\}$$

where $\mathbf{Z}_0 = \text{diag}(Z_{0,1}, \ldots, Z_{0,n})$.

Proof

By definition, $\boldsymbol{a}^{\pm} = \frac{1}{2}\mathbf{Z}_0^{-1}\{\boldsymbol{V} \pm \mathbf{Z}_0\boldsymbol{I}\}$. Thus, from $\boldsymbol{V} = \boldsymbol{Z}\boldsymbol{I}$ and $\boldsymbol{a}^- = \mathbf{S}\boldsymbol{a}^+$,

$$\mathbf{Z}_0^{-1}\{\mathbf{V} - \mathbf{Z}_0\boldsymbol{I}\} = \mathbf{S}\mathbf{Z}_0^{-1}\{\boldsymbol{V} + \mathbf{Z}_0\boldsymbol{I}\} \Rightarrow \left\{\mathbf{I} - \mathbf{Z}_0\mathbf{S}\mathbf{Z}_0^{-1}\right\}\boldsymbol{V} = \mathbf{Z}_0\left\{\mathbf{I} + \mathbf{S}\right\}\boldsymbol{I}$$

and hence, $\boldsymbol{V} = \{\mathbf{I} - \mathbf{Z}_0\mathbf{S}\mathbf{Z}_0^{-1}\}^{-1}\mathbf{Z}_0\{\mathbf{I} + \mathbf{S}\}\boldsymbol{I}$, as desired. □

6.5 Worked Problems

6.5.1 Exercise

Establish Equation (6.1) from Theorem 6.1.1.

Solution

Let $C = \bigcup_k (C_k \cup C_k')$. Then, setting $\Omega = \mathbf{R}^3 - C$, and noting trivially that $\varphi_i \mid C_i = V_i$ is a constant for each i, it follows at once that $\sum_i \int_{\partial C_i} \rho_i \varphi_i' \mathrm{d}^2\boldsymbol{x} = \sum_i V_i' \int_{\partial C_i} \rho_i \mathrm{d}^2\boldsymbol{x} = V_i' Q_i$, where $Q_i = \int_{\partial C_i} \rho_i$ (as charges within a conductor will quickly diffuse to the boundary of the conductor). Likewise, $\sum_i \int_{\partial C_i} \rho_i' \varphi_i \mathrm{d}^2\boldsymbol{x} = \sum_i V_i Q_i'$, and Theorem 6.1.1 thus yields the desired result. □

6.5.2 Exercise

For a system of thin conductors $\{(C_i, V_i, I_i)\}$ and $\{(C_i', V_i', I_i')\}$ for $i = 1, \ldots, n$, establish that $\sum_{i=1}^n V_i I_i' = \sum_{i=1}^n V_i' I_i$.

Solution

From Theorem 6.1.3, $\int_\Omega \boldsymbol{E} \cdot \boldsymbol{J}' \mathrm{d}^2 x = \int_\Omega \boldsymbol{E}' \cdot \boldsymbol{J} \mathrm{d}^2 x$, where $\Omega = \mathbf{R}^3 - \bigcup_k (C_k \cup C_k')$. Under the assumption of thin conductor, restricting to the boundary of the conductors leads to

$$\int_\Omega \boldsymbol{E} \cdot \boldsymbol{J}' \mathrm{d}^3 x \to \sum_k \int_{\partial(C_k \cup C_k')} \boldsymbol{E}_k \cdot \boldsymbol{J}_k' \mathrm{d}^2 x = -\sum_k \int_{\partial(C_k \cup C_k')} \nabla\varphi_k \cdot \boldsymbol{J}_k' \mathrm{d}^2 x$$

where the current density becomes surface current density. Furthermore, motivated by approximating a thin conductor as a line conductor, by appealing to integration by parts, $\int \partial_\xi \varphi_k J_{k,\xi} \mathrm{d}x = \varphi_k J_{k,\xi} - \int \varphi_k \, \partial_\xi J_{k,\xi} \mathrm{d}x$, it follows clearly that

$$\sum_k \int_{\partial(C_k \cup C_k')} \nabla\varphi_k \cdot J_k' \mathrm{d}^2x = \sum_k \left\{ \int \varphi_k J_k' \cdot n_k' \mathrm{d}^2x - \int_{\partial(C_k \cup C_k')} \varphi_k \nabla \cdot J' \mathrm{d}^2x \right\},$$

where n_k' is the unit normal to the cross-section of the conductor C_k'. However, under electrostatic conditions, $\nabla \cdot J' = 0$, hence, $\varphi_k \mid \partial C_k = V_k$ is a constant, implies that $\int_{\partial(C_k \cup C_k')} \nabla\varphi_k \cdot J_k' \mathrm{d}x = \sum_k \varphi_k I_k'$, where $\int_{\partial C_k} J_k' \cdot n_k' \mathrm{d}^2x = I_k'$. Hence, by symmetry, the equality $\sum_k V_k I_k' = \sum_k V_k' I_k$ follows. □

6.5.3 Exercise

Establish Corollary 6.3.2.

Solution

From Equation (6.21), substituting (6.22) into $\partial_t(6.21)$:

$$\begin{aligned}
\partial_z^2 V &= -\mathbf{R}\partial_z I - \mathbf{L}\partial_t \partial_z I \\
&= -\mathbf{R}\{-\mathbf{Y}V - \mathbf{C}\partial_t V\} - \mathbf{L}\partial_t\{-\mathbf{Y}V - \mathbf{C}\partial_t V\} \\
&= \mathbf{RY}V + \mathbf{RC}\partial_t V + \mathbf{LY}\partial_t V + \mathbf{LC}\partial_t^2 V \\
&= \mathbf{RY}V + (\mathbf{RC} + \mathbf{LY})\partial_t V + \mathbf{LC}\partial_t^2 V
\end{aligned}$$

Similarly, substituting Equation (6.21) into $\partial_t(6.22)$ yields (6.24). To establish the second part of the corollary, compare Equation (4.10) with (6.23), and note that under perfect TEM conditions, (6.23) reduces to $\partial_z^2 V = \mathbf{RY}V + \mathbf{LY}\partial_t V + \mathbf{LC}\partial_t^2 V$, and hence, it is clear that $\mathbf{LY} = \mu\sigma\mathbf{I}$ and $\mathbf{LC} = \mu\varepsilon\mathbf{I}$.

Likewise, under perfect TEM conditions, Equation (6.24) reduces to $\partial_z^2 I = \mathbf{YR}I + \mathbf{YL}\partial_t I + \mathbf{CL}\partial_t^2 I$ and hence, comparing with (4.11) yields: $\mathbf{YL} = \mu\sigma\mathbf{I}$ and $\mathbf{CL} = \mu\varepsilon\mathbf{I}$. Whence, both $\mathbf{Y}$ and $\mathbf{C}$ commute with $\mathbf{L}$ by inspection. Hence, from the definition of the wave equation, the waves propagate along the conductors at speed $\frac{1}{\sqrt{\mu\varepsilon}}$. □

6.5.4 Exercise

Prove Lemma 6.3.8.

Solution

From $\mathbf{N}(\ell) = \mathbf{M}_{12}(\ell) - \mathbf{M}_{11}(\ell)\mathbf{Z}_S - \mathbf{Z}(\ell)\{\mathbf{M}_{22}(\ell) - \mathbf{M}_{21}(\ell)\mathbf{Z}_S\}$, under the assumption that $\ell \ll \lambda$, it follows from Lemma 6.3.7, and noting via Taylor expansion that

$$\cosh(\mathrm{i}a) = \cos a \approx 1 - \tfrac{a^2}{2} + o(a^4) \quad \text{and} \quad \sinh(\mathrm{i}a) = \mathrm{i}\sin a \approx \mathrm{i}\left\{a - \tfrac{a^3}{3!} + o(a^5)\right\},$$

it follows that

$$\mathbf{M}_{11}(\ell) = \mathbf{Z}_0\mathbf{P}^t\left\{\mathbf{I} + \tfrac{(\beta\ell)^2}{2}\mathbf{I} + \cdots\right\}(\mathbf{P}^t)^{-1}\mathbf{Z}_0^{-1} \approx \mathbf{I}$$

$$\mathbf{M}_{12}(\ell) = \mathrm{i}\mathbf{Z}_0\mathbf{P}^t\left\{\beta\ell\mathbf{I} - \tfrac{(\beta\ell)^3}{3!}\mathbf{I} + \cdots\right\}(\mathbf{P}^t)^{-1} \approx \mathrm{i}\beta\ell\mathbf{Z}_0$$

$$\mathbf{M}_{21}(\ell) = \mathrm{i}\mathbf{P}^t\left\{\beta\ell\mathbf{I} - \tfrac{(\beta\ell)^3}{3!}\mathbf{I} + \cdots\right\}(\mathbf{P}^t)^{-1}\mathbf{Z}_0^{-1} \approx \mathrm{i}\beta\ell\mathbf{Z}_0^{-1}$$

$$\mathbf{M}_{22}(\ell) = \mathbf{P}^t\left\{\mathbf{I} + \tfrac{(\beta\ell)^2}{2}\mathbf{I} + \cdots\right\}(\mathbf{P}^t)^{-1} \approx \mathbf{I}$$

Substituting these values into $\mathbf{N}(\ell)$ yields

$$\begin{aligned}\mathbf{N}(\ell) &\approx -(\mathbf{Z}_S + \mathbf{Z}(\ell)) + \mathrm{i}\beta\ell\left(\mathbf{Z}_0 + \mathbf{Z}(\ell)\mathbf{Z}_0^{-1}\mathbf{Z}_S\right)\\ &= -\left\{\mathbf{I} - \mathrm{i}\beta\ell\left(\mathbf{Z}_0 + \mathbf{Z}(\ell)\mathbf{Z}_0^{-1}\mathbf{Z}_S\right)(\mathbf{Z}_S + \mathbf{Z}(\ell))^{-1}\right\}(\mathbf{Z}_S + \mathbf{Z}(\ell)).\end{aligned}$$

Furthermore, noting that $\|\mathbf{Z}_0^{-1}\mathbf{Z}_S - \mathbf{Z}^{-1}(\ell)(\mathbf{Z}_S - \mathbf{Z}_0)\| \le 1$ implies, via the triangle inequality, that the following holds: $1 + \|\mathbf{Z}^{-1}(\ell)(\mathbf{Z}_S - \mathbf{Z}_0)\| \ge \|\mathbf{Z}_0^{-1}\mathbf{Z}_S\| \Rightarrow 1 + \|\mathbf{Z}^{-1}(\ell)(\mathbf{Z}_S - \mathbf{Z}_0)\| - \|\mathbf{Z}_0^{-1}\mathbf{Z}_S\| \ge 0$, hence, $\beta\ell \ll 1$ implies that

$$\begin{aligned}\mathbf{N}^{-1}(\ell) &\approx -(\mathbf{Z}_S + \mathbf{Z}(\ell))^{-1}\left\{\mathbf{I} - \mathrm{i}\beta\ell\left(\mathbf{Z}_0 + \mathbf{Z}(\ell)\mathbf{Z}_0^{-1}\mathbf{Z}_S\right)(\mathbf{Z}_S + \mathbf{Z}(\ell))^{-1}\right\}^{-1}\\ &\approx -(\mathbf{Z}_S + \mathbf{Z}(\ell))^{-1}\left\{\mathbf{I} + \mathrm{i}\beta\ell\left(\mathbf{Z}_0 + \mathbf{Z}(\ell)\mathbf{Z}_0^{-1}\mathbf{Z}_S\right)(\mathbf{Z}_S + \mathbf{Z}(\ell))^{-1}\right\},\end{aligned}$$

as asserted. □

6.5.5 Exercise

Establish Proposition 6.3.9.

Solution

By the assumption, Lemma 6.3.8 applies. Whence, from Corollary 6.3.7,

$$\begin{aligned}I(0) &\approx \mathbf{N}^{-1}(\ell)\left\{\mathrm{i}\beta\ell\mathbf{Z}(\ell)\mathbf{Z}_0^{-1} - \mathbf{I}\right\}V_S\\ &\approx (\mathbf{Z}_S + \mathbf{Z}(0))^{-1}(\mathbf{I} + \mathrm{i}\beta\ell\Phi(\ell))\left(\mathbf{I} - \mathrm{i}\beta\ell\mathbf{Z}(\ell)\mathbf{Z}_0^{-1}\right)V_S\\ &\approx (\mathbf{Z}_S + \mathbf{Z}(0))^{-1}\left\{\mathbf{I} + \mathrm{i}\beta\ell\left(\Phi(\ell) - \mathbf{Z}(\ell)\mathbf{Z}_0^{-1}\right)\right\}V_S,\end{aligned}$$

where $\Phi(\ell) = (\mathbf{Z}_0 + \mathbf{Z}(\ell)\mathbf{Z}_0^{-1}\mathbf{Z}_S)(\mathbf{Z}_S + \mathbf{Z}(\ell))^{-1}$. Next,

$$I(\ell) \approx i\beta\ell\mathbf{Z}_0^{-1}\left\{\mathbf{I} - \mathbf{Z}_S\left(\mathbf{Z}_S + \mathbf{Z}(\ell)\right)^{-1}\left(\mathbf{I} + i\beta\ell\Phi(\ell)\right)\right\}V_S$$

$$\approx i\beta\ell\mathbf{Z}_0^{-1}\left\{\mathbf{I} - \mathbf{Z}_S\left(\mathbf{Z}_S + \mathbf{Z}(\ell)\right)^{-1}\right\}V_S.$$

Similarly,

$$V(0) \approx \left\{\mathbf{I} - \mathbf{Z}_S\left(\mathbf{Z}_S + \mathbf{Z}\right)^{-1}\left(\mathbf{I} + i\beta\ell\Phi(\ell)\right)\left(\mathbf{I} - i\beta\ell\mathbf{Z}(\ell)\mathbf{Z}_0^{-1}\right)\right\}V_S$$

$$\approx \left\{\mathbf{I} - \mathbf{Z}_S\left(\mathbf{Z}_S + \mathbf{Z}\right)^{-1}\left(\mathbf{I} + i\beta\ell\left(\Phi(\ell) - \mathbf{Z}(\ell)\mathbf{Z}_0^{-1}\right)\right)\right\}V_S,$$

and

$$V(\ell) \approx \left\{\mathbf{I} + (i\beta\ell\mathbf{Z}_0 - \mathbf{Z}_S)\left(\mathbf{Z}_S + \mathbf{Z}(\ell)\right)^{-1}\left(\mathbf{I} + i\beta\ell\Phi(\ell)\right)\left(\mathbf{I} - i\beta\ell\mathbf{Z}(\ell)\mathbf{Z}_0^{-1}\right)\right\}V_S$$

$$\approx \left\{\mathbf{I} + (i\beta\ell\mathbf{Z}_0 - \mathbf{Z}_S)(\mathbf{Z}_S + \mathbf{Z})^{-1}\left(\mathbf{I} + i\beta\ell\left(\Phi(\ell) - \mathbf{Z}(\ell)\mathbf{Z}_0^{-1}\right)\right)\right\}V_S$$

$$\approx \left\{\mathbf{I} + (i\beta\ell\mathbf{Z}_0 - \mathbf{Z}_S)(\mathbf{Z}_S + \mathbf{Z})^{-1}\right\}V_S + i\beta\ell(i\beta\ell\mathbf{Z}_0 - \mathbf{Z}_S)(\mathbf{Z}_S + \mathbf{Z})^{-1} \times$$

$$\left(\Phi(\ell) - \mathbf{Z}(\ell)\mathbf{Z}_0^{-1}\right)V_S$$

$$\approx \left\{\mathbf{I} - \mathbf{Z}_S(\mathbf{Z}_S + \mathbf{Z})^{-1} + i\beta\ell\left(\mathbf{Z}_0\left(\mathbf{Z}_S + \mathbf{Z}\right)^{-1} - \mathbf{Z}_S\left(\mathbf{Z}_S + \mathbf{Z}\right)^{-1}\left(\Phi(\ell) - \mathbf{Z}(\ell)\mathbf{Z}_0^{-1}\right)\right)\right\}V_S,$$

as required. □

6.5.6 Exercise

Verify Equation (6.29).

Solution

By (6.28), $V_k^- = S_{k1}V_1^+ + \cdots + S_{kj}V_j^+ + \cdots + S_{kn}V_n^+$. Inasmuch as S_{ij} are constants, it follows clearly that on setting $V_i^+ = 0\ \forall i \neq j$, $V_k^- = S_{kj}V_j^+ \Rightarrow S_{kj} = \frac{V_k^-}{V_j^+}$, as asserted. □

References

1. Balanis, C. 1982. *Antenna Theory: Analysis and Design*. New York: John Wiley & Sons.
2. Carson, J. 1924. A generalization of the reciprocal theorem. *Bell Sys. Tech. J.* **3**(3) July: 393–399.

3. Carson, J. 1930. The reciprocal energy theorem. Bell Sys. Tech. J. **9**(2) April: 325–331.
4. Chang, D. 1992. *Field and Wave Electromagnetics*. Reading, MA: Addison-Wesley.
5. Orfanidis, O. 2002. *Electromagnetic Waves and Antenna*. Rutgers University, ECE Dept., http://www.ece.rutgers.edu/~orfanidi/ewa/.
6. Paul, C. 1994. *Analysis of Multiconductor Transmission Lines*. New York, John Wiley & Sons.
7. Plonsey, R. and Collin, R. 1961. *Principles and Applications of Electromagnetic Fields*. New York: McGraw-Hill.
8. Pozar, D. 2005. *Microwave Engineering*. New York: John Wiley & Sons.
9. Rothwell, E. and Cloud, M. 2001. *Electromagnetics*. Boca Raton, FL: CRC Press.
10. Smythe, W. 1950. *Static and Dynamic Electricity*. New York: McGraw-Hill.

7

Waveguides and Cavity Resonance

It was established in Chapter 1 via Theorem 1.4.1 that TEM waves cannot be sustained on a single conductor. In contrast, it is established below that TE and TM waves can propagate within a single hollow conductor. Applications of these principles can be found in radars and microwave ovens, to name a few.

By way of introduction, waves propagating between two disjoint conductors are investigated, followed by attaching the conductors together to form a single hollow conducting structure. Finally, the hollow conductor is closed at both ends to form an enclosed cavity. Resonance within a cavity forms as a result of standing waves sustained within the cavity. Readers who are interested in pursuing the topic of dielectric waveguides (which unfortunately is not covered here) or more advanced theory and applications may consult References [1,4,6,8]. The elementary theory can be found in most references on electromagnetic theory; for example, see References [2,5,7].

7.1 Parallel Plate Guides

By way of motivation, consider waves propagating between two large parallel plates. A pair of transmission lines is a special case of two distinct conductors. It is thus clear that TEM can be sustained between two parallel plates. However, can two parallel plates sustain TE and TM modes? More precisely, consider two infinite parallel planes separated by a distance $x = a$, and for simplicity, suppose that an incident electromagnetic wave is propagating in the z-direction. Can the parallel planes support TE and TM modes?

Initially, the transverse electric field propagation is considered. The transverse magnetic field propagation will conclude this introduction. Without loss of generality, in rectangular coordinates, suppose the direction of propagation is $\mathbf{e}_z$. Then, by definition, $E_z = 0$ for TE propagation.

Explicitly, consider the space $(\Omega, \mu, \varepsilon)$, where

$$\Omega = \{(x, y, z) \in \mathbf{R}^3 : -\infty < x, z < \infty, 0 < y < a\}$$

and suppose that the current density $J = 0$ on Ω, the charge density $\rho = 0$ on Ω, and $\partial\Omega$ is a perfect electrical conductor. So, consider Maxwell's first equation (1.15). Solving for it yields:

$$\nabla \times \boldsymbol{E} = \begin{vmatrix} \mathbf{e}_x & \mathbf{e}_y & \mathbf{e}_z \\ \partial_x & \partial_y & \partial_z \\ E_x & E_y & 0 \end{vmatrix} = \begin{pmatrix} -\partial_z E_y \\ \partial_z E_x \\ \partial_x E_y - \partial_y E_x \end{pmatrix} = -\begin{pmatrix} \partial_t B_x \\ \partial_t B_y \\ \partial_t B_z \end{pmatrix}$$

Clearly, for TE mode, B_z is not constrained to be zero on Ω. Thus,

$$\partial_z E_y = \partial_t B_x \tag{7.1}$$

$$\partial_z E_x = -\partial_t B_y \tag{7.2}$$

$$\partial_x E_y - \partial_y E_x = -\partial_t B_z \tag{7.3}$$

Next, solving for Equation (1.17),

$$\nabla \times \boldsymbol{B} = \begin{vmatrix} \mathbf{e}_x & \mathbf{e}_y & \mathbf{e}_z \\ \partial_x & \partial_y & \partial_z \\ B_x & B_y & B_z \end{vmatrix} = \begin{pmatrix} \partial_y B_z - \partial_z B_y \\ -\partial_x B_z + \partial_z B_x \\ \partial_x B_y - \partial_y B_x \end{pmatrix} = \mu\sigma \begin{pmatrix} E_x \\ E_y \\ 0 \end{pmatrix} + \mu\varepsilon \begin{pmatrix} \partial_t E_x \\ \partial_t E_y \\ 0 \end{pmatrix}$$

That is,

$$\partial_y B_z - \partial_z B_y = \mu\sigma E_x + \mu\varepsilon\,\partial_t E_x \tag{7.4}$$

$$-\partial_x B_z + \partial_z B_x = \mu\sigma E_y + \mu\varepsilon\,\partial_t E_y \tag{7.5}$$

$$\partial_x B_y - \partial_y B_x = 0 \tag{7.6}$$

Equations (7.1)–(7.6) can now be used to establish the properties of TE wave propagation.

7.1.1 Theorem

Suppose $(\boldsymbol{E},\boldsymbol{B})|_{E_z=0}$ propagates in the z-direction in some homogeneous domain $(\Omega,\mu,\varepsilon,\sigma)$, where $\Omega \subset \mathbf{R}^3$. Suppose that the charge density $\rho|\Omega = 0$. Then, the TE to z-mode is completely characterized by

$$-\Delta B_z + \mu\sigma\,\partial_t B_z + \mu\varepsilon\,\partial_t^2 B_z = 0 \tag{7.7}$$

$$-\Delta \boldsymbol{E}_\perp + \mu\sigma\,\partial_t \boldsymbol{E}_\perp + \mu\varepsilon\,\partial_t^2 \boldsymbol{E}_\perp = 0 \tag{7.8}$$

Proof

Now, $\partial_y(7.4) - \partial_x(7.5)$ yields

$$\partial_y^2 B_z + \partial_x^2 B_z + 2\partial_z^2 B_z + \partial_z(\partial_x B_x + \partial_y B_y)$$
$$= \mu\sigma\,\partial_t(\partial_y E_x - \partial_x E_y) + \mu\varepsilon\,\partial_t^2(\partial_y E_x - \partial_x E_y)$$

Thus, invoking Gauss' law $\nabla \cdot \boldsymbol{E} = 0$ together with Equation (7.3) give

$$-\Delta B_z + \mu\sigma\,\partial_t B_z + \mu\varepsilon\,\partial_t^2 B_z = 0$$

Next, $\partial_x(7.3) + \partial_z(7.1)$ together with $\nabla \cdot \boldsymbol{E} = 0$ yields

$$\partial_y^2 E_y + \partial_x^2 E_y + \partial_z^2 E_y = -\partial_t(\partial_x B_z - \partial_z B_x) = \mu\sigma\,\partial_t E_y + \mu\varepsilon\,\partial_t^2 E_y$$

and likewise, $\partial_y(7.3) + \partial_z(7.2)$ together with Gauss' law yields

$$\partial_y^2 E_x + \partial_x^2 E_x + \partial_z^2 E_x = -\partial_t(\partial_y B_z - \partial_z B_y) = \mu\sigma\,\partial_t E_x + \mu\varepsilon\,\partial_t^2 E_x$$

Hence, along a similar vein, together with $\partial_y(7.3) + \partial_z(7.1)$, $-\Delta \boldsymbol{E}_\perp + \mu\sigma\,\partial_t \boldsymbol{E}_\perp + \mu\varepsilon\,\partial_t^2 \boldsymbol{E}_\perp = 0$, as required. □

7.1.2 Proposition

A time-harmonic TE to *z*-mode wave propagation satisfies $\boldsymbol{E}_\perp = (\eta/\mu)\mathbf{e}_z \times \boldsymbol{B}_\perp$, where η is some frequency-dependent parameter.

Proof

A general technique is employed to establish the assertion (see Section 1.4). First, decompose the fields and operators into transverse and longitudinal components, and then equate the transverse (longitudinal) components with the transverse (longitudinal) components. This is done as follows. Set $\nabla = \nabla_\perp + \mathbf{e}_z\partial_z$, $\boldsymbol{B} = \boldsymbol{B}_\perp + \mathbf{e}_z B_z$, and by assumption, $\boldsymbol{E} = \boldsymbol{E}_\perp$. Then, from Maxwell's equations and using the convention $e^{-i\omega t}$ instead of $e^{i\omega t}$ for time harmonicity so that $\partial_t \leftrightarrow -i\omega$,

$$\nabla \times \boldsymbol{E} = i\omega \boldsymbol{B} \Rightarrow \nabla_\perp \times \boldsymbol{E}_\perp + \mathbf{e}_z \times \partial_z \boldsymbol{E}_\perp = i\omega \boldsymbol{B}_\perp + i\omega \mathbf{e}_z B_z$$

That is,

$$\nabla_\perp \times \boldsymbol{E}_\perp = i\omega \mathbf{e}_z B_z \tag{7.9}$$

$$\mathbf{e}_z \times \partial_z \boldsymbol{E}_\perp = i\omega \boldsymbol{B}_\perp \tag{7.10}$$

From Equation (7.10), $\mathbf{e}_z \times (\mathbf{e}_z \partial_z \times \boldsymbol{E}_\perp) = -\partial_z \boldsymbol{E}_\perp \Rightarrow \partial_z \boldsymbol{E}_\perp = -\mathrm{i}\omega \mathbf{e}_z \times \boldsymbol{B}_\perp$. As a warm-up exercise, verify this in Exercise 7.4.2. Next, noting that $\partial_z e^{-\gamma z} = -\gamma e^{-\gamma z} \Rightarrow \partial_z \leftrightarrow -\gamma$, it follows that

$$\gamma \boldsymbol{E}_\perp = \mathrm{i}\omega \mathbf{e}_z \times \boldsymbol{B}_\perp \Rightarrow \boldsymbol{E}_\perp = \frac{1}{\mu}\frac{\mathrm{i}\omega\mu}{\gamma} \mathbf{e}_z \times \boldsymbol{B}_\perp \tag{7.11}$$

as required. □

7.1.3 Definition

The *TE to z-mode wave impedance* η_{TE} for a time-harmonic wave $(\boldsymbol{E},\boldsymbol{B})|_{E_z=0}$ is defined by $\eta_{\mathrm{TE}} = \frac{\mathrm{i}\omega\mu}{\gamma}$, where γ depends on the boundary conditions.

Now, consider an electromagnetic field propagating between two infinite parallel plates illustrated in Figure 7.1.

7.1.4 Proposition

Referring to Figure 7.1, suppose that $(\boldsymbol{E},\boldsymbol{B})|_{E_z=0}$ is a time-harmonic wave propagating in the $\mathbf{e}_z$-direction on (Ω,μ,ε), and suppose further that $\rho = 0$ on Ω and there is no variation along the $\mathbf{e}_x$-direction. Then, the *n*-mode wave propagation $(\boldsymbol{E},\boldsymbol{B})|_{E_z=0}$ satisfies

$$E_{x,n} = E_{0,n} \sin\left(\tfrac{n\pi}{a} y\right) e^{-\gamma_n z} \tag{7.12}$$

$$B_{y,n} = \tfrac{\gamma_n}{\omega} E_{0,n} \sin\left(\tfrac{n\pi}{a} y\right) e^{-\gamma_n z} \tag{7.13}$$

$$B_{z,n} = \tfrac{\mathrm{i}}{\omega}\tfrac{n\pi}{a} E_{0,n} \cos\left(\tfrac{n\pi}{a} y\right) e^{-\gamma_n z} \tag{7.14}$$

for some constant $E_{0,n}$, and $\gamma_n^2 = \left(\frac{n\pi}{a}\right)^2 - \omega^2\mu\varepsilon - \mathrm{i}\omega\mu\sigma$. Furthermore, if $\sigma = 0$, there exists a real sequence $(\omega_n)_n$ of *cut-off* angular frequencies such that

(a) $$\omega > \omega_n \Rightarrow e^{-\gamma_n z} = e^{-\mathrm{i}\beta_n z}, \beta_n = \Im m \gamma_n$$

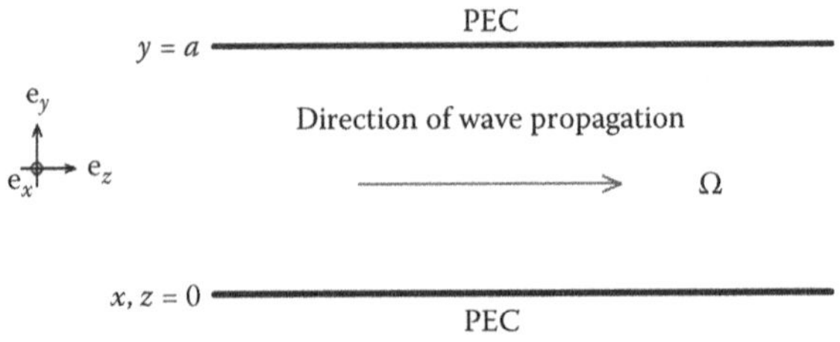

FIGURE 7.1
Wave propagation between two infinite parallel planes.

that is, the waves propagate without attenuation,

(b) $$\omega < \omega_n \Rightarrow e^{-\gamma_n z} = e^{-\alpha_n z}, \alpha_n = \Re e\gamma_n$$

that is, the waves propagate with exponential attenuation.

Finally, for $\omega = \omega_n$, standing waves exist between the two parallel planes, and no wave propagates along the $\mathbf{e}_z$-direction.

Proof

Now, applying the assumption $\partial_x = 0$ to Equations (7.3) and (7.6), respectively, yields

$$\partial_y E_x = \partial_t B_z \Rightarrow \partial_y E_x = -i\omega B_z \tag{7.15}$$

$$\partial_y B_x = 0 \tag{7.16}$$

Next, noting that $\partial_x\partial_y = \partial_y\partial_x$ as $\boldsymbol{E}, \boldsymbol{B}$ are of class C^2, it follows from Equation (7.16) that $\partial_y(7.1) \Rightarrow 0 = \partial_y\partial_z E_y = \partial_z\partial_y E_y \Rightarrow E_y$ is independent of y. Because $E_y \mid \partial\Omega = 0$, it follows that $E_y = 0$ on Ω in order to satisfy the boundary conditions. In particular, (7.1) implies at once that $B_x = 0$ on Ω. Hence, (7.8) reduces to $\Delta E_x + (i\omega\mu\sigma + \omega^2\mu\varepsilon)E_x = 0$. For notational simplicity, set $k^2 = \omega^2\mu\varepsilon + i\omega\mu\sigma$.

The solution of Equation (7.8) can be easily determined via the separation of variables. Now, by assumption, $\partial_x = 0 \Rightarrow \Delta = \partial_y^2 + \partial_z^2$. Hence, set $E_x(y,z) = \Phi(y)\Psi(z)$. Then, from Proposition 7.1.2,

$$\tfrac{\partial_y^2\Phi}{\Phi} + \tfrac{\partial_z^2\Psi}{\Psi} + k^2 = 0 \Rightarrow -k_y^2 + \gamma^2 + k^2 = 0$$

where $\tfrac{\partial_y^2\Phi}{\Phi} + k_y^2 = 0$ and $\tfrac{\partial_z^2\Psi}{\Psi} - \gamma^2 = 0$. The general solution of $\tfrac{\partial_y^2\Phi}{\Phi} + k_y^2 = 0$ is $\Phi(y) = A\cos k_y y + B\sin k_y y$, and that of $\tfrac{\partial_z^2\Psi}{\Psi} + \gamma^2 = 0$ is given by $\Psi = A'e^{-\hat{\gamma}z} + B'e^{\gamma z} = A'e^{-\gamma z}$ due to the absence of boundary as $z \to \infty$; that is, $B' \equiv 0$ as there can be no reflected waves. Thus, $\Psi = e^{-\gamma z}$ corresponds to the fundamental solution.

The boundary condition $E_x|\partial\Omega = 0$ implies that $\Phi(0) = 0 = \Phi(a)$ and hence, set $A = 0$ and $k_y a = \pi n$, for $= 0, 1, 2, \ldots$. Thus, the fundamental solution is $\Phi_n(y) = \sin k_n y \equiv \sin\tfrac{n\pi}{a}y$. In particular, denoting $\gamma_n = \gamma$,

$$\gamma_n^2 = k_n^2 - k^2 = \left(\tfrac{n\pi}{a}\right)^2 - \omega^2\mu\varepsilon - i\omega\mu\sigma \tag{7.17}$$

So, setting $\gamma_n = \alpha_n + i\beta_n$, it follows that

$$\begin{cases} \alpha_n^2 - \beta_n^2 = \left(\tfrac{n\pi}{a}\right)^2 - \omega^2\mu\varepsilon \\ 2\alpha_n\beta_n = -\omega\mu\sigma \end{cases}$$

Next, noting that $(\alpha_n^2+\beta_n^2)^2=(\beta_n^2-\alpha_n^2)^2+4\alpha_n^2\beta_n^2$, it follows that

$$\alpha_n^2+\beta_n^2=\sqrt{\left(\left(\tfrac{n\pi}{a}\right)^2-\omega^2\mu\varepsilon\right)^2+(\omega\mu\sigma)^2}$$

and hence, adding and subtracting from $\alpha_n^2-\beta_n^2$ yields

$$\begin{cases}\alpha_n=\frac{1}{\sqrt{2}}\left\{\sqrt{\left(\left(\tfrac{n\pi}{a}\right)^2-\omega^2\mu\varepsilon\right)^2+(\omega\mu\sigma)^2}-\left(\omega^2\mu\varepsilon-\left(\tfrac{n\pi}{a}\right)^2\right)\right\}^{\frac{1}{2}}\\ \beta_n=\frac{1}{\sqrt{2}}\left\{\sqrt{\left(\left(\tfrac{n\pi}{a}\right)^2-\omega^2\mu\varepsilon\right)^2+(\omega\mu\sigma)^2}+\omega^2\mu\varepsilon-\left(\tfrac{n\pi}{a}\right)^2\right\}^{\frac{1}{2}}\end{cases} \tag{7.18}$$

From Equation (7.18), it is clear that $\alpha_n,\beta_n\geq 0\ \forall\omega\geq 0$ as $|a+b|+a\geq 0\ \forall a\in\mathbf{R}, b\geq 0$ In particular, $\sigma>0\Rightarrow\alpha_n,\beta_n>0$. If $\sigma=0$, then (7.18) reduces to the form $f_\pm(\xi)=\sqrt{|\xi|\mp\xi}$, where $f_+\leftrightarrow\alpha_n$ and $f_-\leftrightarrow\beta_n$. Indeed, it is evident that

(a) $$\omega^2\mu\varepsilon-\left(\tfrac{n\pi}{a}\right)^2>0\Rightarrow\alpha_n=0\Rightarrow e^{-\gamma_n z}=e^{-i\beta_n z}$$

(b) $$\omega^2\mu\varepsilon-\left(\tfrac{n\pi}{a}\right)^2<0\Rightarrow\beta_n=0\Rightarrow e^{-\gamma_n z}=e^{-\alpha_n z}$$

Thus, scenario (a) corresponds to forward wave propagation without attenuation, and scenario (b) corresponds to wave attenuation; these waves are called *evanescent waves*, as they die out exponentially for $z>\frac{1}{\beta_n}$.

In summary, $E_{x,n}=E_{0,n}\sin(\frac{n\pi}{a}y)e^{i\gamma_n z}$, where $E_{0,n}$ is some constant depending on initial conditions. And from Equation (7.2), via $\partial_z\leftrightarrow-\gamma_n$, $B_{y,n}=\frac{\gamma_n}{\omega}E_{0,n}$ $\sin(\frac{n\pi}{a}y)e^{-\gamma_n z}$. Moreover, the solution for (7.14) can be easily solved via (7.14):

$$\partial_y E_x=-i\omega B_z\Rightarrow B_{z,n}=\tfrac{in\pi}{\omega a}E_{0,n}^+\cos\left(\tfrac{n\pi}{a}y\right)e^{i\gamma_n z}$$

Last, observe from (a) and (b) that $\omega^2\mu\varepsilon-\left(\frac{n\pi}{a}\right)^2=0\Leftrightarrow\omega_n=\frac{1}{\sqrt{\mu\varepsilon}}\frac{n\pi}{a}$ implies the existence of a sequence $(\omega_n)_n$ of critical frequencies determining whether waves are attenuated. Indeed, as $\omega=\omega_n\Rightarrow\alpha_n,\beta_n=0\Rightarrow e^{-\gamma_n z}=1\ \forall z$, it is obvious that there is no wave propagation along the z-direction as the field is independent of z. Thus, the solution $(E_{x,n},B_{y,n},B_{z,n})$ shows that standing waves exist between the two parallel planes when $\omega=\omega_n$, as required. □

7.1.5 Corollary

The wave impedance for the TE to *z*-mode of Proposition 7.1.4 is given by

$$\eta_{\mathrm{TE},n}=\eta\left\{1-\left(\tfrac{\omega_n}{\omega}\right)^2+i\tfrac{\sigma}{\omega\varepsilon}\right\}^{-\frac{1}{2}} \tag{7.19}$$

where $\eta = \sqrt{\mu/\varepsilon}$. In particular, when the dielectric is lossless, that is, $\sigma = 0$, then, for each fixed $n \in \mathbf{N}$,

(a) $$\omega < \omega_n \Rightarrow \eta_{\mathrm{TE},n} = -\mathrm{i}\eta\left\{\left(\tfrac{\omega_n}{\omega}\right)^2 - 1\right\}^{-\frac{1}{2}}$$

(b) $$\omega > \omega_n \Rightarrow \eta_{\mathrm{TE},n} = \eta\left\{1 - \left(\tfrac{\omega_n}{\omega}\right)^2\right\}^{-\frac{1}{2}}$$

Finally, $\omega >> \omega_n \Rightarrow \eta_{\mathrm{TE},n} \to \eta$. In particular, $\eta_{\mathrm{TE},n} > \eta$ whenever $\omega > \omega_n$.

Proof

From Proposition 7.1.2, $\eta_{\mathrm{TE},n} = \frac{\mathrm{i}\omega\mu}{\gamma_n}$. Furthermore, from (7.17),

$$\gamma_n = \sqrt{\left(\tfrac{n\pi}{a}\right)^2 - \omega^2\mu\varepsilon - \mathrm{i}\omega\mu\sigma} = \mathrm{i}\omega\sqrt{\mu\varepsilon}\sqrt{1 - \left(\tfrac{\omega_n}{\omega}\right)^2 + \mathrm{i}\tfrac{\sigma}{\omega\varepsilon}}$$

where $\omega_n = \frac{1}{\sqrt{\mu\varepsilon}}\frac{n\pi}{a}$. Hence,

$$\eta_{\mathrm{TE},n} = \sqrt{\tfrac{\mu}{\varepsilon}}\left\{1 - \left(\tfrac{\omega_n}{\omega}\right)^2 + \mathrm{i}\tfrac{\sigma}{\omega\varepsilon}\right\}^{-\frac{1}{2}}$$

Next, on setting $\sigma = 0$ and $\hat{\gamma}_n = \sqrt{1 - \left(\frac{\omega_n}{\omega}\right)^2}$, it is obvious that $\omega < \omega_n \Rightarrow \hat{\gamma}_n$ is purely imaginary, whereas $\omega > \omega_n \Rightarrow \hat{\gamma}_n$ is real. To complete the proof, it suffices to note that $\omega_n << \omega \Rightarrow \hat{\gamma}_n \to 1 \Rightarrow \eta_{\mathrm{TE},n} \to \eta$, and $\omega > \omega_n \Rightarrow 1 - \left(\frac{\omega_n}{\omega}\right)^2 < 1 \Rightarrow \eta_{\mathrm{TE},n} > \eta$. □

It is clear from Corollary 7.1.5, even when the dielectric is lossless, $\hat{\gamma}_n \in \mathbf{R} \Rightarrow \gamma_n$ is purely imaginary and hence $\mathrm{e}^{-\gamma_n z} \nrightarrow 0$ as $z \to \infty$; that is, when the TE mode impedance is real, the fields are traveling waves and will thus propagate without attenuation. That is, for real TE mode impedance, power can be transferred.

On the other hand, still considering the lossless dielectric case for simplicity, $\hat{\gamma}_n$ is imaginary implying that $\gamma_n > 0$ and hence, $\mathrm{e}^{-\gamma_n z} \to 0$ as $z \to \infty$. From a circuit analogy, the impedance is imaginary and hence reactive; thus, no power is transferred. Physically, the fields are evanescent waves and are thus attenuated as they propagate between the parallel planes. Wave propagation cannot be sustained. In particular, when the waves are at the cut-off frequency $\omega = \omega_n$, $\hat{\gamma}_n = 0 \Rightarrow \gamma_n = 0 \Rightarrow \mathrm{e}^{-\gamma_n z} \equiv 1\ \forall z > 0$ and hence, waves cannot be sustained at the cut-off frequency. Physically, via Corollary 7.1.5, the wave impedance approaches infinity. See Figure 7.2 for the variation of the TE mode impedance as the angular frequency varies. The plot is based on the normalized n-mode TE wave impedance $\frac{\eta_{\mathrm{TE},n}}{\eta}$ for simplicity, where η is the wave impedance of the dielectric, against the normalized angular frequency $\frac{\omega_n}{\omega}$ for some fixed ω_n. Lastly, note that $\frac{\omega_n}{\omega} = 0 \Leftrightarrow \omega \to \infty$, for any fixed $\omega_n > 0$.

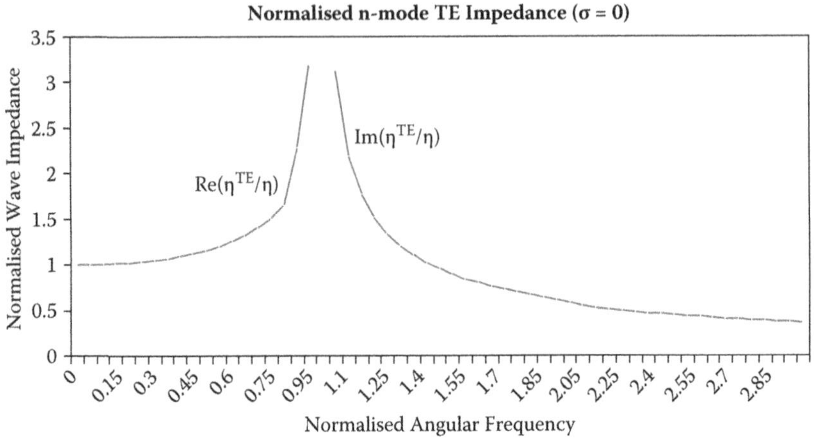

FIGURE 7.2
Normalised n-mode TE wave impedance in a lossless dielectric.

7.1.6 Theorem

Suppose $(\boldsymbol{E},\boldsymbol{B})|_{B_z=0}$ propagates in the z-direction in some homogeneous domain $(\Omega,\mu,\varepsilon,\sigma)$, where $\Omega\subset\mathbf{R}^3$. Suppose that the charge density $\rho|\Omega=0$. Then, the TM to z-mode is completely characterized by

$$-\Delta E_z+\mu\sigma\,\partial_t E_z+\mu\varepsilon\,\partial_t^2 E_z=0 \tag{7.20}$$

$$-\Delta \boldsymbol{B}_\perp+\mu\sigma\,\partial_t \boldsymbol{B}_\perp+\mu\varepsilon\,\partial_t^2 \boldsymbol{B}_\perp=0 \tag{7.21}$$

Proof

There is a quick way to establish Equations (7.20) and (7.21) via duality sketched in Section 3.6. Expressing (7.7) as $-\Delta H_z+\mu\sigma\,\partial_t H_z+\mu\varepsilon\,\partial_t^2 H_z=0$, where $B_z=\mu H_z$, and invoking the duality transformation $\boldsymbol{H}\to-\boldsymbol{E}$ yields $\Delta E_z-\mu\sigma\,\partial_t E_z-\mu\varepsilon\,\partial_t^2\ E_z=0$. Likewise, via $\boldsymbol{E}\to H$, (7.8) transforms into $-\Delta\boldsymbol{H}_\perp+\mu\sigma\,\partial_t\boldsymbol{H}_\perp+\mu\varepsilon\,\partial_t^2\boldsymbol{H}_\perp=0$, yielding (7.21) after utilizing $\boldsymbol{B}=\mu\boldsymbol{H}$. □

7.1.7 Proposition

A time-harmonic TM to z-mode wave propagation satisfies $\boldsymbol{B}_\perp=(\mu/\eta)\mathbf{e}_z\times\boldsymbol{E}_\perp$, where η is some frequency-dependent parameter.

Proof

The proof follows Proposition 7.1.2 *mutatis mutandis*. From

$$\nabla\times\boldsymbol{B}=(\mu\sigma-i\omega\mu\varepsilon)\boldsymbol{E}\Rightarrow\nabla_\perp\times\boldsymbol{B}_\perp+\mathbf{e}_z\times\partial_z\boldsymbol{B}_\perp=\kappa\boldsymbol{E}_\perp+\kappa\mathbf{e}_z E_z$$

where $\kappa = \mu\sigma - i\omega\mu\varepsilon$. That is,

$$\nabla_\perp \times \boldsymbol{B}_\perp = \kappa \mathbf{e}_z E_z \tag{7.22}$$

$$\mathbf{e}_z \times \partial_z \boldsymbol{B}_\perp = \kappa \boldsymbol{E}_\perp \tag{7.23}$$

From Equation (7.23), $\mathbf{e}_z \times (\mathbf{e}_z \,\partial_z \times \boldsymbol{B}_\perp) = -\partial_z \boldsymbol{B}_\perp \Rightarrow -\partial_z \boldsymbol{B}_\perp = \kappa \mathbf{e}_z \times \boldsymbol{E}_\perp$, whence $\partial_z \leftrightarrow -\gamma$ gives

$$\gamma \boldsymbol{B}_\perp = \kappa \mathbf{e}_z \times \boldsymbol{E}_\perp \Rightarrow \boldsymbol{B}_\perp = \mu \tfrac{\kappa}{\mu\gamma} \mathbf{e}_z \times \boldsymbol{E}_\perp$$

as required. □

7.1.8 Remark

Clearly, from Proposition 7.1.2, $\boldsymbol{E}_\perp = (\eta/\mu)\mathbf{e}_z \times \boldsymbol{B}_\perp = \eta \mathbf{e}_z \times \boldsymbol{H}_\perp$, invoking duality, $\boldsymbol{E} \to \boldsymbol{H}$ and $\boldsymbol{H} \to -\boldsymbol{E}$, yields $\boldsymbol{H}_\perp = -\tilde{\eta}\mathbf{e}_z \times \boldsymbol{E}_\perp \Rightarrow \boldsymbol{B}_\perp = -(\mu/\eta)\mathbf{e}_z \times \boldsymbol{E}_\perp$, for some frequency dependent parameter $\hat{\eta}$. However, this would not have yielded the explicit expression for $\hat{\eta}$.

7.1.9 Definition

The *TM to z-mode wave impedance* η_{TM} for a time-harmonic wave $(\boldsymbol{E},\boldsymbol{B})|_{B_z=0}$ is defined by $\eta_{TM} = -(\mu\gamma/\kappa)$, where γ depends on the boundary conditions and $\kappa = \mu\sigma - i\omega\mu\varepsilon$.

7.1.10 Proposition

Referring to Figure 7.1, suppose that $(\boldsymbol{E},\boldsymbol{B})|_{B_z=0}$ is a time-harmonic wave propagating in the $\mathbf{e}_z$-direction on (Ω,μ,ε), and suppose further that $\rho = 0$ on Ω and there is no variation along the $\mathbf{e}_x$-direction. Then, the n-mode wave propagation $(\boldsymbol{E},\boldsymbol{B})|_{B_z=0}$ satisfies

$$B_{x,n} = B_{0,n}^{+} \cos\left(\tfrac{n\pi}{a} y\right) e^{-\gamma_n z} \tag{7.24}$$

$$E_y = -\tfrac{\gamma_n}{\kappa} B_{0,n}^{+} \cos\left(\tfrac{n\pi}{a} y\right) e^{-\gamma_n z} \tag{7.25}$$

$$E_z = \tfrac{1}{\kappa} B_{0,n}^{+} \tfrac{n\pi}{a} \sin\left(\tfrac{n\pi}{a} y\right) e^{-\gamma_n z} \tag{7.26}$$

for some constant $B_{0,n}^{+}$ with $\kappa = \mu\sigma - i\omega\mu\varepsilon$. Furthermore, if $\sigma = 0$, there exists a real sequence $(\omega_n)_n$ of *cut-off* angular frequencies such that

(a) $\omega > \omega_n \Rightarrow e^{-\gamma_n z} = e^{-i\beta_n z}$, $\beta_n = \Im m \gamma_n$; that is, the waves propagate without attenuation

(b) $\omega < \omega_n \Rightarrow e^{-\gamma_n z} = e^{-\alpha_n z}$, $\alpha_n = \Re e \gamma_n$; that is, the waves propagate with attenuation

Finally, for $\omega = \omega_n$, standing waves exist between the two parallel planes, and no waves propagate along the $\mathbf{e}_z$-direction.

Proof

The proof mimics that of Proposition 7.1.4; see Exercise 7.4.3. □

7.1.11 Corollary

The wave impedance for the TM to the z-mode of Proposition 7.1.10 is given by

$$\eta_{TM,n} = \eta\left\{1 + i\tfrac{\sigma}{\omega\varepsilon}\right\}^{-1}\sqrt{1 - \left(\tfrac{\omega_n}{\omega}\right)^2 + i\tfrac{\sigma}{\omega\varepsilon}} \tag{7.27}$$

where $\eta = \sqrt{\frac{\mu}{\varepsilon}}$. In particular, when the dielectric is lossless (i.e., $\sigma = 0$) then, for each fixed $n \in \mathbf{N}$,

(a) $$\omega < \omega_n \Rightarrow \eta_{TM,n} = i\eta\sqrt{\left(\tfrac{\omega_n}{\omega}\right)^2 - 1}$$

(b) $$\omega > \omega_n \Rightarrow \eta_{TM,n} = \eta\sqrt{1 - \left(\tfrac{\omega_n}{\omega}\right)^2}$$

Finally, $\omega >> \omega_n \Rightarrow \eta_{TM,n} \to \eta$. In particular, $\eta_{TM,n} < \eta$ whenever $\omega > \omega_n$, and hence, $\eta_{TM,n} < \eta_{TE,n}\ \forall \omega > \omega_n$.

Proof

From Definition 7.1.9, $\eta_{TM,n} = -\frac{\mu\gamma_n}{\kappa}$, where $\kappa = \mu\sigma - i\omega\mu\varepsilon$. Hence, following the proof of Corollary 7.1.5 *mutatis mutandis*, and on setting $\gamma_n = i\omega\sqrt{\mu\varepsilon}\sqrt{1 - \left(\frac{\omega_n}{\omega}\right)^2 + i\frac{\sigma}{\omega\varepsilon}} \equiv i\omega\sqrt{\mu\varepsilon}\hat{\gamma}_n$, it can be shown (see Exercise 7.4.4) that

$$\eta_{TM,n} = \eta\left\{1 + i\tfrac{\sigma}{\omega\varepsilon}\right\}^{-1}\sqrt{1 - \left(\tfrac{\omega_n}{\omega}\right)^2 + i\tfrac{\sigma}{\omega\varepsilon}}$$

The two cases (a) and (b) are self-evident and (b) implies $\eta_{TM,n} < \eta\ \forall \omega > \omega_n$. Lastly, $\eta_{TM,n} < \eta_{TE,n}\ \forall \omega > \omega_n$ follows directly from Corollary 7.1.5. □

The plot of the normalized TM wave impedance $\frac{\eta_{TM,n}}{\eta}$ versus the normalized angular frequency $\frac{\omega_n}{\omega}$ is shown in Figure 7.3, where ω_n is assumed fixed.

7.1.12 Theorem

Given the parallel wave guide illustrated in Figure 7.1, suppose a time-harmonic n-mode TE or TM wave is propagating at some fixed frequency $\omega > \omega_n$. Then, the wavelength of the waves within Ω is given by

$$\lambda = \tfrac{2\sqrt{2}\pi}{\omega} v \left\{\sqrt{\left(\left(\tfrac{\omega_n}{\omega}\right)^2 - 1\right)^2 + \left(\tfrac{\sigma}{\omega\varepsilon}\right)^2} + 1 - \left(\tfrac{\omega_n}{\omega}\right)^2\right\}^{-\frac{1}{2}} \tag{7.28}$$

In particular, for a lossless medium,

$$\lambda = \tfrac{2\pi v}{\omega}\left\{1 - \left(\tfrac{\omega_n}{\omega}\right)^2\right\}^{-\frac{1}{2}}$$

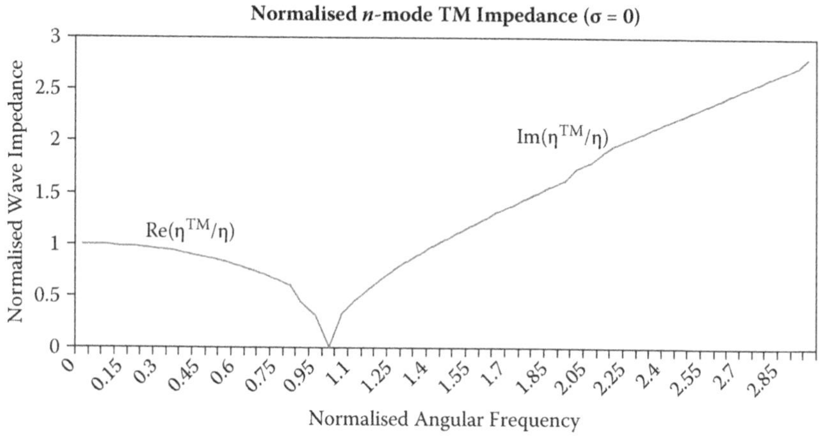

FIGURE 7.3
Normalised n-mode TM wave impedance in a lossless dielectric.

Proof

Now, an admissible TE or TM wave propagating between two parallel plates Ω has the form $f(y)e^{-\gamma_n z}e^{-i\omega t}$. Setting $\gamma_n = \alpha_n + i\beta_n$ defined by Equation (7.18), it is clear that the phase of the wave is defined by $e^{-i(\omega t+\beta_n z)}$. Thus, the point of constant phase is $\Theta = \omega t + \beta_n z$ and the phase velocity is thus

$$\tfrac{d}{dt}\Theta = \omega + \beta_n \tfrac{dz}{dt} = 0 \Rightarrow \tfrac{dz}{dt} = -\tfrac{\omega}{\beta_n}$$

whence, the n-mode phase speed $\hat{u}_n$ of the wave in Ω is

$$\hat{u}_n = \tfrac{\omega}{\beta_n} = f\lambda_n \Rightarrow \lambda_n = \tfrac{2\pi}{\beta_n} \tag{7.29}$$

Rearranging β_n slightly as

$$\tfrac{1}{\sqrt{2}}\left\{\sqrt{\left(\left(\tfrac{n\pi}{a}\right)^2 - \omega^2\mu\varepsilon\right)^2 + (\omega\mu\sigma)^2} + \omega^2\mu\varepsilon - \left(\tfrac{n\pi}{a}\right)^2\right\}^{\frac{1}{2}} =$$

$$\tfrac{1}{\sqrt{2}}\omega\sqrt{\mu\varepsilon}\left\{\sqrt{\left(\left(\tfrac{\omega_n}{\omega}\right)^2 - 1\right)^2 + \left(\tfrac{\sigma}{\omega\varepsilon}\right)^2} + 1 - \left(\tfrac{\omega_n}{\omega}\right)^2\right\}^{\frac{1}{2}}$$

it is clear that

$$\lambda_n = \tfrac{\sqrt{2}}{f}v\left\{\sqrt{\left(\left(\tfrac{\omega_n}{\omega}\right)^2 - 1\right)^2 + \left(\tfrac{\sigma}{\omega\varepsilon}\right)^2} + 1 - \left(\tfrac{\omega_n}{\omega}\right)^2\right\}^{-\frac{1}{2}}$$

where $v = \frac{1}{\sqrt{\mu\varepsilon}}$ is the phase velocity in $(\mathbf{R}^3, \mu, \varepsilon)$. For the lossless case, $\sigma = 0$ and the result thus follows. □

7.1.13 Example

Determine the time-average power transmitted by an admissible n-mode TE wave propagating in $\Omega|_{\sigma=0}$ defined in Figure 7.1, and hence, deduce the velocity of the energy flow. To determine the power flow, it suffices to consider the time-average power transmitted across a finite cross-section $R = [0,1] \times [0,a]$. By definition,

$$\langle P\rangle_R = \tfrac{1}{2}\int_R \Re\mathrm{e}(\boldsymbol{E}\times\boldsymbol{H}^*)\cdot\boldsymbol{n}\,\mathrm{d}^2x = \tfrac{1}{2}\int_0^1\int_0^a \Re\mathrm{e}(\boldsymbol{E}\times\boldsymbol{H}^*)\cdot\mathbf{e}_z\,\mathrm{d}y\,\mathrm{d}x$$

From Proposition 7.1.4, $\Re\mathrm{e}(\boldsymbol{E}\times\boldsymbol{H}^*)\cdot\mathbf{e}_z = \frac{1}{\mu}E_{x,n}B_{y,n}^* = \frac{1}{\mu}|E_{0,n}|^2\frac{\beta_n}{\omega}\sin^2\left(\frac{n\pi}{a}y\right)$, as $\gamma_n = \mathrm{i}\beta_n$ for lossless Ω. Hence,

$$\langle P\rangle_R = \tfrac{a}{4}\tfrac{1}{\mu}|E_{0,n}|^2\tfrac{\beta_n}{\omega}$$

is the time-average power transmitted across the cross-section R.

Next, to determine the velocity of energy flow, recall first by definition that power is the time-rate of the flow of energy. Second, the time-average energy density per unit length along the direction of wave propagation is given by $\langle w\rangle_C = \frac{1}{2}\int_C \Re\mathrm{e}\left(\frac{1}{2}\boldsymbol{D}\cdot\boldsymbol{E}^* + \frac{1}{2}\boldsymbol{B}\cdot\boldsymbol{H}^*\right)\mathrm{d}^3x$, where $C = R \times [0,1]$. Observe also that the assumption of $\partial_x = 0$ renders the choice of R,C meaningful and hence, $\langle P\rangle_R, \langle w\rangle_C$ meaningful.

By definition, $\langle w\rangle_C = \frac{1}{4}\int_0^1\int_0^1\int_0^a(\varepsilon|\boldsymbol{E}|^2 + \mu|\boldsymbol{H}|^2)\mathrm{d}y\,\mathrm{d}x\,\mathrm{d}z$. From Proposition 7.1.4,

$$\varepsilon|\boldsymbol{E}|^2 + \mu|\boldsymbol{H}|^2 = \varepsilon(E_{0,n})^2\sin^2\left(\tfrac{n\pi}{a}y\right) + \tfrac{1}{\mu}\left(E_{0,n}^{+}\right)^2\left\{\left(\tfrac{\beta_n}{\omega}\right)^2\sin^2\left(\tfrac{n\pi}{a}y\right) + \left(\tfrac{n\pi}{\omega a}\right)^2\cos^2\left(\tfrac{n\pi}{a}y\right)\right\}$$

and hence,

$$\langle w\rangle_C = \tfrac{a}{8}(E_{0,n})^2\left\{\varepsilon + \tfrac{1}{\mu}\left(\left(\tfrac{\beta_n}{\omega}\right)^2 + \left(\tfrac{n\pi}{\omega a}\right)^2\right)\right\}$$

Thus, note that if $\tilde{u}_n$ is the n-mode velocity of energy propagation, then $\tilde{u}_n\langle w\rangle_C$ has precisely the dimensions of power. It is intuitively clear that $\tilde{u}_n\langle w\rangle_C$ corresponds to the time-average power: $\langle P\rangle_R = \tilde{u}_n\langle w\rangle_C$. Hence, the velocity of energy propagation along $\Omega|_{\sigma=0}$ is given by

$$\tilde{u}_n = \tfrac{\langle P\rangle_R}{\langle w\rangle_C} = 2\beta_n\omega\left\{(\omega^2+\omega_n^2)\mu\varepsilon + \beta_n^2\right\}^{-1} = \tfrac{2\beta_n\omega}{(\omega^2+\omega_n^2+\omega^2-\omega_n^2)\mu\varepsilon} = \tfrac{\beta_n}{\omega\mu\varepsilon}$$

However, for $\sigma = 0$, $\omega > \omega_n \Rightarrow \beta_n = \omega\sqrt{\mu\varepsilon}\sqrt{1-\left(\frac{\omega_n}{\omega}\right)^2}$, as can be easily verified. Hence,

$$\tilde{u}_n = \frac{1}{\sqrt{\mu\varepsilon}}\sqrt{1-\left(\frac{\omega_n}{\omega}\right)^2} \equiv v\sqrt{1-\left(\frac{\omega_n}{\omega}\right)^2} \tag{7.30}$$

It is clear from the equation that $\tilde{u} < v \; \forall \omega > \omega_n$.

Finally, from the proof of Proposition 7.1.12, $\hat{u}_n = \frac{\omega}{\beta_n} \Rightarrow \hat{u}_n\tilde{u}_n = \frac{1}{\mu\varepsilon}$. That is, the product of the phase velocity and the energy propagation velocity (i.e., group velocity) yields precisely the velocity of propagation of a plane wave in $(\mathbf{R}^3, \mu, \varepsilon)$. A plot of the normalized phase $\frac{\hat{u}_n}{v}$ velocity and the normalized velocity of energy propagation $\frac{\tilde{u}_n}{v}$ as a function of the normalized angular frequency $\frac{\omega_n}{\omega}$ is shown in Figure 7.4.

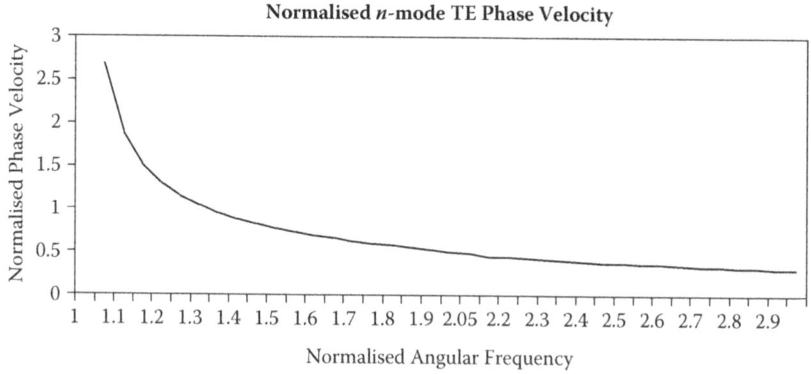

FIGURE 7.4 A
Normalised phase velocity vs. normalised angular frequency.

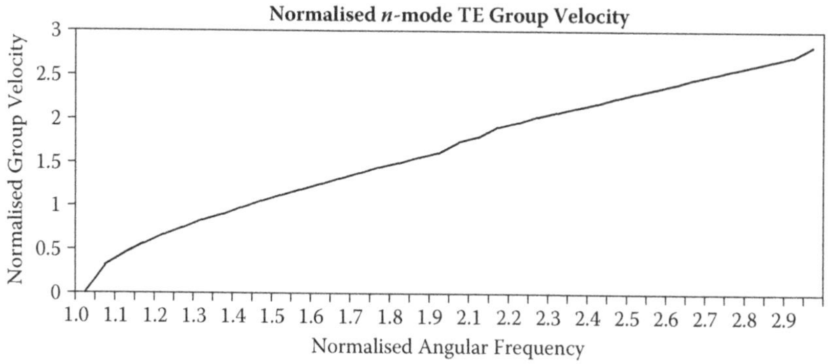

FIGURE 7.4B
Normalised group velocity vs. normalised angular frequency.

7.2 Rectangular Waveguides

In Section 7.1, the concept of TE/TM mode propagation was introduced between two parallel planes. The parallel planes constitute a *waveguide*. In particular, a pair of transmission lines constitutes a waveguide for TEM propagation. In short, a waveguide is any structure* that supports TE, TM, or TEM modes. In what follows, let $\Omega = (0,b) \times (0,a) \times [0,\infty)$ denote a semi-infinite rectangular waveguide, and $\partial\Omega$ denote its boundary.

7.2.1 Proposition

Suppose that $(\boldsymbol{E},\boldsymbol{B})|_{E_z=0}$ is a time-harmonic wave propagating in the e_z-direction on $(\Omega,\mu,\varepsilon,\sigma)$, and suppose further that $\rho = 0$ on Ω and $\partial\Omega$ is a perfect electrical conductor. Then, $(\boldsymbol{E},\boldsymbol{B})|_{E_z=0}$ satisfies

$$E_{x,mn} = E_{0,mn}\cos\left(\tfrac{m\pi}{b}x\right)\sin\left(\tfrac{n\pi}{a}y\right)\mathrm{e}^{-\gamma_{mn}z} \tag{7.31}$$

$$E_{y,mn} = -\tfrac{ma}{nb}E_{0,mn}\sin\left(\tfrac{m\pi}{b}x\right)\cos\left(\tfrac{n\pi}{a}y\right)\mathrm{e}^{-\gamma_{mn}z} \tag{7.32}$$

$$B_{x,mn} = \tfrac{\mathrm{i}\gamma_{mn}}{\omega}\tfrac{ma}{nb}E_{0,mn}\sin\left(\tfrac{m\pi}{b}x\right)\cos\left(\tfrac{n\pi}{a}y\right)\mathrm{e}^{-\gamma_{mn}z} \tag{7.33}$$

$$B_{y,mn} = \tfrac{\mathrm{i}\gamma_{mn}}{\omega}E_{0,mn}\cos\left(\tfrac{m\pi}{b}x\right)\sin\left(\tfrac{n\pi}{a}y\right)\mathrm{e}^{-\gamma_{mn}z} \tag{7.34}$$

$$B_{z,mn} = \tfrac{\mathrm{i}\pi}{\omega}E_{0,mn}\left\{1+\left(\tfrac{ma}{nb}\right)^2\right\}\cos\left(\tfrac{m\pi}{b}x\right)\cos\left(\tfrac{m\pi}{a}y\right)\mathrm{e}^{-\gamma_{mn}z} \tag{7.35}$$

for some constant $E_{0,mn}$ and $\gamma_{mn}^2 = \left(\tfrac{m\pi}{b}\right)^2 + \left(\tfrac{n\pi}{a}\right)^2 - \omega^2\mu\varepsilon - \mathrm{i}\omega\mu\sigma$. Furthermore, if $\sigma = 0$, there exists a real sequence $(\omega_{mn})_{m,n}$ of *cut-off* angular frequencies such that

(a) $\omega > \omega_{mn} \Rightarrow \mathrm{e}^{-\gamma_{mn}z} = \mathrm{e}^{-\mathrm{i}\beta_{mn}z}, \beta_{mn} = \Im\mathrm{m}\gamma_{mn}$, and hence waves propagate without attenuation

(b) $\omega < \omega_{mn} \Rightarrow \mathrm{e}^{-\gamma_{mn}z} = \mathrm{e}^{-\alpha_{mn}z}, \alpha_{mn} = \Re\mathrm{e}\gamma_{mn}$ and hence waves propagate with exponential attenuation

Proof

From Theorem 7.1.1, $\Delta B_z + k^2 B_z = 0$ subject to the boundary condition

$$\begin{cases} \partial_x B_z = 0 \text{ if } x = 0, b \\ \partial_y B_z = 0 \text{ if } y = 0, a \end{cases}$$

* More precisely, a single conductor or a set of disjoint conductors.

as $\partial\Omega$ is a perfect electrical conductor. So, to begin, consider $\Delta E_x + k^2 E_x = 0$. Then, via the separation of variables

$$B_x(x,y,z) = \Phi(x)\Psi(y)\Theta(z) \Rightarrow \tfrac{\partial_x^2 \Phi}{\Phi} + \tfrac{\partial_y^2 \Psi}{\Psi} + \tfrac{\partial_z^2 \Theta}{\Theta} + k^2 = 0$$

Set $\frac{\partial_x^2 \Phi}{\Phi} = -k_x^2$, $\frac{\partial_y^2 \Psi}{\Psi} = -k_y^2$, and $\frac{\partial_z^2 \Theta}{\Theta} = \gamma^2$. Thus,

$$\Phi = A\cos k_x x + B\sin k_x x \Rightarrow \partial_x \Phi = -Ak_x \sin k_x x + Bk_x \cos k_x x$$

yielding, via the boundary condition, $B = 0 \Rightarrow \Phi = \cos\left(\frac{m\pi}{b}x\right)$. Similarly, $\Psi = \cos\left(\frac{n\pi}{a}y\right)$ and $\Theta = e^{-\gamma z}$. That is, the fundamental solution is $B_x(x,y,z) = \cos\left(\frac{m\pi}{b}x\right)\cos\left(\frac{m\pi}{a}y\right)e^{-\gamma z}$.

Next, invoking Theorem 7.1.1 again, $\Delta \boldsymbol{E}_\perp + k^2 \boldsymbol{E}_\perp = 0$. Thus, appealing to the boundary condition for $\boldsymbol{E}_\perp$,

$$\begin{cases} E_y = 0 \text{ if } x = 0, b \\ E_x = 0 \text{ if } y = 0, a \end{cases}$$

the separation of variables applied to the pair E_x, E_y yields:

$$E_x \sim \sin\left(\tfrac{n\pi}{a}y\right)\left\{A'\cos k_x' x + B'\sin k_x' x\right\}e^{-\gamma' z}$$

$$E_y \sim \sin\left(\tfrac{m\pi}{b}x\right)\left\{A''\cos k_y'' y + B''\sin k_y'' y\right\}e^{-\gamma'' z}$$

Now, appealing to Equation (7.3), $\partial_x E_y - \partial_y E_x = i\omega B_z$, it follows at once by invoking the boundary condition for B_z, to wit,

$$\partial_x\left\{\partial_x E_y - \partial_y E_x\right\}\Big|_{x=0,b} = 0 = \partial_y\left\{\partial_x E_y - \partial_y E_x\right\}\Big|_{y=0,a}$$

that the following must hold,

$$\gamma' = \gamma = \gamma''$$

$$A'\cos k_x' x + B'\sin k_x' x \sim \cos\left(\tfrac{m\pi}{b}x\right) \Rightarrow B' = 0, k_x' = \tfrac{m\pi}{b}$$

$$A''\cos k_y'' y + B''\sin k_y'' y \sim \cos\left(\tfrac{n\pi}{a}y\right) \Rightarrow B'' = 0, k_y'' = \tfrac{n\pi}{a}$$

The results give the fundamental solutions:

$$E_x = A' \cos\left(\tfrac{m\pi}{b}x\right)\sin\left(\tfrac{n\pi}{a}y\right)e^{-\gamma' z}$$

$$E_y = A'' \sin\left(\tfrac{m\pi}{b}x\right)\cos\left(\tfrac{n\pi}{a}y\right)e^{-\gamma' z}$$

In order to determine the pair A', A'', it suffices to appeal to Gauss' law: $\nabla \cdot \boldsymbol{E}_\perp = 0$ (by assumption). Hence, $\partial_x E_x = -\partial_y E_y \Rightarrow A''\tfrac{n\pi}{a} = -A'\tfrac{m\pi}{b} \Rightarrow A'' = -\tfrac{ma}{nb}A'$. Finally, from Equation (7.3), it is clear that $i\omega B^{+}_{0,nm} = A''\tfrac{m\pi}{b} - A'\tfrac{n\pi}{a} = -A'\pi\left\{1+\left(\tfrac{ma}{nb}\right)^2\right\} \Rightarrow B^{+}_{0,nm} = \tfrac{i\pi}{\omega}A'\left\{1+\left(\tfrac{ma}{nb}\right)^2\right\}$. From this, it follows that on setting $A' = E_{0,mn}$, some constant,

$$E_x = E_{0,mn}\cos\left(\tfrac{m\pi}{b}x\right)\sin\left(\tfrac{n\pi}{a}y\right)e^{-\gamma_{mn}z}$$

$$E_{y,mn} = -\tfrac{ma}{nb}E_{0,mn}\sin\left(\tfrac{m\pi}{b}x\right)\cos\left(\tfrac{n\pi}{a}y\right)e^{-\gamma_{mn}z}$$

$$B_{z,mn} = \tfrac{i\pi}{\omega}E_{0,mn}\left\{1+\left(\tfrac{ma}{nb}\right)^2\right\}\cos\left(\tfrac{m\pi}{b}x\right)\cos\left(\tfrac{m\pi}{a}y\right)e^{-\gamma_{mn}z}$$

where $\gamma_{mn}^2 = k_m^2 + k_n^2 - k^2 = \left(\tfrac{m\pi}{b}\right)^2 + \left(\tfrac{n\pi}{a}\right)^2 - \omega^2\mu\varepsilon - i\omega\mu\sigma$. From Equation (7.18), it is obvious that

$$\begin{cases} \alpha_{mn} = \tfrac{1}{\sqrt{2}}\left\{\sqrt{\left(\left(\tfrac{m\pi}{b}\right)^2+\left(\tfrac{n\pi}{a}\right)^2-\omega^2\mu\varepsilon\right)^2+(\omega\mu\sigma)^2} - \left(\omega^2\mu\varepsilon-\left(\tfrac{m\pi}{b}\right)^2-\left(\tfrac{n\pi}{a}\right)^2\right)\right\}^{\frac{1}{2}} \\ \beta_{mn} = \tfrac{1}{\sqrt{2}}\left\{\sqrt{\left(\left(\tfrac{m\pi}{b}\right)^2+\left(\tfrac{n\pi}{a}\right)^2-\omega^2\mu\varepsilon\right)^2+(\omega\mu\sigma)^2} + \omega^2\mu\varepsilon-\left(\tfrac{m\pi}{b}\right)^2-\left(\tfrac{n\pi}{a}\right)^2\right\}^{\frac{1}{2}} \end{cases}$$

This is easily obtained via the replacement: $\left(\tfrac{n\pi}{a}\right)^2 \to \left(\tfrac{m\pi}{b}\right)^2 + \left(\tfrac{n\pi}{a}\right)^2$. Hence, for $\sigma = 0$, and setting $\omega_{mn}^2\mu\varepsilon = \left(\tfrac{m\pi}{b}\right)^2 + \left(\tfrac{n\pi}{a}\right)^2$, it is clear that $\alpha_{mn} = 0$ if $\omega > \omega_{mn}$, whereas $\beta_{mn} = 0$ if $\omega < \omega_{mn}$.

The explicit expressions for the remaining magnetic field density components follow directly from Equations (7.1) and (7.2), respectively:

$$B_{x,mn} = -\tfrac{i\gamma_{mn}}{\omega}E_{y,mn} = \tfrac{i\gamma_{mn}}{\omega}\tfrac{ma}{nb}E^{+}_{0,mn}\sin\left(\tfrac{m\pi}{b}x\right)\cos\left(\tfrac{n\pi}{a}y\right)e^{-\gamma_{mn}z}$$

$$B_{y,mn} = \tfrac{i\gamma_{mn}}{\omega}E_{x,mn} = \tfrac{i\gamma_{mn}}{\omega}E^{+}_{0,mn}\cos\left(\tfrac{m\pi}{b}x\right)\sin\left(\tfrac{n\pi}{a}y\right)e^{-\gamma_{mn}z}$$

The proof that resonance occurs when $\omega = \omega_{mn}$, the *cut-off angular frequency*, follows that of Proposition 7.1.4 *mutatis mutandis*. In particular, no waves propagate along the waveguide. □

7.2.2 Corollary

The TE_{mn} to z-mode wave impedance $\eta_{\mathrm{TE},mn}$ for a time-harmonic $(\boldsymbol{E},\boldsymbol{B})|_{E_z=0}$ is given by $\eta_{\mathrm{TE},mn} = \frac{i\omega\mu}{\gamma_{mn}} = \eta\left\{1-\left(\frac{\omega_{mn}}{\omega}\right)^2 + i\frac{\sigma}{\omega\varepsilon}\right\}^{-\frac{1}{2}}$. In particular, when $\sigma = 0$, for each fixed $(m,n) \in \mathrm{N} \times \mathrm{N}$,

(a) $$\omega < \omega_{mn} \Rightarrow \eta_{\mathrm{TE},mn} = -i\eta\left\{\left(\tfrac{\omega_{mn}}{\omega}\right)^2 - 1\right\}^{-1}$$

(b) $$\omega > \omega_{mn} \Rightarrow \eta_{\mathrm{TE},mn} = \eta\left\{1-\left(\tfrac{\omega_{mn}}{\omega}\right)^2\right\}^{-1}$$

Finally, $\omega >> \omega_{mn} \Rightarrow \eta_{\mathrm{TE},mn} \to \eta$. In particular, $\eta_{\mathrm{TE},mn} > \eta$ whenever $\omega > \omega_{mn}$.

Proof

Noting that $\gamma_{mn} = i\omega\sqrt{\mu\varepsilon}\sqrt{1-\left(\frac{\omega_{mn}}{\omega}\right)^2 + i\frac{\sigma}{\omega\varepsilon}}$, Definition 7.1.3 leads at once to

$$\eta_{\mathrm{TE},mn} = \frac{i\omega\mu}{\gamma_{mn}} = \eta\left\{1-\left(\tfrac{\omega_{mn}}{\omega}\right)^2 + i\tfrac{\sigma}{\omega\varepsilon}\right\}^{-\frac{1}{2}}$$

The remaining assertions follow the proof of Corollary 7.1.5 *mutatis mutandis*. □

It is evident from the previous section that traveling waves are sustained if the frequency is greater than the cut-off frequency. From Corollary 7.2.2, waves will propagate when the wave impedance is real; when the wave impedance is imaginary, waves are not sustained—they are attenuated—as no power is transferred.

7.2.3 Proposition

Suppose that $(\boldsymbol{E},\boldsymbol{B})|_{B_z=0}$ is a time-harmonic wave propagating in the $\mathbf{e}_z$-direction on $(\Omega,\mu,\varepsilon,\sigma)$, and suppose further that $\rho = 0$ on Ω and $\partial\Omega$ is a perfect electrical conductor. Then, $(\boldsymbol{E},\boldsymbol{B})|_{B_z=0}$ satisfies

$$E_{z,mn} = E_0 \sin\left(\tfrac{m\pi}{b}x\right)\sin\left(\tfrac{n\pi}{a}y\right)e^{-\gamma_{mn}z} \tag{7.36}$$

$$E_{x,mn} = -\gamma_{mn}\left\{\left(\tfrac{m\pi}{b}\right)^2 + \left(\tfrac{n\pi}{a}\right)^2\right\}^{-1} \tfrac{m\pi}{b} E_0 \cos\left(\tfrac{m\pi}{b}x\right)\sin\left(\tfrac{n\pi}{a}y\right)e^{-\gamma_{mn}z} \tag{7.37}$$

$$E_{y,mn} = -\gamma_{mn}\left\{\left(\tfrac{m\pi}{b}\right)^2 + \left(\tfrac{n\pi}{a}\right)^2\right\}^{-1} \tfrac{n\pi}{a} E_0 \sin\left(\tfrac{m\pi}{b}x\right)\cos\left(\tfrac{n\pi}{a}y\right)e^{-\gamma_{mn}z} \tag{7.38}$$

$$B_{x,mn} = \kappa\left\{\left(\tfrac{m\pi}{b}\right)^2 + \left(\tfrac{n\pi}{a}\right)^2\right\}^{-1} \tfrac{n\pi}{a} E_0^+ \sin\left(\tfrac{m\pi}{b}x\right)\cos\left(\tfrac{n\pi}{a}y\right)e^{-\gamma_{mn}z} \tag{7.39}$$

$$B_{y,mn} = -\kappa\left\{\left(\tfrac{m\pi}{b}\right)^2 + \left(\tfrac{n\pi}{a}\right)^2\right\}^{-1} \tfrac{m\pi}{b} E_0^+ \cos\left(\tfrac{m\pi}{b}x\right)\sin\left(\tfrac{n\pi}{a}y\right)e^{-\gamma_{mn}z} \tag{7.40}$$

for some constant E_0, $\kappa = \mu\sigma - i\omega\mu\varepsilon$, and $\gamma_{mn}^2 = \left(\frac{m\pi}{b}\right)^2 + \left(\frac{n\pi}{a}\right)^2 - \omega^2\mu\varepsilon - i\omega\mu\sigma$. Furthermore, if $\sigma = 0$, there exists a real sequence $(\omega_{mn})_{m,n}$ of cut-off angular frequencies such that

a) $\omega > \omega_{mn} \Rightarrow e^{-\gamma_{mn}z} = e^{-i\beta_{mn}z}, \beta_{mn} = \Im m\gamma_{mn}$, and hence waves propagate without attenuation

b) $\omega < \omega_{mn} \Rightarrow e^{-\gamma_{mn}z} = e^{-\alpha_{mn}z}, \alpha_{mn} = \Re e\gamma_{mn}$ and hence waves propagate with exponential attenuation

Proof

From Exercise 7.4.2, the TM mode reduces Maxwell's equations to

$$\partial_y E_z + \gamma_{mn} E_y = i\omega B_x, -\partial_x E_z - \gamma_{mn} E_x = i\omega B_y, \partial_y E_x = \partial_x E_y$$

$$\gamma_{mn} B_y = \kappa E_x, -\gamma_{mn} B_x = \kappa E_y, \partial_x B_y - \partial_y B_x = \kappa E_z$$

By Theorem 7.1.6, $-\Delta E_z + \mu\sigma\partial_t E_z + \mu\varepsilon\partial_t^2 E_z = 0$. So, once again, appealing to the separation of variables, set $E_z = \Phi(x)\Psi(y)\Theta(z)$, subject to the boundary condition:

$$E_z = 0 \text{ for } \begin{cases} x = 0, b \\ y = 0, a \end{cases}$$

By now, it should be obvious that the fundamental solution is $E_z = \sin\left(\frac{m\pi}{b}x\right)\sin\left(\frac{n\pi}{a}y\right)e^{-\gamma z}$. To see this, it suffices to set $\Phi = A\cos k_x x + B\sin k_x x$, $\Psi = A'\cos k_y y + B'\sin k_y y$, and $\Theta = A''e^{-\gamma z} + B''e^{\gamma z}$. Then, the boundary conditions and the requirement that the solution be finite yield the desired fundamental solution. Thus, set $E_{z,mn} = E_0 \sin\left(\frac{m\pi}{b}x\right)\sin\left(\frac{n\pi}{a}y\right)e^{-\gamma_{mn}z}$.

Next, substituting $E_x = \frac{\gamma_{mn}}{\kappa} B_y$ into $-\partial_x E_z - \gamma_{mn} E_x = i\omega B_y$ yields:

$$B_y = -\left\{\frac{\gamma_{mn}^2}{\kappa} + i\omega\right\}^{-1} \partial_x E_z = -\left\{\frac{\gamma_{mn}^2}{\kappa} + i\omega\right\}^{-1} \frac{m\pi}{b} E_0^+ \cos\left(\frac{m\pi}{b}x\right)\sin\left(\frac{n\pi}{a}y\right)e^{-\gamma_{mn}z}$$

Substituting $E_y = -\frac{\gamma_{mn}}{\kappa} B_x$ into $\partial_y E_z + \gamma_{mn} E_y = i\omega B_x$ leads to:

$$B_x = \left\{\frac{\gamma_{mn}^2}{\kappa} + i\omega\right\}^{-1} \partial_y E_z = \left\{\frac{\gamma_{mn}^2}{\kappa} + i\omega\right\}^{-1} \frac{n\pi}{a} E_0^+ \sin\left(\frac{m\pi}{b}x\right)\cos\left(\frac{n\pi}{a}y\right)e^{-\gamma_{mn}z}$$

and finally, back-substituting yields

$$E_x = -\gamma_{mn}\left\{\left(\frac{m\pi}{b}\right)^2 + \left(\frac{n\pi}{a}\right)^2\right\}^{-1} \frac{m\pi}{b} E_0^+ \cos\left(\frac{m\pi}{b}x\right)\sin\left(\frac{n\pi}{a}y\right)e^{-\gamma_{mn}z}$$

$$E_y = -\gamma_{mn}\left\{\left(\frac{m\pi}{b}\right)^2 + \left(\frac{n\pi}{a}\right)^2\right\}^{-1} \frac{n\pi}{a} E_0^+ \sin\left(\frac{m\pi}{b}x\right)\cos\left(\frac{n\pi}{a}y\right)e^{-\gamma_{mn}z}$$

And the results thus follow from the simplification: $-\frac{\gamma_{mn}}{\kappa}\left\{\frac{\gamma_{mn}^2}{\kappa}+i\omega\right\}^{-1}=-\frac{\gamma_{mn}}{\gamma_{mn}^2-k^2}$ and by substituting $i\omega\kappa=k^2$. So, by definition, $\gamma_{mn}^2-k^2=\left(\frac{m\pi}{b}\right)^2+\left(\frac{n\pi}{a}\right)^2$ yields the desired results. Lastly, the remaining assertions follow from the proof of Proposition 7.2.1 *mutatis mutandis* via the replacement: $\left(\frac{n\pi}{a}\right)^2\to\left(\frac{m\pi}{b}\right)^2+\left(\frac{n\pi}{a}\right)^2$. □

7.2.4 Corollary

The TM_{mn} to *z*-mode wave impedance $\eta_{\text{TM},mn}$ for a time-harmonic $(\boldsymbol{E},\boldsymbol{B})|_{E_z=0}$ is given by $\eta_{\text{TM},mn}=\eta\{1+i\frac{\sigma}{\omega\varepsilon}\}^{-1}\sqrt{1-\left(\frac{\omega_{mn}}{\omega}\right)^2+i\frac{\sigma}{\omega\varepsilon}}$. In particular, when $\sigma=0$, for each fixed $(m,n)\in\mathbf{N}\times\mathbf{N}$,

a) $$\omega<\omega_{mn}\Rightarrow\eta_{\text{TM},mn}=i\eta\sqrt{\left(\frac{\omega_{mn}}{\omega}\right)^2-1}$$

b) $$\omega>\omega_{mn}\Rightarrow\eta_{\text{TM},mn}=\eta\sqrt{1-\left(\frac{\omega_{mn}}{\omega}\right)^2}$$

Finally, $\omega>>\omega_{mn}\Rightarrow\eta_{\text{TM},mn}\to\eta$. In particular, $\eta_{\text{TM},mn}<\eta$ whenever $\omega>\omega_{mn}$.

Proof

Noting that $\gamma_{mn}=i\omega\sqrt{\mu\varepsilon}\sqrt{1-\left(\frac{\omega_{mn}}{\omega}\right)^2+i\frac{\sigma}{\omega\varepsilon}}$ and $\frac{\mu}{\kappa}=\frac{i}{\omega\varepsilon}\{1+i\frac{\sigma}{\omega\varepsilon}\}^{-1}$, where we recall that $\kappa=\mu\sigma-i\omega\mu\varepsilon$ and $\omega_{mn}=\frac{1}{\sqrt{\mu\varepsilon}}\sqrt{\left(\frac{m\pi}{b}\right)^2+\left(\frac{n\pi}{a}\right)^2}$, Definition 7.1.9 leads at once to

$$\eta_{\text{TM},mn}=-\frac{\mu\gamma_{mn}}{\kappa}=\eta\{1+i\tfrac{\sigma}{\omega\varepsilon}\}^{-1}\sqrt{1-\left(\tfrac{\omega_{mn}}{\omega}\right)^2+i\tfrac{\sigma}{\omega\varepsilon}}$$

The remaining assertions are obvious; see the proof of Corollary 7.1.11. □

Thus, for a lossless rectangular waveguide, $\eta_{\text{TM},mn}<\eta<\eta_{\text{TE},mn}$ whenever $\omega>\omega_{mn}$ (see Section 7.1 for a pair of parallel plane waveguides). Intuitively, this is expected, as a pair of parallel plane waveguides is merely a special case of the rectangular waveguide wherein one dimension (here, the *x*-direction) extends out to infinity instead of being bounded. The following result, in view of Theorem 7.1.12 is obvious via the replacement $\omega_n\to\omega_{mn}$.

7.2.5 Theorem

Given an admissible time-harmonic TE_{mn} or TM_{mn} wave is propagating at some fixed frequency $\omega>\omega_{mn}$, the wavelength of the waves within $(\Omega,\mu,\varepsilon,\sigma)$ is given by

$$\lambda=\frac{2\sqrt{2}\pi}{\omega}v\left\{\sqrt{\left(\left(\frac{\omega_{mn}}{\omega}\right)^2-1\right)^2+\left(\frac{\sigma}{\omega\varepsilon}\right)^2}+1-\left(\frac{\omega_{mn}}{\omega}\right)^2\right\}^{-\frac{1}{2}}\tag{7.41}$$

In particular, for a lossless medium, $\lambda=\frac{2\pi v}{f}\left\{1-\left(\frac{\omega_{mn}}{\omega}\right)^2\right\}^{-\frac{1}{2}}$ □

7.2.6 Remark

It follows directly from the proof of Theorem 7.1.12 and Example 7.1.13 that the phase velocity $\hat{u}_{mn}$ of an admissible TE_{mn} or TM_{mn} mode is given by

$$\hat{u}_{mn} = \sqrt{2}v\left\{\sqrt{\left(\left(\tfrac{\omega_{mn}}{\omega}\right)^2 - 1\right)^2 + \left(\tfrac{\sigma}{\omega\varepsilon}\right)^2} + 1 - \left(\tfrac{\omega_{mn}}{\omega}\right)^2\right\}^{-\frac{1}{2}}$$

and the velocity of energy propagation $\tilde{u}_{mn}$ satisfies $\hat{u}_{mn}\tilde{u}_{mn} = v^2$ and hence,

$$\tilde{u}_{mn} = \tfrac{1}{\sqrt{2}}v\left\{\sqrt{\left(\left(\tfrac{\omega_{mn}}{\omega}\right)^2 - 1\right)^2 + \left(\tfrac{\sigma}{\omega\varepsilon}\right)^2} + 1 - \left(\tfrac{\omega_{mn}}{\omega}\right)^2\right\}^{\frac{1}{2}}$$

In particular, because the sequence $(\omega_{mn})_{m,n}$ of cut-off angular frequencies is discrete and hence monotonic, the sequence can be ordered such that $\{\omega_{m_0n_0} < \omega_{m_1n_1} < \cdots < \omega_{m_kn_k} < \cdots\}$. Hence, given $\omega > 0$ such that $\omega_{m_kn_k} > \omega > \omega_{m_{k-1}n_{k-1}}$, for some $(m_{k-1}, n_{k-1}), (m_k, n_k) \in \mathbf{N} \times \mathbf{N}$, the TE or TM wave propagation is of (m_{k-1}, n_{k-1})-mode. To see this, it suffices to assume that the wave is of mode (m_{k-1}, n_{k-1}) and (m', n'), where $\omega_{m'n'} < \omega_{m_{k-1},n_{k-1}}$. Then, $\tilde{u}_{m_{k-1}n_{k-1}} < \tilde{u}_{m'n'}$ and $\hat{u}_{m_{k-1}n_{k-1}} > \hat{u}_{m'n'}$. However, the respective velocities of wave propagation for a fixed mode are unique. Hence, the wave can only be of mode (m_{k-1}, n_{k-1}) and higher mode waves* and thus transfers energy (or information) at a slower speed with respect to lower mode waves in a fixed waveguide.

7.2.7 Example

Consider the semi-infinite rectangular waveguide $(\Omega, \mu, \varepsilon, \sigma)$ such that $\partial\Omega$ is a perfect electrical conductor. Observe that for $(m,0)$-mode, where $m \neq 0$, the TE propagation leads, via Equations (7.23)–(7.27) to:

$$E_{x,m0} = 0 = B_{y,m0} \quad \text{and} \quad (E_{y,m0}, B_{x,m0}, B_{z,m0}) \neq 0$$

and hence, TE_{m0} exists. On the other hand, via Equations (7.28)–(7.32) for TM mode,

$$E_{z,m0} = 0 = E_{x,m0} = B_{y,m0} \quad \text{and} \quad (E_{y,m0}, B_{x,m0}) \neq 0$$

Inasmuch as TEM cannot exist on Ω, it follows at once that TM_{m0} does not exist. By symmetry, it can be easily seen that for $(0,n)$-mode, where $n \neq 0$, TE_{0n} exists but TM_{0n} does not exist. □

* In the sense of a higher admissible angular frequency.

7.2.8 Example

Suppose $\Omega = (\Omega', \mu', \varepsilon') \cup (\Omega'', \mu'', \varepsilon'')$, where $\Omega' = (0,b) \times (0,a) \times (0, z_0]$ and $\Omega'' = (0,b) \times (0,a) \times [z_0, \infty)$. Is there a frequency $\omega > 0$ such that there is no reflection of a TE mode at the boundary interface at $z = z_0$, where it is assumed that $\{\mu', \varepsilon'\} \neq \{\mu'', \varepsilon''\}$ (as sets)? If so, derive the conditions under which this can occur by modifying (μ'', ε'') in Ω''.

Now, in order for the coefficient of reflection to be zero, it is necessary and sufficient that $\eta'_{\mathrm{TE},mn} = \eta''_{\mathrm{TE},mn}$, the wave impedances in Ω', Ω'', respectively. Hence, by Corollary 7.2.2,

$$\eta'_{\mathrm{TE},mn} = \eta''_{\mathrm{TE},mn} \Leftrightarrow \frac{\eta'}{\eta''} = \left\{1 - \left(\frac{\omega'_{mn}}{\omega_{mn}}\right)^2\right\}\left\{1 - \left(\frac{\omega''_{mn}}{\omega_{mn}}\right)^2\right\}^{-1} = \frac{\omega^2 - \omega'^2_{mn}}{\omega^2 - \omega''^2_{mn}}$$

where $\eta' = \sqrt{\frac{\mu'}{\varepsilon'}}$ and $\eta'' = \sqrt{\frac{\mu''}{\varepsilon''}}$, whence, rearranging yields

$$\omega^2 = \left\{\frac{\eta'}{\eta''} - 1\right\}^{-1}\left\{\frac{1}{\mu'\varepsilon'}\frac{\eta'}{\eta''} - \frac{1}{\mu''\varepsilon''}\right\} \Rightarrow \omega = \left\{\frac{\eta'}{\eta''} - 1\right\}^{-\frac{1}{2}}\sqrt{\frac{1}{\mu'\varepsilon'}\frac{\eta'}{\eta''} - \frac{1}{\mu''\varepsilon''}}$$

Thus, a unique solution $\omega > 0$ exists if and only if the following two criteria are satisfied,

(a) $$\eta' > \eta'' \Leftrightarrow \mu'\varepsilon'' > \mu''\varepsilon'$$

(b) $$\sqrt{\frac{1}{\mu'\varepsilon'}\frac{\eta'}{\eta''} - \frac{1}{\mu''\varepsilon''}} > 1 \Leftrightarrow \frac{\mu''\varepsilon''}{\mu'\varepsilon'}\frac{\eta'}{\eta''} = \frac{\varepsilon''}{\varepsilon'}\sqrt{\frac{\mu''\varepsilon''}{\mu'\varepsilon'}} > 1 \Leftrightarrow \varepsilon''^3\mu'' > \varepsilon'^3\mu'$$

Now, it is clear that conditions (a) and (b) can be satisfied in two ways. First, suppose that $\mu', \mu'' > 0$ are fixed. Then, (a) implies that $\varepsilon'' > \frac{\mu''}{\mu'}\varepsilon'$ and (b) implies that $\varepsilon'' > \left(\frac{\mu'}{\mu''}\right)^{\frac{1}{3}}\varepsilon'$, whence, set $\varepsilon'' > \max\left\{\frac{\mu''}{\mu'}\varepsilon', \left(\frac{\mu'}{\mu''}\right)^{\frac{1}{3}}\varepsilon'\right\}$. Then, criteria (a) and (b) are automatically satisfied by construction. Similarly, if $\varepsilon', \varepsilon''$ are fixed, then (a) implies that $\mu'' < \frac{\varepsilon''}{\varepsilon'}\mu'$, and (b) implies that $\mu'' > \left(\frac{\varepsilon'}{\varepsilon''}\right)^3\mu'$. Therefore, μ'' must satisfy $\left(\frac{\varepsilon'}{\varepsilon''}\right)^3\mu' < \mu'' < \frac{\varepsilon''}{\varepsilon'}\mu'$. In particular, this can only hold if $\varepsilon' < \varepsilon''$. Thus, $\left(\frac{\varepsilon'}{\varepsilon''}\right)^3\mu' < \mu'' < \frac{\varepsilon''}{\varepsilon'}\mu'$ and $\varepsilon' < \varepsilon''$ together will ensure that criteria (a) and (b) are preserved.

Notice the major difference between the two solutions, to wit, either choosing ε'' or μ'' for impedance matching. In the former, $\mu', \mu'' > 0$ are completely arbitrary, whereas the latter requires that the pair $(\varepsilon', \varepsilon'')$ satisfies the constraint $\varepsilon' < \varepsilon''$ in order for a solution to exist.

7.3 Cavity Resonance

This chapter closes with a quantitative description of cavity resonance and its properties. Again, for simplicity, consider a closed rectangular cavity surrounded by perfect electrical conducting boundaries. More precisely, let

$\Omega=(0,b)\times(0,a)\times(0,z_0)$, where $\partial\Omega$ is a perfect electrical conductor. Notice that in this final scenario, the waves are confined in a compact space. Admissible frequencies are investigated, together with the wave impedance and cavity wavelength.

7.3.1 Proposition

Given a homogeneous bounded space $(\Omega,\mu,\varepsilon,\sigma)$, where $\partial\Omega$ is a perfect electrical conductor, an admissible time-harmonic TE to z-mode wave propagation $(\boldsymbol{E},\boldsymbol{B})|_{E_z=0}$ satisfies:

$$B_{z,mnp}=B_0\cos\left(\tfrac{m\pi}{b}x\right)\cos\left(\tfrac{n\pi}{a}y\right)\sin\left(\tfrac{p\pi}{z_0}z\right) \tag{7.42}$$

$$B_{x,mnp}=-\left\{k^2-\left(\tfrac{p\pi}{z_0}\right)^2\right\}^{-1}\tfrac{m\pi}{b}\tfrac{p\pi}{z_0}B_0\sin\left(\tfrac{m\pi}{b}x\right)\cos\left(\tfrac{n\pi}{a}y\right)\sin\left(\tfrac{p\pi}{z_0}z\right) \tag{7.43}$$

$$B_{y,mnp}=-\left\{k^2-\left(\tfrac{p\pi}{z_0}\right)^2\right\}^{-1}\tfrac{n\pi}{a}\tfrac{p\pi}{z_0}B_0\cos\left(\tfrac{m\pi}{b}x\right)\sin\left(\tfrac{n\pi}{a}y\right)\sin\left(\tfrac{p\pi}{z_0}z\right) \tag{7.44}$$

$$E_{x,mnp}=-i\omega\left\{k^2-\left(\tfrac{p\pi}{z_0}\right)^2\right\}^{-1}\tfrac{n\pi}{a}B_0\cos\left(\tfrac{m\pi}{b}x\right)\sin\left(\tfrac{n\pi}{a}y\right)\sin\left(\tfrac{p\pi}{z_0}z\right) \tag{7.45}$$

$$E_{y,mnp}=i\omega\left\{k^2-\left(\tfrac{p\pi}{z_0}\right)^2\right\}^{-1}\tfrac{m\pi}{b}B_0\sin\left(\tfrac{m\pi}{b}x\right)\cos\left(\tfrac{n\pi}{a}y\right)\sin\left(\tfrac{p\pi}{z_0}z\right) \tag{7.46}$$

where $k^2=\omega^2\mu\varepsilon+i\mu\omega\sigma$. Moreover, if $\sigma=0$, there exists a real sequence $(\omega_{mnp})_{m,n,p}$ of angular frequencies such that the TE_{mnp} mode *resonates* in Ω, where $\omega_{mnp}=\frac{1}{\sqrt{\mu\varepsilon}}\left\{\left(\frac{m\pi}{b}\right)^2+\left(\frac{n\pi}{a}\right)^2+\left(\frac{n\pi}{a}\right)^2\right\}^{\frac{1}{2}}$.

Proof

The proof should be routine by now via the separation of variables. From Equation (7.7), let $B_z=\Phi(x)\Psi(y)\Theta(z)$. Then, imposing the boundary conditions on (7.7) yields

$$B_z=0\quad\text{for}\quad z=0,z_0,\ \partial_x B_z=0\quad\text{for}\quad x=0,b,\ \partial_y B_z=0\quad\text{for}\quad y=0,a$$

Thus, the fundamental solutions are: $\Phi(x)=\cos\left(\frac{m\pi}{b}x\right)$, $\Psi(x)=\cos\left(\frac{n\pi}{a}y\right)$ and $\Theta(z)=\sin\left(\frac{p\pi}{z_0}z\right)$. That is, $B_{z,mnp}=B_0\cos\left(\frac{m\pi}{b}x\right)\cos\left(\frac{n\pi}{a}y\right)\sin\left(\frac{p\pi}{z_0}z\right)$, for some constant B_0. Therefore Equation (7.1) yields $B_{x,mnp}=-\frac{i}{\omega}\partial_z E_{y,mnp}$, and substituting this into (7.5) gives $-\partial_x B_{z,mnp}=\kappa E_{y,mnp}-\frac{i}{\omega}\partial_z^2 E_{y,mnp}$. However, noting that $\partial_z^2\leftrightarrow-\gamma^2\equiv\left(\frac{m\pi}{b}\right)^2+\left(\frac{n\pi}{a}\right)^2-k^2$, it follows that

$$-\partial_x B_{z,mnp}=\left(\kappa+\tfrac{i\gamma^2}{\omega}\right)E_{y,mnp}=\tfrac{i}{\omega}\left(k^2-\gamma^2\right)E_{y,mnp}=\tfrac{i}{\omega}\left\{k^2-\left(\tfrac{p\pi}{z_0}\right)^2\right\}E_{y,mnp}$$

and hence,

$$E_{y,mnp} = \mathrm{i}\omega\left\{k^2 - \left(\tfrac{p\pi}{z_0}\right)^2\right\}^{-1}\partial_x B_{z,mnp}$$

$$= -\mathrm{i}\omega\left\{k^2 - \left(\tfrac{p\pi}{z_0}\right)^2\right\}^{-1}\tfrac{m\pi}{b}B_0\sin\left(\tfrac{m\pi}{b}x\right)\cos\left(\tfrac{n\pi}{a}y\right)\sin\left(\tfrac{p\pi}{z_0}z\right)$$

Similarly, Equation (7.2) leads to $B_{y,mnp} = -\frac{\mathrm{i}}{\omega}\partial_z E_{x,mnp}$, and hence, via (7.4),

$$\partial_y B_{z,mnp} = \left\{\kappa + \tfrac{\mathrm{i}\gamma^2}{\omega}\right\}E_{x,mnp} = \tfrac{\mathrm{i}}{\omega}\left\{k^2 - \left(\tfrac{p\pi}{z_0}\right)^2\right\}E_{x,mnp}$$

Thus,

$$E_{x,mnp} = \mathrm{i}\omega\left\{k^2 - \left(\tfrac{p\pi}{z_0}\right)^2\right\}^{-1}\tfrac{n\pi}{a}B_0\cos\left(\tfrac{m\pi}{b}x\right)\sin\left(\tfrac{n\pi}{a}y\right)\sin\left(\tfrac{p\pi}{z_0}z\right)$$

and finally,

$$B_{x,mnp} = -\left\{k^2 - \left(\tfrac{p\pi}{z_0}\right)^2\right\}^{-1}\tfrac{m\pi}{b}\tfrac{p\pi}{z_0}B_0\sin\left(\tfrac{m\pi}{b}x\right)\cos\left(\tfrac{n\pi}{a}y\right)\sin\left(\tfrac{p\pi}{z_0}z\right)$$

$$B_{y,mnp} = -\left\{k^2 - \left(\tfrac{p\pi}{z_0}\right)^2\right\}^{-1}\tfrac{n\pi}{a}\tfrac{p\pi}{z_0}B_0\cos\left(\tfrac{m\pi}{b}x\right)\sin\left(\tfrac{n\pi}{a}y\right)\sin\left(\tfrac{p\pi}{z_0}z\right)$$

Lastly, from $k^2 = \left(\frac{m\pi}{b}\right)^2 + \left(\frac{n\pi}{a}\right)^2 + \left(\frac{p\pi}{z_0}\right)^2, \sigma=0 \Rightarrow \omega_{mnp}^2 = \frac{1}{\mu\varepsilon}\left\{\left(\frac{m\pi}{b}\right)^2 + \left(\frac{n\pi}{a}\right)^2 + \left(\frac{p\pi}{z_0}\right)^2\right\}$. Observe that there are no cut-off frequencies, as the finite boundary allows the waves to reflect back and forth. The fields attain a state of resonance when $\omega = \omega_{mnp}$. At resonance, there is no power transmitted as they are stored reactively for the case wherein $\sigma = 0$. □

7.3.2 Proposition

Given a homogeneous bounded space $(\Omega, \mu, \varepsilon, \sigma)$, where $\partial\Omega$ is a perfect electrical conductor, an admissible time-harmonic TM to z-mode wave propagation $(\boldsymbol{E}, \boldsymbol{B})|_{B_z=0}$ satisfies:

$$E_{z,mnp} = E_0\sin\left(\tfrac{m\pi}{b}x\right)\sin\left(\tfrac{n\pi}{a}y\right)\cos\left(\tfrac{p\pi}{z_0}z\right) \tag{7.47}$$

$$E_{x,mnp} = -\left\{k^2 - \left(\tfrac{p\pi}{z_0}\right)^2\right\}^{-1}\tfrac{m\pi}{b}\tfrac{p\pi}{z_0}E_0\cos\left(\tfrac{m\pi}{b}x\right)\sin\left(\tfrac{n\pi}{a}y\right)\sin\left(\tfrac{p\pi}{z_0}z\right) \tag{7.48}$$

$$E_{y,mnp} = -\left\{k^2 - \left(\tfrac{p\pi}{z_0}\right)^2\right\}^{-1}\tfrac{n\pi}{a}\tfrac{p\pi}{z_0}E_0\sin\left(\tfrac{m\pi}{b}x\right)\cos\left(\tfrac{n\pi}{a}y\right)\sin\left(\tfrac{p\pi}{z_0}z\right) \tag{7.49}$$

$$B_{x,mnp} = \kappa\left\{k^2 - \left(\tfrac{p\pi}{z_0}\right)^2\right\}^{-1} \tfrac{n\pi}{a} B_0 \sin\left(\tfrac{m\pi}{b}x\right)\cos\left(\tfrac{n\pi}{a}y\right)\cos\left(\tfrac{p\pi}{z_0}z\right) \tag{7.50}$$

$$B_{y,mnp} = -\kappa\left\{k^2 - \left(\tfrac{p\pi}{z_0}\right)^2\right\}^{-1} \tfrac{m\pi}{b} B_0 \cos\left(\tfrac{m\pi}{b}x\right)\sin\left(\tfrac{n\pi}{a}y\right)\cos\left(\tfrac{p\pi}{z_0}z\right) \tag{7.51}$$

where $k^2 = \omega^2\mu\varepsilon + i\mu\omega\sigma$ and $\kappa = \mu\sigma - i\omega\mu\varepsilon$. Moreover, if $\sigma = 0$, there exists a real sequence $(\omega_{mnp})_{m,n,p}$ of angular frequencies such that the TM$_{mnp}$ mode resonates in Ω, where $\omega_{mnp} = \frac{1}{\sqrt{\mu\varepsilon}}\left\{\left(\frac{m\pi}{b}\right)^2 + \left(\frac{n\pi}{a}\right)^2 + \left(\frac{n\pi}{a}\right)^2\right\}^{\frac{1}{2}}$.

Proof

The proof is left for the reader to establish; see Exercise 7.4.6. □

It ought to be pointed out that within the rectangular cavity, only standing waves exist, as the waves are reflected from the walls of the cavity, and hence, a cavity stores electromagnetic energy. However, if there exists an aperture in the wall, current is induced around the aperture, and, as shown in the following chapter, the aperture will radiate. Thus, from an EMC perspective, cavity resonance can be a source of annoyance. Clearly, if the walls have finite conductivity, then the energy is lost over time as ohmic heat. This leads to the concept of quantifying resonance within a cavity.

Before proceeding to quantify cavity resonance, the concept of surface impedance is required. First, recall from (1.40), that $\boldsymbol{B}_\perp = -\frac{i\gamma}{\omega}\mathbf{e}_z \times \boldsymbol{E}_\perp \Rightarrow \boldsymbol{H}_\perp = -\frac{i\gamma}{\mu\omega}\mathbf{e}_z \times \boldsymbol{E}_\perp$. Hence, set $\eta = \frac{i\omega\mu}{\gamma}$, where $\gamma = i\sqrt{\omega^2\mu\varepsilon + i\omega\mu\sigma}$ from (1.32). Then, η defines the wave impedance for a TEM wave propagating in the z-direction, and $\eta = \frac{\omega\mu}{\sqrt{\omega^2\mu\varepsilon + i\omega\mu\sigma}}$.

That η defines the wave impedance is not difficult to see via the following equivalent expression. Given a TEM to z-direction wave (E_x, H_y), from Chapter 1,

$$E_x = E^+ e^{-\gamma z} + E^- e^{\gamma z} \Rightarrow H_y = \tfrac{E^+}{\eta} e^{-\gamma z} - \tfrac{E^-}{\eta} e^{\gamma z}$$

Intuitively, $E_x \leftrightarrow V, H_y \leftrightarrow I \Rightarrow \eta \leftrightarrow R$, where $V = -\int \boldsymbol{E}\cdot d\boldsymbol{l}$, $I = \oint \boldsymbol{H}\cdot d\boldsymbol{l}$ and $R = \frac{V}{I}$.

Denote $\eta = R_S + iX_S$. As a particular instance, this applies to fields penetrating a conductor, and it is called the *surface impedance* of the conductor. Indeed, noting trivially that

$$\sqrt{\omega^2\mu\varepsilon + i\omega\mu\sigma} = \omega\sqrt{\mu\varepsilon}\sqrt{1 + \tfrac{i\sigma}{\omega\varepsilon}}$$

setting $\sqrt{1 + \frac{i\sigma}{\omega\varepsilon}} = \xi_+ + i\xi_-$ yields $\xi_\pm = \frac{1}{\sqrt{2}}\left\{\sqrt{1 + \left(\frac{\sigma}{\omega\varepsilon}\right)^2} \pm 1\right\}^{\frac{1}{2}}$. Thus, putting $\eta_0 = \sqrt{\frac{\mu}{\varepsilon}}$,

$$\eta = \eta_0 \tfrac{\xi_+ - i\xi_-}{\xi_+^2 + \xi_-^2} \Rightarrow R_S = \eta_0\left\{1 + \left(\tfrac{\sigma}{\omega\varepsilon}\right)^2\right\}^{-\frac{1}{2}}\xi_+,\ X_S = -\eta_0\left\{1 + \left(\tfrac{\sigma}{\omega\varepsilon}\right)^2\right\}^{-\frac{1}{2}}\xi_-$$

7.3.3 Lemma

The time-average power density loss as a TEM wave transmits into a conductive boundary at $z = z_0$ is given by $\langle \tilde{W} \rangle = -\frac{1}{2\sqrt{2}\eta_0} \mathbf{e}_z \left\{ |E_x|^2 + |E_y|^2 \right\}_{z=z_0} \left\{ \sqrt{1+\left(\frac{\sigma}{\omega\varepsilon}\right)^2} + 1 \right\}^{\frac{1}{2}}$.

Proof

The time-average Poynting vector is $\langle \boldsymbol{S} \rangle_{z=z_0} = \frac{1}{2} \Re\mathrm{e}\left(\boldsymbol{E}_\perp \times \boldsymbol{H}_\perp^* \right)_{z=z_0}$. Thus, the time-average power density loss $\langle \tilde{W} \rangle_{z=z_0} = -\langle \boldsymbol{S} \rangle_{z=z_0}$ via the conservation of energy. Evaluating this expression yields:

$$
\begin{aligned}
\langle \tilde{W} \rangle &= -\tfrac{1}{2} \mathbf{e}_z \left\{ |E_x|^2 + |E_y|^2 \right\}_{z=z_0} \Re\mathrm{e}\left(\tfrac{1}{\eta^*} \right) \\
&= -\tfrac{1}{2} \mathbf{e}_z \left\{ |E_x|^2 + |E_y|^2 \right\}_{z=z_0} \tfrac{R_S}{R_S^2 + X_S^2} \\
&= -\tfrac{1}{2\sqrt{2}} \mathbf{e}_z \left\{ |E_x|^2 + |E_y|^2 \right\}_{z=z_0} \tfrac{1}{\eta_0} \left\{ \sqrt{1+\left(\tfrac{\sigma}{\omega\varepsilon}\right)^2} + 1 \right\}^{\frac{1}{2}}.
\end{aligned}
$$

□

From the above lemma, it is clear that for a rectangular cavity Ω, the total time-average power loss is over the sum of the boundaries: $\langle W \rangle = \int_{\partial\Omega} \langle \tilde{W} \rangle \cdot \boldsymbol{n} \mathrm{d}^2 x$. That is,

$$
\langle W \rangle = \sum_i \int_{\partial\Omega_i} \langle \tilde{W} \rangle \cdot \boldsymbol{n}_i \mathrm{d}^2 x = -\tfrac{1}{2\sqrt{2}\eta_0} \left\{ |E_x|^2 + |E_y|^2 \right\}_{z=z_0} \left\{ \sqrt{1+\left(\tfrac{\sigma}{\omega\varepsilon}\right)^2} + 1 \right\}
$$

where $\partial\Omega_i$ is the ith-face of $\partial\Omega$.

7.3.4 Definition

Let ω_{mnp} be some fixed resonant angular frequency in a rectangular cavity $(\Omega, \mu, \varepsilon, \sigma)$ with $(\partial\Omega, \sigma_{\partial\Omega})$, and suppose that $\sigma = 0$ with $0 < \sigma_{\partial\Omega} < \infty$. Furthermore, let $\langle U \rangle = \langle U_E \rangle + \langle U_B \rangle$ denote the time-average energy stored in Ω, where $U_E = \frac{1}{2} \boldsymbol{D} \cdot \boldsymbol{E}^*$ ($U_B = \frac{1}{2} \boldsymbol{B} \cdot \boldsymbol{H}^*$) is the electric (magnetic) energy, and $\langle W \rangle$ denotes the time-average power dissipated in Ω, with $W = \frac{1}{2} |\boldsymbol{J}_S|^2 R_S = \frac{1}{2} |\boldsymbol{H}_\perp|^2 R_S$. Then, the *quality factor* of (m, n, p)-mode is defined by $Q_{mnp} = \omega_{mnp} \frac{\langle U \rangle}{\langle W \rangle}$.

7.3.5 Lemma

Given a rectangular cavity $(\Omega, \mu, \varepsilon)$ with $(\partial\Omega, \sigma_{\partial\Omega})$, the time-average power loss satisfies $\langle W \rangle = \langle U_0 \rangle \mathrm{e}^{-\frac{\omega_0}{Q} t}$, where ω_0 is some fixed resonant angular frequency. In particular, the frequency distribution of the magnitude of the electric field within the cavity in frequency domain is

$$\left|\tilde{E}(\omega)\right| = \tfrac{1}{2\pi}\left|E_0\right|\left\{\left(\tfrac{\omega_0}{2Q}\right)^2 + (\omega-\omega_0)^2\right\}^{-\frac{1}{2}}$$

for some initial value E_0.

Proof

By the conservation of energy, the negative time rate of change of stored electromagnetic energy U is precisely the energy loss on the cavity walls W. Whence, from Definition 7.3.4, $Q = \omega_0 \frac{\langle U\rangle}{\langle W\rangle} \Rightarrow -\frac{d}{dt}\langle U\rangle = \langle W\rangle = \frac{\omega_0}{Q}\langle U\rangle \Rightarrow \langle U\rangle = \langle U_0\rangle e^{-\frac{\omega_0}{Q}t}$. Thus, the energy decays exponentially in time. In particular, if the electric field $E(t) = E_0 e^{i\omega_0 t}$ initially, it will decay in the cavity according to $E(t) \to E_0 e^{-\frac{\omega_0}{2Q}t} e^{i\omega_0 t}$, as $E \propto \sqrt{U}$, whence, taking the Fourier transform of $E(t)$,

$$E(t) = \int_0^\infty \tilde{E}(\omega) e^{i\omega t}\, dt \Rightarrow \tilde{E}(\omega) = \tfrac{1}{2\pi}\int_0^\infty E(t) e^{-i\omega t}\, dt$$

Substituting $E(t) = E_0 e^{-\frac{\omega_0}{2Q}t} e^{i\omega_0 t}$ into the integral to obtain the electric field magnitude in the frequency domain yields: $\left|\tilde{E}(\omega)\right|^2 = \left|\frac{E_0}{2\pi}\right|^2 \left\{\left(\frac{\omega_0}{2Q}\right)^2 + (\omega-\omega_0)^2\right\}^{-1}$, as required. Show this in Exercise 7.4.7. □

7.3.6 Example

Given a lossless rectangular cavity (Ω,μ,ε) and $(\partial\Omega,\sigma)$, determine the quality factor Q for an arbitrary TE_{mnp} mode at resonance.

Now, the time-average electric energy stored in Ω is

$$\langle U_E\rangle = \tfrac{1}{2}\varepsilon\int_\Omega |\mathbf{E}|^2\, d^3x = \tfrac{1}{2}\varepsilon\left(\tfrac{\omega_{mnp}}{\omega_{mn}^2}\right)^2 B_0^2 \left\{\left(\tfrac{n\pi}{a}\right)^2 \int_0^b \cos^2\left(\tfrac{m\pi}{b}x\right)dx \int_0^a \sin^2\left(\tfrac{n\pi}{a}y\right)dy \right.$$

$$\left. \int_0^{z_0} \sin^2\left(\tfrac{p\pi}{z_0}z\right)dz + \left(\tfrac{m\pi}{b}\right)^2 \int_0^b \sin^2\left(\tfrac{m\pi}{b}x\right)dx \int_0^a \cos^2\left(\tfrac{n\pi}{a}y\right)dy \int_0^{z_0} \sin^2\left(\tfrac{p\pi}{z_0}z\right)dz \right\}$$

and hence,

$$\langle U_E\rangle = \tfrac{1}{2}\varepsilon\left(\tfrac{\omega_{mnp}}{\omega_{mn}^2}\right)^2 B_0^2 \left\{\left(\tfrac{n\pi}{a}\right)^2 \tfrac{abz_0}{8} + \left(\tfrac{m\pi}{b}\right)^2 \tfrac{abz_0}{8}\right\} = \tfrac{1}{\mu}\left(\tfrac{\omega_{mnp}}{\omega_{mn}}\right)^2 B_0^2 \tfrac{abz_0}{16}$$

at resonance, where for notational convenience, $\omega_{mn}^2 = \frac{1}{\mu\varepsilon}\left\{\left(\frac{m\pi}{b}\right)^2 + \left(\frac{n\pi}{a}\right)^2\right\}$.

The time-average magnetic energy stored in Ω is

$$\langle U_B\rangle = \tfrac{1}{2\mu}\int_\Omega |\boldsymbol{B}|^2\,\mathrm{d}^3x = \tfrac{1}{2\mu\omega_{mn}^4(\mu\varepsilon)^2}B_0^2\left\{\left(\tfrac{m\pi}{a}\tfrac{p\pi}{z_0}\right)^2\int_0^b \sin^2\left(\tfrac{m\pi}{b}x\right)\mathrm{d}x\right.$$

$$\int_0^a \cos^2\left(\tfrac{n\pi}{a}y\right)\mathrm{d}y\int_0^{z_0}\sin^2\left(\tfrac{p\pi}{z_0}z\right)\mathrm{d}z +$$

$$\left(\tfrac{n\pi}{a}\tfrac{p\pi}{z_0}\right)^2\int_0^b \cos^2\left(\tfrac{m\pi}{b}x\right)\mathrm{d}x\int_0^a \sin^2\left(\tfrac{n\pi}{a}y\right)\mathrm{d}y\int_0^{z_0}\sin^2\left(\tfrac{p\pi}{z_0}z\right)\mathrm{d}z +$$

$$\left.\omega_{mn}^4(\mu\varepsilon)^2\int_0^b \cos^2\left(\tfrac{m\pi}{b}x\right)\mathrm{d}x\int_0^a \sin^2\left(\tfrac{n\pi}{a}y\right)\mathrm{d}y\int_0^{z_0}\sin^2\left(\tfrac{p\pi}{z_0}z\right)\mathrm{d}z\right\}$$

and hence, noting trivially by definition that $\omega_{mnp}^2\mu\varepsilon = \omega_{mn}^2\mu\varepsilon + \left(\tfrac{p\pi}{z_0}\right)^2$,

$$\langle U_B\rangle = \tfrac{1}{2\mu\omega_{mn}^4\mu\varepsilon}B_0^2\left\{\tfrac{1}{\mu\varepsilon}\left(\left(\tfrac{n\pi}{a}\right)^2+\left(\tfrac{m\pi}{b}\right)^2\right)\left(\tfrac{p\pi}{z_0}\right)^2\tfrac{abz_0}{8} + \omega_{mn}^4\mu\varepsilon\tfrac{abz_0}{8}\right\} = \tfrac{1}{\mu}\left(\tfrac{\omega_{mnp}}{\omega_{mn}}\right)^2 B_0^2\tfrac{abz_0}{16}$$

Thus, $\langle U\rangle = \tfrac{1}{\mu}\left(\tfrac{\omega_{mnp}}{\omega_{mn}}\right)^2 B_0^2\tfrac{abz_0}{8}$ is the total stored electromagnetic energy in the cavity.

To complete the problem, the loss must be computed. From $W = \tfrac{1}{2}|\boldsymbol{H}_\perp|^2 R_S$, where $R_S = \Re\mathrm{e}\eta_S$ is the resistance of the surface $\partial\Omega$, it follows that the time-average energy loss by the conductive walls is $\langle W\rangle = \tfrac{1}{2}\int_{\partial\Omega}|\boldsymbol{H}_\perp|^2 R_S\mathrm{d}^2x$. Next, noting that the energy loss on $\partial\Omega|_{x=0}$ is identical to that of $\partial\Omega|_{x=b}$, and likewise for $\{y=0, y=a\}$ and for $\{z=0, z=z_0\}$, it follows that

$$\langle W\rangle = \int_{\partial\Omega|_{x=0}}|\boldsymbol{H}_\perp|^2 R_S\mathrm{d}^2x + \int_{\partial\Omega|_{y=0}}|\boldsymbol{H}_\perp|^2 R_S\mathrm{d}^2x + \int_{\partial\Omega|_{z=0}}|\boldsymbol{H}_\perp|^2 R_S\mathrm{d}^2x$$

Now, $\partial\Omega|_{x=0} = [0,a]\times[0,z_0] \Rightarrow \boldsymbol{H}_\perp = (H_y, H_z)_{x=0}$. Hence,

$$\langle W\rangle|\partial\Omega_{x=0} = R_S\tfrac{1}{\mu^2}B_0^2\tfrac{1}{\left(\omega_{mn}^2\mu\varepsilon\right)^2}\left(\tfrac{n\pi}{a}\tfrac{p\pi}{z_0}\right)^2\int_0^a\int_0^{z_0}\sin^2\left(\tfrac{n\pi}{a}y\right)\sin^2\left(\tfrac{p\pi}{z_0}z\right)\mathrm{d}y\,\mathrm{d}z +$$

$$R_S\tfrac{1}{\mu^2}B_0^2\int_0^a\int_0^{z_0}\cos^2\left(\tfrac{n\pi}{a}y\right)\sin^2\left(\tfrac{p\pi}{z_0}z\right)\mathrm{d}y\,\mathrm{d}z$$

$$= R_S\tfrac{1}{\mu^2}B_0^2\tfrac{az_0}{4}\left\{\tfrac{1}{\left(\omega_{mn}^2\mu\varepsilon\right)^2}\left(\tfrac{n\pi}{a}\tfrac{p\pi}{z_0}\right)^2+1\right\}.$$

On the wall $\partial\Omega|_{y=0} = [0,b]\times[0,z_0] \Rightarrow \boldsymbol{H}_\perp = (H_x, H_z)_{y=0} = (0, H_z)_{x=0}$. Hence,

$$\langle W\rangle|\partial\Omega_{y=0} = R_S \tfrac{1}{\mu^2} B_0^2 \int_0^b \int_0^{z_0} \cos^2\left(\tfrac{n\pi}{a} y\right)\sin^2\left(\tfrac{p\pi}{z_0} z\right) dy\, dz = R_S \tfrac{1}{\mu^2} B_0^2 \tfrac{b z_0}{4}.$$

Lastly, $\partial\Omega|_{z=0} = [0,b]\times[0,a] \Rightarrow \boldsymbol{H}_\perp = (H_x, H_y)_{z=0} = (0,0)$. Hence, $\langle W\rangle|\partial\Omega_{z=0} = 0$. Thus,

$$\langle W\rangle = \tfrac{R_S}{\mu^2} B_0^2 \tfrac{z_0}{4}\left\{ a\left(\left(\tfrac{1}{\omega_{mn}^2 \mu\varepsilon} \tfrac{n\pi}{a} \tfrac{p\pi}{z_0}\right)^2 + 1\right) + b\right\}$$

yielding the desired quality factor at resonance:

$$Q_{mnp} = \omega_{mnp} \tfrac{\langle U\rangle}{\langle W\rangle} = \omega_{mnp} \tfrac{\mu}{R_S}\left(\tfrac{\omega_{mnp}}{\omega_{mn}}\right)^2 \tfrac{b}{4}\left\{\left(\tfrac{1}{\omega_{mn}^2 \mu\varepsilon} \tfrac{n\pi}{a} \tfrac{p\pi}{z_0}\right)^2 + 1 + \tfrac{b}{a}\right\}^{-1} \qquad \square$$

Now, for very good conductors, recall from Chapter 1 for a TEM wave that

$$\gamma = \alpha + i\beta \approx \tfrac{1}{\delta}(1+i)$$

where $\delta = \sqrt{\tfrac{2}{\omega\mu\sigma}}$ is the skin depth of the conductor, whence, from Definition 7.1.3,

$$\eta = \tfrac{i\omega\mu}{\gamma} = i\omega\mu\delta \tfrac{1-i}{2} = \tfrac{1}{2}\omega\mu\delta(1+i) \Rightarrow R_S = \tfrac{1}{2}\omega\mu\delta$$

and so,

$$Q_{mnp} = \sqrt{\tfrac{\omega_{mnp}\mu\sigma}{2}}\left(\tfrac{\omega_{mnp}}{\omega_{mn}}\right)^2 \tfrac{b}{2}\left\{\left(\tfrac{1}{\omega_{mn}^2 \mu\varepsilon} \tfrac{n\pi}{a} \tfrac{p\pi}{z_0}\right)^2 + 1 + \tfrac{b}{a}\right\}^{-1}$$

7.3.7 Remark

This chapter closes with a brief word on dielectric waveguides. A brief account can be found in References [2,4]. For simplicity, consider a TM wave propagating within a lossless infinite rectangular dielectric slab $(\Omega,\mu,\varepsilon)\subset(\mathbf{R}^3, \mu_0,\varepsilon_0)$ of finite thickness a: $\Omega = \{(x,y,z): x,z\in\mathbf{R}, 0<y<a\}$ with $\partial_x = 0$ assumed. Recall that a TM to z mode is characterized by $E_{z,n}$. Then, it is clear that it must satisfy these equations simultaneously:

(a) $$\Delta_\perp E_z + (\omega^2\mu\varepsilon - \beta^2)E_z = 0 \quad \text{on } \Omega$$

(b) $$\Delta_\perp E_z + (\omega^2\mu_0\varepsilon_0 - \beta^2)E_z = 0 \quad \text{on } \mathbf{R}^3 - \bar{\Omega}$$

where $\bar{\Omega} = \Omega\cup\partial\Omega$.

For propagation to occur on Ω, it follows that the waves must be evanescent on $\mathbf{R}^3 - \bar{\Omega}$, and hence, via the separation of variables, $E_z = \Phi(y)e^{-i\beta z}$ with $\Phi(y) = E_+ \cos k_y y + E_- \sin k_y y$ on Ω, where $k_y^2 = \omega^2\mu\varepsilon - \beta^2$, whereas on $\mathbf{R}^3 - \bar{\Omega}$,

$$\Phi(y) \sim \begin{cases} e^{-\alpha y} & \text{for } y \geq a \\ e^{\alpha y} & \text{for } y \leq 0 \end{cases}$$

where $\alpha^2 = \beta^2 - \omega^2\mu_0\varepsilon_0 > 0$. The following pair defines the *dispersion relations* for the dielectric slab:

$$\begin{cases} k_y^2 = \omega^2\mu\varepsilon - \beta^2 \\ \alpha^2 = \beta^2 - \omega^2\mu_0\varepsilon_0 \end{cases}$$

□

7.4 Worked Problems

7.4.1 Exercise

(a) Establish that $\mathbf{e}_z \times (\mathbf{e}_z\, \partial_z \times \boldsymbol{E}_\perp) = -\partial_z \boldsymbol{E}_\perp \Rightarrow \partial_z \boldsymbol{E}_\perp = -i\omega \mathbf{e}_z \times \boldsymbol{B}_\perp$.

(b) Show that a time-harmonic TE to z-mode wave propagation satisfies $\boldsymbol{E}_\perp = \frac{i\omega}{\gamma^2 + \omega^2\mu\varepsilon + i\omega\mu\sigma} \nabla_\perp \times \mathbf{e}_z B_z$.

Solution

(a) This is just a trivial exercise of directly evaluating the expression by brute force:

$$\mathbf{e}_z\, \partial_z \times \boldsymbol{E}_\perp = \begin{vmatrix} \mathbf{e}_x & \mathbf{e}_y & \mathbf{e}_z \\ 0 & 0 & \partial_z \\ E_x & E_y & 0 \end{vmatrix} = \begin{pmatrix} -\partial_z E_y \\ \partial_z E_x \\ 0 \end{pmatrix}$$

$$\mathbf{e}_z \times (\mathbf{e}_z\, \partial_z \times \boldsymbol{E}_\perp) = \begin{vmatrix} \mathbf{e}_x & \mathbf{e}_y & \mathbf{e}_z \\ 0 & 0 & 1 \\ -\partial_z E_y & \partial_z E_x & 0 \end{vmatrix} = \begin{pmatrix} -\partial_z E_x \\ -\partial_z E_y \\ 0 \end{pmatrix} = -\partial_z \boldsymbol{E}_\perp$$

Therefore, from Equation (7.10), $\mathbf{e}_z \times (7.10) \Rightarrow -\partial_z \boldsymbol{E}_\perp = i\omega \mathbf{e}_z \times \boldsymbol{B}_\perp$, as desired.

(b) From Maxwell's equations: $\nabla \times \boldsymbol{B} = \kappa \boldsymbol{E} \Rightarrow \nabla_\perp \times \boldsymbol{B}_\perp + \mathbf{e}_z \times \partial_z \boldsymbol{B}_\perp + \nabla_\perp \times \mathbf{e}_z B_z = \kappa \boldsymbol{E}_\perp$, where $\kappa = \mu\sigma - i\omega\mu\varepsilon$. Because $\nabla_\perp \times \boldsymbol{B}_\perp = \kappa E_z \mathbf{e}_z = 0$, it follows that

$$\mathbf{e}_z \times \partial_z \boldsymbol{B}_\perp + \nabla_\perp \times \mathbf{e}_z B_z = \kappa \boldsymbol{E}_\perp$$

Upon substituting the result from (a) into the above equation and noting trivially that $\partial_z \mathbf{e}_z \times \boldsymbol{B}_\perp = \mathbf{e}_z \times \partial_z \boldsymbol{B}_\perp$ as $\mathbf{e}_z$ is a constant, it is evident that

$$-\partial_z^2 \boldsymbol{E}_\perp = i\omega \mathbf{e}_z \times \partial_z \boldsymbol{B}_\perp = i\omega\{\kappa \boldsymbol{E}_\perp - \nabla_\perp \times \mathbf{e}_z B_z\} \Rightarrow \partial_z^2 \boldsymbol{E}_\perp + i\omega\kappa \boldsymbol{E}_\perp = i\omega \nabla_\perp \times \mathbf{e}_z B_z$$

Finally, observing that $\partial_z e^{-\gamma z} = -\gamma e^{-\gamma z} \Rightarrow \partial_z \leftrightarrow -\gamma$, it follows that

$$\boldsymbol{E}_\perp = \frac{i\omega}{i\omega\kappa + \gamma^2} \nabla_\perp \times \mathbf{e}_z B_z = \frac{1}{\mu} \frac{i\omega\mu}{i\omega\kappa + \gamma^2} \nabla_\perp \times \mathbf{e}_z B_z$$ □

7.4.2 Exercise

Fill in the details for the proof of Proposition 7.1.2.

Solution

Now, $\partial_z \boldsymbol{E}_\perp = -i\omega \mathbf{e}_z \times \boldsymbol{B}_\perp$ was established in Exercise 7.4.1(a). To complete the details of the remaining proof, observe that $\partial_z^2 e^{-\gamma z} = \gamma^2 e^{-\gamma z}$; that is, $\partial_z^2 \leftrightarrow \gamma^2$, whence,

$$\partial_z^2 \boldsymbol{E}_\perp - i\omega\kappa \boldsymbol{E}_\perp = -i\omega \nabla_\perp \times \mathbf{e}_z B_z \Leftrightarrow \gamma^2 \boldsymbol{E}_\perp - i\omega\kappa \boldsymbol{E}_\perp = -i\omega \nabla_\perp \times \mathbf{e}_z B_z$$

$$\Leftrightarrow (\gamma^2 - i\omega\kappa) \boldsymbol{E}_\perp = -i\omega \nabla_\perp \times \mathbf{e}_z B_z$$

So, recalling that $\kappa = \mu\sigma - i\omega\mu\varepsilon$, it is clear that $-i\omega\kappa = \omega^2\mu\varepsilon - i\omega\mu\sigma$, yielding

$$\boldsymbol{E}_\perp = -\frac{i\omega}{-i\omega\kappa + \gamma^2} \nabla_\perp \times \mathbf{e}_z B_z = -\frac{1}{\mu} \frac{i\omega\mu}{-i\omega\kappa + \gamma^2} \nabla_\perp \times \mathbf{e}_z B_z = -\frac{1}{\mu} \frac{i\omega\mu}{\gamma^2 + \omega^2\mu\varepsilon - i\omega\mu\sigma} \nabla_\perp \times \mathbf{e}_z B_z$$

as required. □

7.4.3 Exercise

Prove Proposition 7.1.10.

Proof

For TM mode, $B_z = 0$. Thus, invoking the assumption that $\partial_x = 0$,

$$\nabla \times \boldsymbol{E} = \begin{vmatrix} \mathbf{e}_x & \mathbf{e}_y & \mathbf{e}_z \\ 0 & \partial_y & \partial_z \\ E_x & E_y & E_z \end{vmatrix} = \begin{pmatrix} \partial_y E_z - \partial_z E_y \\ \partial_z E_x \\ -\partial_y E_x \end{pmatrix} = -\partial_t \begin{pmatrix} B_x \\ B_y \\ 0 \end{pmatrix}$$

$$\nabla \times \boldsymbol{B} = \begin{vmatrix} \mathbf{e}_x & \mathbf{e}_y & \mathbf{e}_z \\ 0 & \partial_y & \partial_z \\ B_x & B_y & 0 \end{vmatrix} = \begin{pmatrix} -\partial_z B_y \\ \partial_z B_x \\ -\partial_y B_x \end{pmatrix} = \mu\sigma \begin{pmatrix} E_x \\ E_y \\ E_z \end{pmatrix} - \mu\varepsilon\partial_t \begin{pmatrix} E_x \\ E_y \\ E_z \end{pmatrix}$$

Noting that $\nabla \cdot \boldsymbol{B} = 0 \Rightarrow \partial_y B_y = 0$ as $\partial_x = 0$ and $B_z \equiv 0$, whence, ∂_y (fourth equation) leads to $0 = \mu\sigma\partial_y E_x + \mu\varepsilon\partial_t\partial_y E_x \Rightarrow E_x$ is independent of y. Hence, in order to satisfying the boundary condition $E_x \mid \partial\Omega = 0 \Rightarrow E_x \equiv 0$. In particular, the fourth equation leads immediately to $\partial_z B_y = 0 \Rightarrow B_y$ is independent of z. Hence, the only solution to satisfy this is $B_y \equiv 0$.

To summarize for ease of reference, on setting $\kappa = \mu\sigma - \mathrm{i}\omega\mu\varepsilon$, $\partial_z \to -\gamma$ and $\partial_t \to -\mathrm{i}\omega$, Maxwell's equations reduce to

$$\partial_y E_z + \gamma E_y = \mathrm{i}\omega B_x$$

$$\partial_z B_x = \kappa E_y$$

$$-\partial_y B_x = \kappa E_z$$

Therefore, ∂_z (second equation) $-\partial_z$ (third equation) yields

$$\partial_z^2 B_x + \partial_y^2 B_x = \kappa(\partial_z E_y - \partial_y E_z) = -\mathrm{i}\omega\kappa B_x \Rightarrow -\Delta B_x + k^2 B_x = 0$$

where $k^2 = \omega^2\mu\varepsilon + \mathrm{i}\omega\mu\sigma$. The boundary condition is

$$\partial_y B_x \mid \partial\Omega = 0 \Leftrightarrow \partial_y B_x \mid_{y=0} = 0 = \partial_y B_x \mid_{y=a}$$

Thus, via the separation of variables with $B_x(y,z) = \Phi(y)\Psi(z)$ and imposing the boundary conditions, the fundamental solutions are: $\Phi_n(y) = \cos\frac{n\pi}{a}y$ and $\Psi(z) = \mathrm{e}^{-\gamma_n z}$, where $\gamma_n^2 = \left(\frac{n\pi}{a}\right)^2 - \omega^2\mu\varepsilon - \mathrm{i}\omega\mu\sigma$ satisfies Equation (7.18). So, the solution is thus $B_{x,n} = B_{0,n}^{+}\cos\left(\frac{n\pi}{a}y\right)\mathrm{e}^{-\gamma_n z}$.

Finally, the latter two equations yield, respectively,

$$E_y = -\tfrac{\gamma_n}{\kappa} B_x = -\tfrac{\gamma_n}{\kappa} B_{0,n}^{+} \cos\left(\tfrac{n\pi}{a} y\right) e^{-\gamma_n z} \quad \text{and} \quad E_z = -\tfrac{1}{\kappa} \partial_y B_x = \tfrac{1}{\kappa} B_{0,n}^{+} \tfrac{n\pi}{a} \sin\left(\tfrac{n\pi}{a} y\right) e^{-\gamma_n z}$$

To complete the proof, it suffices to observe that as γ_n for TM is identical with that of TE, the properties derived for TE apply equally to TM. □

7.4.4 Exercise

Establish $\eta_{\mathrm{TM},n} = \eta\left\{1 + i\tfrac{\sigma}{\omega\varepsilon}\right\}^{-1} \sqrt{1 - \left(\tfrac{\omega_n}{\omega}\right)^2 + i\tfrac{\sigma}{\omega\varepsilon}}$ of Corollary 7.1.11.

Solution

From Definition 7.1.9, $\eta_{\mathrm{TM},n} = -\tfrac{\mu\gamma_n}{\kappa}$, where $\kappa = \mu\sigma - i\omega\mu\varepsilon$. Next, set $\gamma_n = i\omega\sqrt{\mu\varepsilon}\sqrt{1 - \left(\tfrac{\omega_n}{\omega}\right)^2 + i\tfrac{\sigma}{\omega\varepsilon}} \equiv i\omega\sqrt{\mu\varepsilon}\hat{\gamma}_n$ and noting that $\sigma - i\omega\varepsilon = -i\omega\varepsilon\left(1 + \tfrac{i\sigma}{\omega\varepsilon}\right)$,

$$\eta_{\mathrm{TM},n} = -\tfrac{\mu\gamma_n}{\kappa} = i\omega\sqrt{\mu\varepsilon}\hat{\gamma}_n \tfrac{1}{i\omega\varepsilon}\left(1 + \tfrac{i\sigma}{\omega\varepsilon}\right)^{-1} = \sqrt{\tfrac{\mu}{\varepsilon}}\left(1 + \tfrac{i\sigma}{\omega\varepsilon}\right)^{-1} \sqrt{1 - \left(\tfrac{\omega_n}{\omega}\right)^2 + i\tfrac{\sigma}{\omega\varepsilon}} \quad \square$$

7.4.5 Exercise

Derive the TE modes for a semi-infinite hollow cylinder $\Omega = B_a(0) \times [0, \infty)$, where $B_a(0) = \{(x, y) \in \mathbf{R}^2 : x^2 + y^2 < a^2\}$, and assume that $\partial\Omega$ is a perfect electrical conductor. Furthermore, determine the wave impedance and waveguide wavelength for each fixed mode.

Solution

The appropriate coordinate system to be invoked here is the cylindrical coordinate system. As an aside, Equations (7.7) and (7.8) apply in all coordinate systems. Hence, it suffices to express the Laplacian Δ in cylindrical coordinates. Recalling that

$$\Delta B_z = \left(\tfrac{1}{r}\partial_r + \partial_r^2 + \tfrac{1}{r^2}\partial_\phi^2\right) B_z + \partial_z^2 B_z \equiv \Delta_{r\phi} B_z + \partial_z^2 B_z$$

it follows that (7.7) becomes $0 = \Delta_{r\phi} B_z + (\gamma^2 + k^2) B_z$. So, attempt the separation of variables method once again: set $B_z = \Psi(r)\Phi(\phi)$. Then, upon expanding out the expression, dividing by $\Psi(r)\Phi(\phi)$ and multiplying by r^2,

$$\tfrac{1}{r}\tfrac{\partial_r \Psi}{\Psi} + \tfrac{\partial_r^2 \Psi}{\Psi} + \tfrac{1}{r^2}\tfrac{\partial_\phi^2 \Phi}{\Phi} + (\gamma^2 + k^2) = 0 \Rightarrow r\tfrac{\partial_r \Psi}{\Psi} + r^2\tfrac{\partial_r^2 \Psi}{\Psi} + r^2(\gamma^2 + k^2) = -\tfrac{\partial_\phi^2 \Phi}{\Phi} = n^2$$

for some constant n^2, as the left side of the equation is solely a function of r and the center equation is solely a function of ϕ.

The general solution for Φ is obvious by now: $\Phi = A\cos n\phi + B\sin n\phi$. The differential equation

$$r\frac{\partial_r\Psi}{\Psi} + r^2\frac{\partial_r^2\Psi}{\Psi} + r^2(\gamma^2+k^2) - n^2 = 0 \Leftrightarrow \frac{1}{r}\frac{\partial_r\Psi}{\Psi} + \frac{\partial_r^2\Psi}{\Psi} + (\gamma^2+k^2) - \frac{n^2}{r^2} = 0$$

is known as *Bessel's equation*. The general solution for Bessel's equation is well known:

$$\Psi = CJ_n((\gamma^2+k^2)r) + DN_n((\gamma^2+k^2)r)$$

where

$$J_n(qr) = \sum_{m\geq 0}\frac{(-1)^m(qr)^{n+2m}}{m!(n+m)!2^{n+2m}}$$

is *Bessel's function of the first kind* (there are other equivalent representations), and

$$N_n(qr) = \frac{\cos(n\pi)J_n(qr) - J_{-n}(qr)}{\sin n\pi}$$

For more details, see References [2,3,7,9].

Now, Φ is periodic in ϕ and hence, $n \in \mathbf{Z}$ for each n. In particular, $\sin n\pi = 0$ implies at once that $D \equiv 0 \Rightarrow \Psi_n = J_n((\gamma^2+k^2)r)$ is a fundamental solution. Because the coefficients A,B of Φ depend upon the choice of an arbitrary reference angle ϕ, it follows for simplicity that either $A = 0$ or $B = 0$ may be chosen without any loss of generality. So, let $\Phi_n = \cos n\phi$ denote the fundamental solution. Then, the general solution is: $B_{z,n} = B_0 J_n((\gamma^2+k^2)r)\cos n\phi$.

Imposing the boundary condition on $B_{z,n} \Rightarrow \partial_r B_{z,n}|_{r=a} = 0$. This is equivalent to:

$$\partial_r J_n((\gamma_{mn}^2+k^2)a) = 0 \Leftrightarrow \gamma_{mn}^2 = p_{mn} - k^2$$

where $p_{mn}a$ are the roots of Bessel's function $\partial_r J_n(p_{mn}a) = 0\ \forall m,n$. Thus,

$$\gamma_{mn}^2 = p_{mn} - \omega^2\mu\varepsilon - i\omega\mu\sigma$$

and hence, setting $\gamma_{mn} = \alpha_{mn} + i\beta_{mn}$, following the proof of Proposition 7.1.4 *mutatis mutandis* yields

$$\begin{cases} \alpha_{mn} = \frac{1}{\sqrt{2}}\left\{\sqrt{(p_{mn}-\omega^2\mu\varepsilon)^2 + (\omega\mu\sigma)^2} - (\omega^2\mu\varepsilon - p_{mn})\right\}^{\frac{1}{2}} \\ \beta_{mn} = \frac{1}{\sqrt{2}}\left\{\sqrt{(p_{mn}-\omega^2\mu\varepsilon)^2 + (\omega\mu\sigma)^2} + \omega^2\mu\varepsilon - p_{mn}\right\}^{\frac{1}{2}} \end{cases}$$

In particular, for $\sigma = 0$,

(a) $\omega^2\mu\varepsilon - p_{mn} > 0 \Rightarrow \alpha_{mn} = 0 \Rightarrow e^{-\gamma_{mn}z} = e^{-i\beta_{mn}z}$ (traveling wave),

(b) $\omega^2\mu\varepsilon - p_{mn} < 0 \Rightarrow \beta_{mn} = 0 \Rightarrow e^{-\gamma_{mn}z} = e^{-\alpha_{mn}z}$ (evanescent wave).

Finally, to determine the remaining fields, consider Maxwell's equations for TE mode:

$$\nabla \times \mathbf{E} = \tfrac{1}{r}\begin{vmatrix} \mathbf{e}_r & r\mathbf{e}_\phi & \mathbf{e}_z \\ \partial_r & \partial_\phi & \partial_z \\ E_r & rE_\phi & 0 \end{vmatrix} = \tfrac{1}{r}\begin{pmatrix} -r\partial_z E_\phi \\ r\partial_z E_r \\ \partial_r(rE_\phi) - \partial_\phi E_r \end{pmatrix} = i\omega\begin{pmatrix} B_r \\ B_\phi \\ B_z \end{pmatrix}$$

$$\nabla \times \mathbf{B} = \tfrac{1}{r}\begin{vmatrix} \mathbf{e}_r & r\mathbf{e}_\phi & \mathbf{e}_z \\ \partial_r & \partial_\phi & \partial_z \\ B_r & rB_\phi & B_z \end{vmatrix} = \tfrac{1}{r}\begin{pmatrix} \partial_\phi B_z - r\partial_z B_\phi \\ -r\partial_r B_z + r\partial_z B_r \\ \partial_r(rB_\phi) - \partial_\phi B_r \end{pmatrix} = \kappa\begin{pmatrix} E_r \\ E_\phi \\ 0 \end{pmatrix}$$

whence, via the correspondence $\partial_z \leftrightarrow -\gamma_{mn}$ and observing that $p_{mn} = i\omega\kappa + \gamma_{mn}^2$,

$$\gamma_{mn}E_{\phi,mn} = i\omega B_{r,mn} \Rightarrow B_{r,mn} = -\tfrac{i\gamma_{mn}}{\omega}E_{\phi,mn}$$

$$-\gamma_{mn}E_{r,mn} = i\omega B_{\phi,mn} \Rightarrow B_{\phi,mn} = \tfrac{i\gamma_{mn}}{\omega}E_{r,mn}$$

and substituting the first equation into $-\partial_r B_z - \gamma_{mn}B_{r,mn} = \kappa E_{\phi,mn}$ leads to:

$$-\partial_r B_z = \kappa E_{\phi,mn} + \gamma_{mn}B_{r,mn} = \left\{\kappa - \tfrac{i\gamma_{mn}^2}{\omega}\right\}E_{\phi,mn} = -\tfrac{i}{\omega}\left\{i\omega\kappa + \gamma_{mn}^2\right\}E_{\phi,mn} \Rightarrow$$

$$E_{\phi,mn} = -i\omega\left\{i\omega\kappa + \gamma_{mn}^2\right\}^{-1}\partial_r B_z = -\tfrac{i\omega}{p_{mn}}\partial_r B_z$$

Similarly, substituting the second equation into $\tfrac{1}{r}\partial_\phi B_{z,mn} + \gamma_{mn}B_{\phi,mn} = \kappa E_{r,mn}$ leads to:

$$\tfrac{1}{r}\partial_\phi B_{z,mn} = \kappa E_{r,mn} - \gamma_{mn}B_{\phi,mn} = \left\{\kappa - \tfrac{i\gamma_{mn}^2}{\omega}\right\}E_{r,mn} = -\tfrac{i}{\omega}\left\{i\omega\kappa + \gamma_{mn}^2\right\}E_{r,mn} \Rightarrow$$

$$E_{r,mn} = i\omega\left\{i\omega\kappa + \gamma_{mn}^2\right\}^{-1}\partial_r B_z = \tfrac{i\omega}{p_{mn}}\partial_r B_z$$

In summary,

$$B_{z,mn} = B_0 J_n(p_{mn}r)\cos(n\phi)$$

$$B_{r,mn} = -\tfrac{\gamma_{mn}}{p_{mn}} B_0(\partial_r J_n(p_{mn}r))\cos(n\phi)e^{-\gamma_{mn}z}$$

$$B_{\varphi,mn} = \tfrac{\gamma_{mn}}{p_{mn}} B_0(\partial_r J_n(p_{mn}r))\cos(n\phi)e^{-\gamma_{mn}z}$$

$$E_{\phi,mn} = -\tfrac{i\omega}{p_{mn}} B_0(\partial_r J_n(p_{mn}r))\cos(n\phi)e^{-\gamma_{mn}z}$$

$$E_{r,mn} = \tfrac{i\omega}{p_{mn}} B_0(\partial_r J_n(p_{mn}r))\cos(n\phi)e^{-\gamma_{mn}z}$$

Regarding the wave impedance, by appealing to Definition 7.1.3, and defining $\omega_{mn}^2\mu\varepsilon = p_{mn}$,

$$\eta_{\mathrm{TE},mn} = \tfrac{i\mu\omega}{\gamma_{mn}} = \sqrt{\tfrac{\mu}{\varepsilon}}\left\{1-\left(\tfrac{\omega_{mn}}{\omega}\right)^2 + \tfrac{i\sigma}{\omega\varepsilon}\right\}^{-\frac{1}{2}}$$

where $\gamma_{mn} = \sqrt{p_{mn} - \omega^2\mu\varepsilon - i\omega\mu\sigma} = i\omega\sqrt{\mu\varepsilon}\sqrt{1-\left(\tfrac{\omega_{mn}}{\omega}\right)^2 + \tfrac{i\sigma}{\omega\varepsilon}}$; see the wave impedance for a rectangular waveguide in Corollary 7.2.2. Specifically, Corollary 7.2.2 applies equally to a cylindrical waveguide, and it is clear that ω_{mn} defines the cut-off angular frequencies for a lossless cylindrical waveguide.

Lastly, the determination of the wavelength within the guide follows that of Theorem 7.1.12 identically. Set $\Theta = \omega t + \beta_{mn}z$. Then, $\hat{u}_{mn} \equiv \tfrac{dz}{dt} = \left|\tfrac{\omega}{\beta_{mn}}\right| = f\lambda_{mn} \Rightarrow \lambda_{mn} = \tfrac{2\pi}{\beta_{mn}}$. Thus,

$$\lambda_{mn} = 2\sqrt{2}\pi\left\{\sqrt{\left(p_{mn} - \omega^2\mu\varepsilon\right)^2 + \left(\omega\mu\sigma\right)^2} + \omega^2\mu\varepsilon - p_{mn}\right\}^{-\frac{1}{2}}$$

$$= \tfrac{\sqrt{2}}{f}v\left\{\sqrt{\left(p_{mn} - \omega^2\mu\varepsilon\right)^2 + \left(\omega\mu\sigma\right)^2} + \omega^2\mu\varepsilon - p_{mn}\right\}^{-\frac{1}{2}}$$

whenever $\omega > \omega_{mn}$, where $v = \tfrac{1}{\sqrt{\mu\varepsilon}}$ and $\omega = 2\pi f$. □

7.4.6 Exercise

Establish Proposition 7.3.2.

Solution

Invoking the separation of variables, Equation (7.20) leads to $E_z = \Phi(x)\Psi(y)\Theta(z)$, subject to the boundary condition $\Phi|_{x=0,b} = 0 = \Psi|_{y=0,a}$, whence, $\Phi = \sin\left(\tfrac{m\pi}{b}x\right)$ and $\Psi = \sin\left(\tfrac{n\pi}{a}y\right)$ are the desired fundamental solutions, as should be apparent from the proofs given variously in Sections 7.2 and 7.3. In particular, this leads to: $-k_x^2 - k_y^2 + \gamma^2 + k^2 = 0$, where $k_x = \frac{m\pi}{b}$, $k_y = \frac{n\pi}{a}$. Furthermore, as E_z is reflected between $z = 0$ and $z = z_0$, and $\Theta|_{z=0,z_0} \neq 0$ (as there exists a surface current induced on the boundary), it follows that $\Theta = Ae^{-\gamma z} + Be^{\gamma z} \Rightarrow B \neq -A$. That is, $\Theta \neq \sin\left(\frac{p\pi}{z_0}\right)$. Indeed, as the reflection coefficient of a perfect conductor is −1, it follows that in a lossless medium, $B \equiv A$ and hence, yielding the fundamental solution $\Theta = \cos\left(\frac{p\pi}{z_0}z\right)$. Thus, $E_{z,mnp} = E_0 \sin\left(\frac{m\pi}{b}x\right)\sin\left(\frac{n\pi}{a}y\right)\cos\left(\frac{p\pi}{z_0}z\right)$.

For ease of reference, Maxwell's six equations are rewritten below for TM mode (see Exercise 7.4.3) and labeled, respectively, from (1)–(6):

$$\partial_y E_z - \partial_z E_y = -\partial_t B_x,\ -\partial_x E_z + \partial_z E_x = -\partial_t B_y,\ \partial_x E_y - \partial_y E_x = 0$$

$$-\partial_z B_y = \kappa E_x,\ \partial_z B_x = \kappa E_y,\ \partial_x B_y - \partial_y B_x = \kappa E_z$$

Then, following the proof of Proposition 7.3.1, and noting that $\partial_z^2 \leftrightarrow -\gamma^2 = \left(\frac{p\pi}{z_0}\right)^2$, substituting (5) into (1) leads to: $\partial_y E_z = i\omega B_x + \partial_z\left(\frac{1}{\kappa}\partial_z B_x\right) = \frac{1}{\kappa}\left(i\omega\kappa - \left(\frac{p\pi}{z_0}\right)^2\right)$ $B_x = \frac{1}{\kappa}\left(\kappa^2 - \left(\frac{p\pi}{z_0}\right)^2\right)B_x$. Thus,

$$B_{x,mnp} = \kappa\left\{k^2 - \left(\frac{p\pi}{z_0}\right)^2\right\}^{-1} \frac{n\pi}{a} E_0 \sin\left(\frac{m\pi}{b}x\right)\cos\left(\frac{n\pi}{a}y\right)\cos\left(\frac{p\pi}{z_0}z\right)$$

Likewise, substituting (4) into (2) yields:

$$-\partial_x E_z = i\omega B_y + \partial_z\left(\frac{1}{\kappa}\partial_z B_y\right) = \frac{1}{\kappa}\left(k^2 - \left(\frac{p\pi}{z_0}\right)^2\right)B_y$$

That is,

$$B_{y,mnp} = -\kappa\left\{k^2 - \left(\frac{p\pi}{z_0}\right)^2\right\}^{-1} \frac{m\pi}{b} E_0 \cos\left(\frac{m\pi}{b}x\right)\sin\left(\frac{n\pi}{a}y\right)\cos\left(\frac{p\pi}{z_0}z\right)$$

It is now trivial, via (4) and (5), respectively, to evaluate $\boldsymbol{E}_\perp$:

$$E_{y,mnp} = -\left\{k^2 - \left(\frac{p\pi}{z_0}\right)^2\right\}^{-1} \frac{m\pi}{b}\frac{p\pi}{z_0} E_0 \cos\left(\frac{m\pi}{b}x\right)\sin\left(\frac{n\pi}{a}y\right)\sin\left(\frac{p\pi}{z_0}z\right)$$

$$E_{x,mnp} = -\left\{k^2 - \left(\frac{p\pi}{z_0}\right)^2\right\}^{-1} \frac{n\pi}{a}\frac{p\pi}{z_0} E_0 \sin\left(\frac{m\pi}{b}x\right)\cos\left(\frac{n\pi}{a}y\right)\sin\left(\frac{p\pi}{z_0}z\right)$$

Finally, observe that for a lossless medium, $\sigma = 0$, and $k^2 - \left(\frac{p\pi}{z_0}\right)^2$ reduces to $\left(\frac{m\pi}{b}\right)^2 + \left(\frac{n\pi}{a}\right)^2$, and in particular, the resonant angular frequencies are

$$\omega_{mnp}^2 = \frac{1}{\mu\varepsilon}\left\{\left(\frac{m\pi}{b}\right)^2 + \left(\frac{n\pi}{a}\right)^2 + \left(\frac{p\pi}{z_0}\right)^2\right\}$$

by construction. □

7.4.7 Exercise

Establish that $\left|\tilde{E}(\omega)\right|^2 = \left|\frac{E_0}{2\pi}\right|^2 \left\{\left(\frac{\omega_0}{2Q}\right)^2 + (\omega - \omega_0)^2\right\}^{-1}$, for some fixed resonance frequency ω_0, given that $E(t) = E_0 e^{-\frac{\omega_0}{2Q}t} e^{i\omega_0 t}$.

Proof

From $\tilde{E}(\omega) = \frac{1}{2\pi}\int_0^\infty E(t) e^{-i\omega t} dt$, it is clear that

$$\tilde{E}(\omega) = \frac{1}{2\pi} E_0 \int_0^\infty e^{-\frac{\omega_0}{2Q}t} e^{i\omega_0 t} e^{-i\omega t}\, dt$$

$$= -\frac{1}{2\pi} E_0 \left\{\frac{\omega_0}{2Q} + i(\omega - \omega_0)\right\}^{-1} e^{-Kt}\Big|_{t=0}^{\infty} = \frac{1}{2\pi} E_0 \left\{\frac{\omega_0}{2Q} + i(\omega - \omega_0)\right\}^{-1}$$

where $K = \frac{\omega_0}{2Q} + i(\omega - \omega_0)$. Hence,

$$\left|\tilde{E}(\omega)\right|^2 = \left|\frac{1}{2\pi} E_0\right|^2 \left\{\frac{\omega_0}{2Q} - i(\omega - \omega_0)\right\}\left\{\frac{\omega_0}{2Q} + i(\omega - \omega_0)\right\}\left\{\left(\frac{\omega_0}{2Q}\right)^2 + (\omega - \omega_0)^2\right\}^{-2}$$

$$= \left|\frac{1}{2\pi} E_0\right|^2 \left\{\left(\frac{\omega_0}{2Q}\right)^2 + (\omega - \omega_0)^2\right\}^{-1}$$

as desired. □

References

1. Balanis, C. 1989. *Advanced Engineering Electromagnetics*. New York: John Wiley & Sons.
2. Chang, D. 1992. *Field and Wave Electromagnetics*. Reading, MA: Addison-Wesley.
3. Farlow, S. 1993. *Partial Differential Equations for Scientists and Engineers*. New York: Dover.
4. Jackson, J. 1962. *Classical Electrodynamics*. New York: John Wiley & Sons.
5. Neff, H. 1981. *Basic Electromagnetic Fields*. New York: Harper & Row.
6. Orfanidis, S. 2002. *Electromagnetic Waves and Antenna*. Rutgers University, ECE Dept., http://www.ece.rutgers.edu/~orfanidi/ewa/.
7. Plonsey, R. and Collin, R. 1961. *Principles and Applications of Electromagnetic Fields*. New York: McGraw-Hill.
8. Silver, S. 1949. *Microwave Antenna Theory and Design*. New York: McGraw-Hill.
9. Wylie, C., Jr., 1960. *Advanced Engineering Mathematics*. New York: McGraw-Hill.

8

Basic Antenna Theory

An antenna is essentially a mechanism by which electromagnetic waves are transmitted or received. It operates on the principle that an accelerating charge particle radiates. A broadcasting antenna comprises a waveguide that carries electromagnetic energy from the source, generating fields that radiate out into space. The notion of electric dipoles and the magnetic dipoles introduced in Chapter 1 comprise the building blocks in antenna analysis [3, 6–8]. More advanced theory on antennae can be found in References [2, 9].

An important aspect of antenna theory is the radiation field from apertures. This has direct relevance to EMC engineers from a compliance perspective, to wit, radiation escaping from apertures in chassis enclosing printed circuit boards. The analysis essentially follows from the application of diffraction theory. Excellent accounts of diffraction theory can be found in References [4, 5, 10].

8.1 Radiation from a Charged Particle

In Chapter 1, the existence of electromagnetic waves was established. However, the question regarding their generation was left unanswered. How is electromagnetic radiation generated? To understand the concept at an intuitive level, consider a point charge in vacuum. An electric field is generated by the presence of the point charge via Coulomb's law: $\boldsymbol{E}(r) = \frac{1}{4\pi\varepsilon_0}\frac{q}{r^3}\boldsymbol{r}$, where q, r are the electric charge and distance away from the point charge, respectively (taken to be located at the origin).

On the other hand, via Biot–Savart's law, the magnetic field generated by a moving point charge is $\boldsymbol{B} = \mu\varepsilon \boldsymbol{v} \times \boldsymbol{E}$, where $\boldsymbol{v}$ is the velocity of the point charge in $(\Omega, \mu, \varepsilon)$, $\Omega \subseteq \mathbf{R}^3$. Because a magnetic field forms loops (*cf.* Chapter 1) and the magnetic field vector is normal to both the direction of propagation and the electric field according to $\boldsymbol{v} \times \boldsymbol{E}$, magnetic loops can be envisaged as circulating around the velocity vector.

The above discussion suggests that in order for radiation to be generated, a charge particle must undergo an acceleration, as (i) a static magnetic field does not propagate in space, (ii) a static particle cannot generate a magnetic field (relative to some fixed reference frame), and (iii) a charged particle traveling at some constant velocity only generates a static magnetic field. Thus, by elimination, acceleration appears to be the only candidate. Before proceeding further with the discussion, a fundamental result from Chapter 1 is recalled below.

8.1.1 Lemma

The electromagnetic field of a charged particle moving at some fixed velocity $\boldsymbol{v}$ can be completely characterized by the pair $(\boldsymbol{A},\varphi)$, where $\boldsymbol{A}$ defines a vector potential and φ a scalar potential associated with the charged particle.

Proof

The assertion follows trivially from Definition 1.2.3 and Theorem 1.3.1. To wit, given $(\boldsymbol{A},\varphi)$, $(\boldsymbol{E},\boldsymbol{B})$ can be uniquely derived via $\boldsymbol{E} = -\nabla\varphi - \partial_t \boldsymbol{A}$ and $\boldsymbol{B} = \nabla \times \boldsymbol{A}$. □

8.1.2 Proposition

An accelerating charged particle generates electromagnetic radiation.

Proof

From $\boldsymbol{B} = \mu\varepsilon\boldsymbol{v} \times \boldsymbol{E}$, it follows that $0 \neq \mu\varepsilon\left(\frac{d}{dt}\boldsymbol{v}\right)\times\boldsymbol{E} = \partial_t\boldsymbol{B} - \mu\varepsilon\boldsymbol{v}\times\partial_t\boldsymbol{E} \Rightarrow \partial_t\boldsymbol{A} \neq 0$. The result is now evident from Equations (1.15) and (1.17); that is, a time-varying electric field resulting from an accelerating charge is related to a time-varying magnetic field and vice versa, yielding electromagnetic waves; see Equations (1.28) and (1.29). Indeed, it is obvious that $\partial_t\boldsymbol{A} \neq 0 \Leftrightarrow \frac{d}{dt}\boldsymbol{v} \neq 0$ and hence, $\frac{d}{dt}\boldsymbol{v} = 0 \Leftrightarrow \boldsymbol{B}$ is a static field. □

Upon examining the proof of Proposition 8.1.2 more carefully, it is clear that $\partial_t\boldsymbol{A} \neq 0$ generates both a time-varying electric field and a time-varying magnetic field. When a charged particle is not accelerating, the fields are attached to the particle. On the other hand, upon accelerating, the fields are detached from the charged particle; the detached field manifests as radiation! In this sense, it is not precisely correct to think of a time-varying electric field generating a time-varying magnetic field and vice versa. The correct perspective is the following: the fields are generated by in accelerating charge* via $\partial_t\boldsymbol{A} \neq 0$.

In order to consider radiation, relativistic effects must be taken into account. Without going into detail, with respect to a fixed laboratory frame,

* Strictly speaking, the transformation between an electric field and a magnetic field is a relativistic effect; recall from a footnote in Chapter 1 that the electric and magnetic fields comprise an entity called the electromagnetic field.

the scalar potential and the vector potential of a moving charged particle of charge q and velocity v are, respectively [11],

$$\varphi = \tfrac{1}{4\pi\varepsilon_0}\tfrac{1}{1-\frac{v}{c}}\tfrac{q}{|r-r_0|} \quad \text{and} \quad A = \tfrac{1}{4\pi\varepsilon_0}\tfrac{1}{1-\frac{v}{c}}\tfrac{q}{|r-r_0|}v$$

These potentials are called the *Liénard–Wiechert* potentials.

Now, recalling that $E = -\nabla\varphi - \partial_t A$, it is can be shown (*cf.* Exercise 8.6.1) that the electric field can be decomposed into a term involving the acceleration and a term involving the static term:

$$E = -\tfrac{q}{4\pi\varepsilon_0}\left(1-\tfrac{v}{c}\right)^{-2}\tfrac{1}{cr}\tfrac{d}{dt}v + \tfrac{q}{4\pi\varepsilon_0}\left(1-\tfrac{v}{c}\right)^{-1}\tfrac{v}{r^2}v - \nabla\varphi \equiv E_{\text{accel}} + E_{\text{static}}$$

where

$$E_{\text{accel}} = -\tfrac{q}{4\pi\varepsilon_0}\left(1-\tfrac{v}{c}\right)^{-2}\tfrac{1}{cr}\tfrac{d}{dt}v \quad \text{and} \quad E_{\text{static}} = \tfrac{q}{4\pi\varepsilon_0}\left(1-\tfrac{v}{c}\right)^{-1}\tfrac{v}{r^2}v - \nabla\varphi$$

The key point to note here is that $E_{\text{accel}} \propto \frac{1}{r}$ whereas $E_{\text{static}} \propto \frac{1}{r^2}$. Hence, for large $r > 0$, $E_{\text{accel}} >> E_{\text{static}}$. Once again, in the absence of acceleration, only the static term remains. In the presence of nonzero acceleration, in the far-field regime, the field detaches, as it were, from the particle and propagates outward. Indeed, it can also be shown [8,11] for a radiation field (E, B), $E \perp B$ and the field is thus TEM, and hence, the saying that an accelerating charged particle radiates.

8.2 Hertzian Dipole Antenna

It was demonstrated above that an accelerating charge generates electromagnetic waves. Using this principle, an antenna can be constructed by considering an antenna to be made up of differential antenna elements; each element approximates a charged dipole called a *Hertzian dipole*.

8.2.1 Definition

Consider two charged point particles of charge $+q$ and $-q$ separated by a constant distance d. The pair of charged particles defined by $(\pm q, d)$ constitutes an *electrostatic dipole* (or more simply an *electric dipole*), where d is the vector pointing from $-q$ to $+q$ and $\|d\| = d$.

The electrostatic field profile for an electric dipole was worked out in Example 1.1.1. Another useful concept to know is the following. Given an electric dipole $(\pm q, d)$, define an *electric dipole moment* by $p = qd$. As an example, suppose a dipole oscillates at an angular frequency ω; then $q(t) = |q|\,e^{i\omega t}$, whence,

$i(t) = \frac{d}{dt}q(t) = i\omega|q|e^{i\omega t}$ and $p = -i\frac{I}{\omega}d$, where $i(t) = Ie^{i\omega t}$. Indeed, the potential of an electric dipole can now be expressed in terms of the dipole moment as

$$V = \frac{1}{4\pi\varepsilon}\frac{1}{r^2}p \cdot r$$

where the origin is taken to be the center of the dipole axis and r is the displacement of an arbitrary point from the origin.

8.2.2 Definition

A *Hertzian dipole* is an electric dipole $(\pm q, d, \omega)$, where $d \ll 1$, such that the charge oscillates at an angular frequency ω along the distance d between $\pm q$. If $\delta z n$ denotes the differential length along a thin straight conductor* carrying a time-harmonic current per unit length $J = J(z;\omega)$, where $n = \frac{J}{|J|}$ is a unit vector, then, a *Hertzian antenna* is defined to be the triple $(\pm q, \delta z n, J(z;\omega))$, where the oscillating current element $\delta I(z;\omega) = J(z;\omega)\delta z$.

In what follows, suppose without loss of generality that the current $I = I(z;\omega)$ is time harmonic because in most physical applications, the current has a Fourier series which is, in essence, the infinite sum of time-harmonic currents. A physical antenna may be approximated by the sum of Hertzian antennae, or more precisely, may be expressed as the integral over the Hertzian antennae. Indeed, Definition 8.2.2 may be extended to a conductor with a finite (nonvanishing) cross-section. Then, the current per unit length is replaced more generally by a current density over a surface area or volume.

8.2.3 Lemma

Suppose without loss of generality that the center of a Hertzian dipole $(\pm q, \delta z e_z, J(z;\omega))$ is the origin and the current is time harmonic. Then, the vector potential $A = A(r)$ of the Hertzian dipole is given by

$$A(r) = \frac{\mu}{4\pi}\frac{I\delta z}{r}e^{-i\beta r}e_z \tag{8.1}$$

where r is the distance of an arbitrary point from the center of the dipole.

Proof

From $A(r) = \frac{\mu}{4\pi}\int_V J(r')e^{-i\beta|r-r'|}\frac{1}{|r-r'|}d^3r'$, set $J(r') = I(z')\delta(x')\delta(y')e_z$ with $r' = (0,0,z)$ along the z-axis connecting the two charges $\pm q$; $\delta(u)$ is the Dirac-delta distribution satisfying:

$$\delta(t) = \begin{cases} 0 & \text{if } t \neq 0, \\ \infty & \text{if } t = 0, \end{cases}$$

* Here, *thin* approximates a one-dimensional conductor, that is, a conductor with vanishingly small cross-sectional area.

and $\int_{-\infty}^{\infty}\delta(t)\mathrm{d}t = 1$. Furthermore, note that the integral $\int_V \mathrm{d}^3r' = \int_{-\varepsilon}^{\varepsilon}\mathrm{d}x' \int_{-\varepsilon}^{\varepsilon}\mathrm{d}y' \int_{-\frac{1}{2}\delta z}^{\frac{1}{2}\delta z}\mathrm{d}z'$, where V is the 1-dimensional current element of length δz and $\varepsilon > 0$ is arbitrary. Indeed, because of the definition of the Dirac-delta distribution, the integral can also be defined by $\int_V \mathrm{d}^3r' = \int_{-\infty}^{\infty}\mathrm{d}x' \int_{-\infty}^{\infty}\mathrm{d}y' \int_{-\frac{1}{2}\delta z}^{\frac{1}{2}\delta z}\mathrm{d}z'$. Finally observe that on setting $\|\boldsymbol{r}\| = r = \sqrt{x^2+y^2+z^2}$ (in rectangular coordinates) and noting that

$$\begin{aligned}\|\boldsymbol{r}-\boldsymbol{r}'\| &= \sqrt{(x-x')^2+(y-y')^2+(z-z')^2}\\ &= \sqrt{x^2+y^2+z^2+\left(x'^2+y'^2+z'^2-2(xx'+yy'+zz')\right)}\\ &= r\sqrt{1-\tfrac{2(xx'+yy'+zz')}{r^2}+\left(\tfrac{r'}{r}\right)^2}\end{aligned}$$

it follows that

$$r' << r \Rightarrow \|\boldsymbol{r}-\boldsymbol{r}'\| \approx r\sqrt{1-\tfrac{2(xx'+yy'+zz')}{r^2}} \approx r\left\{1-\tfrac{xx'+yy'+zz'}{r^2}\right\}$$

where the binomial expansion $\sqrt{1+\varepsilon} \approx 1+\frac{1}{2}\varepsilon$ for $|\varepsilon| << 1$ was applied.

Thus, substituting the above approximation into the vector potential integral yields

$$\begin{aligned}\boldsymbol{A}(\boldsymbol{r}) &= \mathbf{e}_z \tfrac{\mu}{4\pi}\int_{-\infty}^{\infty}\int_{-\infty}^{\infty}\int_{-\frac{1}{2}\delta z}^{\frac{1}{2}\delta z} I\delta(x')\delta(y')\mathrm{e}^{-\mathrm{i}\beta\sqrt{x^2+y^2+(z-z')^2}}\left\{x^2+y^2+(z-z')^2\right\}^{-\frac{1}{2}}\mathrm{d}x'\mathrm{d}y'\mathrm{d}z'\\ &\approx \mathbf{e}_z \tfrac{\mu}{4\pi} I\int_{-\frac{1}{2}\delta z}^{\frac{1}{2}\delta z}\mathrm{d}z'\int_{-\infty}^{\infty}\delta(x')\mathrm{d}x'\int_{-\infty}^{\infty}\delta(y')\mathrm{e}^{-\mathrm{i}\beta r\sqrt{1-\frac{2zz'}{r^2}}}\tfrac{1}{r}\left\{1-\tfrac{2zz'}{r^2}\right\}^{-1}\mathrm{d}y'\\ &\approx \mathbf{e}_z \tfrac{\mu}{4\pi} I\int_{-\frac{1}{2}\delta z}^{\frac{1}{2}\delta z}\tfrac{1}{r}\left\{1+\tfrac{2zz'}{r^2}\right\}\mathrm{e}^{-\mathrm{i}\beta r\left\{1-\frac{zz'}{r^2}\right\}}\mathrm{d}z'\\ &\approx \mathbf{e}_z \tfrac{\mu}{4\pi}\tfrac{I}{r}\int_{-\frac{1}{2}\delta z}^{\frac{1}{2}\delta z}\mathrm{e}^{-\mathrm{i}\beta r\left\{1-\frac{zz'}{r^2}\right\}}\mathrm{d}z'\\ &= \mathbf{e}_z \tfrac{\mu}{4\pi}\tfrac{I}{r}\tfrac{2r}{\beta z}\mathrm{e}^{-\mathrm{i}\beta r}\sin\tfrac{\beta z\delta z}{2r} \approx \tfrac{\mu}{4\pi}\tfrac{I\delta z}{r}\mathrm{e}^{-\mathrm{i}\beta r}\mathbf{e}_z\end{aligned}$$

up to first order in $\frac{1}{r}$, where the approximation $\sin\theta \approx \theta$ for $|\theta| << 1$ was used. □

8.2.4 Theorem

Suppose without loss of generality that the center of a Hertzian dipole $(\pm q, \delta z\mathbf{e}_z, \boldsymbol{J}(z;\omega))$ is the origin and the current is time harmonic, where

$J(r') = I(z')\delta(x')\delta(y')\mathbf{e}_z$. If the dipole is placed in a homogeneous dielectric medium $(\mathbf{R}^3, \mu, \varepsilon)$, then the electric field and magnetic field of the dipole are given in spherical coordinates by

$$E_r(r,\theta) = -\tfrac{i}{2\pi\omega\varepsilon}\tfrac{I\delta z}{r^2}e^{-i\beta r}\cos\theta\left\{\tfrac{1}{r} + i\beta\right\} \tag{8.2}$$

$$E_\theta(r,\theta) = -\tfrac{i}{4\pi\omega\varepsilon}\tfrac{I\delta z}{r}e^{-i\beta r}\sin\theta\left\{-\beta^2 + \tfrac{1}{r^2} + i\tfrac{\beta}{r}\right\} \tag{8.3}$$

$$B_\phi(r,\theta) = \tfrac{\mu}{4\pi}\tfrac{I\delta z}{r}e^{-i\beta r}\left\{\tfrac{1}{r} + i\beta\right\}\sin\theta \tag{8.4}$$

Proof

Figure 8.1 shows a vector $\boldsymbol{r}$ represented in two different coordinate systems: the rectangular coordinates and spherical polar coordinates. The transformations between rectangular and spherical coordinates are given by

$$x = r\sin\theta\cos\phi$$

$$y = r\sin\theta\sin\phi$$

$$z = r\cos\theta$$

Hence, $dz = \partial_r z dr + \partial_\theta z d\theta = \cos\theta dr - \sin\theta r d\theta$. That is, $\mathbf{e}_z = \cos\theta\mathbf{e}_r - \sin\theta\mathbf{e}_\theta$. From Proposition 8.2.3, expressing the vector potential in terms of spherical coordinates yields:

$$A_r = A_z\cos\theta = \tfrac{\mu}{4\pi}\tfrac{I\delta z}{r}e^{-i\beta r}\cos\theta$$

$$A_\theta = -A_z\sin\theta = -\tfrac{\mu}{4\pi}\tfrac{I\delta z}{r}e^{-i\beta r}\sin\theta$$

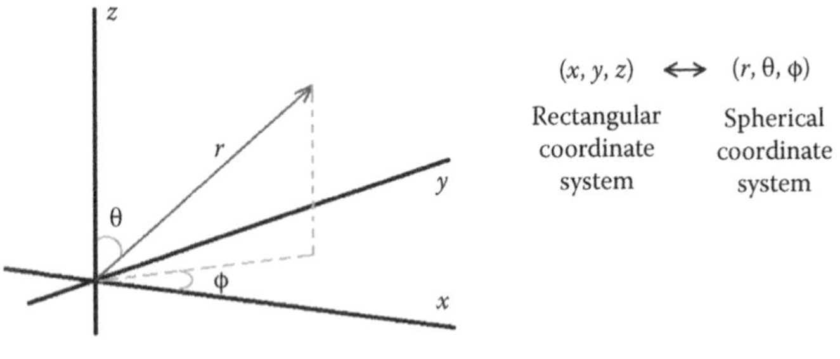

FIGURE 8.1.
Transformation between rectangular and spherical coordinates.

Observe that $A_\phi = 0$ as there is no ϕ-dependency. Moreover, note that in spherical polar coordinates, $\nabla = (\partial_r, \frac{1}{r}\partial_\theta, \frac{1}{r\sin\theta}\partial_\phi)$. Hence,

$$\nabla \times \boldsymbol{A} = \begin{vmatrix} \mathbf{e}_r & \mathbf{e}_\theta & \mathbf{e}_\phi \\ \partial_r & \frac{1}{r}\partial_\theta & \frac{1}{r\sin\theta}\partial_\phi \\ A_r & A_\theta & 0 \end{vmatrix} = \frac{1}{r}\begin{vmatrix} \mathbf{e}_r & r\mathbf{e}_\theta & \mathbf{e}_\phi \\ \partial_r & \partial_\theta & \frac{1}{r\sin\theta}\partial_\phi \\ A_r & rA_\theta & 0 \end{vmatrix}$$

$$= \frac{1}{r^2\sin\theta}\begin{vmatrix} \mathbf{e}_r & r\mathbf{e}_\theta & r\sin\theta\mathbf{e}_\phi \\ \partial_r & \partial_\theta & \partial_\phi \\ A_r & rA_\theta & 0 \end{vmatrix}$$

$$= \frac{1}{r}\{\partial_r(rA_\theta) - \partial_\theta A_r\}\mathbf{e}_\phi$$

Now,

$$\partial_\theta A_r = \frac{\mu}{4\pi}\frac{I\delta z}{r}\mathrm{e}^{-\mathrm{i}\beta r}\cos\theta = -\frac{\mu}{4\pi}\frac{I\delta z}{r}\mathrm{e}^{-\mathrm{i}\beta r}\sin\theta \quad \text{and} \quad \partial_r(rA_\theta) = \mathrm{i}\beta\frac{\mu}{4\pi}I\delta z\mathrm{e}^{-\mathrm{i}\beta r}\sin\theta$$

So, this gives

$$\nabla \times \boldsymbol{A} = \frac{\mu}{4\pi}\frac{I\delta z}{r}\mathrm{e}^{-\mathrm{i}\beta r}\left\{\frac{1}{r} + \mathrm{i}\beta\right\}\sin\theta\mathbf{e}_\phi.$$

Hence, from $\boldsymbol{B} = \nabla \times \boldsymbol{A}$,

$$B_\phi = \frac{\mu}{4\pi}\frac{I\delta z}{r}\mathrm{e}^{-\mathrm{i}\beta r}\left\{\frac{1}{r} + \mathrm{i}\beta\right\}\sin\theta\mathbf{e}_\phi$$

Next, via $\nabla \times \boldsymbol{B} = \mathrm{i}\omega\mu\varepsilon E$ as the medium is lossless, it follows that $\boldsymbol{E} = -\frac{\mathrm{i}}{\mu\omega\varepsilon}\nabla \times \boldsymbol{B}$. Hence,

$$\nabla \times \boldsymbol{B} = \begin{vmatrix} \mathbf{e}_r & \mathbf{e}_\theta & \mathbf{e}_\phi \\ \partial_r & \frac{1}{r}\partial_\theta & \frac{1}{r\sin\theta}\partial_\varphi \\ 0 & 0 & B_\phi \end{vmatrix} = \frac{1}{r^2\sin\theta}\begin{vmatrix} \mathbf{e}_r & r\mathbf{e}_\theta & r\sin\theta\mathbf{e}_\varphi \\ \partial_r & \partial_\theta & \partial_\varphi \\ 0 & 0 & (r\sin\theta)B_\phi \end{vmatrix}$$

$$= \frac{1}{r^2\sin\theta}\{\partial_\theta((r\sin\theta)B_\phi)\mathbf{e}_r - r\partial_r((r\sin\theta)B_\phi)\mathbf{e}_\theta\}$$

So, substituting

$$\partial_\theta(r\sin\theta H_\phi) = \frac{2\sin\theta\cos\theta}{4\pi}I_0\delta z\mathrm{e}^{-\mathrm{i}\beta r}\left\{\frac{1}{r} + \mathrm{i}\beta\right\}$$

$$\partial_r(r\sin\theta H_\phi) = \frac{\sin^2\theta}{4\pi}I_0\delta z\mathrm{e}^{-\mathrm{i}\beta r}\left\{-\beta^2 + \frac{1}{r^2} + \mathrm{i}\frac{\beta}{r}\right\}$$

results in

$$E = -\frac{i}{\omega\mu\varepsilon}\nabla\times B = -\frac{i}{4\pi\omega\mu\varepsilon}\frac{I\delta z}{r}e^{-i\beta r}\begin{pmatrix} \frac{2\cos\theta}{r}\left\{\frac{1}{r}+i\beta\right\} \\ \sin\theta\left\{-\beta^2+\frac{1}{r^2}+i\frac{\beta}{r}\right\} \\ 0 \end{pmatrix}$$

as required. □

Now, observe trivially that $\beta r << 1 \Leftrightarrow \frac{\beta}{r} << \frac{1}{r^2}$ and $\beta r << 1 \Rightarrow e^{-i\beta r} \approx 1 - i\beta r + o(r^2)$. Hence,

$$E_r(r,\theta) \approx -\frac{i}{2\pi\omega\varepsilon}\frac{I\delta z}{r^2}\{1-i\beta r\}\cos\theta\left\{\frac{1}{r}+i\beta\right\} \approx -\frac{i}{2\pi\omega\varepsilon}\frac{I\delta z}{r^3}\cos\theta$$

$$E_\theta(r,\theta) = -\frac{i}{4\pi\omega\varepsilon}\frac{I\delta z}{r}\{1-i\beta r\}\sin\theta\left\{-\beta^2+\frac{1}{r^2}+i\frac{\beta}{r}\right\} \approx -\frac{i}{4\pi\omega\varepsilon}\frac{I\delta z}{r^3}\sin\theta$$

$$B_\phi(r,\theta) = \frac{\mu}{4\pi}\frac{I\delta z}{r}\{1-i\beta r\}\left\{\frac{1}{r}+i\beta\right\}\sin\theta \approx \frac{\mu}{4\pi}\frac{I\delta z}{r^2}\sin\theta$$

Conversely, $\beta r << 1 \Leftrightarrow \frac{\beta}{r} << \frac{1}{r^2}$, and so,

$$E_r(r,\theta) = -\frac{i}{2\pi\omega\varepsilon}\frac{I\delta z}{r^2}e^{-i\beta r}\cos\theta\left\{\frac{1}{r}+i\beta\right\} \approx \frac{\beta}{2\pi\omega\varepsilon}\frac{I\delta z}{r^2}e^{-i\beta r}\cos\theta$$

$$E_\theta(r,\theta) = -\frac{i}{4\pi\omega\varepsilon}\frac{I\delta z}{r}e^{-i\beta r}\sin\theta\left\{-\beta^2+\frac{1}{r^2}+i\frac{\beta}{r}\right\} \approx \frac{i\beta^2}{4\pi\omega\varepsilon}\frac{I\delta z}{r}e^{-i\beta r}\sin\theta$$

$$B_\phi(r,\theta) = \frac{\mu}{4\pi}\frac{I\delta z}{r}e^{-i\beta r}\left\{\frac{1}{r}+i\beta\right\}\sin\theta \approx i\beta\frac{\mu}{4\pi}\frac{I\delta z}{r}e^{-i\beta r}\sin\theta$$

8.2.5 Definition

Given a Hertzian dipole $(\pm q, \delta z\mathbf{e}_z, J(z;\omega))$ where $J(r') = I(z')\delta(x')\delta(y')\mathbf{e}_z$, the *near field* (or *near zone*) is defined by the criterion $\beta r \ll 1$. In particular, the electromagnetic field in the near-field regime to first order in $\frac{1}{r}$ satisfies:

$$E_r(r,\theta) \approx -\frac{i}{2\pi\omega\varepsilon}\frac{I\delta z}{r^3}\cos\theta \tag{8.5}$$

$$E_\theta(r,\theta) \approx -\frac{i}{4\pi\omega\varepsilon}\frac{I\delta z}{r^3}\sin\theta \tag{8.6}$$

$$B_\phi(r,\theta) \approx \frac{\mu}{4\pi}\frac{I\delta z}{r^2}\sin\theta \tag{8.7}$$

The *far field* (or *far zone*) is defined by the criterion $\beta r \gg 1$. In particular, the electromagnetic field in the far-field regime, up to first order in $\frac{1}{r}$, satisfies:

$$E_\theta(r,\theta) \approx \frac{i\beta^2}{4\pi\omega\varepsilon}\frac{I\delta z}{r}e^{-i\beta r}\sin\theta \tag{8.8}$$

$$B_\phi(r,\theta) \approx \frac{i\beta\mu}{4\pi}\frac{I\delta z}{r}e^{-i\beta r}\sin\theta \tag{8.9}$$

Finally, a zone that is neither a near zone nor far zone is called an *intermediate zone.*

Notice from Equations (8.5)–(8.7) in the near zone that $\boldsymbol{E} \propto \frac{1}{r^3}$ and $\boldsymbol{B} \propto \frac{1}{r^2}$. Thus, the electric field is much stronger than the magnetic field in the near zone. Specifically, $\frac{1}{\omega\varepsilon r^3} >> \frac{\mu}{r^2} \Rightarrow \frac{1}{\omega\mu\varepsilon} >> r$; the electric field is much stronger than the magnetic field, where $\omega,\mu,\varepsilon > 0$ are fixed. On the other hand, in the far zone, (8.8) and (8.9) reveal that the fields resemble that of a plane wave: the electric field and magnetic field for a plane wave are related by $B_\phi(r,\theta) = \frac{\mu}{\eta}E_\theta$, where $\beta = \omega\sqrt{\mu\varepsilon}$ was utilized.

Now, recall that the time-average power absorbed by a resistor is given by $\langle P\rangle = \frac{1}{2}|I|^2 R$. This is the energy dissipated by the load. In an analogous manner, the power absorbed by the intervening medium (dielectric) as radiation propagates across the medium is the energy dissipated from the source. Call this equivalent load the *radiation resistance*. In particular, this means that more power is dissipated by the medium if its radiation resistance is high.

Intuitively, an antenna may be thought of as an energy storage device inasmuch as it supports standing waves; the radiation resistance then affords a means whereby the energy can be extracted from the standing waves by transforming them into traveling waves. In the case of a Hertzian dipole, let $\langle P\rangle$ denote the time-average outward power flow from a Hertzian dipole. If I is the time-harmonic current flowing along the dipole, then the radiation resistance of the Herzian dipole is given by $R = 2\frac{\langle P\rangle}{|I|^2}$.

8.2.6 Proposition

The radiation resistance of a Hertzian dipole $(\pm q, \delta z\mathbf{e}_z, \boldsymbol{J}(z;\omega))$, where $\boldsymbol{J}(\boldsymbol{r}') = I(z')\delta(x')\delta(y')\mathbf{e}_z$, in a homogeneous medium (μ,ε) is given by $R = \frac{\eta}{6\pi}(\beta\delta z)^2$, where δz is the length of the Hertzian dipole and $\eta = \sqrt{\frac{\mu}{\varepsilon}}$.

Proof

First, recall that the time-average power is defined via the Poynting vector as $\langle S\rangle = \frac{1}{2}\text{Re}(\boldsymbol{E}\times\boldsymbol{H}^*)$. Explicitly,

$$\boldsymbol{E}\times\boldsymbol{H}^* = -\frac{i}{\omega\varepsilon}\left(\frac{1}{4\pi}\frac{\delta z}{r}\right)^2|I|^2\begin{vmatrix} \mathbf{e}_r & \mathbf{e}_\theta & \mathbf{e}_\varphi \\ 2\cos\theta\left\{\frac{1}{r}+i\beta\right\} & \sin\theta\left\{-\beta^2+\frac{1}{r^2}+i\frac{\beta}{r}\right\} & 0 \\ 0 & 0 & \left\{\frac{1}{r}-i\beta\right\}\sin\theta \end{vmatrix}$$

$$= -\frac{i}{\omega\varepsilon}\left(\frac{1}{4\pi}\frac{\delta z}{r}\right)^2|I|^2\begin{pmatrix} \sin^2\theta\left\{\frac{1}{r}-i\beta\right\}\left\{-\beta^2+\frac{1}{r^2}+i\frac{\beta}{r}\right\} \\ \sin 2\theta\left\{\frac{1}{r}-i\beta\right\}\left\{\frac{1}{r}+i\beta\right\} \\ 0 \end{pmatrix}$$

After some tedious simplification, the components of the Poynting vector are:

$$\begin{aligned}(\boldsymbol{E}\times\boldsymbol{H}^*)_r &= -\tfrac{\sin^2\theta}{\omega\varepsilon}\left(\tfrac{\delta z}{4\pi}\right)^2\tfrac{1}{r^2}|I|^2\left\{i\tfrac{1}{r^3}-\beta^3\right\}\\ (\boldsymbol{E}\times\boldsymbol{H}^*)_\theta &= -i\tfrac{\sin 2\theta}{\omega\varepsilon}\left(\tfrac{\delta z}{4\pi}\right)^2\tfrac{1}{r^2}|I|^2\left\{\tfrac{1}{r^2}+\beta^2\right\}\end{aligned} \tag{8.10}$$

From this,

$$\langle \boldsymbol{S}\rangle = \tfrac{\sin^2\theta}{2\omega\varepsilon}\left(\tfrac{\delta z}{4\pi}\right)^2\tfrac{1}{r^2}|I|^2\begin{pmatrix}\beta^3\\ 0\\ 0\end{pmatrix} \tag{8.11}$$

The time-average power $\langle P\rangle$ of the dipole is the power radiated radially outwards. That is,

$$\begin{aligned}\langle P\rangle &= \int_0^{2\pi}\int_0^{\pi}\langle \boldsymbol{S}\rangle\cdot\hat{\mathbf{r}}r^2\sin\theta\, d\theta\, d\phi\\ &= \tfrac{1}{2\omega\varepsilon}\left(\tfrac{\delta z}{4\pi}\right)^2|I|^2\beta^3\int_0^{2\pi}\int_0^{\pi}\sin^3\theta\, d\theta\, d\phi\\ &= \tfrac{1}{2\omega\varepsilon}\left(\tfrac{\delta z}{4\pi}\right)^2|I|^2\beta^3\tfrac{2\pi}{12}\left[\cos 3\theta - 9\cos\theta\right]_0^{\pi}\\ &= \tfrac{1}{12\pi\omega\varepsilon}\delta z^2|I|^2\beta^3\end{aligned}$$

Therefore, $R = 2\frac{\langle P\rangle}{|I|^2} = \frac{1}{6\pi\omega\varepsilon}\beta^3\delta z^2 = \frac{\eta}{6\pi}(\beta\delta z)^2$, where $\eta = \sqrt{\mu/\varepsilon}$, as required. □

It is clear that the radiation resistance of a Hertzian dipole is very small as $\delta z \ll 1$. In particular, a Hertzian dipole is a poor radiator unless $\beta\delta z \gg 1$; that is, $\omega \gg \frac{1}{\delta z\sqrt{\mu\varepsilon}}$. Hence, in the microwave range, a Hertzian dipole makes a poor antenna.

8.3 Magnetic Dipole Antenna

In Section 8.2, an elementary open-ended antenna in terms of an electric dipole was considered. A closed-ended antenna forms a loop: indeed, this forms the basis for a loop antenna. An elementary loop antenna is defined via a magnetic dipole. Informally, this can be viewed as a charged particle of charge q traversing around a small loop γ with a constant velocity v, where γ

spans a differential surface area $\delta S(\gamma) \subset \mathbf{R}^2$ and $\boldsymbol{n}_\gamma$ is the unit vector normal to the area $\delta S(\gamma)$; see Example 1.2.4.

8.3.1 Definition

A *magnetic dipole* is the triple $(q, \gamma, \omega \boldsymbol{n}_\gamma)$, where $\boldsymbol{n}_\gamma = \frac{1}{\omega} \boldsymbol{r} \times \boldsymbol{v}$ is the unit normal on the surface $\delta S(\gamma) \subset \mathbf{R}^2$ spanned by γ, and ω is the angular velocity around γ. The *magnetic moment* of a magnetic dipole is defined by $\boldsymbol{m} = q\omega\,|\delta S(\gamma)|\,\boldsymbol{n}_\gamma$, where $|\delta S(\gamma)|$ denotes the surface area of $\delta S(\gamma)$.

By way of an example, consider an ideal current loop γ of radius δr with a time-harmonic current $I = I(\omega)$ flowing around the loop, where $\delta S(\gamma) \subset \mathbf{R}^2$. Then, the magnetic dipole moment of the current loop is $\boldsymbol{m} = I\pi\delta r^2 \mathbf{e}_z$. This follows from the fact that $I = q\omega$, $|\delta S(\gamma)| = \pi \delta r^2$ and $\boldsymbol{n}_\gamma = \mathbf{e}_z$. In view of this result, a magnetic dipole for a current I flowing around an arbitrary loop γ is defined by the triple $(I, \gamma, \delta S(\gamma))$, where $\delta S(\gamma) = \frac{1}{2} \oint_\gamma \boldsymbol{r} \times \mathrm{d}\boldsymbol{r}$. This is also called a *magnetic dipole antenna*.

8.3.2 Proposition

Given a differential magnetic dipole $(I, \gamma, \delta S(\gamma))$ in an isotropic homogeneous medium (μ, ε), that is, $|\delta S(\gamma)| << 1$, the vector potential in spherical coordinates may be approximated by $\boldsymbol{A}(r,\theta) = \frac{\mu m}{4\pi} \mathrm{e}^{-\mathrm{i}\beta r} \sin\theta \frac{1}{r^2}\{1 + \mathrm{i}\beta r\}\mathbf{e}_\phi$, where $m = I|\delta S(\gamma)|$.

Proof

Recall that $\boldsymbol{A}(r,\theta) = \frac{\mu}{4\pi} I_0 \oint_\gamma \frac{1}{R} \mathrm{e}^{-\mathrm{i}\beta R}\, \mathrm{d}\boldsymbol{l}$ for a one-dimensional current-carrying loop. Next, consider γ to be a planar loop of radius δr, that is, a loop embedded in $\mathbf{R}^2$. Without loss of generality, one may assume that the center of γ coincides with the origin. In Cartesian coordinates,

$$\mathrm{d}\boldsymbol{l} = \delta r \mathrm{d}\phi \mathbf{e}_\phi = (-\mathbf{e}_x \sin\phi + \mathbf{e}_y \cos\phi)\delta r \mathrm{d}\phi$$

$$\boldsymbol{R} = \boldsymbol{r} - \delta r \mathbf{e}_r = (x - \delta r \cos\phi)\mathbf{e}_x + (y - \delta r \sin\phi)\mathbf{e}_y + z\mathbf{e}_z$$

Now, noting that $r^2 = x^2 + y^2 + z^2$, and using the fact that $r >> \delta r$,

$$\begin{aligned} R &= \sqrt{(x - \delta r\cos\phi)^2 + (y - \delta r \sin\phi)^2 + z^2} \\ &= \sqrt{r^2 - 2x\delta r\cos\phi - 2y\delta r \sin\phi + o(\delta r^2)} \\ &\approx r\sqrt{1 - \tfrac{2x\delta r}{r}\cos\phi - \tfrac{2y\delta r}{r}\sin\phi} \end{aligned}$$

Therefore,

$$\frac{1}{R} \approx \frac{1}{r}\left\{1 + \frac{x\delta r}{r^2}\cos\phi + \frac{y\delta r}{r^2}\sin\phi\right\}$$

and hence

$$\tfrac{\mathrm{d}l}{R} \approx \tfrac{1}{r}\left\{1+\tfrac{x\delta r}{r^2}\cos\phi+\tfrac{y\delta r}{r^2}\sin\phi\right\}\left\{-\mathbf{e}_x\sin\phi+\mathbf{e}_y\cos\phi\right\}\delta r\mathrm{d}\phi$$

Hence, evaluating the following integrals:

$$\int_0^{2\pi}\left\{\sin\phi+\tfrac{x\delta r}{r^2}\sin\phi\cos\phi+\tfrac{y\delta r}{r^2}\sin^2\phi\right\}\mathrm{d}\phi=\tfrac{y\delta r}{r^2}\pi$$

$$\int_0^{2\pi}\left\{\cos\phi+\tfrac{x\delta r}{r^2}\cos^2\phi+\tfrac{y\delta r}{r^2}\sin\phi\cos\phi\right\}\mathrm{d}\phi=\tfrac{x\delta r}{r^2}\pi$$

In terms of spherical coordinates (*cf.* Example A.1.6 of the Appendix),

$$x = r\sin\theta\cos\phi,\ y = r\sin\theta\sin\phi,\ z = r\cos\theta$$

it follows that

$$\oint_\gamma \tfrac{\mathrm{d}l}{R} \approx -\tfrac{\pi\delta r^2}{r^2}\sin\theta\sin\phi\mathbf{e}_x+\tfrac{\pi\delta r^2}{r^2}\sin\theta\cos\phi\mathbf{e}_y$$

It thus remains to transform $(\mathbf{e}_x,\mathbf{e}_y)$ into spherical coordinates. So, referring to the Appendix, given any vector v in some coordinate basis $(\mathbf{e}_1,\mathbf{e}_2,\mathbf{e}_3)$, that is, $v=v^1\mathbf{e}_1+v^2\mathbf{e}_2+v^3\mathbf{e}_3$, expressing v in terms of another coordinate basis $(\mathbf{e}_1',\mathbf{e}_2',\mathbf{e}_3')$ is given simply by

$$v=(\mathbf{v}\cdot\mathbf{e}_1')\mathbf{e}_1'+(\mathbf{v}\cdot\mathbf{e}_2')\mathbf{e}_2'+(\mathbf{v}\cdot\mathbf{e}_3')\mathbf{e}_3'$$

For simplicity, set $v=-\tfrac{\pi\delta r^2}{r^2}\sin\theta\sin\phi\mathbf{e}_x+\tfrac{\pi\delta r^2}{r^2}\sin\theta\cos\phi\mathbf{e}_y$. Then,

$$v\cdot\mathbf{e}_r=-\tfrac{\pi\delta r^2}{r^2}\{\sin^2\theta\sin\phi\cos\phi-\sin^2\theta\sin\phi\cos\phi\}=0$$

$$v\cdot\mathbf{e}_\theta=-\tfrac{\pi\delta r^2}{r^2}\{\sin\theta\cos\theta\sin\phi\cos\phi-\sin\theta\cos\theta\sin\phi\cos\phi\}=0$$

$$v\cdot\mathbf{e}_\phi=\tfrac{\pi\delta r^2}{r^2}\{\sin\theta\sin^2\phi+\sin\theta\cos^2\phi\}=\tfrac{\pi\delta r^2}{r^2}\sin\theta$$

Furthermore, observe that

$$\mathrm{e}^{-\mathrm{i}\beta R}=\mathrm{e}^{-\mathrm{i}\beta(r-r+R)}\approx\mathrm{e}^{-\mathrm{i}\beta r}(1-\mathrm{i}\beta(R-r))=\mathrm{e}^{-\mathrm{i}\beta r}((1+\mathrm{i}\beta r)-\mathrm{i}\beta R)$$

whence,

$$\oint_\gamma e^{-i\beta R}\tfrac{dl}{R} \approx e^{-i\beta r}\oint_\gamma (1+\beta r)\tfrac{dl}{R} - e^{-i\beta r}\oint_\gamma i\beta\, dl = e^{-i\beta r}\oint_\gamma (1+\beta r)\tfrac{dl}{R}$$

as

$$\int_0^{2\pi}\left\{\begin{array}{c}\sin\phi\\ \cos\phi\end{array}\right\}d\phi = 0$$

That is, $\oint_\gamma i\beta\, dl = 0$. Thus, it follows at once that

$$A(r,\theta) = \tfrac{\mu I}{4\pi}\tfrac{\pi\delta r^2}{r^2}(1+i\beta r)e^{-i\beta r}\sin\theta\,\mathbf{e}_\phi = \tfrac{\mu}{4\pi} I\,|\delta S(\gamma)|\,\tfrac{1+i\beta r}{r^2}e^{-i\beta r}\sin\theta\,\mathbf{e}_\phi$$

as required. □

8.3.3 Theorem

Given a differential magnetic dipole $(I,\gamma,\delta S(\gamma))$, that is, $|\delta S(\gamma)| << 1$, the electric field and magnetic field arising from the magnetic dipole are given by

$$E_\phi = -\tfrac{i}{\omega\varepsilon}\tfrac{m}{4\pi}e^{-i\beta r}\sin\theta\tfrac{\beta^2}{r^2}\{1+i\beta r\} \quad (8.12)$$

$$B_r = \tfrac{\mu}{2\pi}me^{-i\beta r}\cos\theta\tfrac{1}{r^3}\{1+i\beta r\} \quad (8.13)$$

$$B_\theta = \tfrac{\mu}{4\pi}me^{-i\beta r}\sin\theta\tfrac{1}{r^3}\{-(\beta r)^2+1+i\beta r\} \quad (8.14)$$

Proof

From $\boldsymbol{B} = \nabla\times\boldsymbol{A}$, where

$$\nabla\times\boldsymbol{A} = \tfrac{1}{r^2\sin\theta}\begin{vmatrix}\mathbf{e}_r & r\mathbf{e}_\theta & r\sin\theta\mathbf{e}_\phi\\ \partial_r & \partial_\theta & \partial_\phi\\ 0 & 0 & r\sin\theta A_\phi\end{vmatrix} = \tfrac{1}{r^2\sin\theta}\begin{pmatrix}\partial_\theta(r\sin\theta A_\phi)\\ -r\partial_r(r\sin\theta A_\phi)\\ 0\end{pmatrix}$$

with

$$\partial_\theta(r\sin\theta A_\phi) = \tfrac{\mu}{4\pi}m\tfrac{1+i\beta r}{r}e^{-i\beta r}\sin 2\theta$$

and

$$r\partial_r(r\sin\theta A_\phi) = \tfrac{\mu}{4\pi} m e^{-i\beta r}\left\{r\beta^2 - \tfrac{1}{r} - i\beta\right\}\sin^2\theta.$$

Thus, it follows that

$$B_r = \tfrac{\mu}{2\pi} m e^{-i\beta r} \tfrac{1+i\beta r}{r^3}\cos\theta$$

$$B_\theta = \tfrac{\mu}{4\pi} m e^{-i\beta r}\left\{-\tfrac{\beta^2}{r} + \tfrac{1}{r^3} + i\tfrac{\beta}{r^2}\right\}\sin\theta$$

Similarly, the electric field follows from $\nabla \times \boldsymbol{B} = i\omega\mu\varepsilon \boldsymbol{E}$, where the medium is lossless, and

$$\nabla\times\boldsymbol{B} = \tfrac{1}{r^2\sin\theta}\begin{vmatrix} \mathbf{e}_r & r\mathbf{e}_\theta & r\sin\theta\mathbf{e}_\phi \\ \partial_r & \partial_\theta & \partial_\phi \\ B_r & rB_\theta & 0 \end{vmatrix} = \tfrac{1}{r^2\sin\theta}\begin{pmatrix} -\partial_\phi(rB_\theta) \\ r\partial_\phi B_r \\ r\sin\theta(\partial_r(rB_\theta) - \partial_\theta B_r) \end{pmatrix}$$

Then, noting that $\partial_\phi B_\theta = 0 = \partial_\phi B_r$ as there are no ϕ-dependencies,

$$\partial_r B_\theta = -\tfrac{\mu}{4\pi} m \sin\theta e^{-i\beta r}\left\{-\tfrac{2\beta^2}{r^2} + \tfrac{3}{r^4} - i\tfrac{\beta^3}{r} + i\tfrac{3\beta}{r^3}\right\}$$

$$\partial_\theta B_r = -\tfrac{\mu}{4\pi} m e^{-i\beta r}\sin\theta\left\{\tfrac{2}{r^3} + i\tfrac{2\beta}{r^2}\right\}$$

whence, $\partial_r(rB_\theta) - \partial_\theta B_r = \tfrac{\mu}{4\pi} m e^{-i\beta r}\sin\theta\left\{\tfrac{\beta^2}{r} + i\beta^3\right\} = \tfrac{m}{4\pi} e^{-i\beta r}\sin\theta \tfrac{\beta^2}{r}\{1 + i\beta r\}$, and the electric field is thus $E_\phi = -\tfrac{i}{\omega\varepsilon}\tfrac{m}{4\pi} e^{-i\beta r}\sin\theta\tfrac{\beta^2}{r^2}\{1+i\beta r\}$, completing the proof. □

Once again, observe that in the far zone, $(E_\phi, B_\theta) \sim o\left(\tfrac{1}{r}\right)$ and $B_\phi \sim o\left(\tfrac{1}{r^2}\right)$, hence, only (E_ϕ, B_θ) are dominant. Explicitly, for $\beta r >> 1$, $\tfrac{\beta^2}{r^2}\{1+i\beta r\} \approx i\tfrac{\beta^3}{r}$, $\tfrac{1+i\beta r}{r^3} \approx i\tfrac{\beta}{r^2}$ and finally, $-\tfrac{\beta^2}{r} + \tfrac{1}{r^3} + i\tfrac{\beta}{r^2} = -\tfrac{\beta^2}{r}\left\{1 - \tfrac{1}{(\beta r)^2} - i\tfrac{1}{\beta r}\right\} \approx -\tfrac{\beta^2}{r}$, yielding the result given below.

8.3.4 Corollary

Given a differential magnetic dipole $(I, \gamma, \delta S(\gamma))$, that is, $|\delta S(\gamma)| << 1$, the electric and magnetic far fields, to first order in $\tfrac{1}{r}$, are given by

$$E_\phi = \tfrac{1}{4\pi}\tfrac{m}{\omega\varepsilon}\tfrac{\beta^3}{r} e^{-i\beta r}\sin\theta \tag{8.15}$$

$$B_\theta = -\tfrac{\mu}{4\pi} m \tfrac{\beta^2}{r} e^{-i\beta r}\sin\theta \tag{8.16}$$

□

In particular, the wave impedance in the far zone is $Z_{\text{far}} \approx \mu \frac{E_\phi}{B_\theta} = \sqrt{\frac{\mu}{\varepsilon}} = \eta$, as expected, where $\beta = \omega\sqrt{\mu\varepsilon}$ was utilized.

8.3.5 Corollary

Given a differential magnetic dipole $(I, \gamma, \delta S(\gamma))$, that is, $|\delta S(\gamma)| << 1$, the near-field electric and magnetic fields are given by

$$E_\phi = -\frac{i}{\omega\varepsilon}\frac{1}{4\pi} m \frac{\beta^2}{r^2} e^{-i\beta r} \sin\theta \tag{8.17}$$

$$B_r = \frac{\mu}{2\pi} m \frac{1}{r^3} e^{-i\beta r} \cos\theta \tag{8.18}$$

$$B_\theta = \frac{\mu}{4\pi} m \frac{1}{r^3} e^{-i\beta r} \sin\theta \tag{8.19}$$

provided that $|\delta S(\gamma)| << r^2 < 1$ continues to hold in the limit as r approaches the loop γ.

Proof

In the near zone, $\beta r << 1$. Hence, as long as the loop γ is sufficiently small so that $|\delta S(\gamma)| << r^2 < 1$, then it is clear that $1 + i\beta r \approx 1$ and $-(\beta r)^2 + 1 + i\beta r \approx 1$. The result thus follows. □

Some remarks regarding Corollary 8.3.5 are due. For a sufficiently small loop γ such that $|\delta S(\gamma)| << r^2 << 1$ holds, the magnetic field dominates the electric field: $|\boldsymbol{H}| >> |\boldsymbol{E}|$ as $\boldsymbol{E} \propto \frac{1}{r^2}$ whereas $\boldsymbol{B} \propto \frac{1}{r^3}$. Hence the saying that the magnetic field dominates in the near field for magnetic loops. More important, observe that in the near field, the magnitude of the magnetic field is independent of the wavelength of the electromagnetic field whereas the magnitude of the electric field is directly proportional to its frequency (modulo the phase $e^{-i\beta r}$):

$$E_\phi = -\frac{i}{\omega\varepsilon}\frac{m}{4\pi} e^{-i\beta r} \sin\theta \frac{\beta^2}{r^2} = -i\frac{\mu}{4\pi} m \frac{\omega}{r^2} e^{-i\beta r} \sin\theta$$

Finally, recall that for a fixed distance r, near-zone approximation implies that $r << \lambda$, where λ is the wavelength of the electromagnetic field; this is independent of $|\delta S(\gamma)|$. The requirement that $|\delta S(\gamma)| << r^2$ follows from the derivation of Proposition 8.3.2.

8.3.6 Definition

Given a fixed Hertzian or magnetic loop antenna, let $P_s(\theta, \phi) \equiv r^2 \langle \boldsymbol{S} \rangle \cdot \mathbf{e}_r$ define the power per unit solid angle, and set $\bar{P}_r = \int_0^{2\pi} \int_0^{2\pi} P_s(\theta, \phi) \sin\theta \, d\theta d\phi$ to be the spatial average power radiated across a sphere. Then, the antenna *gain function* (or the *directive gain*) is given by

$$G(\theta,\phi) = \frac{4\pi P_s(\theta,\phi)}{\bar{P}_r} \tag{8.20}$$

Finally, the antenna *directivity* is defined by $D = \max_{\theta,\phi} G(\theta,\phi)$.

8.3.7 Example

Compare the gain and directivity between a Hertzian antenna and a loop antenna in the far zone. First, consider a Hertzian antenna $(J, \delta z \mathbf{e}_z)$. From Equations (8.8) and (8.9),

$$r^2\langle \mathbf{S}\rangle = \frac{r^2}{2\mu}\Re e(\mathbf{E}\times\mathbf{B}^*) = \mathbf{e}_z\frac{\beta^3}{(4\pi)^2\omega\varepsilon}(I\delta z)^2 e^{-i2\beta r}\sin^2\theta$$

From this, $\bar{P}_r = \int_0^{2\pi}\int_0^{\pi} P_s(\theta,\phi)\sin\theta\, d\theta d\phi = 2\pi\frac{\beta^3}{(4\pi)^2\omega\varepsilon}(I\delta z)^2 e^{-i2\beta r}\int_0^{\pi}\sin^3\theta d\theta$. Noting that $\int_0^{\pi}\sin^3\theta d\theta = \frac{4}{3}$, it follows that the Hertzian gain is $G_H(\theta,\phi) = \frac{3}{2}\sin^2\theta$. The Hertzian directivity is clearly $D_H = \frac{3}{2}$, as $\max_{\theta}\{\sin^2\theta\} = 1$.

To complete the example, the gain and directivity of a magnetic loop antenna $(I, \gamma, \delta S(\gamma))$ are evaluated below. However, by comparing Equations (8.15) and (8.16) with (8.8) and (8.9), it is obvious that their gains and directivities are identical. Explicitly, from (8.15) and (8.16),

$$r^2\langle \mathbf{S}\rangle = \frac{r^2}{2\mu}\Re e(\mathbf{E}\times\mathbf{B}^*) = \mathbf{e}_z\frac{m^2\beta^5}{(4\pi)^2 2\omega\varepsilon}e^{-i2\beta r}\sin^2\theta$$

and hence, $\bar{P}_r = \int_0^{2\pi}\int_0^{\pi} P_s(\theta,\phi)\sin\theta\, d\theta d\phi = 2\pi\frac{m^2\beta^5}{(4\pi)^2 2\omega\varepsilon}e^{-i2\beta r}\int_0^{\pi}\sin^3\theta d\theta$. Thus, the magnetic loop gain is $G_M(\theta,\phi) = \frac{3}{2}\sin^2\theta$ and the directivity $D_M = \frac{3}{2}$, as claimed. □

8.4 Microstrip Antenna: A Qualitative Overview

A thin electrical conductor on a printed circuit board is called a *trace*. Recall from Chapter 5 that a microstrip is defined to be a trace lying on the top layer of a PCB and bounded from below by a grounded plane (assumed here to be a perfect electrical conductor), whereas a stripline is defined to be a trace that is bounded from above and below by ground planes (or power planes). The dielectric medium above a microstrip is typically air. Figure 8.2 illustrates the difference between a microstrip and a stripline.

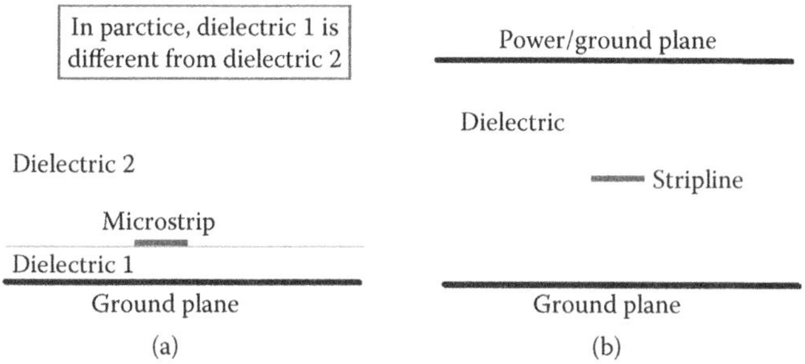

FIGURE 8.2.
Two kinds of traces on printed circuit boards: (a) microstrip, (b) stripline.

The far-field and near-field effects of a microstrip are analyzed below. The microstrip is assumed to be perfectly terminated for simplicity. This criterion can clearly be generalized to an arbitrary load. The length of the microstrip is ℓ and its height above its ground plane is h. In the interest of simplicity, it is further assumed that the dielectrics above and below the microstrip are identical and the trace is a thin solid cylinder of length ℓ.

The method of images is used to solve the far-field effects of a microstrip (*cf.* Figure 8.3). Let θ denote the angle between $\boldsymbol{r}$ and the microstrip at $z = 0$. Likewise, let $\boldsymbol{r}'$ denote the vector of the image microstrip at $z = 0$, and θ' the angle between $\boldsymbol{r}'$ and the image microstrip. Finally, the current $I = I_0 e^{-i\beta z}$ is assumed to travel in the +z-direction toward the termination (load). By definition, the image current travels in the opposite direction.

8.4.1 Proposition

Consider a matched microstrip of length ℓ shown in Figure 8.3 and suppose some current $I(z) = I_0 e^{-i\beta z}$ is flowing along the conductor, where the conductor is surrounded by a homogeneous dielectric medium (μ, ε), with I_0 being a constant. Then, in the far field, the electric field and magnetic field generated

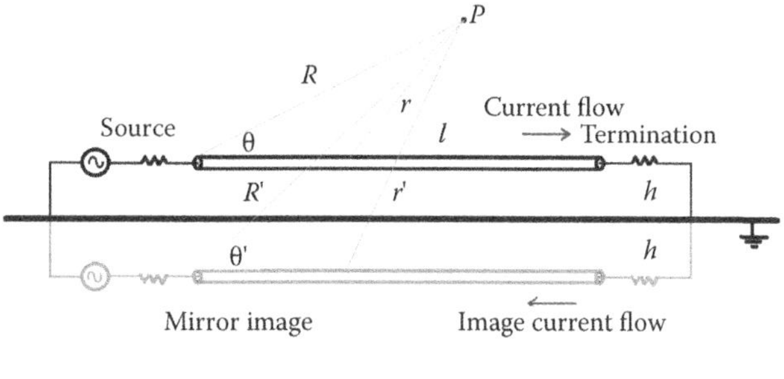

FIGURE 8.3.
The electric field of a microstrip in a homogeneous dielectric medium.

by the trace at a distance $r >> \ell$ such that $\frac{\ell}{r} << \sin^2\theta$, where the pair (r,θ) as defined in Figure 8.3 is fixed, are

$$\delta E_\theta(r,\theta) \approx \frac{I_0\beta}{\pi\varepsilon}\frac{\sin\theta}{r(\cos\theta-1)}e^{-i\beta r}e^{-i\frac{h\beta}{r}\sin\theta}e^{i\frac{1}{2}\beta(\cos\theta-1)\ell}\cos\left(\frac{\ell}{2}\beta(\cos\theta-1)\right)\sin\left(\frac{h\beta}{r}\sin\theta\right) \tag{8.21}$$

$$\delta B_\phi(r,\theta) \approx \frac{iI_0\beta}{4\pi}\eta\frac{\sin\theta}{r(\cos\theta-1)}e^{-i\beta r}\left(1-e^{i\beta(\cos\theta-1)\ell}\right)\left(1-e^{-i\frac{2h\beta}{r}\sin\theta}\right) \tag{8.22}$$

where $\eta = \sqrt{\mu/\varepsilon}$.

Proof

Without loss of generality, consider the top trace depicted in Figure 8.3 (modulo the ground plane). That is, consider the electric field from the microstrip, where the pair (r,θ) is fixed. From Equation (8.8), $dE_\theta(r,\theta) \approx \frac{i\beta^2}{4\pi\omega\varepsilon}\frac{Idz}{r}e^{-i\beta r}\sin\theta$, whence, $E_\theta(r,\theta) \approx \frac{i\beta^2 I_0}{4\pi\omega\varepsilon}\int_0^\ell \frac{1}{r}e^{-i\beta r}e^{-i\beta z}\sin\Theta dz$. To evaluate this integral, let $R^2 = r^2 + z^2 - 2rz\cos\theta$, where $0 \le z \le \ell$, and set $L = r\cos\theta >> \ell$. Then, by construction,

$$\cos\Theta \equiv \frac{L-z}{R} \Rightarrow \Theta = \arccos\left\{\frac{L-z}{\sqrt{r^2+z^2-2rz\cos\theta}}\right\}$$

Here, Θ is the angle between $\boldsymbol{R}$ and the microstrip. Then, recalling the identity $\sin(\arccos\phi) = \sqrt{1-\phi^2}$, it follows that

$$\sin\Theta = \sqrt{1-\frac{(L-z)^2}{r^2+z^2-2rz\cos\theta}} \approx \sqrt{1-\left(\frac{L}{r}\right)^2\left\{1-\frac{2z}{r}(1-\cos\theta)\right\}}$$

$$= \sqrt{1-\cos^2\theta\left\{1-\frac{2z}{r}(1-\cos\theta)\right\}}$$

to first order in $\frac{z}{r}$. In particular, by assumption,

$$\frac{z}{r} << \sin^2\theta \Rightarrow \sin\Theta \approx \sin\theta\left\{1+\frac{z}{r}(1-\cos\theta)\cot^2\theta\right\}$$

where the binomial approximation and the identity $1 = \cos^2\theta + \sin^2\theta$ were invoked.

Hence,

$$\int_0^\ell \frac{1}{R}e^{-i\beta(R+z)}\sin\Theta dz \approx \sin\theta\int_0^\ell \frac{1}{R}e^{-i\beta(R+z)}dz + \frac{\cos^2\theta}{r\sin\theta}(1-\cos\theta)\int_0^\ell \frac{z}{R}e^{-i\beta(R+z)}dz$$

up to first order in $\frac{z}{r}$. Moreover, appealing to the binomial approximation again,

$$R \approx r\left(1-\frac{z}{r}\cos\theta + o\left(\left(\frac{z}{r}\right)^2\right)\right) \quad \text{and} \quad \frac{1}{R} \approx \frac{1}{r}\left(1+\frac{z}{r}\cos\theta\right) + o\left(\left(\frac{z}{r}\right)^2\right)$$

Thus, evaluating the integrals, and noting that $\int z e^{az} dz = e^{az}\frac{1-az}{a^2}$,

$$\int_0^\ell \tfrac{1}{R} e^{-i\beta(R+z)}\,dz \approx \tfrac{1}{r}\int_0^\ell e^{-i\beta(R+z)}\left\{1+\tfrac{z}{r}\cos\theta\right\}dz \approx \tfrac{1}{r}e^{-i\beta r}\int_0^\ell e^{-i\beta(\cos\theta-1)z}\left\{1+\tfrac{z}{r}\cos\theta\right\}dz$$

$$\int_0^\ell \tfrac{z}{R} e^{-i\beta(R+z)}\,dz \approx \tfrac{1}{r}e^{-i\beta r}\int_0^\ell z e^{-i\beta(\cos\theta-1)z}\left\{1+\tfrac{z}{r}\cos\theta\right\}dz$$

Taking the approximation up to $\frac{1}{r^2}$,

$$\int_0^\ell \tfrac{1}{R} e^{-i\beta(R+z)}\sin\Theta dz \approx \tfrac{\sin\theta}{\beta r(\cos\theta-1)} e^{-i\beta r}\left\{1-e^{i\beta(\cos\theta-1)\ell}\right\}+$$

$$\left\{\sin\theta\cos\theta+(\cos\theta-1)\tfrac{\cos^2\theta}{\sin\theta}\right\}\tfrac{1}{\beta^2(\cos\theta-1)^2 r^2}e^{-i\beta r}\left\{e^{i\beta(\cos\theta-1)\ell}(1-i\beta(\cos\theta-1)\ell)-1\right\}$$

That is, the electric field in the far zone for the microstrip is approximated by

$$E_\theta(r,\theta) \approx \tfrac{iI_0\beta}{4\pi\varepsilon}\tfrac{\sin\theta}{r(\cos\theta-1)}e^{-i\beta r}\left\{1-e^{i\beta(\cos\theta-1)\ell}\right\}+$$

$$\tfrac{iI_0}{4\pi\varepsilon}\left\{\sin\theta\cos\theta+(\cos\theta-1)\tfrac{\cos^2\theta}{\sin\theta}\right\}\tfrac{1}{\beta^2(\cos\theta-1)^2 r^2}e^{-i\beta r}\left\{e^{i\beta(\cos\theta-1)\ell}(1-i\beta(\cos\theta-1)\ell)-1\right\}$$

Similarly, to determine the field from the image trace, it suffices to note that

$$r' = \sqrt{r^2+4h^2+4rh\sin\theta} \quad \text{and} \quad \theta' = \arccos\tfrac{L}{r'}$$

the image field can be obtained via: $r \to r', \theta \to \theta', I_0 \to -I_0$. Hence,

$$E_\theta(r',\theta') \approx -\tfrac{iI_0\beta}{4\pi\varepsilon}\tfrac{\sin\theta'}{r'(\cos\theta'-1)}e^{-i\beta r'}\left\{1-e^{i\beta(\cos\theta'-1)\ell}\right\}+$$

$$-\tfrac{iI_0}{4\pi\varepsilon}\left\{\sin\theta'\cos\theta'+(\cos\theta'-1)\tfrac{\cos^2\theta'}{\sin\theta'}\right\}\tfrac{1}{\beta^2(\cos\theta'-1)^2 r'^2}e^{-i\beta r'}\left\{e^{i\beta(\cos\theta'-1)\ell}(1-i\beta(\cos\theta'-1)\ell)-1\right\}$$

In particular, the resultant electric field at the point P is, to first order in $\frac{1}{r}$, which is equivalent to the condition $\beta r^* >> 1 \Leftrightarrow \frac{\beta}{r^*} >> \frac{1}{(r^*)^2}$, where $r^* \in \{r, r'\}$,

$$\delta E_\theta(r,\theta) = E_\theta(r,\theta) + E_\theta(r',\theta')$$

$$\approx \tfrac{iI_0\beta}{4\pi\varepsilon}\left\{\tfrac{\sin\theta}{r(\cos\theta-1)}e^{-i\beta r}\left(1-e^{i\beta(\cos\theta-1)\ell}\right)-\tfrac{\sin\theta'}{r'(\cos\theta'-1)}e^{-i\beta r'}\left(1-e^{i\beta(\cos\theta'-1)\ell}\right)\right\}$$

In turn, this can be simplified (see Exercise 8.6.4) to first order in $\frac{1}{r}$, to

$$\delta E_\theta(r,\theta) \approx \tfrac{iI_0}{\pi\varepsilon}\tfrac{\beta}{r}\tfrac{\sin\theta}{\cos\theta-1}e^{-i\beta r}e^{-i\frac{h\beta}{r}\sin\theta}e^{i\frac{1}{2}\beta(\cos\theta-1)\ell}\sin\left(\tfrac{\ell}{2}\beta(\cos\theta-1)\right)\sin\left(\tfrac{h\beta}{r}\sin\theta\right)$$

Finally, to complete the proof, it suffices to recall that $H_\phi = \frac{1}{\eta}E_\theta$, and hence, $\frac{1}{\eta}\frac{\mu}{\varepsilon} = \eta$ yields

$$\delta B_\theta(r,\theta) \approx \frac{iI_0}{\pi}\eta\frac{\beta}{r}\frac{\sin\theta}{\cos\theta-1}e^{-i\beta r}e^{-i\frac{h\beta}{r}\sin\theta}e^{i\frac{1}{2}\beta(\cos\theta-1)\ell}\sin\left(\tfrac{\ell}{2}\beta(\cos\theta-1)\right)\sin\left(\tfrac{h\beta}{r}\sin\theta\right) \quad \square$$

8.4.2 Remark

Indeed, for $0 < h \ll 1$, the pair $(\delta E_\theta, \delta B_\phi)$ can be further approximated by

$$\delta E_\theta(r,\theta) \approx \frac{iI_0 h}{\pi\varepsilon}\frac{\beta^2}{r^2}\frac{\sin^2\theta}{\cos\theta-1}e^{-i\beta r}e^{-i\frac{h\beta}{r}\sin\theta}e^{i\frac{1}{2}\beta(\cos\theta-1)\ell}\sin\left(\tfrac{1}{2}\beta(\cos\theta-1)\ell\right)$$

via $\sin\left(\frac{h\beta}{r}\sin\theta\right) \approx \frac{h\beta}{r}\sin\theta + o\left(\left(\frac{\beta}{r}\right)^3\right)$. Likewise,

$$\delta B_\theta(r,\theta) \approx \frac{iI_0 h}{\pi}\eta\frac{\beta^2}{r^2}\frac{\sin^2\theta}{\cos\theta-1}e^{-i\beta r}e^{-i\frac{h\beta}{r}\sin\theta}e^{i\frac{1}{2}\beta(\cos\theta-1)\ell}\sin\left(\tfrac{1}{2}\beta(\cos\theta-1)\ell\right)$$

From this, it is clear that a microstrip is a poor radiator, as the field falls off as $\frac{1}{r^2}$; and by inference, a differential pair is also a poor radiator. Furthermore, it is clear that along the axis of the microstrip, the fields are zero, that is, when $\theta = 0$, whereas the fields are maximal when $\theta = \frac{\pi}{2}$. Finally, observe that for $\theta \neq 0$, $\sin\left(\frac{1}{2}\beta(\cos\theta-1)\ell\right) = 0 \Leftrightarrow \frac{1}{2}\beta(\cos\theta-1)\ell = n\pi$ for n = 1,2, … . That is, from $\beta = \omega\sqrt{\mu\varepsilon}$, $\delta E_\theta = 0 = \delta B_\phi \Leftrightarrow \omega = \frac{2\pi n}{\sqrt{\mu\varepsilon}(\cos\theta-1)\ell}$ $\forall n = 0,1,2,\ldots$.

8.5 Array Antenna and Aperture Antenna

This chapter ends with a brief account of the array antenna and aperture antenna. From Section 8.4, it is clear that a linear antenna has the form:

$$\boldsymbol{E} = \boldsymbol{E}_0 F(\theta,\phi)\frac{e^{-i\beta r}}{r} \tag{8.23}$$

where $F(\theta,\phi)$ is the *antenna factor*, and $\boldsymbol{E}_0$ is the initial field strength. The antenna factor determines the antenna characteristics that depend solely upon the antenna geometry. This forms the basis for analyzing an antenna array structure.

Now, Figure 8.4 illustrates a *linear* $(n + 1)$-array antenna, where the k-antenna has current $I_k = C_k I_0 e^{ik\xi}$ $\forall k = 0,1,2,\ldots$, with some fixed phase $\xi > 0$, constant $C_k > 0$, with $C_0 = 1$, and the antennae are separated by a constant distance d. In particular, Equation (8.23) becomes $\boldsymbol{E}_k = \boldsymbol{E}_{0,k}F(\theta,\phi)\frac{e^{-i\beta r}}{r}$, where $\boldsymbol{E}_{0,k} = C_k \boldsymbol{E}_0 e^{ik\xi}$. Denote a linear $(n + 1)$-array antenna of length ℓ by $\{(I_k, \theta_k, d, \xi, \ell)\}_n$.

8.5.1 Theorem

Given a linear $(n+1)$-antenna array $\{(I_k, \theta_k, d, \xi, \ell)\}_n$, suppose that $L = r_0 \cos\theta_0 >> nd$. Then, the cross-sectional far-field profile of the electric field, to second order in $\frac{1}{r_0}$, satisfies

$$\boldsymbol{E} \approx E_0 F(\theta, \phi) \frac{e^{-i\beta r_0}}{r_0} \sum_{k=0}^{n} C_k e^{ik(\beta d \cos\theta_0 + \xi)} \left\{1 + \frac{kd}{r_0} \cos\theta_0\right\} \tag{8.24}$$

Proof

From Figure 8.4, it is clear that $r_k^2 = r_0^2 + (kd)^2 - 2kr_0 d \cos\theta_0 \; \forall k$ and hence,

$$r_k \approx r_0 \sqrt{1 - \frac{2kd}{r_0} \cos\theta_0} \approx r_0 \left(1 - \frac{kd}{r_0} \cos\theta_0\right)$$

and

$$\frac{1}{r_k} \approx \frac{1}{r_0} \left(1 - \frac{2kd}{r_0} \cos\theta_0\right)^{-\frac{1}{2}} \approx \frac{1}{r_0} \left(1 + \frac{kd}{r_0} \cos\theta_0\right)$$

Likewise, by assumption,

$$L >> nd \Rightarrow \cos\theta_k = \frac{r_k}{L - kd} \approx \frac{r_k}{L} \left(1 + \frac{kd}{L}\right) \approx \cos\theta_0 \left\{1 + kd\left(\frac{1}{L} - \frac{\cos\theta_0}{r_0}\right)\right\}$$

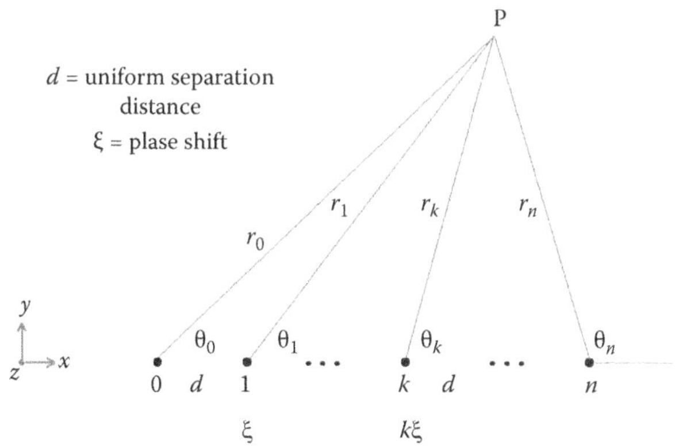

FIGURE 8.4.
Linear array of n-antenna.

Thus, $L = r_0 \cos\theta_0 \Rightarrow \cos\theta_k \approx \cos\theta_0 \left\{1 + \frac{kd}{r_0}\sin\theta_0 \tan\theta_0\right\}$ as

$$\frac{1}{\cos\theta_0} - \cos\theta_0 = \frac{1-\cos^2\theta_0}{\cos\theta_0} = \frac{\sin^2\theta_0}{\cos\theta_0} = \sin\theta_0 \tan\theta_0$$

Moreover, noting that

$$\beta r_k - k\xi \approx \beta r_0 - \beta kd\cos\theta_0 - k\xi = \beta r_0 - k(\beta d\cos\theta_0 + \xi)$$

it follows at once that on setting $\boldsymbol{E}_k = C_k E_0 \boldsymbol{n}_k \equiv E_0 \boldsymbol{C}_k$ with $\boldsymbol{E}_0 = E_0 \boldsymbol{n}_0 \equiv E_0 \boldsymbol{C}_0$, where $\boldsymbol{n}_k = \frac{1}{r_k}\boldsymbol{r}_k$ are unit vectors,

$$\boldsymbol{E} = \sum_{k=0}^{n} \boldsymbol{E}_k \approx E_0 F(\theta,\phi) \sum_{k=0}^{n} \boldsymbol{C}_k \frac{1}{r_k} e^{-i\beta r_0} e^{ik(\beta d\cos\theta_0 + \xi)}$$

$$\approx E_0 F(\theta,\phi) \frac{e^{-i\beta r_0}}{r_0} \sum_{k=0}^{n} \boldsymbol{C}_k e^{ik(\beta d\cos\theta_0 + \xi)} \left\{1 + \frac{kd}{r_0}\cos\theta_0\right\}$$

□

8.5.2 Remark

Observe that because $\min r_k >> nd$, it follows that $\boldsymbol{n}_k \approx \boldsymbol{n}_0$ and hence, Equation (8.24) reduces to $\boldsymbol{E} \approx \boldsymbol{E}_0 F(\theta,\phi) \frac{e^{-i\beta r_0}}{r_0} \sum_{k=0}^{n} C_k e^{ik(\beta d\cos\theta_0 + \xi)} \left\{1 + \frac{kd}{r_0}\cos\theta_0\right\}$. The quantity

$$|A| = \left| \sum_{k=0}^{n} C_k e^{ik(\beta d\cos\theta_0 + \xi)} \left\{1 + \frac{kd}{r_0}\cos\theta_0\right\} \right|$$

is called the *array factor* up to first order in $\frac{1}{r_0}$. More commonly, the array factor is often defined as the zeroth order of $\frac{1}{r_0}$:

$$|A| = \left| \sum_{k=0}^{n} C_k e^{ik(\beta d\cos\theta_0 + \xi)} \right|$$

8.5.3 Corollary

Under the conditions of Theorem 8.5.1, if $C_k = C_0\ \forall k$ and $e^{ik(\beta d\cos\theta_0 + \xi)} \neq 1\ \forall k > 0$, then to first order in $\frac{1}{r_0}$,

$$\boldsymbol{E} \approx C_0 \boldsymbol{E}_0 F(\theta,\varphi) \frac{e^{-i\beta r_0}}{r_0} e^{i\frac{n}{2}(\beta d\cos\theta_0 + \xi)} \frac{\sin\left\{\frac{1}{2}(n+1)(\beta d\cos\theta_0 + \xi)\right\}}{\sin\left\{\frac{1}{2}(\beta d\cos\theta_0 + \xi)\right\}} \tag{8.25}$$

Proof

From Theorem 8.5.1 and Remark 8.5.2, $\boldsymbol{E} \approx C_0 E_0 F(\theta,\phi)\frac{e^{-i\beta r_0}}{r_0}\sum_{k=0}^{n} e^{ik(\beta d\cos\theta_0+\xi)}$. Next, noting that $\sum_{k=0}^{n} e^{ik(\beta d\cos\theta_0+\xi)} = 1 + e^{i(\beta d\cos\theta_0+\xi)} + \cdots + e^{in(\beta d\cos\theta_0+\xi)}$ defines a geometric series: $1 + z \cdots + z^{n-1} = \frac{z^n-1}{z-1}$ $\forall z \in \mathbf{C}$, $z \neq 1, n < \infty$, it follows that

$$\sum_{k=0}^{n} e^{ik(\beta d\cos\theta_0+\xi)} = \frac{e^{i(n+1)(\beta d\cos\theta_0+\xi)} - 1}{e^{i(\beta d\cos\theta_0+\xi)} - 1}$$

$$= \frac{e^{i(n+1)(\beta d\cos\theta_0+\xi)/2}}{e^{i(\beta d\cos\theta_0+\xi)/2}} \frac{e^{i(n+1)(\beta d\cos\theta_0+\xi)/2} - e^{-i(n+1)(\beta d\cos\theta_0+\xi)/2}}{e^{i(\beta d\cos\theta_0+\xi)/2} - e^{-i(\beta d\cos\theta_0+\xi)/2}}$$

$$= e^{i\frac{n}{2}(\beta d\cos\theta_0+\xi)} \frac{\sin\left\{\frac{1}{2}(n+1)(\beta d\cos\theta_0+\xi)\right\}}{\sin\left\{\frac{1}{2}(\beta d\cos\theta_0+\xi)\right\}}$$

and hence,

$$\boldsymbol{E} \approx C_0 E_0 F(\theta,\varphi)\frac{e^{-i\beta r_0}}{r_0} e^{i\frac{n}{2}(\beta d\cos\theta_0+\xi)} \frac{\sin\left\{\frac{1}{2}(n+1)(\beta d\cos\theta_0+\xi)\right\}}{\sin\left\{\frac{1}{2}(\beta d\cos\theta_0+\xi)\right\}} \qquad \square$$

Note finally from Corollary 8.5.3 that by measuring the distance from the center of the linear array to the point of interest, that is, on setting $\bar{r} = r_0 - \frac{1}{2} nd\cos\theta_0$ (*cf.* Figure 8.5), Equation (8.25) simplifies to

$$\boldsymbol{E} \approx C_0 E_0 F(\theta,\phi)\frac{e^{-i\beta\bar{r}}}{\bar{r}} e^{i\frac{n}{2}\xi} \frac{\sin\left\{\frac{1}{2}(n+1)(\beta d\cos\theta_0+\xi)\right\}}{\sin\left\{\frac{1}{2}(\beta d\cos\theta_0+\xi)\right\}} \tag{8.26}$$

Lastly, the magnetic field follows immediately from Maxwell's equation: $\nabla \times \boldsymbol{E} = i\omega \boldsymbol{B}$, where the convention $e^{-i\omega t}$ is adopted here for time harmonicity instead of $e^{i\omega t}$.

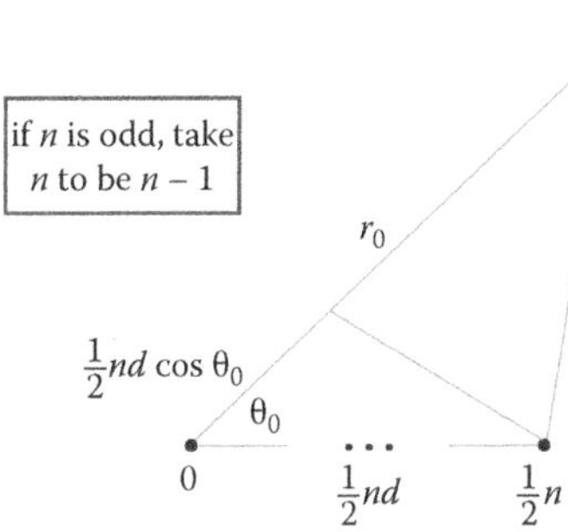

FIGURE 8.5.
Approximating distance in a linear array antenna.

8.5.4 Example

Determine the array factor of a linear 2-antenna array, where we assume for simplicity that $C_0 = C_1$. By Remark 8.5.2,

$$
\begin{aligned}
|A| &= C_0\left|\sum_{k=0}^{1} e^{ik(\beta d\cos\theta_0+\xi)}\right| = C_0\left|1+e^{i(\beta d\cos\theta_0+\xi)}\right| \\
&= C_0\left|2e^{i\frac{1}{2}(\beta d\cos\theta_0+\xi)}\,\tfrac{1}{2}\left\{e^{-i\frac{1}{2}(\beta d\cos\theta_0+\xi)}+e^{i\frac{1}{2}(\beta d\cos\theta_0+\xi)}\right\}\right| \\
&= 2C_0\left|\cos\left(\tfrac{1}{2}(\beta d\cos\theta_0+\xi)\right)\right|
\end{aligned}
$$

It is clear from the above that $A = 0 \Leftrightarrow \beta d\cos\theta_0 + \xi = (2n-1)\pi\ \forall n = 1,2,\ldots$. Thus, if ω is fixed, then $\xi = (2n-1)\pi - \beta d\cos\theta_0$ leads to far-field cancellation. Likewise, if the phase ξ is fixed, then transmitting at

$$\omega = \tfrac{(2n-1)\pi-\xi}{\sqrt{\mu\varepsilon}\,d\cos\theta_0}$$

will also lead to far-field cancellation. Conversely, A is maximal if and only if $\beta d\cos\theta_0 + \xi = 2n\pi\ \forall n = 1,2,\ldots$. □

8.5.5 Definition

The *irradiance* (or *intensity*) of an electromagnetic field (**E**,**B**) is defined by

$$I \equiv \langle S\rangle = \tfrac{1}{2\mu_0}\,|\mathbf{E}\times\mathbf{B}^*| = \tfrac{1}{2\mu_0}\,|E|^2 \tag{8.27}$$

Thus, in view of Example 8.5.4, the intensity is inversely proportional to r^2 and directly proportional to $|A|^2$, the array factor. In contrast, the electric field and the magnetic field fall off—in the far-field regime—as $\frac{1}{r}$ and their magnitudes are directly proportional to $|A|$.

8.5.6 Proposition

Given a linear $(n + 1)$-antenna array $\{(I_k,\theta_k,d,\xi,\ell)\}_n$ with $L = r_0\cos\theta_0 >> nd$, up to first order in $\frac{1}{\bar{r}}$, where $\bar{r} = r_0 - \frac{1}{2}nd\cos\theta_0$, set $\psi = \beta d\cos\theta_0 + \xi$. If $I_k = I_0\ \forall k$, then

(a) The maximal field strength occurs when $\psi = 0$.
(b) The minimal field strength occurs when $\psi = \frac{2k\pi}{n+1}$, for all $k > 0$.

Proof

To establish the assertions, it suffices to consider the array factor $|A|$. To simplify the analysis, note first of all that $I_k = I_0\ \forall k \Rightarrow C_k = C_0\ \forall k$. Hence, the normalized array factor $\tilde{A}$ can be defined from Equation (8.26):

$$|\tilde{A}| = \frac{1}{n+1}\left|\frac{\sin\{\frac{1}{2}(n+1)\psi\}}{\sin\{\frac{1}{2}\psi\}}\right|$$

Then, via Taylor expansion,

$$\sin x = x - \tfrac{1}{3!}x^3 + o(x^5)$$

it follows at once that

$$\lim_{\psi\to 0}|\tilde{A}| = \lim_{\psi\to 0}\frac{1}{n+1}\left|\frac{\sin\{\frac{1}{2}(n+1)\psi\}}{\sin\{\frac{1}{2}\psi\}}\right| = \lim_{\psi\to 0}\frac{1}{n+1}\left|\frac{(n+1)\{\frac{1}{2}\psi+o(\psi^3)\}}{\frac{1}{2}\psi+o(\psi^3)}\right| = 1$$

and hence, yielding the maximal field strength as claimed in (a). Regarding (b), it suffices to note that

$$|\tilde{A}| = 0 \Leftrightarrow \tfrac{1}{2}(n+1)\psi = k\pi \Rightarrow \psi = \tfrac{2k\pi}{n+1} \text{ for all } k > 0$$

as required. □

A linear antenna array factor, for $n = 10$, is plotted in Figure 8.6 to illustrate the maxima and minima.

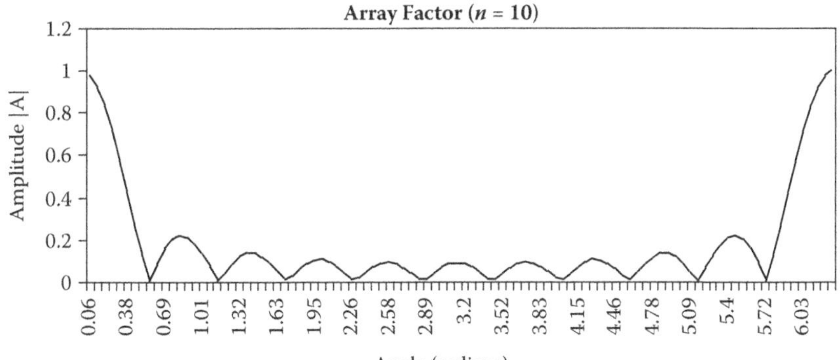

FIGURE 8.6.
The antenna array factor for 10 linear radiators.

The remainder of this section is devoted to a short survey on aperture antennae. As chassis for electronic devices often have apertures, it is important for EMC engineers to understand the basics of aperture radiators. Indeed, the essence of aperture radiators originates from the application of scalar diffraction theory; specifically, the Kirchhoff integral theorem, which is a particular case of Green's theorem, stated below without proof [4].

8.5.7 Theorem (Kirchhoff)

Given some compact neighborhood $V \subset \mathbf{R}^3$ of an arbitrary point r_0, suppose a scalar field $U: V \to \mathbf{R}$ satisfies the *homogeneous Helmholtz* equation $\Delta U + k^2 U = 0$ on V. Set $G(\boldsymbol{r}) = \frac{1}{r} e^{ik\cdot r}$ and let $\boldsymbol{n}$ be a unit normal vector field on ∂V. Then,

$$U(\boldsymbol{r}) = \frac{1}{4\pi} \int_{\partial V} \{G(\boldsymbol{r}')\partial_n U(\boldsymbol{r}') - U(\boldsymbol{r}')\partial_n G(\boldsymbol{r}')\} d^2\boldsymbol{r}' \tag{8.28}$$

Moreover, if the requirement that $V \subset \mathbf{R}^3$ be compact is relaxed, then U must satisfy the boundary condition known as the *Sommerfeld radiation condition* $\lim_{r\to\infty} r\{\partial_n U - ikU\} = 0$, in order for the integral to converge. □

An immediate application of Theorem 8.5.7 is to consider an infinite conductive plane $\mathbf{R}^2$ at $z = 0$, with a single rectangular aperture $\Omega = \left[-\frac{1}{2}\delta, \frac{1}{2}\delta\right] \times \left[-\frac{1}{2}\ell, \frac{1}{2}\ell\right]$, where $\delta \ll \ell$, that is, a *slit*. For simplicity, set $\mathbf{R}^3_+ = \{(x,y,z) \in \mathbf{R}^3 : z \geq 0\}$ and $\mathbf{R}^3_- = \mathbf{R}^3 - \mathbf{R}^3_+$. And as a simple example, consider a single point source (and the field is thus necessarily spherical) at $\boldsymbol{R} \in \mathbf{R}^3_-$ and an arbitrary point $\boldsymbol{r}' \in \Omega$, where by an abuse of notation, Ω is identified with $\Omega \times \{0\}$. Finally, for notational simplicity, let U be a scalar field satisfying the Helmholtz equation, $\Delta U + k^2 U = 0$, where $k = \omega\sqrt{\mu\varepsilon}$ is the wave number, and let $\boldsymbol{n}$ denote a unit normal vector field on Ω directed into $\mathbf{R}^3_-$.

8.5.8 Lemma (Fresnel–Kirchhoff Diffraction Formula)

Given a homogeneous medium $(\mathbf{R}^3, \mu, \varepsilon)$, a unit point source at $\boldsymbol{R} \in \mathbf{R}^3_-$ and a slit $\Omega \subset \partial\mathbf{R}^3_+$, suppose that $\partial\mathbf{R}^3_+ - \Omega$ is a perfect electrical conductor. Then, the field U at an arbitrary point $\boldsymbol{r} \in \mathbf{R}^3_+$ satisfying $kr, kR \gg 1$* may be approximated by

$$U(\boldsymbol{r}) = -\frac{i}{\lambda} \int_{\Omega} \frac{1}{\bar{r}R} e^{ik(\bar{r}+\bar{R})} \frac{1}{2}(\cos\theta - \cos\Theta) dx' dy' \tag{8.29}$$

where $\lambda = \frac{2\pi}{\omega\sqrt{\mu\varepsilon}}$, $\bar{\boldsymbol{r}} = \boldsymbol{r} - \boldsymbol{r}' = (x - x', y - y', z)$ with $\boldsymbol{r}' = (x', y', 0) \in \Omega$ being an arbitrary point, with $\cos\theta = \frac{1}{r}(\bar{\boldsymbol{r}} \cdot \boldsymbol{n})$ and $\cos\Theta = \frac{1}{R}(\bar{\boldsymbol{R}} \cdot \boldsymbol{n})$.

* That is, in the far-field regime.

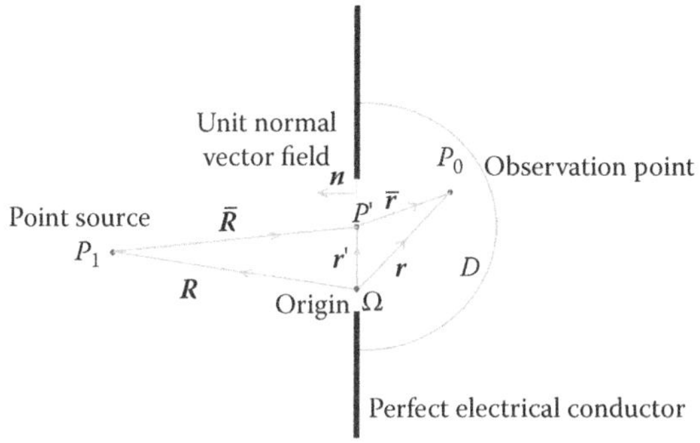

FIGURE 8.7.
Derivation of the Fresnel–Kirchhoff diffraction formula.

Proof

Theorem 8.5.7 is applied twice to obtain Equation (8.29); see Figure 8.7 for details. Here, let the point source be at $P_1 = (X, Y, Z)$ and an arbitrary observation point $P_0 = (x, y, z)$. Finally, let $P' = (x', y', 0)$ be an arbitrary point on Ω, $D \subset \mathbf{R}_+^3$ an arbitrary bounded neighborhood of P_0 such that $\Omega \subset \partial D \cap \partial \mathbf{R}_+^3$, and $\boldsymbol{n}$ a unit normal vector field on $\partial D \cap \Omega$.

Now, appealing to Equation (8.28), the field at P_0 from a point source at P' is:

$$U(\boldsymbol{r}) = \tfrac{1}{4\pi}\int_{\partial D}\left\{\tfrac{e^{ik\bar{r}}}{\bar{r}}\partial_n U(\boldsymbol{r}') - U(\boldsymbol{r}')\partial_n \tfrac{e^{ik\bar{r}}}{\bar{r}}\right\} d^2\boldsymbol{r}'$$

Now, by definition, $\partial_n G \equiv \nabla G \cdot \boldsymbol{n}$, and

$$\nabla \tfrac{1}{\bar{r}} e^{ik\bar{r}} = e^{ik\bar{r}} \nabla \tfrac{1}{\bar{r}} + \tfrac{1}{\bar{r}} \nabla e^{ik\bar{r}} = -e^{ikr} \tfrac{1}{\bar{r}^2} \tfrac{1}{\bar{r}} \bar{\boldsymbol{r}} + \tfrac{ik}{\bar{r}} e^{ik\bar{r}} \tfrac{1}{\bar{r}} \bar{\boldsymbol{r}} = \tfrac{1}{\bar{r}} e^{ik\bar{r}} \left(-\tfrac{1}{\bar{r}} + ik\right) \tfrac{1}{\boldsymbol{r}} \bar{\boldsymbol{r}}$$

whence $k\bar{r} >> 1 \Leftrightarrow k >> \frac{1}{\bar{r}}$ implies that $\nabla \frac{1}{\bar{r}} e^{ik\bar{r}} \approx ik \frac{1}{\bar{r}} e^{ik\bar{r}} \frac{1}{\bar{r}} \bar{\boldsymbol{r}}$, yielding

$$\begin{aligned} U(\boldsymbol{r}) &\approx \tfrac{1}{4\pi}\int_{\partial D - \Omega}\left\{\tfrac{e^{ik\bar{r}}}{\bar{r}}\partial_n U(\bar{\boldsymbol{r}}) - U(\bar{\boldsymbol{r}}) ik \tfrac{1}{\bar{r}} e^{ik\bar{r}} \tfrac{1}{\bar{r}} \bar{\boldsymbol{r}} \cdot \boldsymbol{n}\right\} d^2\bar{r} \\ &= \tfrac{1}{4\pi}\int_{\partial D - \Omega} \tfrac{e^{ik\bar{r}}}{\bar{r}} \left\{\partial_n U(\bar{\boldsymbol{r}}) - U(\bar{\boldsymbol{r}}) ik \tfrac{1}{\bar{r}} \bar{\boldsymbol{r}} \cdot \boldsymbol{n}\right\} \bar{r}^2 \sin\bar{\theta}\, d\bar{\theta}\, d\bar{\phi} \\ &= \tfrac{1}{4\pi}\int_{\partial D - \Omega} e^{ik\bar{r}} \left\{\partial_n U(\bar{\boldsymbol{r}}) - U(\bar{\boldsymbol{r}}) ik \tfrac{1}{\bar{r}} \bar{\boldsymbol{r}} \cdot \boldsymbol{n}\right\} \bar{r} \sin\bar{\theta}\, d\bar{\theta}\, d\bar{\phi} \end{aligned}$$

Now, invoking the Sommerfeld radiation condition,

$$\lim_{\bar{r}\to\infty} U(r) = \lim_{\bar{r}\to\infty} \tfrac{1}{4\pi}\int_{\partial D - \Omega} e^{ik\bar{r}} \bar{r} \left\{\partial_n U(\bar{\boldsymbol{r}}) - U(\bar{\boldsymbol{r}}) ik \tfrac{1}{\bar{r}} \bar{\boldsymbol{r}} \cdot \boldsymbol{n}\right\} \sin\bar{\theta}\, d\bar{\theta}\, d\bar{\phi} = 0$$

That is, the integral on the boundary at infinity vanishes, and hence, the only field contribution arises on Ω. Thus, the surface integral reduces to integration on the aperture Ω:

$$U(\boldsymbol{r}) \approx \tfrac{1}{4\pi}\int_{\Omega} \tfrac{1}{r'}e^{ikr'}\left\{\partial_n U(\boldsymbol{r}') - U(\boldsymbol{r}')ik\tfrac{1}{r'}\boldsymbol{r}'\cdot\boldsymbol{n}\right\}d^2r$$

Set $\frac{1}{r'}\boldsymbol{r}'\cdot\boldsymbol{n} = \cos\theta$. Then, $U(\boldsymbol{r}) \approx \frac{1}{4\pi}\int_{\Omega}\frac{1}{r'}e^{ikr'}\{\partial_n U(\boldsymbol{r}') - ikU(\boldsymbol{r}')\cos\theta\}d^2r$.

Next, given that a point source originates at P_1, it follows that $U(\bar{\boldsymbol{R}}) = \frac{1}{R}e^{ik\bar{R}}$ as a point source generates spherical waves. Therefore substituting this into the integral yields:

$$U(\boldsymbol{r}) \approx \tfrac{1}{4\pi}\int_{\Omega}\tfrac{1}{r'}e^{ikr'}\left\{\partial_n\left(\tfrac{1}{\bar{R}}e^{ik\bar{R}}\right) - ik\tfrac{1}{\bar{R}}e^{ik\bar{R}}\cos\theta\right\}d^2r$$

Moreover, applying the assumption that $k\bar{R} >> 1$, $\nabla\frac{1}{\bar{R}}e^{ik\bar{R}} \approx ik\frac{1}{\bar{R}}e^{ik\bar{R}}\frac{1}{\bar{R}}\bar{\boldsymbol{R}}$ and on setting $\frac{1}{\bar{R}}\bar{\boldsymbol{R}}\cdot\boldsymbol{n} = \cos\Theta$, it follows at once that $U(\boldsymbol{r}) \approx \frac{ik}{4\pi}\int_{\Omega}\frac{1}{\bar{r}}e^{ik\bar{r}}\frac{1}{\bar{R}}e^{ik\bar{R}}$ $\{\cos\Theta - \cos\theta\}d^2r$, as desired, where $k = \frac{2\pi}{\lambda}$. □

It is clear that the above result yields good approximation to radiation wavelengths that are at least as short as the optical wavelengths. That is, it will not yield good results for radiation in the microwave regime. Unfortunately, the microwave range is often the range of interest for EMC engineers; in particular, the source is often close to apertures and hence, the criterion $kR >> 1$ is violated. However, this can be partially rectified from the proof of Lemma 8.5.8.

8.5.9 Lemma

Given a homogeneous medium $(\mathbf{R}^3,\mu,\varepsilon)$, a point source of magnitude U_0 at $R \in \mathbf{R}^3_-$, and a slit $\Omega \subset \partial\mathbf{R}^3_+$, suppose that $\partial\mathbf{R}^3_+ - \Omega$ is a perfect electrical conductor. Then, the field U at an arbitrary point $\boldsymbol{r} \in \mathbf{R}^3_+$ satisfying $kr >> 1$ may be approximated by

$$U(\boldsymbol{r}) = -\tfrac{iU_0}{\lambda}\int_{\Omega}\tfrac{1}{\bar{r}\bar{R}}e^{ik(\bar{r}+\bar{R})}\tfrac{1}{2}\left\{\cos\theta - \left(\tfrac{i}{k\bar{R}}+1\right)\cos\Theta\right\}dx'dy' \tag{8.29}$$

Proof

$$\partial_n\left(\tfrac{1}{\bar{R}}e^{ik\bar{R}}\right) = \tfrac{1}{\bar{R}}e^{ik\bar{R}}\left(-\tfrac{1}{\bar{R}} + ik\right)\tfrac{1}{\bar{R}}\bar{\boldsymbol{R}}\cdot\boldsymbol{n} = \tfrac{ik}{\bar{R}}e^{ik\bar{R}}\left(\tfrac{i}{k\bar{R}}+1\right)\cos\Theta$$

and replace $\frac{e^{ik\xi}}{\xi} \to U_0\frac{e^{ik\xi}}{\xi}$. □

Clearly, if the source is not a spherical radiator, then the integral $U(\boldsymbol{r}) \approx \frac{1}{4\pi}\int_{\Omega}\frac{1}{r'}e^{ikr'}\{\partial_n U - ikU\cos\theta\}d^2r$ must be evaluated for arbitrary source U. An example is if a chassis possesses a slit aperture, and the source is nonspherical. Then, this integral must be evaluated instead of appealing to Equation (8.29); often the integral can only be solved numerically.

However, there is another formalism that lends itself more accessiblly to computation. The proof can be found, for example, in [4], or more informally in [5]. This is the *Huygens–Fresnel principle* and it essentially states every point on an unobstructed wavefront is a secondary spherical source with the same frequency.* The precise statement is given without proof below.

8.5.10 Theorem (Huygens–Fresnel)

Given a homogeneous medium $(\mathbf{R}^3, \mu, \varepsilon)$ and a slit $\Omega \subset \partial\mathbf{R}^3_+$, suppose that $\partial\mathbf{R}^3_+ - \Omega$ is a perfect electrical conductor. Then, the field U at an arbitrary point $\boldsymbol{r} \in \mathbf{R}^3_+$ satisfying $kr >> 1$ may be approximated by

$$U(\boldsymbol{r}) = -\tfrac{\mathrm{i}}{\lambda}\int_{\Omega}\tfrac{1}{\bar{r}}\mathrm{e}^{\mathrm{i}k\bar{r}}U(x', y')\cos\theta\,\mathrm{d}x'\mathrm{d}y' \tag{8.30}$$

where $U|\Omega$ is the secondary source at each point on Ω, and $\cos\theta = \frac{1}{\bar{r}}\bar{\boldsymbol{r}}\cdot\boldsymbol{n}$. □

8.5.11 Example

Suppose a constant plane wave $U = U_0\mathrm{e}^{-\mathrm{i}\beta z}$ in $\mathbf{R}^3_-$ is incident on Ω.† Determine the field strength at $P_0 = (r_0, \theta_0, z_0)$, if $r_0 >> r' = \sqrt{x'^2 + y'^2}$ with $(x', y', 0) \in \Omega$, where $\delta, \ell > 0$ are arbitrary, that is, not necessarily a slit. From Figure 8.7,

$$\bar{r} = \sqrt{(x_0 - x')^2 + (y_0 - y')^2 + z_0^2} = r_0\sqrt{1 + \frac{2(x_0x' + y_0y')}{r_0^2} + \frac{x'^2 + y'^2}{r_0^2}} \approx r_0\sqrt{1 + \frac{2(x_0x' + y_0y')}{r_0^2}}$$

whence

$$\bar{r} \approx r_0\left(1 + \frac{x_0x' + y_0y'}{r_0^2}\right) \quad \text{and} \quad \frac{1}{\bar{r}} \approx \frac{1}{r_0}\left(1 - \frac{x_0x' + y_0y'}{r_0^2}\right)$$

Lastly, noting that $\cos\bar{\theta} = \frac{z_0}{\bar{r}}$ and $\cos\theta_0 = \frac{z_0}{r_0}$, it follows immediately that $\cos\bar{\theta} = \cos\theta_0\frac{r_0}{\bar{r}} \approx \cos\theta_0\left(1 - \frac{x_0x' + y_0y'}{r_0^2}\right)$. So, substituting these approximations into Equation (8.30),

$$U(r_0) = -\tfrac{\mathrm{i}}{\lambda}U_0\mathrm{e}^{-\mathrm{i}\beta z}\int_{\Omega}\tfrac{1}{\bar{r}}\mathrm{e}^{\mathrm{i}k\bar{r}}\cos\bar{\theta}\,\mathrm{d}x'\mathrm{d}y'$$

$$\approx -\tfrac{\mathrm{i}}{\lambda}U_0\mathrm{e}^{-\mathrm{i}\beta z}\tfrac{1}{r_0}\mathrm{e}^{\mathrm{i}kr_0}\cos\theta_0\int_{\Omega}\mathrm{e}^{\mathrm{i}kx_0x'/r_0}\mathrm{e}^{\mathrm{i}ky_0y'/r_0}\left(1 - \frac{x_0x' + y_0y'}{r_0^2}\right)^2\mathrm{d}x'\mathrm{d}y'$$

* Physically, this assertion is clearly false as waves cannot generate more waves on their own; however, the principle yields a decent approximation for most engineering purposes.

† This is equivalent to a point source at infinity.

$$\approx -\tfrac{i}{\lambda}U_0 e^{-i\beta z}\tfrac{1}{r_0}e^{ikr_0}\cos\theta_0\int_\Omega e^{ikx_0x'/r_0}e^{iky_0y'/r_0}\left(1-\tfrac{2(x_0x'+y_0y')}{r_0^2}\right)dx'dy'$$

$$\approx -\tfrac{i}{\lambda}U_0 e^{-i\beta z}\tfrac{1}{r_0}e^{ikr_0}\cos\theta_0\int_\Omega e^{ikx_0x'/r_0}e^{iky_0y'/r_0}\,dx'dy'+$$

$$\tfrac{i}{\lambda}U_0 e^{-i\beta z}\tfrac{1}{r_0}e^{ikr_0}\cos\theta_0\int_\Omega e^{ikx_0x'/r_0}e^{iky_0y'/r_0}\,\tfrac{2(x_0x'+y_0y')}{r_0^2}dx'dy'$$

$$= -\tfrac{i}{\lambda}U_0 e^{-i\beta z}\tfrac{1}{r_0}e^{ikr_0}\cos\theta_0\int_\Omega e^{ikx_0x'/r_0}e^{iky_0y'/r_0}\,dx'dy'+$$

$$\tfrac{i}{\lambda}U_0 e^{-i\beta z}\tfrac{1}{r_0}e^{ikr_0}\cos\theta_0\tfrac{2x_0}{r_0^2}\int_\Omega e^{ikx_0x'/r_0}e^{iky_0y'/r_0}\,x'dx'dy'+$$

$$\tfrac{i}{\lambda}U_0 e^{-i\beta z}\tfrac{1}{r_0}e^{ikr_0}\cos\theta_0\tfrac{2y_0}{r_0^2}\int_\Omega e^{ikx_0x'/r_0}e^{iky_0y'/r_0}\,y'dx'dy'$$

Now, for simplicity, set

$$U(\mathbf{r}_0) = -\tfrac{i}{\lambda}U_0 e^{-i\beta z}\tfrac{1}{r_0}e^{ikr_0}\cos\theta_0\left\{I_1(\mathbf{r}_0)-\tfrac{2x_0}{r_0^2}I_2(\mathbf{r}_0)-\tfrac{2y_0}{r_0^2}I_3(\mathbf{r}_0)\right\}.$$

Then, employing the identity $\sin\phi = \frac{1}{2i}(e^{i\phi}-e^{-i\phi})$ and defining $\mathrm{sinc}\phi = \frac{\sin\phi}{\phi}$,

$$I_1(\mathbf{r}_0) = \int_\Omega e^{ikx_0x'/r_0}e^{iky_0y'/r_0}\,dx'dy' = \int_{-\frac{1}{2}\delta}^{\frac{1}{2}\delta} e^{ikx_0x'/r_0}\,dx'\int_{-\frac{1}{2}\ell}^{\frac{1}{2}\ell} e^{iky_0y'/r_0}\,dy'$$

$$= \left(\tfrac{r_0}{k}\right)^2\tfrac{4}{x_0y_0}\sin\left(\tfrac{1}{2}k\tfrac{x_0\delta}{r_0}\right)\sin\left(\tfrac{1}{2}k\tfrac{y_0\ell}{r_0}\right)$$

$$I_2(\mathbf{r}_0) = \int_\Omega e^{ikx_0x'/r_0}e^{iky_0y'/r_0}\,x'dx'dy'$$

$$= \tfrac{i\delta}{k}\left\{\mathrm{sinc}\left(\tfrac{1}{2}k\tfrac{x_0\delta}{r_0}\right)-\cos\left(\tfrac{1}{2}k\tfrac{x_0\delta}{r_0}\right)\right\}\tfrac{2r_0}{ky_0}\sin\left(\tfrac{1}{2}k\tfrac{y_0\ell}{r_0}\right)$$

$$= \tfrac{i2\delta}{y_0x_0}\left(\tfrac{r_0}{k}\right)^2\left\{\mathrm{sinc}\left(\tfrac{1}{2}k\tfrac{x_0\delta}{r_0^2}\right)-\cos\left(\tfrac{1}{2}k\tfrac{x_0\delta}{r_0^2}\right)\right\}\sin\left(\tfrac{1}{2}k\tfrac{y_0\ell}{r_0^2}\right)$$

$$I_3(\mathbf{r}_0) = \int_\Omega e^{ikx_0x'/r_0^2}e^{iky_0y'/r_0^2}\,y'dx'dy'$$

$$= \tfrac{i\ell}{k}\left\{\mathrm{sinc}\left(\tfrac{1}{2}k\tfrac{y_0\ell}{r_0^2}\right)-\cos\left(\tfrac{1}{2}k\tfrac{y_0\ell}{r_0^2}\right)\right\}\tfrac{2r_0^2}{kx_0}\sin\left(\tfrac{1}{2}k\tfrac{x_0\delta}{r_0^2}\right)$$

$$= \tfrac{i2\ell}{x_0y_0}\left(\tfrac{r_0}{k}\right)^2\left\{\mathrm{sinc}\left(\tfrac{1}{2}k\tfrac{y_0\ell}{r_0^2}\right)-\cos\left(\tfrac{1}{2}k\tfrac{y_0\ell}{r_0^2}\right)\right\}\sin\left(\tfrac{1}{2}k\tfrac{x_0\delta}{r_0^2}\right)$$

whence substituting $k = \frac{2\pi}{\lambda}$ yields

$$U(\boldsymbol{r}_0) = -\mathrm{i}U_0 \mathrm{e}^{-\mathrm{i}\beta z} \mathrm{e}^{ikr_0} \tfrac{\cos\theta_0}{\pi^2} \tfrac{\lambda r_0}{x_0 y_0} \Big\{ \sin\left(\pi \tfrac{x_0 \delta}{\lambda r_0}\right) \sin\left(\pi \tfrac{y_0 \ell}{\lambda x_0}\right) +$$

$$-\mathrm{i}\tfrac{x_0\delta}{r_0^2}\left(\mathrm{sinc}\left(\pi \tfrac{x_0\delta}{\lambda r_0}\right) - \cos\left(\pi \tfrac{x_0\delta}{\lambda r_0}\right)\right)\sin\left(\pi \tfrac{y_0\ell}{\lambda x_0}\right) - \mathrm{i}\tfrac{x_0\delta}{r_0^2}\left(\mathrm{sinc}\left(\pi \tfrac{y_0\ell}{\lambda x_0}\right) - \cos\left(\pi \tfrac{y_0\ell}{\lambda x_0}\right)\right)\sin\left(\pi \tfrac{x_0\delta}{\lambda r_0}\right)\Big\}.$$

Now, noting that

$$\lim_{x\to\infty}\left\{x \sin\tfrac{1}{x} \sin\tfrac{1}{x}\right\} = x\left\{\tfrac{1}{x} - \tfrac{1}{3!}\tfrac{1}{x^3} + o\left(\tfrac{1}{x^5}\right)\right\}\sin\tfrac{1}{x} \to 0$$

and $\lim_{x\to 0}\frac{\sin x}{x} \to 0$, it follows clearly that

$$\lim_{\lambda\to\infty} \lambda \sin\left(\pi \tfrac{x_0\delta}{\lambda r_0}\right)\sin\left(\pi \tfrac{y_0\ell}{\lambda x_0}\right) = 0$$

$$\lim_{\lambda\to\infty} \lambda\left(\mathrm{sinc}\left(\pi \tfrac{x_0\delta}{\lambda r_0}\right) - \cos\left(\pi \tfrac{x_0\delta}{\lambda r_0}\right)\right)\sin\left(\pi \tfrac{y_0\ell}{\lambda x_0}\right) = 0$$

$$\lim_{\lambda\to\infty} \lambda\left(\mathrm{sinc}\left(\pi \tfrac{y_0\ell}{\lambda x_0}\right) - \cos\left(\pi \tfrac{y_0\ell}{\lambda x_0}\right)\right)\sin\left(\pi \tfrac{x_0\delta}{\lambda r_0}\right) = 0$$

That is, for any fixed δ, ℓ, r_0, $\lim_{\lambda\to\infty} U(\boldsymbol{r}_0) = 0$. The implication is the following: for any fixed rectangular aperture, longer wavelength radiation implies less radiation escaping from the aperture.

Next, for $r_0 >> 1$, it is clear that

$$U(\boldsymbol{r}_0) \approx -\mathrm{i}U_0 \mathrm{e}^{-\mathrm{i}\beta z} \mathrm{e}^{ikr_0} \tfrac{\cos\theta_0}{\pi^2} \tfrac{\lambda r_0}{x_0 y_0} \sin\left(\pi \tfrac{x_0\delta}{\lambda r_0}\right)\sin\left(\pi \tfrac{y_0\ell}{\lambda x_0}\right)$$

to first order in $\frac{1}{r_0}$. In particular, the field is zero precisely for each fixed δ, ℓ, r_0 when

$$\tfrac{x_0\delta}{\lambda r_0} = 2n \Rightarrow \lambda = \tfrac{x_0\delta}{2nr_0} \quad \text{or} \quad \tfrac{y_0\ell}{\lambda r_0} = 2n \Rightarrow \lambda = \tfrac{x_0\delta}{2nr_0}$$

for all $n = 0, 1, 2, \ldots$. This yields the *diffraction fringes* observed on a screen placed at z_0 away from the aperture. Indeed, $U(\boldsymbol{r}_0)$ can be expressed more intuitively:

$$U(\boldsymbol{r}_0) \approx -\mathrm{i}U_0 \mathrm{e}^{-\mathrm{i}\beta z} \mathrm{e}^{ikr_0} \tfrac{\cos\theta_0}{\pi^2} \tfrac{\pi^2 \ell\delta}{\lambda r_0} \mathrm{sinc}\left(\pi \tfrac{x_0\delta}{\lambda r_0}\right)\mathrm{sinc}\left(\pi \tfrac{y_0\ell}{\lambda x_0}\right)$$

$$= -\mathrm{i}U_0 \mathrm{e}^{-\mathrm{i}\beta z} \mathrm{e}^{ikr_0} \tfrac{\cos\theta_0}{\pi^2} \tfrac{k}{r_0} \tfrac{\pi \ell\delta}{2} \mathrm{sinc}\left(\pi \tfrac{x_0\delta}{\lambda r_0}\right)\mathrm{sinc}\left(\pi \tfrac{y_0\ell}{\lambda x_0}\right)$$

It is thus obvious when expressed in this form that long wavelength radiation yields a lower intensity for a given point away from a fixed aperture, a result obtained less elegantly from the previous paragraph. □

8.5.12 Proposition

Consider n identical slits adjacent to one another, each slit separated from the other by some distance $h > 0$:

$$\Omega_j = \left[-\tfrac{L}{2}+(j-1)(h+\delta), -\tfrac{L}{2}+(j-1)(h+\delta)+\delta\right]\times\left[-\tfrac{\ell}{2},\tfrac{\ell}{2}\right]\times\{0\}$$

where $L = N(h+\delta)+\delta$, with Ω_j lying on the plane $z = 0$: $\Omega_j \subset \mathbf{R}^2 \times \{0\}\ \forall j \in \mathbf{N}$. Suppose that $\min_j r_j >> \max\{h,\delta\}$, and set $\boldsymbol{r}_{0,j} = (\frac{\delta-L}{2}+(j-1)(h+\delta),0,0)$ to be the center of Ω_j, and let $\boldsymbol{r}_0 = (x_0, y_0, z_0) \in \mathbf{R}^3_+$ be an arbitrary point. Then, the field at $\boldsymbol{r}_0$ is

$$U(\boldsymbol{r}_0) \approx \frac{U_0}{k^2\lambda}\sum_j \left\{ \frac{2}{\Lambda_j(x,X)\Lambda_j(y,Y)}\frac{1}{r_jR_j}e^{ik|\boldsymbol{r}_j+\boldsymbol{R}_j|}\left\{\left(\cos\theta_j - \left(1+\tfrac{i}{kR_j}\right)\cos\Theta_j\right)\right\}\times \right. \tag{8.31}$$

$$e^{ik\Lambda_j(x,X)\left(\Delta_j^- + \frac{\delta}{2}\right)}\sin\left(\tfrac{1}{2}k\Lambda_j(x,X)\delta\right)\cos\left(\tfrac{1}{2}k\Lambda_j(y,Y)\ell\right) -$$

$$\frac{i\delta\cos\left(\frac{1}{2}k\Lambda_j(x,X)\delta\right)}{\Lambda_j(x,X)\Lambda_j(y,Y)}\frac{1}{r_jR_j}e^{ik|\boldsymbol{r}_j+\boldsymbol{R}_j|}\left\{\Lambda_j'(x,X)+\tfrac{i}{kR_j}\Lambda_j''(x,X)\cos\Theta_j\right\}\times$$

$$e^{ik\Lambda_j(x,X)\Delta_j^+}\left\{\left(1-ik\Lambda_j(x,X)\right)\Delta_j^- e^{-i\frac{1}{2}k\Lambda_j(x,X)\delta}\operatorname{sinc}\left(\tfrac{1}{2}k\Lambda_j(x,X)\delta\right)-1\right\}+$$

$$\frac{\ell\sin\left(\frac{1}{2}k\Lambda_j(x,X)\delta\right)}{\Lambda_j(x,X)\Lambda_j(y,Y)}\frac{1}{r_jR_j}e^{ik|\boldsymbol{r}_j+\boldsymbol{R}_j|}\left\{\Lambda_j'(y,Y)+\tfrac{i}{kR_j}\Lambda_j''(y,Y)\cos\Theta_j\right\}\times$$

$$\left. e^{ik\Lambda_j(x,X)\left(\Delta_j^- + \frac{\delta}{2}\right)}\left\{\operatorname{sinc}\left(\tfrac{1}{2}k\Lambda_j(y,Y)\ell\right)-\cos\left(\tfrac{1}{2}k\Lambda_j(y,Y)\ell\right)\right\}\right\}$$

where

$$\Lambda_j(u,U) \equiv \frac{u_j}{r_j}-\frac{U_j}{R_j},\quad \Lambda_j'(u,U) = \frac{u_j}{r_j^2}\cos\theta_j + \frac{U_j}{R_j^2}\cos\Theta_j + (\cos\theta_j - \cos\Theta_j)\left(\frac{u_j}{r_j^2}-\frac{U_j}{R_j^2}\right)$$

and

$$\Lambda_j''(u,U) = \frac{u_j}{r_j^2}-\frac{3U_j}{R_j^2},$$

with

$$u_j \in \{x_j, y_j, z_j\}, U_j \in \{X_j, Y_j, Z_j\}.$$

Proof

For each fixed Ω_j, set $\boldsymbol{r}_j = \boldsymbol{r}_0 - \boldsymbol{r}_{0,j}$ and $\boldsymbol{r}'_j = \boldsymbol{r}' + \boldsymbol{r}_{0,j}$. Then, $\bar{\boldsymbol{r}}_j = \boldsymbol{r}_j - \boldsymbol{r}'$, for each $j = 1,2, \ldots$ Likewise, set $\boldsymbol{R}_j = \boldsymbol{r}_{0,j} - \boldsymbol{R}_0$ and $\bar{\boldsymbol{R}}_j = \boldsymbol{r}'_{0,j} - \boldsymbol{R}_0$. Then, $\bar{\boldsymbol{R}}_j = \boldsymbol{R}_j + \boldsymbol{r}'$. Recall that $\boldsymbol{r}' \in \Omega_{\frac{1}{2}L}$, where for simplicity, it is assumed without loss of generality that $\frac{L}{2} \in \mathbf{N}$, whence

$$\bar{r}_j = \sqrt{(x_j - x')^2 + (y_j - y')^2 + z_j^2} \approx r_j \left\{1 - \frac{x_j x' + y_j y'}{r_j^2}\right\}$$

$$\frac{1}{\bar{r}_j} = \left\{(x_j - x')^2 + (y_j - y')^2 + z_j^2\right\}^{-\frac{1}{2}} \approx \frac{1}{r_j}\left\{1 + \frac{x_j x' + y_j y'}{r_j^2}\right\}$$

$$\bar{R}_j = \sqrt{(X_j + x')^2 + (Y_j + y')^2 + Z_j^2} \approx R_j \left\{1 + \frac{X_j x' + Y_j y'}{R_j^2}\right\}$$

$$\frac{1}{\bar{R}_j} = \left\{(X_j + x')^2 + (Y_j + y')^2 + Z_j^2\right\}^{-\frac{1}{2}} \approx \frac{1}{R_j}\left\{1 - \frac{X_j x' + Y_j y'}{R_j^2}\right\}$$

Moreover, note that $\cos\bar{\theta}_j = \frac{z_0}{\bar{r}_j}$ and $\cos\theta_j = \frac{z_0}{r_j} \Rightarrow \cos\bar{\theta}_j = \cos\theta_j \frac{r_j}{\bar{r}_j}$. Likewise, $\cos\bar{\Theta}_j = \frac{Z_0}{\bar{R}_j}$ and $\cos\Theta_j = \frac{Z_0}{R_j} \Rightarrow \cos\bar{\Theta}_j = \cos\Theta_j \frac{R_j}{\bar{R}_j}$, whence,

$$\cos\bar{\theta}_j \approx \cos\theta_j \left\{1 + \frac{x_j x' + y_j y'}{r_j^2}\right\} \quad \text{and} \quad \cos\bar{\Theta}_j \approx \cos\Theta_j \left\{1 - \frac{X_j x' + Y_j y'}{R_j^2}\right\}$$

Thus after many tedious algebraic manipulations, which the dedicated reader is encouraged to perform, where the binomial approximation is applied,

$$\frac{1}{\bar{r}_j \bar{R}_j} e^{ik|\bar{\boldsymbol{r}}_j + \bar{\boldsymbol{R}}_j|} (\cos\theta_j - \cos\Theta_j)$$

$$\approx \frac{1}{r_j R_j} e^{ik|\boldsymbol{r}_j + \boldsymbol{R}_j|} \left\{1 + \left(\frac{x_j}{r_j^2} - \frac{X_j}{R_j^2}\right)x' + \left(\frac{y_j}{r_j^2} - \frac{Y_j}{R_j^2}\right)y'\right\} \left\{\cos\theta_j - \cos\Theta_j + \right.$$

$$\left. \left(\frac{x_j}{r_j^2}\cos\theta_j + \frac{X_j}{R_j^2}\cos\Theta_j\right)x' + \left(\frac{y_j}{r_j^2}\cos\theta_j + \frac{Y_j}{R_j^2}\cos\Theta_j\right)y'\right\}$$

and

$$\frac{1}{\bar{r}_j \bar{R}_j} e^{ik|\bar{\boldsymbol{r}}_j + \bar{\boldsymbol{R}}_j|} \frac{i}{k\bar{R}_j} \cos\bar{\Theta}_j$$

$$\approx \frac{1}{r_j R_j} e^{ik|\boldsymbol{r}_j + \boldsymbol{R}_j|} \frac{i}{kR_j} \cos\Theta_j \left\{1 + \left(\frac{x_j}{r_j^2} - \frac{3X_j}{R_j^2}\right)x' + \left(\frac{y_j}{r_j^2} - \frac{3Y_j}{R_j^2}\right)y'\right\}$$

Upon expanding and collecting factors for x', y', set

$$\Lambda_j(u,U) \equiv \tfrac{u_j}{r_j} - \tfrac{U_j}{R_j},$$

where $u_j \in \{x_j, y_j, z_j\}, U_j \in \{X_j, Y_j, Z_j\}$

$$\Lambda_j'(u,U) = \tfrac{u_j}{r_j^2}\cos\theta_j + \tfrac{U_j}{R_j^2}\cos\Theta_j + (\cos\theta_j - \cos\Theta_j)\left(\tfrac{u_j}{r_j^2} - \tfrac{U_j}{R_j^2}\right)$$

$$\Lambda_j''(u,U) = \tfrac{u_j}{r_j^2} - \tfrac{3U_j}{R_j^2}$$

$$\Xi_j^{(1)} = \tfrac{1}{2r_jR_j} e^{ik|r_j+R_j|}(\cos\theta_j - \cos\Theta_j),\ \Xi_j^{(2)} = \tfrac{1}{2r_jR_j} e^{ik(r_j+R_j)}\tfrac{i}{kR_j}\cos\Theta_j$$

$$\Xi_j^{(1)}(u,U) = \tfrac{1}{2r_jR_j} e^{ik|r_j+R_j|}\Lambda_j'(u,U),\ \Xi_j^{(2)}(u,U) = \tfrac{1}{2r_jR_j} e^{ik(r_j+R_j)}\tfrac{i}{kR_j}\cos\Theta_j\Lambda_j''(u,U)$$

Then, to first order in x', y', from Equation (8.30),

$$\tfrac{1}{2\bar{r}_j\bar{R}_j} e^{ik|\bar{r}_j+\bar{R}_j|}\left(\cos\bar{\theta}_j - \cos\bar{\Theta}_j - \tfrac{i}{kR_j}\cos\bar{\Theta}_j\right)$$

$$\approx \left(\Xi_j^{(1)} + \Xi_j^{(2)}\right)e^{ik\Lambda_j(x,X)x'}e^{ik\Lambda_j(y,Y)y'} + \left(\Xi_j^{(1)}(x,X) + \Xi_j^{(2)}(x,X)\right)e^{ik\Lambda_j(x,X)x'}e^{ik\Lambda_j(y,Y)y'}x'$$

$$+\left(\Xi_j^{(1)}(y,Y) + \Xi_j^{(2)}(y,Y)\right)e^{ik\Lambda_j(x,X)x'}e^{ik\Lambda_j(y,Y)y'}y'$$

Furthermore, set $I_{1,j} = \int_{\Delta_j^-}^{\Delta_j^+} e^{ik\Lambda_j(x,X)x'}e^{ik\Lambda_j(y,Y)y'}dx'$, where $\Delta_j^- = -\tfrac{L}{2} + (j-1)(h+\delta)$ and $\Delta_j^+ = \Delta_j^- + \delta$. Also, set

$$I_{2,j} = \int_{\Delta_j^-}^{\Delta_j^+} e^{ik\Lambda_j(x,X)x'}x'dx' \int_{-\frac{\ell}{2}}^{\frac{\ell}{2}} e^{ik\Lambda_j(y,Y)y'}dy'$$

and

$$I_{3,j} = \int_{\Delta_j^-}^{\Delta_j^+} e^{ik\Lambda_j(x,X)x'}dx' \int_{-\frac{\ell}{2}}^{\frac{\ell}{2}} e^{ik\Lambda_j(y,Y)y'}y'dy'$$

Then, using the identities

$$\sin x = \tfrac{e^{ix} - e^{-ix}}{2i},\ \cos x = \tfrac{e^{ix} + e^{-ix}}{2},\ \operatorname{sinc} x = \tfrac{\sin x}{x},\ \sin 2x = 2\sin x\cos x,$$

and last, $\int x e^{ikx} dx = \frac{1}{k^2}\{e^{ikx}(1-ikx)\}$, it can be shown after some routine effort that

$$I_{1,j} = \frac{4i}{k^2\Lambda_j(x,X)\Lambda_j(y,Y)} e^{ik\Lambda_j(x,X)\left(\Delta_j^- + \frac{\delta}{2}\right)} \sin\left(\tfrac{1}{2}k\Lambda_j(x,X)\delta\right)\cos\left(\tfrac{1}{2}k\Lambda_j(y,Y)\ell\right)$$

$$I_{2,j} = \frac{2\delta\cos\left(\frac{1}{2}k\Lambda_j(x,X)\delta\right)}{k^2\Lambda_j(x,X)\Lambda_j(y,Y)} e^{ik\Lambda_j(x,X)\Delta_j^+} \left\{(1-ik\Lambda_j(x,X))\Delta_j^- e^{-i\frac{1}{2}k\Lambda_j(x,X)\delta}\operatorname{sinc}\left(\tfrac{1}{2}k\Lambda_j(x,X)\delta\right)-1\right\}$$

$$I_{3,j} = \frac{2i\ell\sin\left(\frac{1}{2}k\Lambda_j(x,X)\delta\right)}{k^2\Lambda_j(x,X)\Lambda_j(y,Y)} e^{ik\Lambda_j(x,X)\left(\Delta_j^- + \frac{\delta}{2}\right)} \left\{\operatorname{sinc}\left(\tfrac{1}{2}k\Lambda_j(y,Y)\ell\right)-\cos\left(\tfrac{1}{2}k\Lambda_j(y,Y)\ell\right)\right\}$$

whence

$$U(r_0) \approx -\frac{iU_0}{\lambda}\sum_j \int_{\Omega_j} \frac{1}{2\bar{r}_j\bar{R}_j} e^{ik|\bar{r}_j+\bar{R}_j|}\left\{\cos\bar{\theta}_j - \left(\frac{i}{k\bar{R}_j}+1\right)\cos\bar{\Theta}_j\right\}dx'dy'$$

$$\approx -\frac{iU_0}{\lambda}\sum_j \frac{4i\left(\Xi_j^{(1)}+\Xi_j^{(2)}\right)}{k^2\Lambda_j(x,X)\Lambda_j(y,Y)} e^{ik\Lambda_j(x,X)\left(\Delta_j^- + \frac{\delta}{2}\right)} \sin\left(\tfrac{1}{2}k\Lambda_j(x,X)\delta\right)\cos\left(\tfrac{1}{2}k\Lambda_j(y,Y)\ell\right)$$

$$+\frac{2\delta\left(\Xi_j^{(1)}(x,X)+\Xi_j^{(2)}(x,X)\right)\cos\left(\frac{1}{2}k\Lambda_j(x,X)\delta\right)}{k^2\Lambda_j(x,X)\Lambda_j(y,Y)} e^{ik\Lambda_j(x,X)\Delta_j^+} \times$$

$$\left\{(1-ik\Lambda_j(x,X))\Delta_j^- e^{-i\frac{1}{2}k\Lambda_j(x,X)\delta}\operatorname{sinc}\left(\tfrac{1}{2}k\Lambda_j(x,X)\delta\right)-1\right\}$$

$$+\frac{2i\ell\left(\Xi_j^{(1)}(y,Y)+\Xi_j^{(2)}(y,Y)\right)\sin\left(\frac{1}{2}k\Lambda_j(x,X)\delta\right)}{k^2\Lambda_j(x,X)\Lambda_j(y,Y)} e^{ik\Lambda_j(x,X)\left(\Delta_j^- + \frac{\delta}{2}\right)} \left\{\operatorname{sinc}(\tfrac{1}{2}k\Lambda_j(y,Y)\ell)-\cos\left(\tfrac{1}{2}k\Lambda_j(y,Y)\ell\right)\right\}$$

as required. □

8.6 Worked Problems

8.6.1 Exercise

Show that

$$E = -\frac{q}{4\pi\varepsilon_0}\left(1-\frac{v}{c}\right)^{-2}\frac{1}{r}\frac{d}{dt}v + \frac{q}{4\pi\varepsilon_0}\left(1-\frac{v}{c}\right)^{-1}\frac{v}{r^2}v - \nabla\varphi$$

given that

$$A = \frac{1}{4\pi\varepsilon_0}\frac{1}{1-\frac{v}{c}}\frac{q}{|r-r_0|}v$$

Solution

By definition, $\boldsymbol{E} = -\nabla\varphi - \partial_t \boldsymbol{A}$. Hence, evaluating $\partial_t \boldsymbol{A}$ yields:

$$\begin{aligned}\tfrac{4\pi\varepsilon_0}{q}\partial_t \boldsymbol{A} &= -\tfrac{1}{r^2}v\left(1-\tfrac{v}{c}\right)^{-1}\boldsymbol{v} + \tfrac{1}{r}\left(1-\tfrac{v}{c}\right)^{-2}\tfrac{v}{c}\tfrac{\mathrm{d}}{\mathrm{d}t}\boldsymbol{v} + \tfrac{1}{r}\left(1-\tfrac{v}{c}\right)^{-1}\tfrac{\mathrm{d}}{\mathrm{d}t}\boldsymbol{v}\\ &= -\tfrac{1}{r^2}v\left(1-\tfrac{v}{c}\right)^{-1}\boldsymbol{v} + \tfrac{1}{r}\left(1-\tfrac{v}{c}\right)^{-1}\left\{\left(1-\tfrac{v}{c}\right)^{-1}\tfrac{v}{c}+1\right\}\tfrac{\mathrm{d}}{\mathrm{d}t}\boldsymbol{v}\\ &= -\tfrac{1}{r^2}v\left(1-\tfrac{v}{c}\right)^{-1}\boldsymbol{v} + \tfrac{1}{r}\left(1-\tfrac{v}{c}\right)^{-2}\tfrac{\mathrm{d}}{\mathrm{d}t}\boldsymbol{v}\end{aligned}$$

where $v = \tfrac{\mathrm{d}}{\mathrm{d}t}|\boldsymbol{r}| = \tfrac{\mathrm{d}}{\mathrm{d}t}r$ is the speed of the particle. Thus,

$$\boldsymbol{E}_{\text{accel}} = -\tfrac{q}{4\pi\varepsilon_0}\left(1-\tfrac{v}{c}\right)^{-2}\tfrac{1}{r}\tfrac{\mathrm{d}}{\mathrm{d}t}\boldsymbol{v} \quad \text{and} \quad \boldsymbol{E}_{\text{static}} = \tfrac{q}{4\pi\varepsilon_0}\left(1-\tfrac{v}{c}\right)^{-1}\tfrac{v}{r^2}\boldsymbol{v} - \nabla\varphi,$$

as required. □

8.6.2 Exercise

Show that in the near zone, the electric field and magnetic field can be expressed in terms of the dipole moment as

$$E_r(r,\theta) \approx \tfrac{1}{2\pi\varepsilon}\tfrac{p}{r^3}\cos\theta$$

$$E_\theta(r,\theta) \approx \tfrac{1}{4\pi\varepsilon}\tfrac{p}{r^3}\sin\theta$$

Solution

Recall from Section 8.2 that $p = -\mathrm{i}\tfrac{I}{\omega}\delta z$. Hence, from Definition 8.2.5, substituting this expression into Equations (8.5) and (8.6) yields the desired results. □

8.6.3 Exercise

Suppose a transmitting antenna and a receiving antenna are separated by a distance much larger than the physical dimensions of either of the two antennae. Model the two antennae in terms of equivalent circuits, and hence derive the power absorbed by the receiver.

Solution

The two antennae can be modeled via the following coupled equations:

$$V_1 = I_1 Z_{11} + I_2 Z_{12}$$

$$V_2 = I_1 Z_{21} + I_2 Z_{22}$$

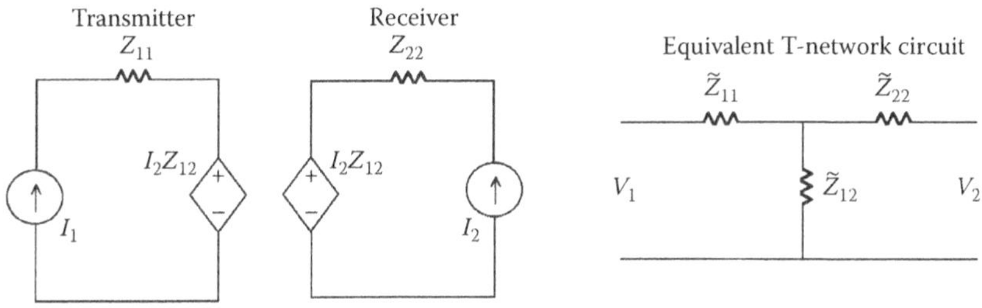

FIGURE 8.8.
Equivalent 2-port network circuit representation.

where the transmitting antenna is modeled by (V_1, Z_{11}), Z_{11} is the load of the transmitter, and V_1 the voltage across the load. The receiving antenna is modeled by (V_2, Z_{22}), where Z_{22} represents the load of the receiver.

By reciprocity, $Z_{12} = Z_{21}$, and hence, the coupled equation, a 2-port network, can be modeled with a 2-port 3-impedance network. That is, the 2-port network can be represented by a T-network or a Π-network. Representing the circuit by a T-network is illustrated in Figure 8.8.

Referring to Figure 8.8, Z_{ii} corresponds to the input impedance under open circuit condition. By Kirchhoff's voltage law,

$$V_1 = \tilde{Z}_{11}I_1 + \tilde{Z}_{12}(I_1 + I_2) \quad \text{and} \quad V_2 = \tilde{Z}_{22}I_2 + \tilde{Z}_{12}(I_1 + I_2)$$

Thus, $I_2 = 0 \Rightarrow V_1 = I_1(\tilde{Z}_{11} + \tilde{Z}_{12})$ and by definition, $Z_{11} = \frac{V_1}{I_1}\Big|_{I_2=0} = \tilde{Z}_{11} + \tilde{Z}_{12}$. Likewise, $I_1 = 0$ implies that $V_2 = I_2(\tilde{Z}_{22} + \tilde{Z}_{12})$ and $Z_{22} = \frac{V_2}{I_2}\Big|_{I_1=0} = \tilde{Z}_{22} + \tilde{Z}_{12}$. Lastly, $V_2 = \tilde{Z}_{12}I_1$ when $I_2 = 0$ implies at once that $Z_{12} = \frac{V_2}{I_1}\Big|_{I_2=0} = \tilde{Z}_{12}$, whence, $\tilde{Z}_{ii} = Z_{ii} - Z_{12}\ \forall i = 1, 2$.

Therefore the antenna transmitter–receiver system may be represented by an equivalent T-network, where the terminal V_2 is attached to a load impedance Z_L of the receiver. Under the assumption of weak coupling, that is, the two antennae are sufficiently far away, $I_2Z_{12} \approx 0$ and hence, $V_1 = I_1Z_{11}$ represents the transmitting antenna.

To determine the receiving antenna equivalent representation, Thévenin's theorem is applied to Z_L. Let V_T denote the open circuit Thévenin voltage and Z_s the Thévenin source impedance. Then, the Thévenin voltage across Z_{12} is obtained by setting $Z_L = \infty \Rightarrow V_T = Z_{12}I_1 = \frac{Z_{12}}{Z_{11}}V_1$. Likewise, under open circuit condition, the Thévenin impedance looking from the receiver end is: $Z_s = Z_{22} - Z_{12} + Z_{12} \parallel (Z_{11} - Z_{12}) = Z_{22} - \frac{Z_{12}^2}{Z_{11}}$. Thus, by Kirchhoff's law, $V_2 = Z_L(-I_2) \Rightarrow -Z_LI_2 = Z_{12}I_1 + Z_{22}I_2 \Rightarrow I_2 = -\frac{Z_{12}}{Z_{22}+Z_L}I_1$. Hence, the time-average power absorbed by the receiving antenna Z_L is

$$\langle P \rangle = -\tfrac{1}{2}\Re\text{e}(V_2I_2^*) = \tfrac{1}{2}|I_2|^2\,\Re\text{e}(Z_L) = \tfrac{1}{2}|I_2|^2\left|\frac{Z_{12}}{Z_{22}+Z_L}\right|^2\Re\text{e}(Z_L). \qquad \square$$

8.6.4 Exercise

Show that

$$\delta E_\theta(r,\theta) \approx \frac{I_0\beta}{\pi\varepsilon}\frac{\sin\theta}{r(\cos\theta-1)}\mathrm{e}^{-\mathrm{i}\beta r}\mathrm{e}^{-\mathrm{i}\frac{h\beta}{r}\sin\theta}\mathrm{e}^{\mathrm{i}\frac{1}{2}\beta(\cos\theta-1)\ell}\cos\left(\tfrac{\ell}{2}\beta(\cos\theta-1)\right)\sin\left(\tfrac{h\beta}{r}\sin\theta\right).$$

Solution

From

$$\delta E_\theta \approx \frac{\mathrm{i}I_0\beta}{4\pi\varepsilon}\left\{\frac{\sin\theta}{r(\cos\theta-1)}\mathrm{e}^{-\mathrm{i}\beta r}\left(1-\mathrm{e}^{\mathrm{i}\beta(\cos\theta-1)\ell}\right)-\frac{\sin\theta'}{r'(\cos\theta'-1)}\mathrm{e}^{-\mathrm{i}\beta r'}\left(1-\mathrm{e}^{\mathrm{i}\beta(\cos\theta'-1)\ell}\right)\right\},$$

observe from $r'=\sqrt{r^2+4h^2+4rh\sin\theta}$ and $\theta'=\arccos\frac{L}{r'}$, that up to first order in $\frac{1}{r}$,

$$r'=\sqrt{r^2+4h^2+4rh\sin\theta}=r\left\{1+\left(\tfrac{2h}{r}\right)^2+\tfrac{4h}{r}\sin\theta\right\}^{\frac{1}{2}}\approx r\left\{1+\tfrac{2h}{r}\sin\theta\right\}$$

$$\frac{1}{r'}\approx\frac{1}{r}\left\{1+\tfrac{4h}{r}\sin\theta\right\}^{-\frac{1}{2}}\approx\frac{1}{r}\left\{1-\tfrac{2h}{r}\sin\theta\right\}$$

$$\theta'=\arccos\tfrac{L}{r'}\approx\arccos\left\{\tfrac{L}{r}\left(1-\tfrac{2h}{r}\sin\theta\right)\right\}\approx\arccos\tfrac{L}{r}=\theta$$

Substituting these approximations into the original first-order expression of δE_θ yields

$$\delta E_\theta(r,\theta)\approx\frac{\mathrm{i}I_0\beta}{4\pi\varepsilon}\frac{\sin\theta}{r(\cos\theta-1)}\mathrm{e}^{-\mathrm{i}\beta r}\left(1-\mathrm{e}^{\mathrm{i}\beta(\cos\theta-1)\ell}\right)\left(1-\mathrm{e}^{-\mathrm{i}\frac{2h\beta}{r}\sin\theta}\right).$$

Finally, to complete the derivation, it suffices to observe that

$$1-\mathrm{e}^{-\mathrm{i}2\Lambda}=2\mathrm{i}\mathrm{e}^{-\mathrm{i}\Lambda}\frac{\mathrm{e}^{\mathrm{i}\Lambda}-\mathrm{e}^{-\mathrm{i}\Lambda}}{2\mathrm{i}}=2\mathrm{i}\mathrm{e}^{-\mathrm{i}\Lambda}\sin\Lambda.$$

Explicitly,

$$\begin{aligned}1-\mathrm{e}^{\mathrm{i}\beta(\cos\theta-1)\ell}&=\mathrm{e}^{\mathrm{i}\frac{1}{2}\beta(\cos\theta-1)\ell}\left(\mathrm{e}^{-\mathrm{i}\frac{1}{2}\beta(\cos\theta-1)\ell}-\mathrm{e}^{\mathrm{i}\frac{1}{2}\beta(\cos\theta-1)\ell}\right)\\&=-2\mathrm{i}\mathrm{e}^{\mathrm{i}\frac{1}{2}\beta(\cos\theta-1)\ell}\sin\left(\tfrac{1}{2}\beta(\cos\theta-1)\ell\right),\end{aligned}$$

$$1-\mathrm{e}^{-\mathrm{i}\frac{2h\beta}{r}\sin\theta}=\mathrm{e}^{-\mathrm{i}\frac{h\beta}{r}\sin\theta}\left(\mathrm{e}^{\mathrm{i}\frac{h\beta}{r}\sin\theta}-\mathrm{e}^{-\mathrm{i}\frac{h\beta}{r}\sin\theta}\right)=2\mathrm{i}\mathrm{e}^{-\mathrm{i}\frac{h\beta}{r}\sin\theta}\sin\left(\tfrac{h\beta}{r}\sin\theta\right),$$

giving the desired result. □

8.6.5 Exercise

Work out Example 8.5.11 for a circular aperture of radius $\delta > 0$ using (8.30) at some observation point $P_0 = (x_0, y_0, z_0) \in \mathbf{R}^3$, where $\delta << r_0 = \sqrt{x_0^2 + y_0^2 + z_0^2}$.

Solution

Fix and suppose that a uniform field $U = U_0 \mathrm{e}^{-\mathrm{i}\beta z}$ impinges upon an aperture $\Omega = \{(x, y, 0) \in \mathbf{R}^3 : x^2 + y^2 \le \delta^2\}$, where $\partial\mathbf{R}_+^3 - \Omega$ is a perfect electrical conductor. Set $R = \sqrt{x_0^2 + y_0^2}$. Then, $\cos\Phi = \frac{x_0}{R}$ and $\sin\Phi = \frac{y_0}{R}$. The scalar field at P_0 via (8.30) as follows. Let $\cos\theta = -\frac{1}{r_0}\boldsymbol{r}_0 \cdot \boldsymbol{n}$, where $\boldsymbol{n}$ is the normal unit vector field on Ω pointing towards $\mathbf{R}_-^3$. For each $\boldsymbol{r}' \in \Omega$, let $\cos\overline{\theta} = -\frac{1}{\overline{r}}\overline{\boldsymbol{r}} \cdot \boldsymbol{n}$, where $\overline{\boldsymbol{r}} = \boldsymbol{r}_0 - \boldsymbol{r}'$. Finally, set $\cos\phi = \frac{x'}{r'}$ and $\sin\phi = \frac{y'}{r'}$.

From (8.30), $U(\boldsymbol{r}) = -\frac{\mathrm{i}}{\lambda}\int_\Omega \frac{1}{\overline{r}}\mathrm{e}^{ik\overline{r}}U(x', y')\cos\overline{\theta}\,\mathrm{d}x'\mathrm{d}y'$. To first order in $\frac{1}{r_0}$,

$$\overline{r} = \sqrt{(x_0 - x')^2 + (y_0 - y')^2 + z_0^2} \approx r_0\left\{1 - \frac{x_0x' + y_0y'}{r_0^2}\right\} \quad \text{and} \quad \frac{1}{\overline{r}} \approx \frac{1}{r_0}\left\{1 + \frac{x_0x' + y_0y'}{r_0^2}\right\}.$$

Furthermore,

$$\cos\overline{\theta} = \cos\theta\frac{r_0}{\overline{r}} \approx \cos\theta\left\{1 + \frac{x_0x' + y_0y'}{r_0^2}\right\}$$

and

$$\overline{r} \approx r_0 - \frac{x_0}{r_0}r'\cos\phi - \frac{y_0}{r_0}r'\sin\phi,$$

whence,

$$\begin{aligned}
\frac{1}{\overline{r}}\mathrm{e}^{ik\overline{r}}\cos\overline{\theta} &\approx \frac{\cos\theta}{r_0}\mathrm{e}^{ik\overline{r}}\left\{1 + \frac{2(x_0x' + y_0y')}{r_0^2}\right\} \\
&= \frac{\cos\theta}{r_0}\mathrm{e}^{ikr_0}\mathrm{e}^{-\mathrm{i}(k/r_0)r'\{x_0\cos\phi + y_0\sin\phi\}}\left\{1 + \frac{2x_0}{r_0^2}r'\cos\phi + \frac{2y_0}{r_0^2}r'\sin\phi\right\} \\
&= \frac{\cos\theta}{r_0}\mathrm{e}^{ikr_0}\mathrm{e}^{-\mathrm{i}(k/r_0)R\{\cos\Phi\cos\phi + \sin\Phi\sin\phi\}r'}\left\{1 + \frac{2x_0}{r_0^2}r'\cos\phi + \frac{2y_0}{r_0^2}r'\sin\phi\right\} \\
&= \frac{\cos\theta}{r_0}\mathrm{e}^{ikr_0}\mathrm{e}^{-\mathrm{i}(k/r_0)R\cos(\phi-\Phi)r'}\left\{1 + \frac{2x_0}{r_0^2}r'\cos\phi + \frac{2y_0}{r_0^2}r'\sin\phi\right\}.
\end{aligned}$$

Next, set $\varphi = \phi - \Phi$ and $\xi = \frac{kR}{r_0}r'$. Then, $\mathrm{d}\varphi = \mathrm{d}\phi$ and $\mathrm{d}\xi = \frac{kR}{r_0}\mathrm{d}r'$. Also, note from the axisymmetry of the problem that the following integral is invariant under angular rotation:

$$\int_0^{2\pi} \mathrm{e}^{-\mathrm{i}(kR/r_0)r'\cos(\phi-\Phi)}\,\mathrm{d}\phi = \int_{-\Phi}^{2\pi-\Phi} \mathrm{e}^{-\mathrm{i}(kR/r_0)r'\cos\varphi}\,\mathrm{d}\varphi = \int_0^{2\pi} \mathrm{e}^{-\mathrm{i}(kR/r_0)r'\cos\varphi}\,\mathrm{d}\varphi.$$

Lastly, noting that the *zero-order Bessel's function of the first kind* may also be defined by

$$J_0(\zeta) = \tfrac{1}{2\pi}\int_0^{2\pi} e^{-i\zeta\cos\varphi}\, d\varphi,$$

it follows that on setting

$$\int_\Omega \tfrac{1}{\bar{r}} e^{ik\bar{r}} \cos\bar{\theta} \approx \tfrac{\cos\theta}{r_0} e^{ikr_0}\left\{I_0 + \tfrac{2x_0}{r_0^2} I_1 + \tfrac{2y_0}{r_0^2} I_2\right\}$$

$$I_0 = \int_0^\delta\int_0^{2\pi} e^{-i(k/r_0)Rr'\cos\varphi} r'\, d\varphi\, dr' = \int_0^\delta J_0\left(\tfrac{kR}{r_0} r'\right) r'\, dr' = \left(\tfrac{r_0}{kR}\right)^2 \int_0^{\bar{\delta}} \xi J_0(\xi)\, d\xi = \left(\tfrac{r_0}{kR}\right)^2 \bar{\delta} J_1(\bar{\delta}),$$

where $\bar{\delta} = \frac{kR}{r_0}\delta$ and $J_1(\xi) = -\frac{i}{2\pi}\int_0^{2\pi} e^{i(\varphi+\zeta\cos\varphi)} d\varphi$ is the *first-order* Bessel's function of the first kind in integral representation. Thus, $I_0 = \frac{r_0\delta}{kR} J_1\left(\frac{kR}{r_0}\delta\right)$.

The second term $I_2 = \int_0^\delta dr' \int_0^{2\pi} r' d\varphi\left\{e^{-i(kR/r_0)r'\cos\varphi} r' \cos(\varphi+\Phi)\right\}$. So, from the expansion

$$\cos(\varphi+\Phi) = \cos\varphi\cos\Phi - \sin\varphi\sin\Phi,$$

$$\int_0^{2\pi} r'\, d\varphi\left\{e^{-i(kR/r_0)r'\cos\varphi} r'\cos(\varphi+\Phi)\right\}$$

$$= (r')^2\cos\Phi\int_0^{2\pi} e^{-i(kR/r_0)r'\cos\varphi}\cos\varphi\, d\varphi - (r')^2\sin\Phi\int_0^{2\pi} e^{-i(kR/r_0)r'\cos\varphi}\sin\varphi\, d\varphi$$

Set $\psi = \cos\varphi \Rightarrow d\psi = -\sin\varphi\, d\varphi$ and $\psi(0) = 1, \psi(2\pi) = 1$. Thus,

$$\int_0^{2\pi} e^{-i(kR/r_0)r'\cos\varphi}\sin\varphi\, d\varphi = -\int_1^1 e^{-i(kR/r_0)r'\psi}\, d\psi = 0$$

yielding

$$\int_0^{2\pi} r'\, d\varphi\left\{e^{-i(kR/r_0)r'\cos\phi} r'\cos(\varphi+\Phi)\right\} = (r')^2\cos\Phi\int_0^{2\pi} e^{-i(kR/r_0)r'\cos\phi}\cos\varphi\, d\varphi$$

However, by definition,

$$J_1(\xi) = -\tfrac{i}{2\pi}\int_0^{2\pi} e^{i(\varphi+\zeta\cos\varphi)}\, d\varphi = -\tfrac{i}{2\pi}\int_0^{2\pi} e^{i\zeta\cos\varphi}\cos\varphi\, d\varphi$$

implies at once that

$$\int_0^{2\pi} r'\, d\varphi\left\{e^{-i(kR/r_0)r'\cos\varphi} r'\cos(\varphi+\Phi)\right\} = i2\pi(r')^2\cos\Phi J_1\left(-\tfrac{kR}{r_0} r'\right)$$

Thus,

$$I_2 = \mathrm{i}2\pi \tfrac{x_0}{R} \int_0^{\delta} \mathrm{d}r' (r')^2 J_1\left(-\tfrac{kR}{r_0} r'\right).$$

The final term $I_3 = \int_0^{\delta} \mathrm{d}r' \int_0^{2\pi} r' \mathrm{d}\varphi \left\{ \mathrm{e}^{-\mathrm{i}(kR/r_0) r' \cos\varphi} r' \sin(\varphi + \Phi) \right\}$. Once again, via the expansion

$$\sin(\varphi + \Phi) = \sin\varphi \cos\Phi + \sin\Phi \cos\varphi$$

$$\int_0^{2\pi} r' \mathrm{d}\varphi \left\{ \mathrm{e}^{-\mathrm{i}(kR/r_0) r' \cos\varphi} r' \sin(\varphi + \Phi) \right\}$$

$$= (r')^2 \cos\Phi \int_0^{2\pi} \mathrm{e}^{-\mathrm{i}(kR/r_0) r' \cos\varphi} \sin\varphi \, \mathrm{d}\varphi + (r')^2 \sin\Phi \int_0^{2\pi} \mathrm{e}^{-\mathrm{i}(kR/r_0) r' \cos\varphi} \cos\varphi \, \mathrm{d}\varphi$$

$$= \mathrm{i}2\pi (r')^2 \tfrac{y_0}{R} J_1\left(-\tfrac{kR}{r_0} r'\right)$$

Thus, from Reference [1, Section 11.1.1],

$$\int_0^z t^p J_q(t) \mathrm{d}t = \frac{z^p \Gamma\left(\frac{p+1+1}{2}\right)}{\Gamma\left(\frac{q-p+1}{2}\right)} \sum_{k=0}^{\infty} \frac{(q+2k+1)\Gamma\left(\frac{q-p+1}{2}+k\right)}{\Gamma\left(\frac{q+p+3}{2}+k\right)} J_{q+2k+1}(z)$$

where $\Gamma(z) = \int_0^{\infty} t^{z-1} \mathrm{e}^{-t} \mathrm{d}t$ is the *Gamma function*, with $\Gamma(n) = (n-1)! \; \forall n \in \mathbf{N}$, it is clear that

$$\int_0^{\delta} (r')^2 J_1(\kappa r') \mathrm{d}r' = \tfrac{1}{\kappa^3} \int_0^{\kappa\delta} t^2 J_1(t) \mathrm{d}t = \tfrac{2\delta^2}{\kappa} \sum_{k=0}^{\infty} \tfrac{2(k+1)\Gamma(1+k)}{(k+1)!} J_{2(k+1)}(\kappa\delta)$$

where $\kappa = -\tfrac{kR}{r_0}$.

The complete solution is thus given by

$$U(r_0) \approx -\tfrac{\mathrm{i}U_0}{\lambda} \mathrm{e}^{-\mathrm{i}\beta z} \cos\theta \tfrac{1}{r_0} \mathrm{e}^{\mathrm{i}kr_0} \left\{ I_0 + \tfrac{2x_0}{r_0^2} I_1 + \tfrac{2y_0}{r_0^2} I_3 \right\}$$

$$= -\tfrac{\mathrm{i}U_0}{\lambda} \mathrm{e}^{-\mathrm{i}\beta z} \cos\theta \tfrac{1}{r_0} \mathrm{e}^{\mathrm{i}kr_0} \left\{ \tfrac{r_0\delta}{kR} J_1\left(\tfrac{kR}{r_0}\delta\right) - \tfrac{\mathrm{i}8\pi R}{r_0^2} \tfrac{r_0\delta^2}{kR} \sum_{k=0}^{\infty} \tfrac{2(k+1)\Gamma(1+k)}{(k+1)!} J_{2(k+1)}\left(-\tfrac{kR}{r_0}\delta\right) \right\}$$

In particular, to first order in $\tfrac{1}{r_0}$,

$$U(r_0) \approx -\tfrac{\mathrm{i}U_0}{\lambda} \mathrm{e}^{-\mathrm{i}\beta z} \mathrm{e}^{\mathrm{i}kr_0} \tfrac{\delta}{kR} J_1\left(\tfrac{kR}{r_0}\delta\right) \cos\theta$$

and the field vanishes at the zeros of $J_1\left(\frac{kR}{r_0}\delta\right)$; that is, the roots of $J_1\left(\frac{kR}{r_0}\delta\right)=0$. As a side remark, Bessel's functions have a countably infinite number of zeros.

Moreover, for $0 < \varepsilon << 1$, $J_1(\varepsilon) \sim \frac{\varepsilon}{2}$ asymptotically, and hence, for λ very large, $\frac{kR}{r_0}\delta << 1 \Rightarrow J_1\left(\frac{kR}{r_0}\delta\right) \approx 0$. In other words, for very large wavelengths, very little radiation is emitted from a small circular aperture. Indeed, a similar result was observed for a rectangular aperture. □

References

1. Abramowitz, M. and Stegun, I. (Eds.) 1972. *Handbook of Mathematical Functions with Formulas, Graphs and Mathematical Tables*. Dept. of Commerce, National Bureau of Standards, AMS 55.
2. Balanis, C. 1982. *Antenna Theory: Analysis and Design*. New York: John Wiley & Sons.
3. Chang, D. 1992. *Field and Wave Electromagnetics*. Reading, MA: Addison-Wesley.
4. Goodman, J. 1996. *Introduction to Fourier Optics*. New York: McGraw-Hill.
5. Hecht, E. 1987. *Optics*. Reading, MA: Addison-Wesley.
6. Neff, H. 1981. *Basic Electromagnetic Fields*. New York: Harper & Row.
7. Plonsey, R. and Collin, R. 1961. *Principles and Applications of Electromagnetic Fields*. New York: McGraw-Hill.
8. Schwartz, M. 1972. *Principles of Electrodynamics*. New York: Dover.
9. Silver, S. 1949. *Microwave Antenna Theory and Design*. New York: McGraw-Hill.
10. Koshlyakov, N., Smirnov, M., and Gliner, E. 1964. *Differential Equations of Mathematical Physics*. Amsterdam: North-Holland.
11. Stratton, J. 1941. *Electromagnetic Theory*. New York: McGraw-Hill.

9

Elements of Electrostatic Discharge

Electrostatic discharge can often be a source of immense annoyance to EMC engineers, and more important, it can also be extremely costly for an electronics manufacturer if countermeasures to mitigate discharge are not implemented. As a simple example, consider a person walking on a carpet during winter when the humidity is low who touches a port on the back of an electronic device. The charge buildup and its subsequent discharge into the port may potentially destroy sensitive electronic components. Hence, electrostatic discharge can be potentially damaging to electronic equipment. This chapter is intended to acquaint the reader with the physics of electrostatic discharge at an informal level.

9.1 Electrostatic Shielding

In order to gain insight into the breakdown of dielectrics under high voltage, it is necessary to understand the microscopic behavior of the dielectrics under the influence of an electric field [5,6,12]. In particular, the application of quantum mechanics in the theory of dielectrics is unavoidable: this leads to the topic of solid-state theory, and the reader unfamiliar with quantum mechanics, but who is nevertheless interested in pursuing more advanced aspects, may consult some introductory references such as [1,13] geared toward solid-state theory. On the other hand, for the more empirically minded reader, Paschen breakdown curves have been studied extensively, and describe the maximal field between two electrodes in a gaseous medium needed to initiate breakdown [7].*

A somewhat simplified view of a dielectric is sketched in Figure A.5 (Appendix). The first question that springs to mind is the following: does breakdown occur within a dielectric when its surface cannot hold any more free charges? This in turn begs another question. How much free charges can a dielectric hold before breakdown occurs? This question implies tacitly that the dielectric material is embedded in some homogeneous (dielectric) medium. The simplest approach to the latter is to consider an isolated conducting sphere embedded in a dielectric medium.

* The original work was investigated by Paschen in the nineteenth century [12].

9.1.1 Proposition

Given a neutral spherical conductor $\bar{S}^2 \subset (\mathbf{R}^3, \mu, \varepsilon)$, where $\bar{S}^2 = \{r \in \mathbf{R}^3 : |r| \le \delta\}$, suppose the electric field strength required to cause dielectric breakdown of $(\mathbf{R}^3, \mu, \varepsilon)$ is E_0. Then, the maximal free charge Q that $\bar{S}^2$ can hold prior to the ionization of the dielectric medium is

$$Q = 4\pi\varepsilon E_0 \delta^2 \tag{9.1}$$

Proof

Recall from Example 3.2.1, it was observed that when a point charge q is sufficiently close to $\bar{S}^2$, the mirror image will attract q to $\bar{S}^2$. In particular, whenever q is sufficiently close to $\bar{S}^2$, it will be attracted to the surface of $\bar{S}^2$. In principle, $\bar{S}^2$ can absorb an arbitrarily large charge, as long as there is a sufficiently large force to drive q sufficiently close to the surface to overcome Coulomb repulsion.

Now, given a charge Q on $\bar{S}^2$, the electric field on the surface is $E = \frac{1}{4\pi\varepsilon}\frac{Q}{\delta^2}$. Because E_0 is the dielectric breakdown of $(\mathbf{R}^3, \mu, \varepsilon)$, it follows that ionization of the medium will begin when

$$E_0 = \frac{1}{4\pi\varepsilon}\frac{Q}{\delta^2} \Rightarrow Q = 4\pi\varepsilon E_0 \delta^2$$

as required. □

Indeed, it is clear from Proposition 9.1.1 that even from a classical analysis perspective, a spherical conductor can hold more charges prior to dielectric breakdown if the electric permittivity of the medium is large or if the radius of the spherical conductor is large.

This section ends with a related topic that is frequently employed in EMC design: electrostatic shielding. This is clearly one way of avoiding dielectric breakdown! Consider Figure 9.1, where $C_i \subset \mathbf{R}^3$ denotes a compact conductive boundary of the ith conductor, and for notational convenience, let $D(C_i) \subset \mathbf{R}^3$ define the compact space bounded by C_i, that is, $C_i = \partial D(C_i)$, and set $D(C_i)^\circ$ to be the interior of $D(C_i)$. Suppose conductor $C_1 \subset D(C_2)$ and $C_i \subset \mathbf{R}^3 - D(C_2)\ \forall i \ge 3$. Finally suppose that C_2 is grounded: $V_2 = 0$.

From Lemma 6.1.3,

$$Q_1 = C_{11}V_1 + \cdots + C_{1n}V_n$$
$$\vdots$$
$$Q_n = C_{n1}V_1 + \cdots + C_{nn}V_n$$

Thus, $V_2 = 0 \Rightarrow V_1$ is completely determined with respect to C_2 inasmuch as all of the electric field from C_1 terminates on C_2, and not on $C_i\ \forall i \ge 3$. Thus, $C_{1i} = 0\ \forall i \ge 3$; that is, the electric field on C_1 does not couple with $C_i\ \forall i \ge 3$.

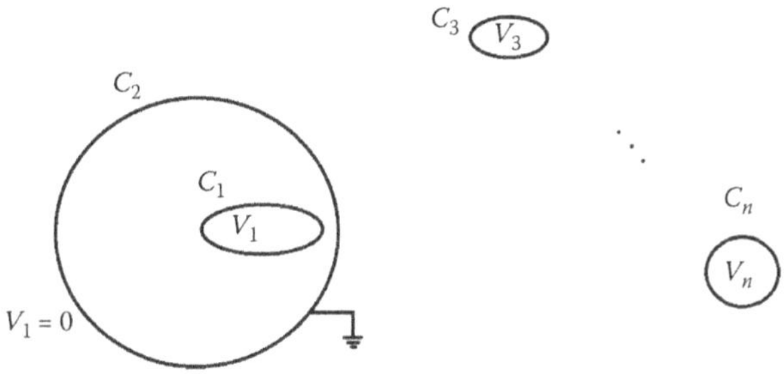

FIGURE 9.1
Electrostatic screening of a conductor.

Hence, the potential on C_1 is independent of the potential on $C_i \ \forall i \geq 3$ as $C_{1i} = C_{i1} \ \forall i$. By reciprocity, $C_i \ \forall i \geq 3$ are shielded from the potential of C_1.

What is more interesting to note—albeit it is intuitively obvious—is that if $V_i = 0 \ \forall i \geq 2$, then $C_{1i} = 0 \ \forall i \geq 3$ implies at once that the system of equations reduces to:

$$Q_1 = C_{11} V_1 \quad \text{and} \quad Q_2 = C_{21} V_1$$

In particular, $Q_i = 0 \ \forall i \geq 3$, and hence, the charge on C_1 will not induce any charge on $C_i \ \forall i \geq 3$. Moreover, because $Q_2 = -Q_1$ as free charges on C_1 will induce the opposite charge on C_2, it follows that $C_{11} = -C_{21} = C_{12}$.

9.2 Dielectric Properties: the Kramers–Kronig Relations

In the first half of this section, the electromagnetic properties of dielectrics are investigated from a classical perspective. Without loss of generality, time harmonicity is assumed: $e^{-i\omega t}$. Recalling that dielectric polarization is the result of bound charges, the rearrangement of bound charges within the dielectric results in "bound" currents, in complete analogy with the movement of free charges resulting in conduction currents. In what follows, define the conduction current by $\boldsymbol{J} = \sigma_1 \boldsymbol{E}$ and the displacement current by $\boldsymbol{D} = \varepsilon_1 \boldsymbol{E}$.

So, from Maxwell's $\nabla \times \boldsymbol{B} = \mu \boldsymbol{J} + \mu\varepsilon_1 \partial_t \boldsymbol{E}$, define $\nabla \times \boldsymbol{B} = -i\omega\mu\varepsilon \boldsymbol{E}$, where

$$\varepsilon = \varepsilon_1 + \tfrac{i\sigma_1}{\omega} \equiv \varepsilon_1 + i\varepsilon_2 \tag{9.2}$$

Furthermore, it can be shown [8] classically that conductivity is complex: $\sigma = \sigma_1 + i\sigma_2$, $\sigma_i \in \mathbf{R}, i = 1, 2$. More specifically, $\sigma_1 = K \frac{g}{g^2 + \omega^2}$ and $\sigma_2 = K \frac{\omega}{g^2 + \omega^2}$, where K is some real constant and $g \gg 1$ is a damping constant that is dependent upon the average collision rate between an electron and the crystal lattice of the conductor. Thus, in view of Equation (9.2), define

$$\varepsilon = \varepsilon_0 + \tfrac{i\sigma}{\omega} \tag{9.3}$$

Then, $\Re e(\varepsilon) = \varepsilon_0 - \frac{K}{g^2+\omega^2}$ and $\Im m(\varepsilon) = \frac{K}{\omega(g^2+\omega^2)}$. For low $\omega > 0$, that is, $g >> \omega$, $\sigma_1 >> \sigma_2 \Rightarrow \sigma \approx \sigma_1$. In the microwave regime, $g >> \omega$ typically holds and hence, the electrical conductivity may be assumed real. In the optical regime, this is false in general.

A more rigorous approach is given below, culminating in the Kramers–Kronig relations, also known as the *dispersion relations*, for the electric permeability and electric conductivity of a dielectric. In what follows, the dielectric is assumed to be isotropic and homogeneous for ease of analysis, and the response of the dielectric under a forcing function to be linear; this guarantees that the Fourier transform may be invoked.

By way of motivation, consider $\boldsymbol{D} = \varepsilon_1 \boldsymbol{E} = \boldsymbol{P} + \varepsilon_0 \boldsymbol{E} \Rightarrow \boldsymbol{P} = (\varepsilon_1 - \varepsilon_0)\boldsymbol{E}$. Recall that the polarization field is the result of displaced electron clouds about the nuclei; see Figure 9.2. The displaced electron clouds generate (atomic or molecular) dipoles within the dielectric subject to an external electric field. That is, the polarization field is the response to a driving electric field, and hence, $\varepsilon_1 - \varepsilon_0$ is the *response function.*

Note that the assumption of response linearity implies that the Fourier transform may be employed:

$$\boldsymbol{P}(\omega) = \tilde{\varepsilon}(\omega)\boldsymbol{E}(\omega) \Rightarrow \boldsymbol{P}(t) = \int \tilde{\varepsilon}(\omega)\boldsymbol{E}(\omega)\mathrm{e}^{-i\omega t}\,\mathrm{d}t$$

where $\tilde{\varepsilon}$ is the response function. Taking the Fourier transform (in frequency domain) leads to

$$\mathcal{F}[\boldsymbol{P}] = \int \boldsymbol{P}(t)\mathrm{e}^{i\omega t}\,\mathrm{d}t = \int \mathrm{e}^{i\omega t}\,\mathrm{d}t \int \tilde{\varepsilon}(t-\tau)\boldsymbol{E}(\tau)\mathrm{d}\tau = \int \mathrm{e}^{i\omega\tau}\,\mathrm{d}\tau \int \tilde{\varepsilon}(t-\tau)\boldsymbol{E}(\tau)\mathrm{e}^{i\omega(t-\tau)}\,\mathrm{d}t.$$

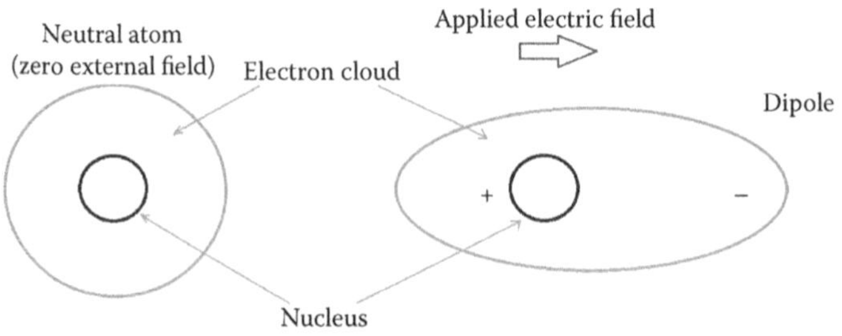

FIGURE 9.2
Electron cloud distortion resulting from an applied electric field.

That is,

$$\mathcal{F}[\boldsymbol{P}](\omega)=\int E(\tau)e^{i\omega\tau}\,d\tau\int\tilde{\varepsilon}(t-\tau)e^{i\omega(t-\tau)}\,dt\equiv\mathcal{F}[\boldsymbol{E}](\omega)\mathcal{F}[\tilde{\varepsilon}](\omega).$$

Last, it is assumed tacitly here that response functions are analytic in ω and in particular, purely on physical grounds, response functions must tend to zero in the limit as $\omega\to\infty$ as a physical system cannot respond quickly enough to an infinitely large frequency. That is, response functions are assumed to be bounded and analytic.

9.2.1 Theorem (Kramers–Kronig Relations)

Suppose a dielectric medium is isotropic, homogeneous, and causal,* then the electric conductivity and permittivity of the medium satisfy

$$\sigma_1(\omega)=\tfrac{2}{\pi}\int_0^\infty \tfrac{\omega' f_2(\omega)}{\omega'^2-\omega^2}\,d\omega' \tag{9.4}$$

$$\sigma_2(\omega)=-\tfrac{2\omega}{\pi}\int_0^\infty \tfrac{f_2(\omega)}{\omega'^2-\omega^2}\,d\omega' \tag{9.5}$$

$$\varepsilon_1(\omega)=\varepsilon_0+\tfrac{2}{\pi}\int_0^\infty \tfrac{\varepsilon_2(\omega')\omega'}{\omega'^2-\omega^2}\,d\omega' \tag{9.6}$$

$$\varepsilon_2(\omega)=-\tfrac{2}{\pi\omega}\int_0^\infty \tfrac{(\varepsilon_1(\omega')-\varepsilon_0)\omega'^2}{\omega'^2-\omega^2}\,d\omega' \tag{9.7}$$

Proof

Consider the Fourier transform $\mathcal{F}[\tilde{\varepsilon}](z)=\int_{-\infty}^{\infty}\tilde{\varepsilon}(t-\tau)e^{iz(t-\tau)}dt$, where $z\in C$. So, setting $z = x + iy$, $x,y\in\mathbf{R}$ (no bearing on the components of the rectangular coordinate system), and invoking causality, that is, $\tilde{\varepsilon}(t-\tau)=0\ \forall t<\tau$, it follows that

$$\mathcal{F}[\tilde{\varepsilon}](z)=\int_{-\infty}^{t}\tilde{\varepsilon}(t-\tau)e^{iz(t-\tau)}\,dt=\int_{-\infty}^{t}\tilde{\varepsilon}(t-\tau)e^{-y(t-\tau)}e^{ix(t-\tau)}\,dt$$

Now, observe that $y>0\Rightarrow e^{-y(t-\tau)}<\infty\ \forall t>\tau$, and the causality assumption implies that $\mathcal{F}[\tilde{\varepsilon}](z)=0\ \forall t-\tau<0\Rightarrow\mathcal{F}[\tilde{\varepsilon}](z)\neq 0\ \forall y>0$. That is, $\mathcal{F}[\tilde{\varepsilon}](z)$ is analytic on the upper half-plane $\mathbf{C}_+=\{z\in\mathbf{C}:\Im m(y)\geq 0\}$ of C. So, consider a closed contour $\Gamma\subset\mathbf{C}_+$ defined by Figure 9.3, where a semi-circular loop $\tilde{\Gamma}=\Gamma*\rho_-*\gamma*\rho_+$ is defined as follows: γ is a semi-circle of radius $\delta>0$ about the point z_0 on the real line, and Γ is a semi-circle of radius $R\gg\delta$, ρ_- is the path from $[-R,z_0-\delta]$, and ρ_+ is the path from $[z_0+\delta,R]$. Then, by definition,

* Informally, causality implies that there is zero response for all times prior to the application of a forcing function; that is, cause and effect.

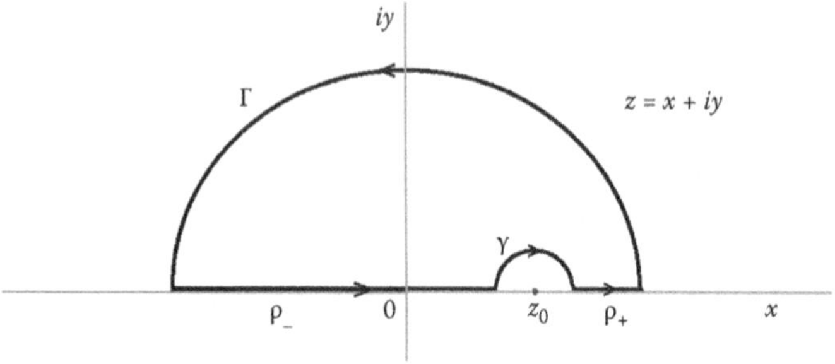

FIGURE 9.3
Contour integral on the upper-half complex plane.

for all points within the space bounded by $\tilde{\Gamma}$ and on $\tilde{\Gamma}$, $\mathcal{F}[\tilde{\varepsilon}](z)$ is analytic and hence, for $z_0 \notin D(\tilde{\Gamma}) = \{z \in \mathbf{C}_+ : |z| < R\} - \{z \in \mathbf{C}_+ : |z - z_0| < \delta\}$, the domain bounded by $\tilde{\Gamma}$, implies that $\frac{\mathcal{F}[\tilde{\varepsilon}](z)}{z-z_0}$ is also analytic on $D(\tilde{\Gamma})$. Hence, by the Cauchy–Goursat theorem,* $\oint_{\tilde{\Gamma}} \frac{\mathcal{F}[\tilde{\varepsilon}](z)}{z-z_0} dz = 0$. Thus,

$$0 = \int_{\Gamma} \frac{\mathcal{F}[\tilde{\varepsilon}](z)}{z-z_0} dz + \int_{\gamma} \frac{\mathcal{F}[\tilde{\varepsilon}](z)}{z-z_0} dz + \int_{\rho_-} \frac{\mathcal{F}[\tilde{\varepsilon}](z)}{z-z_0} dz + \int_{\rho_+} \frac{\mathcal{F}[\tilde{\varepsilon}](z)}{z-z_0} dz$$

From Equation (9.3), $\lim_{R\to\infty} \int_{\Gamma} \frac{\mathcal{F}[\tilde{\varepsilon}](z)}{z-z_0} dz \to 0$. To see this, it suffices to observe that $\mathcal{F}[\tilde{\varepsilon}]$ is analytic and bounded (by assumption) on $D(\tilde{\Gamma})$ and in particular, $\Re e\mathcal{F}[\tilde{\varepsilon}] \le \frac{M'}{R^2}$ and $\Re e\mathcal{F}[\tilde{\varepsilon}] \le \frac{M''}{R}$, for some constants $M', M'' > 0$. Furthermore,

$$\lim_{R\to\infty} \left| \int_{\Gamma} \frac{1}{z-z_0} dz \right| = \lim_{R\to\infty} \left| \int_0^{\pi} \left\{1 - \frac{z_0}{R} e^{-i\theta}\right\}^{-1} d\theta \right| = \pi < \infty$$

where $\int \{1 - ae^{-i\theta}\}^{-1} d\theta = -i\ln(a - e^{i\theta})$ was employed. Thus, $\lim_{R\to\infty} \int_{\Gamma} \frac{\mathcal{F}[\tilde{\varepsilon}](z)}{z-z_0} dz \to 0$, as claimed. Thus, it remains to evaluate

$$I_0 \equiv \lim_{\substack{\delta\to 0 \\ R\to\infty}} \int_{z_0+\delta}^{R} \frac{\mathcal{F}[\tilde{\varepsilon}](z)}{z-z_0} dz + \lim_{\substack{\delta\to 0 \\ R\to\infty}} \int_{-R}^{z_0-\delta} \frac{\mathcal{F}[\tilde{\varepsilon}](z)}{z-z_0} dz = -\lim_{\delta\to 0} \int_{\gamma} \frac{\mathcal{F}[\tilde{\varepsilon}](z)}{z-z_0} dz$$

First, observe that as $\mathcal{F}[\tilde{\varepsilon}]$ is analytic on $\mathbf{C}_+$, it follows that the residue of $\frac{\mathcal{F}[\tilde{\varepsilon}](z)}{z-z_0}$ is $\mathcal{F}[\tilde{\varepsilon}](z_0)$. For notational simplicity, set $f = \mathcal{F}[\tilde{\varepsilon}]$. Then,

$$I_0 = -\lim_{\delta\to 0} \int_{\gamma} \frac{f(z) - f(z_0) + f(z_0)}{z-z_0} = -\lim_{\delta\to 0} \int_{\gamma} \frac{f(z_0)}{z-z_0} - \lim_{\delta\to 0} \int_{\gamma} \frac{f(z) - f(z_0)}{z-z_0} \equiv I_0^{(1)} + I_0^{(2)}$$

* See Reference [3]: if a function is analytic on a complex domain bounded by a simple loop, then the loop integral is zero.

Now, the analyticity of f implies that $f(z) = f(z_0) + f'(z_0)(z - z_0) + \cdots$ and hence, $I_0^{(2)} = 0$. Thus, evaluating $I_0^{(1)}$ completes the initial part of the proof. The method is routine: set $z = \delta e^{i\theta}$. Then, $dz = i\delta e^{i\theta} d\theta$ and evaluating the integral yields

$$\int_\gamma \tfrac{dz}{z - z_0} = i\int_\pi^0 \delta e^{i\theta}\left\{\delta e^{i\theta} - z_0\right\}^{-1} d\theta = -i\int_0^\pi \left\{1 - z_0\delta e^{-i\theta}\right\}^{-1} d\theta = \ln\tfrac{z_0\delta - 1}{z_0\delta + 1}$$

whence,

$$I_0 = -f(z_0)\ln(-1) = -f(z_0)\ln e^{i\pi} = -i\pi f(z_0) = -i\pi\,\mathcal{F}[\tilde{\varepsilon}](z_0)$$

That is, upon replacing $z_0 \to \omega$, it follows at once that

$$\mathcal{F}[\tilde{\varepsilon}](\omega) = -\tfrac{i}{\pi}\mathcal{P}\int_{-\infty}^{\infty} \tfrac{\mathcal{F}[\tilde{\varepsilon}](\omega')}{\omega' - \omega} d\omega' \tag{9.8}$$

where $\mathcal{P}$ denotes the Cauchy principal value and it is defined by

$$\mathcal{P}\int_{-\infty}^{\infty} \tfrac{\mathcal{F}[\tilde{\varepsilon}](\omega')}{\omega' - \omega} d\omega' = \lim_{r\to\infty}\int_{-r}^{r} \tfrac{\mathcal{F}[\tilde{\varepsilon}](\omega')}{\omega' - \omega} d\omega'$$

Next, setting $\mathcal{F}[\tilde{\varepsilon}](\omega) \equiv f = f_1(\omega) + if_2(\omega)$ yields

$$f_1(\omega) = \tfrac{1}{\pi}\mathcal{P}\int_{-\infty}^{\infty} \tfrac{f_2(\omega')}{\omega' - \omega} d\omega' \quad \text{and} \quad f_2(\omega) = -\tfrac{1}{\pi}\mathcal{P}\int_{-\infty}^{\infty} \tfrac{f_1(\omega')}{\omega' - \omega} d\omega'$$

Also, observe that a response function, for $f \in \{\varepsilon, \sigma\}$, satisfies $f(-\omega) = f^*(\omega)$, hence $f_1(-\omega) = f_1(\omega)$ and $f_2(-\omega) = -f_2(\omega)$. This is because

$$\mathcal{F}[\tilde{\varepsilon}](\omega) = \int_{-\infty}^{t} \tilde{\varepsilon}(t - \tau)e^{i\omega(t-\tau)}\, dt \Rightarrow \mathcal{F}[\tilde{\varepsilon}]^*(\omega) = \int_{-\infty}^{t} \tilde{\varepsilon}(t - \tau)e^{-i\omega(t-\tau)}\, dt$$

Thus, the real part is an even function whereas the imaginary part is an odd function of ω, whence,

$$\begin{aligned}
f_1(\omega) &= \tfrac{1}{\pi}\mathcal{P}\int_{-\infty}^{\infty} \tfrac{f_2(\omega')}{\omega' - \omega} d\omega' \\
&= \tfrac{1}{\pi}\mathcal{P}\left\{\int_{-\infty}^{0} \tfrac{f_2(\omega')}{\omega' - \omega} d\omega' + \int_0^{\infty} \tfrac{f_2(\omega')}{\omega' - \omega} d\omega'\right\} = \tfrac{1}{\pi}\mathcal{P}\left\{-\int_0^{-\infty} \tfrac{f_2(\omega')}{\omega' - \omega} d\omega' + \int_0^{\infty} \tfrac{f_2(\omega')}{\omega' - \omega} d\omega'\right\} \\
&= \tfrac{1}{\pi}\mathcal{P}\left\{-\int_0^{\infty} \tfrac{-f_2(\omega')}{-(\omega' + \omega)}(-d\omega') + \int_0^{\infty} \tfrac{f_2(\omega')}{\omega' - \omega} d\omega'\right\} = \tfrac{1}{\pi}\mathcal{P}\left\{\int_0^{\infty} f_2(\omega')\left(\tfrac{1}{\omega' - \omega} + \tfrac{1}{\omega' + \omega}\right) d\omega'\right\} \\
&= \tfrac{2}{\pi}\mathcal{P}\int_0^{\infty} \tfrac{\omega' f_2(\omega)}{\omega'^2 - \omega^2} d\omega'
\end{aligned}$$

and likewise,

$$f_1(\omega) = \tfrac{1}{\pi}\mathcal{P}\int_{-\infty}^{\infty} \tfrac{f_2(\omega')}{\omega'-\omega}\,d\omega'$$

$$= -\tfrac{1}{\pi}\mathcal{P}\left\{-\int_0^{\infty} \tfrac{f_2(\omega')}{-(\omega'+\omega)}(-\,d\omega') + \int_0^{\infty} \tfrac{f_2(\omega')}{\omega'-\omega}\,d\omega'\right\} = \tfrac{1}{\pi}\mathcal{P}\left\{\int_0^{\infty} f_2(\omega')\left(\tfrac{1}{\omega'-\omega} - \tfrac{1}{\omega'+\omega}\right)d\omega'\right\}$$

$$= -\tfrac{2\omega}{\pi}\mathcal{P}\int_0^{\infty} \tfrac{f_2(\omega)}{\omega'^2-\omega^2}\,d\omega'$$

Furthermore, replacing f by σ based on the discussion preceding the theorem, it follows via $\sigma(\omega) = \sigma_1(\omega) + i\sigma_2(\omega)$ that

$$\sigma_1(\omega) = \tfrac{2}{\pi}\mathcal{P}\int_0^{\infty} \tfrac{\omega' f_2(\omega)}{\omega'^2-\omega^2}\,d\omega' \quad \text{and} \quad \sigma_2(\omega) = -\tfrac{2\omega}{\pi}\mathcal{P}\int_0^{\infty} \tfrac{f_2(\omega)}{\omega'^2-\omega^2}\,d\omega'$$

Also, noting from definition $\varepsilon = \varepsilon_0 + i\frac{\sigma}{\omega}$ that $\sigma = \omega\varepsilon_2 - i\omega(\varepsilon - \varepsilon_0)$, and hence, substituting $\omega\varepsilon_1 = \sigma_1$ and $-\omega(\varepsilon - \varepsilon_0) = \sigma_2$ yields

$$\varepsilon_1(\omega) = \varepsilon_0 + \tfrac{2}{\pi}\mathcal{P}\int_0^{\infty} \tfrac{\varepsilon_2(\omega')\omega'}{\omega'^2-\omega^2}\,d\omega' \quad \text{and} \quad \varepsilon_2(\omega) = -\tfrac{2\omega}{\pi}\mathcal{P}\int_0^{\infty} \tfrac{\varepsilon_1(\omega')-\varepsilon_0}{\omega'^2-\omega^2}\,d\omega'$$

Last, noting that as the analytic functions ε_i are bounded in the limit as $\omega \to \infty$, it follows that the integrals are absolutely convergent and hence,

$$\varepsilon_1(\omega) = \varepsilon_0 + \tfrac{2}{\pi}\int_0^{\infty} \tfrac{\varepsilon_2(\omega')\omega'}{\omega'^2-\omega^2}\,d\omega' \quad \text{and} \quad \varepsilon_2(\omega) = -\tfrac{2}{\pi\omega}\int_0^{\infty} \tfrac{(\varepsilon_1(\omega')-\varepsilon_0)\omega'^2}{\omega'^2-\omega^2}\,d\omega'$$

as required. □

The above dispersion relations show how the real and imaginary components are related to one another in the frequency domain.

9.2.2 Corollary

The DC conductivity σ_{DC} of a conductor is given by

$$\sigma_{DC} = \tfrac{2}{\pi}\int_0^{\infty} (\varepsilon_0 - \varepsilon_1(\omega))\,d\omega \tag{9.9}$$

Proof

Because the DC electric conductivity must be real, Equation (9.4) can be invoked. So, setting $\omega = 0$ for the DC case, and substituting $\sigma_2 = \omega(\varepsilon_0 - \varepsilon_1)$, (9.4) leads immediately to

$$\sigma_1(0) = \tfrac{2}{\pi}\int_0^{\infty} \tfrac{\sigma_2(\omega')}{\omega'}\,d\omega' = \tfrac{2}{\pi}\int_0^{\infty} \tfrac{1}{\omega'}(\varepsilon_0 - \varepsilon_1(\omega'))\omega'\,d\omega' = \tfrac{2}{\pi}\int_0^{\infty} (\varepsilon_0 - \varepsilon_1(\omega'))\,d\omega'$$ □

By inspecting Equation (9.6), it is clear that in the limit as $\omega \to 0$, the real part of the electric permittivity becomes $\varepsilon_1(0) = \varepsilon_0 + \frac{2}{\pi}\int_0^\infty \frac{\varepsilon_2(\omega)}{\omega^2} d\omega$. Indeed, these expressions, including those from Theorem 9.2.1 and Corollary 9.2.2, enable the real part of electric permittivity to be determined when the imaginary part of the electric permittivity is known. What is more interesting to note is the following: the Kramers–Kronig relations display how the real and imaginary parts are related; they are not independent of one another!

9.2.3 Remark

It can be shown (see Exercise 9.5.1) that the dielectric constant of a material can be determined classically by considering a collection of oscillating dipoles subject to a time-harmonic monochromatic plane wave:

$$\frac{\varepsilon(\omega)}{\varepsilon_0} = 1 + \frac{Ne^2}{\varepsilon_0 m_0} \sum_j \frac{f_j}{\omega_j^2 - \omega^2 - i\gamma_j \omega} \tag{9.10}$$

where f_i denotes the probability that an electronic transition will take place at the resonant angular frequency ω_j of the jth electronic dipole oscillator. It is a quantum mechanical phenomenon and must clearly satisfy $\sum_j f_j = 1$. Moreover, $|e|$ denotes the electronic charge, m_0 the mass of an electron, and γ_j corresponds to the damping for the jth electronic dipole oscillator.

9.3 Beyond Classical Theory

The dielectric properties of materials cannot be understood completely within the context of classical physics. As such, a brief venture into the world of quantum mechanics is unavoidable. Indeed, quantum mechanics is a necessary evil to understand dielectric breakdown phenomena. Thus, a short summary is provided in order to establish some notations; the interested reader may satisfy his or her intellectual craving with a plethora of books on quantum mechanics and solid-state theory. For example, see also References [2,4] in addition to those given in Section 9.1. Clearly, no pretense is made for completeness; only a brief informal outline is sketched.

9.3.1 Axioms (Quantum Mechanics)

The following axioms describe the basic tenet of quantum mechanics.

(a) The *state* ψ of a (physical) system is completely described by a unit vector $|\psi\rangle$ in a Hilbert space.* That is, the inner product $\langle\psi|\psi\rangle = 1$ for the unit vector.

* See Appendix A.4 for the definition of a Hilbert space. For the present, the reader may view this space as a vector space endowed with an inner product (the "dot" product) structure.

(b) A *physical observable* $\mathbf{A}$ is a *Hermitian* operator on the Hilbert space; that is, $\mathbf{A} = \mathbf{A}^{\dagger}$, where, given any two states ψ, φ, $\mathbf{A}^{\dagger}$ is defined by $\langle\psi|\mathbf{A}\varphi\rangle = \langle\mathbf{A}^{\dagger}\psi|\varphi\rangle$.

(c) Given an observable $\mathbf{A}$, let $\{|\psi_n\rangle\} \subset \mathrm{H}$ be a set of eigenvectors of $\mathbf{A}$, that is, $\mathbf{A}\,|\psi_n\rangle \equiv |\mathbf{A}\psi_n\rangle = a_n\,|\psi_n\rangle\ \forall n = 1,2,\ldots$, where $\mathcal{H}$ denotes the Hilbert space of states and $a_n \in \mathbf{R}\ \forall n$. Then, the probability of measuring the value a_n when the system is in state ψ is precisely $|\langle\psi_n|\psi\rangle|^2$. In particular, the *measurement* of an observable corresponds precisely to one of its eigenvalues. Moreover, the state ψ immediately after the measurement collapses down to the state ψ_n : $|\psi\rangle \to |\psi_n\rangle$, so that the probability of finding the state in $|\psi_n\rangle$ immediately after the measurement is 1.

(d) States are defined by the *Schrödinger equation:* $\frac{d}{dt}|\psi\rangle = -i\frac{2\pi}{h}\mathbf{H}\,|\psi\rangle$, where $h = 6.63 \times 10^{-34}$ J·s is called the *Planck constant* and $\mathcal{H}$ is the *Hamiltonian* operator, and it is associated with the energy of the system; that is, the eigenvalues correspond precisely to the energy states of the system. The Hamiltonian is defined by $\mathbf{H} = -\frac{1}{2m}\hbar^2\Delta + \mathbf{U}$, where $\hbar = \frac{h}{2\pi}$, m is the mass of the particle described by $|\psi\rangle$, U is the potential energy operator, and when it is purely a function of position, it acts upon the state function via multiplication; that is, $\mathbf{U}\psi = U\psi$, where $\mathbf{x}\psi \equiv x\psi$.

9.3.2 Remark

(a) Classically, the state of a system can be described by the positions and momenta of all the particles comprising the system. In quantum mechanics, the state of a closed system can be completely described by a state vector in Hilbert space. (b) A classical observable is any measureable quantity that can be attributed to a system, for example, energy, momentum, position, and the like. (c) The classical Hamiltonian of a system is defined by $H = T + U$, where T is the kinetic energy of the system and U is the potential energy of the system. A critical point to note is the following: the transition from classical to quantum mechanics transforms classical observables into Hermitian operators acting on a Hilbert space. Finally, a wave function is described at an abstract level by a unit vector in Hilbert space, which completely describes the state of a particle.

An important point in introducing the elements of quantum mechanics here is to illustrate the concept of quantum tunneling. This concept has wide applications in engineering, an example of which is tunneling diodes. Tunneling is a purely quantum phenomenon that does not exist in the classical setting. Explicitly, suppose a particle is trapped in a square potential well U with a finite width potential barrier (see Figure 9.4). If the kinetic energy T of the particle satisfies $T < U$, then classically, it is impossible for the particle to escape the potential barrier. However, in quantum mechanics, there is a nontrivial probability that the particle can travel beyond the potential barrier.

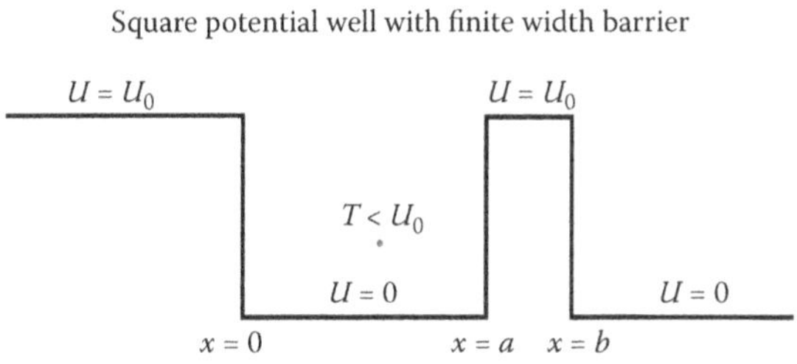

FIGURE 9.4
Finite potential barrier illustrating quantum tunneling phenomenon.

9.3.3 Definition

A wave function ψ representing the state vector $|\psi\rangle$ of a particle is called the *probability amplitude* of the particle, and the probability that the particle exists in some compact space $K \subset \mathbf{R}^3$ is given by $P_\psi(K) = \int_K \psi^*(\boldsymbol{r})\psi(\boldsymbol{r})d^3\boldsymbol{r}$, where $\int_{\mathbf{R}^3} \psi^*(\boldsymbol{r})\psi(\boldsymbol{r})d^3\boldsymbol{r} = 1$.* Moreover, the *probability current density* of the state is defined by $\boldsymbol{j} = -\frac{i}{2m}\hbar(\psi^*\nabla\psi - \psi\nabla\psi^*)$.

Indeed, it can be shown [4] that $\partial_t\rho + \nabla \cdot \boldsymbol{j} = 0$, where $\rho \equiv \psi^*\psi$ defines the *probability density* function. This leads to the conservation of probability. See the charge conservation equation in Maxwell's theory, and hence the respective terminologies defined above. Last, the following definitions are required to describe the phenomenon of tunneling. Suppose ψ_0 is the incident wave, ψ_r the reflected wave, and ψ_t the transmitted wave across a barrier. Then, $R = \left|\frac{j_r}{j_0}\right|$ defines the *reflection coefficient* whereas $S = \left|\frac{j_t}{j_0}\right|$ defines the *transmission coefficient* of the wave function.

9.3.4 Example

Consider a one-dimensional scenario, wherein a particle is trapped in a square potential well of width a as illustrated in Figure 9.4. Let the potential energy of the barrier be defined by

$$U = \begin{cases} U_0 & \text{for } x \in D_1 \cup D_3 \\ 0 & \text{for } x \in D_2 \cup D_4 \end{cases}$$

where $D_1 = (-\infty, 0]$, $D_2 = (0, a)$, $D_3 = [a, b]$ and $D_4 = (a, \infty)$.

Suppose the particle has a kinetic energy $T = E$ and mass m. Inasmuch as the potential energy is independent of time, the particle is in a stationary state and hence, via Exercise 9.5.3, the solution satisfies the time-independent

* The probability that the particle is somewhere in space is unity. Here, it is implicitly assumed that the wave function is normalized.

Schrödinger equation $\mathbf{H}|\psi\rangle = E|\psi\rangle$. That is, the one-dimensional Schrödinger equation for the particle is described by

$$\left\{-\tfrac{1}{2m}\hbar^2\Delta + U\right\}\psi = E\psi \Rightarrow -\tfrac{1}{2m}\hbar^2\tfrac{d^2}{dt^2}\psi = (E-U)\psi$$

where $E = \frac{1}{2}\frac{p^2}{m}$, $p = mv$ is the momentum of the particle.

On $D_1 \cup D_3$, Schrödinger equation becomes

$$-\tfrac{1}{2m}\hbar^2\tfrac{d^2}{dt^2}\psi + U_0\psi = E\psi$$

So, on setting $\beta^2 = 2m\left(\frac{1}{\hbar}\right)^2(U_0 - E)$, the equation becomes $\frac{d^2}{dt^2}\psi - \beta^2\psi = 0$ and hence, the general respective solutions are:

$$\psi_1 = A_1 e^{\beta x} \text{ on } D_1 \tag{9.11}$$

$$\psi_3 = A_3^+ e^{\beta x} + A_3^- e^{-\beta x} \text{ on } D_3 \tag{9.12}$$

where the solution $e^{-\beta x}$ in Equation (9.11) was discarded as it diverges in the limit as $x \to -\infty$ and hence is nonphysical.

However, on $D_2 \cup D_4$, the Schrödinger equation becomes

$$-\tfrac{1}{2m}\hbar^2\tfrac{d^2}{dt^2}\psi = E\psi$$

Thus, set $k^2 = 2m\left(\frac{1}{\hbar}\right)^2 E$. Then, the general solutions for the respective domains are:

$$\psi_2 = A_2^+ e^{ikx} + A_2^- e^{-ikx} \text{ on } D_2 \tag{9.13}$$

$$\psi_4 = A_4 e^{ikx} \text{ on } D_4 \tag{9.14}$$

as there is no boundary for $x > b$ to reflect the waves back toward $x = b$.

To solve the coefficients, it suffices to note that mathematically $\psi, \frac{d}{dx}\psi$ must be continuous at the boundaries: $x = 0,a,b$. So, imposing the boundary conditions at $x = 0$,

$$A_1 = A_2^+ + A_2^- \tag{9.15}$$

$$\beta A_1 = ikA_2^+ - ikA_2^- \tag{9.16}$$

whence, ik(9.15) + (9.16) yields $(\beta + ik)A_1 = ikA_2^+ \Rightarrow \frac{A_1}{A_2^+} = \frac{2ik}{\beta+ik}$.

At $x = a$, the boundary conditions yield

$$A_2^+ e^{ika} + A_2^- e^{-ika} = A_3^+ e^{\beta a} + A_3^- e^{-\beta a} \tag{9.17}$$

$$ikA_2^+ e^{ika} - ikA_2^- e^{-ika} = \beta A_3^+ e^{\beta a} - \beta A_3^- e^{-\beta a} \tag{9.18}$$

Thus, ik(9.17) + (9.18) yields

$$2ikA_2^+ e^{ika} = (ik + \beta) A_3^+ e^{\beta a} + (ik - \beta) A_3^- e^{-\beta a} \tag{9.19}$$

At $x = b$, the boundary conditions yield

$$A_4 e^{ikb} = A_3^+ e^{\beta b} + A_3^- e^{-\beta b} \tag{9.20}$$

$$ikA_4 e^{ikb} = \beta A_3^+ e^{\beta b} - \beta A_3^- e^{-\beta b} \tag{9.21}$$

whence, β(9.20) ± (9.21) lead immediately to

$$A_3^+ = \tfrac{\beta + ik}{2\beta} A_4 e^{ikb - \beta b} \quad \text{and} \quad A_3^- = \tfrac{\beta - ik}{2\beta} A_4 e^{ikb + \beta b}$$

Substituting these expressions into Equation (9.19) gives, after some algebraic manipulation, and using the identities $\sinh x = \frac{1}{2}(e^x - e^{-x})$ and $\cosh x = \frac{1}{2}(e^x + e^{-x})$,

$$2ik\beta A_2^+ e^{-ik(b-a)} = A_4\{-(\beta^2 - k^2)\sinh(\beta(b-a)) + i2k\beta\cosh(\beta(b-a))\}$$

That is, $\frac{A_2^+}{A_4} = e^{ik(b-a)}\left\{\cosh(\beta(b-a)) + i\frac{\beta^2 - k^2}{2k\beta}\sinh(\beta(b-a))\right\}$. In particular,

$$\left|\frac{A_2^+}{A_4}\right|^2 = \cosh^2(\beta(b-a)) + \left(\frac{\beta^2 - k^2}{2k\beta}\right)^2 \sinh^2(\beta(b-a))$$

Finally, observe from Definition 9.3.3 that the current density of the incident wave in D_2 is

$$j_2^+ = -\tfrac{i}{2m}\hbar\left\{\left|A_2^+\right|^2 e^{-ikx} ike^{ikx} + \left|A_2^+\right|^2 e^{ikx} ike^{-ikx}\right\} = -\tfrac{i}{2m}\hbar 2ik\left|A_2^+\right|^2$$

and likewise, on D_4,

$$j_4 = -\tfrac{i}{2m}\hbar\left\{\left|A_4\right|^2 e^{-ikx} ike^{ikx} + \left|A_4\right|^2 e^{ikx} ike^{-ikx}\right\} = -\tfrac{i}{2m}\hbar 2ik\left|A_4\right|^2$$

Hence, the transmission coefficient across the barrier is

$$S_4 = \left|\frac{j_4}{j_2^+}\right| = \left|\frac{A_4}{A_2^+}\right|^2 = \left\{\cosh^2(\beta(b-a)) + \left(\frac{\beta^2 - k^2}{2k\beta}\right)^2 \sinh^2(\beta(b-a))\right\}^{-1}$$

It is evident here that as the width of the potential barrier approaches infinity, the transmission coefficient falls to zero as $\lim_{x\to\infty}\{\cosh x, \sinh x\} \to \infty$. Likewise, if the height of the potential barrier approaches infinity, the transmission coefficient falls to zero.

Furthermore, notice also that on $D_1 \cup D_2$, the probability amplitudes are real and hence, the current density is identically zero by definition. Physically, even though there is a nontrivial probability that the waves will penetrate the potential barrier, the waves are evanescent: they decay exponentially. For instance, it was derived above that $\frac{A_1}{A_2^+} = \frac{2ik}{\beta+ik}$ which is nonzero. This is in sharp distinction with classical physics which prohibits a particle from penetrating a potential barrier that is larger than its inherent kinetic energy. It is easy to see that the penetration is identically zero in quantum mechanics when the potential barrier is infinitely large: $\lim_{U_0\to\infty} \frac{A_1}{A_2^+} \sim \lim_{U_0\to\infty} \frac{1}{\beta} \to 0$. □

In passing, it is known that transistors, FETs, and MOSFETs have leakage currents. Indeed, from Example 9.3.3, it is somewhat obvious that the source of the leakage currents is due to quantum tunneling effects. In particular, quantum tunneling places a lower bound on the gate size in integrated circuits, and hence, Moore's law* will eventually reach an upper bound.

As a prelude to the topic on dielectric breakdown discussed in the following section, this section concludes with a quick review of the band theory of semiconductors. By way of introduction, a comparison between a conductor (i.e., metal) versus semiconductor is drawn. The theory for metals is essentially based on the Drude–Sommerfeld model, and that of semiconductors is based on the Lorentz model.

Informally, the difference between a conductor and an insulator is depicted in Figure 9.5. Basically, electrons in the ground state occupy the valance band. If there is sufficient energy to excite the electrons into the conduction band, and have the electrons remain there, then conduction can take place. Specifically, when the conduction band overlaps the valance band in a solid, and hence, electrons are able to diffuse from the valance band into the conduction band even under low ambient thermal conditions, the material defines a conductor. For a semiconductor, there is a small nonoverlap between the conduction band and the valance band; the separation is usually small enough that thermal energy beyond a small value may be sufficient to excite the valance electrons beyond the forbidden gap, a region wherein electrons states do not exist, into the conduction band.

On the other hand, for insulators, the forbidden zone is large enough that thermal energy alone is insufficient to excite the electrons into the conduction zone. Hence, there are no electrons in the conduction zone—or at least, very few electrons (perhaps due to tunneling) in comparison to a

* *Moore's law* essentially states that the number of transistors on integrated circuits doubles every two years.

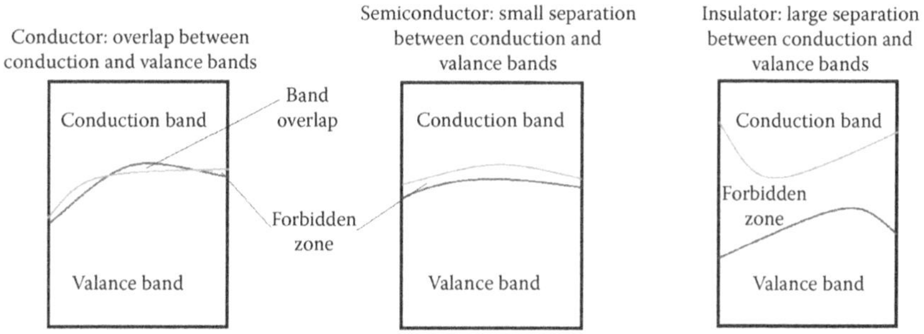

FIGURE 9.5
A simplistic description for (a) conductors, (b) semiconductors, (c) insulators.

semiconductor—rendering the material an insulator. Readers interested in solid-state theory may consult References [1,9,13].

9.3.5 Theorem

Suppose $N > 0$ energy states exist for a system in equilibrium. Then, the probability that the system is in a state with energy E_k is $\mathrm{P}(E_k) = \frac{1}{Z}\mathrm{e}^{-\beta E_k}$, where $\beta = \frac{1}{k_B T}$, with $k_B = 1.38 \times 10^{-23}$ J/K being the Boltzmann constant, T the temperature, and $Z = \sum_{n=1}^{N} \mathrm{e}^{-\beta E_n}$ the *partition function*. In particular, if $|k\rangle$ denotes the quantum state of a system with energy E_k, then the expected value of an observable $\mathbf{A}$ is given by $\langle \mathbf{A} \rangle = \frac{1}{Z}\sum_k \langle k|\mathbf{A}k\rangle \mathrm{e}^{-\beta E_k}$. □

Consider a quantum system comprising $0 < N \leq \infty$ particles. If an arbitrary number of noninteracting particles within the system can occupy any given state, the particles are called *Bose particles* (or *bosons*); if only one particle may occupy a given state,* then the particles are called *Fermi particles* (or *fermions*). Let $\langle n_k \rangle$ denote the expected number of particles occupying a given state with energy E_k. At temperature T = 0 K, electrons in a solid occupy the lowest energy level; the highest energy level occupied by the valance electrons at 0 K is called the *Fermi energy level*.

9.3.6 Theorem

For any fixed temperature T, the probability that fermions exists in a state with energy E_k obeys the *Fermi–Dirac distribution*: $\mathrm{P}(E_k, T) = \left\{\mathrm{e}^{\beta(E_k - \mu)} + 1\right\}^{-1}$, where μ is the *chemical potential* defined by $\mathrm{P}(\mu, T) = 0.5$.† In particular, the probability distribution of holes left below the Fermi level when electrons are thermally excited into the conduction band is $1 - \mathrm{P}(E_k, T) = \left\{\mathrm{e}^{\beta(\mu - E_k)} + 1\right\}^{-1}$. □

* That is, obeying the *Pauli exclusion principle*, such as electrons (on the other hand, photons are Bose particles).

† For temperatures of practical interest, the chemical potential is approximately equal to the Fermi energy [13, p.166].

A solid is often modeled as a crystal lattice with periodic potential wherein the valance electrons are nearly free (i.e., plane waves modulated by the periodicity of the potential energy). In particular, for metal, the electrons freed from the valance band are free to diffuse throughout the metal (via the conduction band). For semiconductors, this holds when the ambient temperature is not too low. A precise statement can be captured below wherein electrons are modeled as an ideal gas.

By way of introduction, given a crystal lattice with periodicity $\hat{\boldsymbol{r}}$, where the lattice is defined by fixed nuclei, define the reciprocal lattice vector $\hat{\boldsymbol{k}}$ by $\hat{\boldsymbol{r}} \cdot \hat{\boldsymbol{k}} = 2\pi l$, for some $l \in \mathbf{N}$. Finally, if $\boldsymbol{p}$ is the momentum of an electron in the crystal lattice, define the electron wave vector $\boldsymbol{k}$ by $\boldsymbol{p} = \hbar \boldsymbol{k}$.

9.3.7 Theorem (Bloch)

Given a crystal lattice with time-independent periodic potential $U(\boldsymbol{r}+\hat{\boldsymbol{r}}) = U(\boldsymbol{r})$, where $\hat{\boldsymbol{r}}$ is the periodicity of the lattice and the nuclei are fixed at the lattice sites, the electron energy eigenfunctions $|\Psi_k\rangle$ of the time-independent Schrödinger equation $|\mathbf{H}\Psi_k\rangle = E(\boldsymbol{k})|\Psi_k\rangle$ satisfy $\Psi_k(\boldsymbol{r}) = e^{i\boldsymbol{k}\cdot\boldsymbol{r}}\psi_k(\boldsymbol{r})$, where $\boldsymbol{k}$ is the wave vector, $\psi_k(\boldsymbol{r}+\overline{\boldsymbol{r}}) = \psi_k(\boldsymbol{r})\ \forall \boldsymbol{k}$ and $E(\boldsymbol{k}+\hat{\boldsymbol{k}}) = E(\boldsymbol{k})\ \forall \boldsymbol{k}$. □

9.3.8 Remark

The eigenvalues $E(\boldsymbol{k})$ define the energy band structure: a surface in $\boldsymbol{k}$-space such that the energy on the surface is the Fermi energy level, called a *Fermi surface* $\partial\tilde{\mathcal{F}}(k_F)$, where $\tilde{\mathcal{F}}(k_F) = \{\boldsymbol{k} \in \mathbf{R}^3 : \boldsymbol{k} \cdot \nabla_k \cdot E(\boldsymbol{k}) = 0, k \le k_F\}$. If $D(\boldsymbol{k}_-, \boldsymbol{k}_+) \subset \mathbf{R}^3$ defines a conduction band in $\boldsymbol{k}$-space, then for a conductor, $D(\boldsymbol{k}_-, \boldsymbol{k}_+) \cap \tilde{\mathcal{F}}(k_F) \ne \varnothing$, whereas for an insulator,

$$d(\partial D(\boldsymbol{k}_-, \boldsymbol{k}_+), \partial\tilde{\mathcal{F}}(k_F)) \equiv \sup\{d(\boldsymbol{k}', \boldsymbol{k}'') : \boldsymbol{k}' \in \partial D(\boldsymbol{k}_-, \boldsymbol{k}_+), \boldsymbol{k}'' \in \partial\tilde{\mathcal{F}}(k_F)\} >> \varepsilon(T)$$

where $\varepsilon(T) > 0$ is some monotonically increasing function of the (ambient) temperature T and $d : \mathbf{R}^3 \times \mathbf{R}^3 \to \mathbf{R}_+$ is the Euclidean metric (defined in Appendix A.2). Thus the minimal energy surface for a conduction band lies well above the Fermi surface for an insulator; see Figure 9.5. Last but not least, the *forbidden zone* is the region in $\boldsymbol{k}$-space wherein electron states cannot exist. It is sometimes called the *band gap.*

9.3.9 Definition

The *Drude model* for a metal assumes that the valance electrons form an ideal classical gas that diffuses throughout the metal in the conduction zone. The *Drude–Sommerfeld model* for a metal is the Drude model wherein the electrons form an ideal Fermi gas, that is, an ideal gas that obeys the Fermi–Dirac statistics. Finally, the *Lorentz model* for semiconductors is the Drude model wherein the valance electrons are bound by some simple harmonic potential.

This section concludes with a brief comment on temperature and material properties. It is intuitively clear from band theory that increasing the thermal temperature of the lattice may potentially impart to the valance electrons sufficient energy to tunnel across the band gap. Hence, the electrical conductivity of a dielectric or semiconductor depends upon the ambient (or operating) temperature. For further studies, the reader is invited to consult any standard references on solid state physics.

9.4 Dielectric Breakdown

A brief exposition is sketched here, based on the previous section, to provide a heuristic understanding of dielectric breakdown, often resulting in electrostatic discharge. Much empirical work has been done on this account, resulting in the Paschen breakdown curves and characteristics thereof for dielectrics; this topic is not pursued here. In what follows, a dielectric is assumed to be an insulator, unless stated otherwise. Furthermore, observe from quantum tunneling effects that an insulator is not a perfect insulator, as electrons from the valance band can always tunnel across the forbidden zone to the conduction band, as the temperature increases. Notwithstanding that the probability is very small. A simple example is worked out below by way of illustration.

9.4.1 Example

Consider a free electron impinging upon a potential barrier of infinite width. Suppose further that upon the application of a constant external electric field E_0, the barrier falls off according to

$$U(x) = \begin{cases} U_0 \quad \text{on } \mathbf{R}_- \cup (a,b) \\ U_0 - eE_0(x-b) \quad \text{on } [b,c] \\ 0 \quad \text{on } [0,a] \cup (c,\infty) \end{cases}$$

where e is the electronic charge. See Figure 9.6.

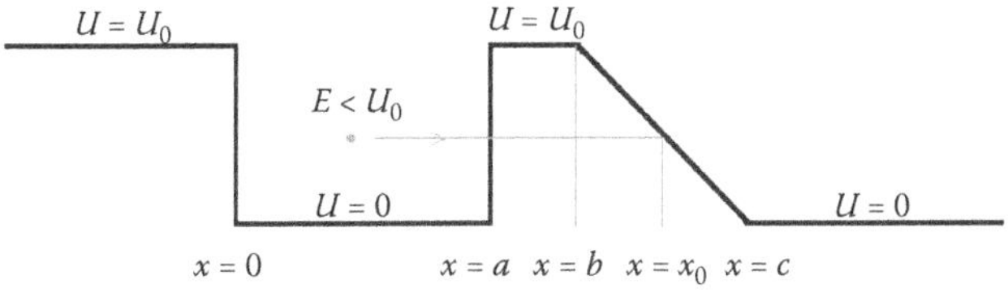

FIGURE 9.6
Potential barrier reduction via an applied electric field.

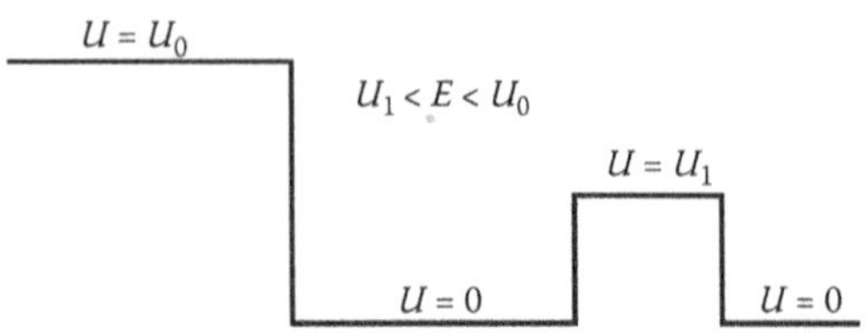

FIGURE 9.7
Potential barrier lower than a particle's kinetic energy.

Now, let the particle kinetic energy $T = E$. Then, by the continuity of U on $[a,b]$, the intermediate value theorem implies $\exists x_0 \in (a,b)$ such that $U(x_0) = E$. In particular, $x \in [a, x_0) \Rightarrow U(x) - E > 0$ and $x \in (x_0, b] \Rightarrow U(x) - E < 0$. From Example 9.3.3, for each fixed $\bar{x} \in [a, x_0)$, the transmission coefficient

$$S_+ = \left\{ \cosh^2(\beta_+(x)(x-a)) + \left(\tfrac{\beta_+^2(x)-k^2}{2k\beta_+(x)}\right)^2 \sinh^2(\beta_+(x)(x-a)) \right\}^{-1}$$

where $\beta_+(x) = \frac{2m}{\hbar^2}(U(x) - E)$ on $[b, x_0)$.

To determine the transmission coefficient on $[x_0, c]$, consider the opposite scenario to Example 9.3.3; see Figure 9.7. In this instance, the particle kinetic energy $E > U_1$, where U_1 is the potential energy of the finite barrier. Then, following Example 9.3.3, *mutatis mutandis*, on $\mathbf{R}_-$,

$$0 = \tfrac{d^2}{dt^2}\psi_1 - \tfrac{2m}{\hbar^2}(U_0 - E)\psi_1 = 0 \Rightarrow \psi_1 = A_1 e^{\beta_1 x}$$

where $\beta_1 = \frac{1}{\hbar}\sqrt{2m(U_0 - E)}$. On the interval $[0,a]$,

$$\tfrac{d^2}{dt^2}\psi_2 + \tfrac{2m}{\hbar^2}E\psi_2 = 0 \Rightarrow \psi_2 = A_2^+ e^{ikx} + A_2^- e^{-ikx}$$

where $k = \frac{1}{\hbar}\sqrt{2mE}$. On the interval (a,b),

$$0 = \tfrac{d^2}{dt^2}\psi_3 + \tfrac{2m}{\hbar^2}(E - U_1)\psi_3 = 0 \Rightarrow \psi_3 = A_3^+ e^{i\beta_2 x} + A_3^- e^{-i\beta_2 x}$$

where $\beta_2 = \frac{1}{\hbar}\sqrt{2m(E - U_1)}$. Finally, on $[b,\infty)$,

$$0 = \tfrac{d^2}{dt^2}\psi_4 + \tfrac{2m}{\hbar^2}E\psi_4 = 0 \Rightarrow \psi_4 = A_4 e^{ikx}$$

The pair of equations at $x = a$ yields

$$2A_2^+ = A_3^+\left(1 + \tfrac{\beta_2}{k}\right)e^{i\beta_2 a} + A_3^-\left(1 - \tfrac{\beta_2}{k}\right)e^{-i\beta_2 a}$$

Likewise, at $x = b$,

$$2A_3^+ e^{i\beta_2 b} = A_4 e^{ikb}\left(1+\tfrac{k}{\beta_2}\right) \quad \text{and} \quad 2A_3^- e^{-i\beta_2 b} = A_4 e^{ikb}\left(1-\tfrac{k}{\beta_2}\right)$$

whence

$$A_2^+ = \tfrac{1}{4}e^{i\beta_2 a}\left(1+\tfrac{\beta_2}{k}\right)A_4 e^{i(k-\beta_2)b}\left(1+\tfrac{k}{\beta_2}\right)+\tfrac{1}{4}e^{-i\beta_2 a}\left(1-\tfrac{\beta_2}{k}\right)A_4 e^{i(k+\beta_2)b}\left(1-\tfrac{k}{\beta_2}\right)$$

and upon some tedious algebraic manipulation,

$$\tfrac{A_2^+}{A_4} = e^{ikb}\left\{\cos(\beta_2(b-a))+\tfrac{i}{2}\tfrac{\beta_2^2+k^2}{2k\beta_2}\sin(\beta_2(b-a))\right\}$$

$$\left(\tfrac{A_2^+}{A_4}\right)^* = e^{-ikb}\left\{\cos(\beta_2(b-a))-\tfrac{i}{2}\tfrac{\beta_2^2+k^2}{2k\beta_2}\sin(\beta_2(b-a))\right\}$$

and hence,

$$\left|\tfrac{A_4}{A_2^+}\right|^2 = \left\{\cos^2(\beta_2(b-a))-\tfrac{1}{4}\left(\tfrac{\beta_2^2+k^2}{2k\beta_2}\right)^2\sin^2(\beta_2(b-a))\right\}^{-1}$$

To complete the solution, it suffices to note that on the closed interval $[x_0, c]$, $E - U(x) > 0$, and hence, the transmission coefficient is

$$S_-(x) = \left\{\cos^2(\beta_-(x)(x-a))-\tfrac{1}{4}\left(\tfrac{\beta_-^2+k^2}{2k\beta_-}\right)^2\sin^2(\beta_-(x)(x-a))\right\}^{-1}$$

where $\beta_-(x) = \tfrac{1}{\hbar}\sqrt{2m(E-U(x))}$. Thus, the average transmission coefficient across the finite width potential barrier is

$$\bar{S} = \tfrac{1}{c-a}\left\{\int_a^{x_0} S_+(x)\,dx + \int_{x_0}^{c} S_-(x)\,dx\right\}$$

Observe by construction that $S_-(c) \le S_-(x) < \bar{S} < S_+(x) \le S_+(a)$ on $[a,c]$. □

It is quite clear from the above example that in the presence of an external field, the potential barrier can be reduced and hence, increase the corresponding transmission coefficient: $S_-(c) < \bar{S}$. This means that electrostatic discharge has a greater probability of occurring when a dielectric is subject to a strong field. In particular, recalling the band structure from the previous section, it is clear that by applying a large enough electric field, the potential barrier may be lowered to a sufficiently small value such that electrons can easily tunnel across the forbidden zone and into the conduction zone. The applied field accelerates the electrons in the conduction zone, and collisions with atomic lattices may potentially excite electrons in the valance band and thereby impart sufficient energy for the valance electrons to jump the forbidden zone and into the conduction band or, if the electron acquires sufficient energy, to leave the dielectric surface (ionization). Thus, at an intuitive level, when a sufficient number of electrons are able to jump to the conduction band, the large current flow in a dielectric results in the phenomenon of breakdown.

As a particular scenario, consider two oppositely charged parallel plate conductors in a homogeneous dielectric medium connected to a common source, that is, a capacitor subject to a fixed potential difference V across the plates. Here, the electric field is just $E = \frac{V}{d}$, where d is the plate separation. For $V >> 0$ such that the potential barrier between the valance band and the conduction band is sufficiently low, the valance electrons can tunnel easily into the conduction band under ambient temperature conditions.

Increasing the electric field further by increasing the potential difference will cause the electrons to accelerate toward the anode; indeed, a sufficiently high field will impart a large enough kinetic energy to the electrons such that their collisions will cause ionization of the dielectric medium. Explicitly,

- The applied electric field will reduce the potential barrier of the dielectric such that the electrons can tunnel across the forbidden zone into the conduction band.
- The electrons in the conduction band will diffuse toward the positively charged conductor (anode).
- The movement of the electrons in the conduction band will generate holes in the valance band (positively charged due to the absence of electrons) which will diffuse toward the negatively charged conductor (cathode).
- The respective charges will accumulate around the boundary of the conductors, forming a space charge that partially shields the conductors and hence, result in a drop in the field.

Heuristically, as more electrons accelerate toward the respective plates, a brief electric arc between the plates may occur. However, note that as the opposite charges form a charge cloud (i.e., space charge) within the vicinity of the plates, the opposing potential—called *screening*—will effectively reduce the applied electric field across the dielectric. The potential is thus modified by the charge cloud. When the electric field falls below some critical level such that the tunneling coefficient becomes negligible, the dielectric ceases to conduct. Thus, electrostatic discharge cannot be a continuous event unless the potential difference between the two fixed conductors increases continuously to supply a strong enough electric field to overcome the screening effects of the ions. The informal picture sketched above can be understood at a more rigorous level by considering a simple example given below: thermionic emission.

9.4.2 Example

Consider a metal conductor surface subject to a high temperature exposure on its surface. Suppose E_F denotes the Fermi energy level of the metal. If W_∞ denotes the work expanded to move an electron from the Fermi level to infinity, then the *work function* ϕ of the metal is defined by $\phi = W_\infty - E_F$. By appealing to the Drude–Sommerfeld model, it can be shown [10 p. 237] that the emission current density from the metal surface is given by the *Richardson–Dushman equation*

$$J = \tfrac{4\pi me}{h^3}(k_B T)^2 e^{-\phi/(k_B T)}$$

where m,e,h,T are, respectively, the mass and absolute charge of an electron, Planck's constant, and temperature. □

9.4.3 Remark

Cold emissions, that is, emissions of electrons under the influence of a strong electric field, from a metal surface are again the result of a quantum tunneling effect. It is called *Fowler–Nordheim tunneling*, and it has applications in metal-oxide semiconductors; the current across the junction is controlled via quantum tunneling through the application of an electric field across the junction.

In closing, the electrostatic discharge can be explained by lowering the potential barrier of a dielectric such that

- The valance electrons can tunnel across the forbidden zone into the conduction band.
- The applied electric field imparts sufficient kinetic energy to the electrons in the conduction band to overcome the work function of the surface.

It is important to note that under the application of a strong electric field, the migrating electrons often acquire large enough kinetic energy such that their collisions with the lattice will impart energy to the lattice; the energy loss manifests as heat. For large enough energy, this leads to the actual breakdown of the dielectric structure within the vicinity of the current flow. At a practical level, an intense discharge can cause vaporization of the material, leading to punctures in the dielectric.

9.4.4 Remark

A rough insight into the above example of a dielectric between two parallel plates can be gained by considering a lossless homogeneous dielectric medium (Ω,μ,ε). Let $W_{\varepsilon,T}$ denote the minimal energy required to allow a valance electron to transition into the conduction band for some fixed temperature T. By modeling the dielectric as a lattice, it may be supposed that an electron has a mean free path $\langle r_\varepsilon \rangle$ between lattice collisions, if sufficient energy is provided to raise the valance electrons into the conduction band. At temperature T, there may only be a small quantity of electrons in the conduction band due to tunneling. A minimal applied uniform electric field E_ε needed to give the electrons in the conduction band an average kinetic energy of $W_{\varepsilon,T}$ can clearly be found from $W_{\varepsilon,T} = F_\varepsilon \langle r_\varepsilon \rangle$, where the required minimal force $F_\varepsilon = |e|\, E_\varepsilon$. Hence, by applying an electric field $E_\varepsilon = \frac{W_{\varepsilon,T}}{|e|\langle r_\varepsilon \rangle}$ across the plates, the conduction electrons will possess enough energy to free the valance electrons into the conduction band. A chain reaction takes place under the applied electric field, causing dielectric ionization to take place and hence a breakdown of the dielectric. Trivially, the required potential difference across the plates is $V_\varepsilon = E_\varepsilon d$. Last, if the electric field along the plates varies, such that $\exists p \in \partial\Omega$ satisfying $E(p) \geq E_\varepsilon$, then breakdown will occur along some sufficiently small tubular neighborhood of p.

9.5 Worked Problems

9.5.1 Exercise

Establish Remark 9.2.3.

Solution

For simplicity, consider a bound electron about some fixed origin subject to a restoring force $m_0\omega_0^2$ and a damping force $m_0\gamma_0 v$, where v is the velocity of the electron. Then, given a plane monochromatic time harmonic electric field $\boldsymbol{E} = \boldsymbol{E}_0 e^{-i(kz+\omega t)}$, the resultant classical expression of the force on the electron is:

$$m_0\ddot{\boldsymbol{r}} + m_0\gamma_0\dot{\boldsymbol{r}} + m_0\omega_0^2\boldsymbol{r} = e\boldsymbol{E}_0 e^{-i\omega t}$$

where $\dot{\boldsymbol{r}} = v$. Then, upon setting $\boldsymbol{r} = \boldsymbol{r}_0 e^{-i\omega t}$, it follows at once that

$$-\omega^2 \boldsymbol{r}_0 - i\omega\gamma_0\boldsymbol{r}_0 + \omega_0^2\boldsymbol{r}_0 = \tfrac{e}{m_0}\boldsymbol{E}_0 \Rightarrow \boldsymbol{r} = \tfrac{e}{m_0}\boldsymbol{E}\frac{1}{\omega_0^2-\omega^2-i\gamma_0\omega}$$

Now, by definition, the dipole moment is $\boldsymbol{p} = e\boldsymbol{r}$, and for a single atomic dipole, $\boldsymbol{p} = (\varepsilon - \varepsilon_0)\boldsymbol{E}$, whence, $\boldsymbol{p} = (\varepsilon - \varepsilon_0)\boldsymbol{E} = e\boldsymbol{r} \Rightarrow \varepsilon = \varepsilon_0 + \frac{e^2}{m_0}\frac{1}{\omega_0^2-\omega^2-i\gamma_0\omega}$. This corresponds precisely to a single mode for an electronic oscillator. If there exist $j = 1,2, \ldots$ modes per electronic oscillator, let f_j denote the probability that an electronic transition will take place at the resonant angular frequency ω_j. Then, by definition, $\sum_j f_j = 1$ (as the total probability must sum to unity) and hence, for a single electronic oscillator,

$$\varepsilon = \varepsilon_0 + \frac{e^2}{m_0} \sum_j \frac{f_j}{\omega_0^2 - \omega^2 - i\gamma_0\omega}$$

Finally, to complete the discussion, if there exist N dipoles per unit volume, then trivially, the electric permittivity of the simple homogeneous dielectric material is given by

$$\varepsilon = \varepsilon_0 + \frac{Ne^2}{m_0} \sum_j \frac{f_j}{\omega_0^2 - \omega^2 - i\gamma_0\omega}$$

as required. □

9.5.2 Exercise

Prove that the eigenvalues of a Hermitian operator are always real.

Solution

Given a Hermitian operator **A**, suppose $|\psi_n\rangle$ is an eigenvector of **A**; that is, $\mathbf{A}|\psi_n\rangle = a_n|\psi_n\rangle$. Then, by definition, $\langle\psi_n|\mathbf{A}\psi_n\rangle = a_n\langle\psi_n|\psi_n\rangle = a_n$, and denoting $*$ to be the complex conjugate, $\langle\mathbf{A}^\dagger\psi_n|\psi_n\rangle = \langle\psi_n|\mathbf{A}^\dagger\psi_n\rangle^* = \langle\psi_n|\mathbf{A}\psi_n\rangle^* = a_n^*$ by the Hermiticity of **A**. Hence, $\langle\mathbf{A}^\dagger\psi_n|\psi_n\rangle = \langle\psi_n|\mathbf{A}\psi_n\rangle \Rightarrow a_n^* = a_n$. Thus, the arbitrariness of $|\psi_n\rangle$ implies that all eigenvalues of **A** must be real.

9.5.3 Exercise

The reason for the Hilbert space formalism of quantum mechanics arises in the following fashion. Consider Axiom 9.3.1(d), $\frac{d}{dt}|\psi\rangle = -i\frac{1}{\hbar}\mathbf{H}|\psi\rangle$. Here,

$$\mathbf{H} = -\frac{1}{2m}\hbar^2\Delta + U(\mathbf{r},t)$$

where Δ is the Laplacian, and in general, the potential energy U depends upon time. In the *Schrödinger (coordinate) representation*, the state vector $|\psi\rangle$ is represented by a complex function $\psi(\mathbf{r},t) \equiv \hat{S}(t)\psi(\mathbf{r})$ such that $\int_{\mathbf{R}^3} \psi(\mathbf{r})\psi^*(\mathbf{r})d^3\mathbf{r} < \infty$ and $\hat{S}$ is a linear operator satisfying $|\hat{S}(t)| = 1$. In particular, the function can be normalized: $\psi \to \tilde{\psi} = \frac{1}{\sqrt{N}}\psi$, where $\int_{\mathbf{R}^3} \psi(\mathbf{r})\psi^*(\mathbf{r})d^3\mathbf{r} = N$. By definition, $\psi \in L^2(\mathbf{R}^3)$, which is the Hilbert space of all square integrable functions: $\langle\psi|\psi\rangle \equiv \int_{\mathbf{R}^3} \psi(\mathbf{r})\psi^*(\mathbf{r})d^3\mathbf{r}$. The wave function $\psi(r)$ is called the *Heisenberg (coordinate) representation* (and is, by definition, independent of time). Finally, recalling that the Fourier transform $\mathcal{F} : L^2(\mathbf{R}^3) \approx L^2(\mathbf{R}^3)$ is a linear isomorphism on a Hilbert space, the *momentum representation* $\psi(\boldsymbol{p})$ is the Fourier transform $\mathcal{F}[\psi](\boldsymbol{p}) = \int_{\mathbf{R}^3} \psi(\boldsymbol{r}) e^{-i\boldsymbol{p}\cdot\boldsymbol{x}} d^3\boldsymbol{r}$ of the coordinate representation $\psi(\boldsymbol{r})$. Lastly, the *Heisenberg uncertainty principle* $\delta p \delta x \geq \frac{1}{2}\frac{h}{2\pi}$, where $\delta p(\delta x)$ is the uncertainty in momentum (position), is a direct consequence of the Hilbert space structure (specifically, the *Cauchy-Schwarz inequality*): $|\psi||\varphi| \geq |\langle\psi|\varphi\rangle|$. This concludes the motivation for formulating quantum mechanics on (an infinite-dimensional) Hilbert space.

Show that the time-independent solution—also called *stationary states*—of the Schrödinger wave equation is given by $\mathbf{H}|\psi\rangle = E|\psi\rangle$, and hence, show explicitly that the time-independent Schrödinger wave equation is $\left\{\frac{-\hbar^2}{2m}\Delta + U(\boldsymbol{r})\right\}\psi(\boldsymbol{r}) = E\psi(\boldsymbol{r})$. In particular, deduce that a solution of the time-independent Schrödinger equation has the form: $\Psi(\boldsymbol{r},t) = \psi(\boldsymbol{r})e^{-i\frac{1}{\hbar}Et}$.

Solution

From Axiom 9.3.1(d), $\frac{d}{dt}|\psi\rangle = -i\frac{1}{h}\mathbf{H}|\psi\rangle$, it follows that in the coordinate representation, $\frac{d}{dt}\psi = -i\frac{1}{h}\mathbf{H}\psi$. Because the Hamiltonian is independent of time, $\frac{d}{dt}\mathbf{H} = 0$ and in particular, $U(r,t) = U(r)$. So, attempt the separation of variables, $\psi(r,t) = \varphi(r)\tau(t)$. Then,

$$\frac{d}{dt}\varphi(\boldsymbol{r})\tau(t) = -i\frac{1}{\hbar}\mathbf{H}\varphi(\boldsymbol{r})\tau(t) \Leftrightarrow -i\frac{1}{\hbar}\frac{1}{\tau(t)}\frac{d}{dt}\tau(t) = \frac{1}{\varphi(\boldsymbol{r})}\mathbf{H}\varphi(\boldsymbol{r})$$

Because the left-hand side of the equality is independent of coordinates by definition, and the right-hand side of the equation is independent of time, they must both be equal to some constant E. That is,

$$-i\frac{1}{\hbar}\frac{1}{\tau(t)}\frac{d}{dt}\tau(t) = E = \frac{1}{\varphi(\boldsymbol{r})}\mathbf{H}\varphi(\boldsymbol{r})$$

whence

$$\frac{1}{\tau(t)}\frac{d}{dt}\tau(t) = i\hbar E \Rightarrow \tau(t) = e^{i\hbar Et}$$

is the fundamental solution, and $\mathbf{H}\varphi(\boldsymbol{r}) = E\varphi(\boldsymbol{r})$ implies immediately that E is an eigenvalue of the Hamiltonian operator $\mathbf{H}$ and $\varphi(r)$ is an eigenfunction of H, whence the stationary Schrödinger wave equation satisfies

$$\left\{\frac{-\hbar^2}{2m}\Delta + U(\boldsymbol{r},t)\right\}\psi = E\psi$$

and the full stationary solution thus has the form $\psi(\boldsymbol{r},t) = \varphi(\boldsymbol{r})e^{\frac{i}{\hbar}Et}$. □

9.5.4. Exercise

Given that the *momentum operator* is defined by $\mathbf{p} = -i\hbar\nabla$, show that for a free particle, that is, potential energy $U = 0$, traveling along the x-axis, the eigenfunction of p is $\psi = e^{\pm ikx}$ with eigenvalue $\pm\hbar k$. In fact, this is just the equation for a plane wave.

Solution

Because the particle is traveling along the x-axis, the variation along the y, z directions are zero and hence, the problem reduces to a one-dimensional scenario. Thus, $\mathbf{p} = -i\bar{h}\frac{d}{dx}$. Moreover, as the potential energy is zero, the time-independent Schrödinger equation defines the state of the particle:

$$\frac{-\hbar^2}{2m}\frac{d^2}{dx^2}\psi = E\psi$$

whence, on setting $k^2 = \frac{2m}{\hbar^2}E$, it follows that

$$0 = \frac{d^2}{dx^2}\psi + k^2\psi \Rightarrow \psi_+ = e^{ikx}, \psi_- = Be^{-ikx}$$

are two possible solutions.

Now, observe that $E = \frac{1}{2m}p^2 \Rightarrow p = \pm\hbar k$, where $p = \hbar k$ denotes the particle traveling along the x-axis in the positive direction whilst $p = -\frac{h}{2\pi}k$ indicates the negative direction. Thus,

$$\mathbf{p}\psi_\pm = -i\frac{h}{2\pi}\frac{d}{dx}\psi_\pm = \pm\hbar k\psi_\pm$$

implies at once that the eigenvalue of p is $p = \hbar k$ for the eigenfunction $\psi_+ = e^{ikx}$, and the eigenvalue is $p = -\hbar k$ for the eigenfunction $\psi_- = Be^{-ikx}$. That is, $\psi_+ = e^{ikx}$ denotes the particle traveling along the positive direction (from left to right) along the x-axis, and $\psi_- = Be^{-ikx}$ denotes the particle traveling along the negative direction (from right to left) along the x-axis, as claimed. □

9.5.5 Exercise

Consider the domains $(\Omega_0, \mu_0, \varepsilon_0, \sigma_0)$ and $(\Omega_1, \mu_1, \varepsilon_1, \sigma_1)$, where $\Omega_i \subset \mathbf{R}^2$ are defined by $\Omega_0 = \left[-\frac{1}{2}a, \frac{1}{2}a\right] \times [0, \delta]$ and $\Omega_1 = [-\frac{1}{2}a, \frac{1}{2}a] \times [\delta, \delta + d]$, and suppose also that

$$C_- = \{(x, 0) \in \mathbf{R}^2 : x \in \mathbf{R}\} \quad \text{and} \quad C_+ = \{(x, \delta + d) \in \mathbf{R}^2 : x \in \mathbf{R}\}$$

are grounded planes that bound Ω_i $\forall i = 0, 1$. Given that $0 < \delta \ll a$, suppose that a constant surface charge density ρ_s exists on $\partial\Omega_0 |_{y=\delta} \equiv \left[-\frac{1}{2}a, \frac{1}{2}a\right] \times \{\delta\}$ at $t = 0$. Determine the charge relaxation time of $\Omega = \Omega_0 \cup \Omega_1$. This question is of particular interest to EMC engineers and material engineers, as occasionally there is a need to know how long it takes to fully discharge a dielectric material.

Solution

Because ρ_s is a constant and $C_\pm$ are grounded conductors, it is clear by construction that $\forall p \in \partial\Omega_0 |_{y=\delta}$ and $\forall p_\pm \in C_\pm$, $-\int_{p_\pm}^{p} \boldsymbol{E}_\pm \cdot d\boldsymbol{l} = V_0$, for some constant V_0. In particular, the assumption $0 < \delta << a$ implies that $\boldsymbol{E}_\pm \approx \pm E_\pm \boldsymbol{e}_y$, where $E_\pm$ are some constants, away from $x = \pm\frac{1}{2}a$. Hence, $\frac{E_+}{d} = V_0 = \frac{E_-}{\delta} \Rightarrow E_+ = \frac{d}{\delta} E_-$ and from Gauss' law, $\rho_1 = \frac{d}{\delta}\frac{\varepsilon_1}{\varepsilon_0}\rho_0$

Now, for simplicity, let $\bar{\sigma}, \bar{\varepsilon}$ denote the electric conductivity and electric permittivity, respectively, for the composite system Ω, whence, from the continuity of charge,

$$\partial_t \rho_s = -\nabla \cdot J_s = -\frac{\bar{\sigma}}{\bar{\varepsilon}}\rho_s \Rightarrow \bar{\tau} = \frac{\bar{\varepsilon}}{\bar{\sigma}}$$

is the composite relaxation time.

Furthermore, observe also that for some small time $t > 0$, $\sigma_i > 0 \, \forall i \Rightarrow \rho_s = \rho_1 + \rho_2$, where ρ_i is in Ω_i. That is, the charge on the boundary will diffuse toward the grounded conductors under the action of $E_\pm$. Hence,

$$\partial_t \rho_s = \partial_t \rho_0 + \partial_t \rho_1 = -\frac{\sigma_0}{\varepsilon_0}\rho_0 - \frac{\sigma_1}{\varepsilon_1}\rho_1 = -\left\{\frac{\sigma_0}{\varepsilon_0} + \frac{\sigma_1}{\varepsilon_0}\frac{d}{\delta}\right\}\rho_0$$

In addition,

$$\left\{\frac{\sigma_0}{\varepsilon_0} + \frac{\sigma_1}{\varepsilon_0}\frac{d}{\delta}\right\}\rho_0 = \frac{\bar{\sigma}}{\bar{\varepsilon}}\rho_s = \frac{\bar{\sigma}}{\bar{\varepsilon}}\left\{1 + \frac{d}{\delta}\frac{\varepsilon_1}{\varepsilon_0}\right\}\rho_0 \Rightarrow \frac{\bar{\sigma}}{\bar{\varepsilon}} = \left\{\sigma_0 + \frac{d}{\delta}\sigma_1\right\}\left\{\varepsilon_0 + \frac{d}{\delta}\varepsilon_1\right\}^{-1}$$

Hence, the charge relaxation time of Ω is $\bar{\tau} = \left\{\varepsilon_0 + \frac{d}{\delta}\varepsilon_1\right\}\left\{\sigma_0 + \frac{d}{\delta}\sigma_1\right\}^{-1}$. Indeed, this can be rewritten in terms of the charge relaxation constant of Ω_0:

$$\bar{\tau} = \tau_0\left\{1 + \frac{d}{\delta}\frac{\varepsilon_1}{\varepsilon_0}\right\}\left\{1 + \frac{d}{\delta}\frac{\sigma_1}{\sigma_0}\right\}^{-1}$$

where $\tau_0 = \frac{\varepsilon_0}{\sigma_0}$ as required. □

References

1. Ahn, D. and Park, S.-H. 2011. *Engineering Quantum Mechanics*. Hoboken, NJ: IEEE Press, John Wiley & Sons.
2. Bohm, D. 1979. *Quantum Theory*. New York: Dover.
3. Churchill, R. and Brown, J. 1990. *Complex Variables and Applications*. New York: McGraw-Hill.

4. Davydov, A.1976. *Quantum Mechanics*. Oxford, UK: Pergamon Press.
5. Dressel, M. and Grüner, G. 2002. *Electrodynamics of Solids: Optical Properties of Electrons in Matter*. Cambridge: University of Cambridge Press, UK.
6. Fox, M. 2001. *Optical Properties of Solids*. Oxford, UK: Oxford University Press.
7. Go, D. and Pohlman, D. (2010). A mathematical model of the modified Paschen's curve for breakdown in microscale gaps. *J. Appl. Phys.* 107(10): 103303.
8. Jackson, D. 1962. *Classical Electrodynamics*. New York: John Wiley & Sons.
9. Jones, W. and March, N. 1973. *Theoretical Solid State Physics. Vol. 1: Perfect Lattices in Equilibrium*. New York: Dover.
10. Kittel, C. 1953. *Introduction to Solid State Physics*. New York: John Wiley & Sons.
11. Paschen, F. (1889). Ueber die zum Funkenübergang in Luft, Wasserstoff und Kohlensäure bei verschiedenen Drucken erforderliche Potentialdifferenz. *Ann. Phys.* 273(5): 69–75.
12. Raju, G. 2003. *Dielectrics in Electric Fields*. New York: Marcel Dekker.
13. Tang, C. 2005. *Fundamentals of Quantum Mechanics for Solid State Electronics and Optics*. Cambridge: Cambridge University Press, UK.

Appendix A

A.1 Coordinate Transformations

Recall that a *vector space* V *over* $\mathbf{R}$ is the quadruple $(V,+,\bullet,\mathbf{R})$ such that the following are preserved

a) Vector addition: $\forall u,v \in V, \quad u+v \in V$
b) *Scalar multiplication:* $\forall v \in V$ and $\forall a \in \mathbf{R}$, $a \bullet v \in V$ (this is often denoted by $av \in V$)

A.1.1 Definition

Given a (real) vector space V, the set of vectors $\{v_1,\ldots,v_k\} \subset V$ is said to be *linearly independent* if $\sum_{i=1}^{k} a^k v_k = 0 \Leftrightarrow a^k = 0\ \forall k$. Furthermore, if $\exists n > 0$ such that $\{v_1,\ldots,v_n\} \subset V$ is linearly independent and $\forall m > n$, $\{u_1,\ldots,u_m\} \subset V$ is not linearly independent, then n is said to be the (real) dimension of V, and $\{v_1,\ldots,v_n\} \subset V$ called a basis of V.

Given a basis $\{v_1,\ldots,v_n\} \subset V$, $\forall v \in V$, $\exists\{a^1,\ldots,a^n\} \subset \mathbf{R}$ such that $v = a^1 v_1 + \cdots + a^n v_n$. Observe that a basis is not unique, and with respect to a fixed basis $\boldsymbol{v} = \{v_1,\ldots,v_n\}$, the set $\{a^1,\ldots,a^n\}$ is the components of v. In particular, the vector v may be represented by $v = (a^1,\ldots,a^n)$ with respect to $\boldsymbol{v}$. The elements of $\mathbf{R}$ are called *scalars*. Denote the dimensions of a vector space V by *dimV*.* Then, a basis $\{v_1,\ldots,v_n\}$ of V is said to *span* V if $n = \dim V$. This is also written as $V = \operatorname{span}\{v_1,\ldots,v_n\}$.

A.1.2 Definition

Let V, W be a pair of finite-dimensional vector spaces. A *linear transformation* (or *operator*) $T: V \to W$ on V is a mapping that satisfies the criteria: $T(av + u) = aT(u) + T(v)\ \forall u,v \in V$ and $a \in \mathbf{R}$. Furthermore, a linear transformation is (a) *injective* (or *one–one*) if $\forall u,v \in V$, $T(u) = T(v) \Rightarrow u = v$, and (b) *surjective* (or *onto*) if $\forall w \in W$, $\exists u \in V$ such that $T(u) = w$; that is, $T(V) = W$. Last, T is said to be a *linear isomorphism* if T is both injective and surjective, and this is denoted by $T: V \approx W$.

By definition, a linear transformation on V is completely defined by its values on a basis $\{v_1,\ldots,v_n\}$ of V. To see this, let $T: V \to W$ be a linear

* Only real vector spaces are considered here; these are vector spaces over $\mathbf{R}$.

transformation and suppose that $T(v_i)=v_i' \ \forall i=1,\ldots,n$. Given any $v \in V$, let $v=\sum_i a^i v_i$ for some scalars $a^i \in \mathbf{R}$. Then, by linearity,

$$T(v)=T(a^1v_1+\cdots+a^nv_n)=\sum_i a^iT(v_i)=\sum_i a^iv_i'.$$

The above linear transformation is *invertible* if $v_i' \neq 0 \ \forall i=1,\ldots,n$, and its *inverse* T^{-1} is defined by $T^{-1}(v_i')=T^{-1}(T(v_i))=v_i$. In other words, T is invertible if $Tu = 0 \Leftrightarrow u = 0$. Equivalently, a linear transformation T is invertible if it is injective.

A basis $\{v_1,\ldots,v_n\}$ of V defines a *coordinate system* for V. A linear isomorphism $T: V \to V$ defines a change in the coordinate system on V; that is, it transforms one coordinate system into another coordinate system. Finally, if W is another vector space such that $W \approx V$, then W may be identified with V.

A.1.3 Theorem

Let V be an n-dimensional (real) vector space, where $n < \infty$. Then, $V \approx \mathbf{R}^n$.

Proof

Let $\{u_i : i=1,\ldots,n\}$ be a basis that spans V and let $\{\mathbf{e}_i : i=1,\ldots,n\}$ denote the standard basis spanning $\mathbf{R}^n$. Define a linear transformation T on V by $T(u_i)=\mathbf{e}_i \ \forall i$. Then, by construction, T is injective. To see this, suppose $u',u'' \in V$. Then, $u'=\sum_i a_i'u_i$ and $u''=\sum_i a_i''u_i$, for some unique $a_i', a_i'' \in \mathbf{R} \ \forall i$, whence,

$$T(u')=T(u'')\Rightarrow \sum_i a_i'\mathbf{e}_i=\sum_i a_i''\mathbf{e}_i \Leftrightarrow \sum_i (a_i'-a_i'')\mathbf{e}_i=0 \Rightarrow a_i'=a_i'' \ \forall i$$

and T is thus injective. Finally, to establish that T is surjective, it suffices to choose any element $w \in \mathbf{R}^n$. Let $w=\sum_i w_ie_i$ and define, $u=\sum_i w_iu_i$. Then, by definition, $u \in V$ and $T(u)=\sum_i w_iT(u_i)=\sum_i w_i\mathbf{e}_i=w$, whence it follows from the arbitrariness of u that T is surjective, as required. □

A.1.4 Example

The space $\mathbf{R}^3$ with respect to the standard basis vectors $\{\mathbf{e}_1,\mathbf{e}_2,\mathbf{e}_3\}$, where $\mathbf{e}_1=(1,0,0), \mathbf{e}_2=(0,1,0), \mathbf{e}_3=(0,0,1)$, is a 3-dimensional vector space. A 3-vector $u \in \mathbf{R}^3$ is expressed interchangeably by its components:

$$u=(a,b,c)=a\mathbf{e}_1+b\mathbf{e}_2+c\mathbf{e}_3$$

Moreover, the *cross-product* of two 3-vectors $u = (u^1,u^2,u^3)$ and $v = (v^1,v^2,v^3)$ is defined by:

$$u \times v=(u^2v^3-u^3v^2,-(u^1v^3-u^3v^1),u^1v^2-u^2v^1)$$

The *inner* (or *scalar*) *product* of two 3-vectors $\boldsymbol{u} = (u^1, u^2, u^3)$ and $\boldsymbol{v} = (v^1, v^2, v^3)$ is given by

$$\boldsymbol{u} \cdot \boldsymbol{v} = u^1 v^1 + u^2 v^2 + u^3 v^3$$

The inner product is also denoted by $\langle \boldsymbol{u} | \boldsymbol{v} \rangle$ or $\langle \boldsymbol{u}, \boldsymbol{v} \rangle$.* Observe that the cross-product of two 3-vectors remains a 3-vector whereas the scalar product of two vectors is a scalar. In particular, $\mathbf{e}_i \cdot \mathbf{e}_j = 0$ for $\forall i \neq j$.

Note as a side remark that more generally, an *orthonormal basis* $\{\mathbf{e}_x, \mathbf{e}_y, \mathbf{e}_z\}$ of $\mathbf{R}^3$ satisfies (i) $\mathbf{e}_{\pi(x)} \times \mathbf{e}_{\pi(y)} = (-1)^{|\pi|} \mathbf{e}_{\pi(z)}$, where $\pi: \{x, y, z\} \to \{x, y, z\}$ is a permutation and $|\pi| = 0$ if the permutation is cyclic, and $|\pi| = 1$ otherwise; (ii) $\mathbf{e}_i \cdot \mathbf{e}_j = \delta_{ij}$, where $i, j \in \{x, y, z\}$.

Now, given a vector space V with basis $v = \{v_1, \ldots, v_n\}$, v is *orthonormal* if $v_i \cdot v_j = \delta_{ij}$, where

$$\delta_{ij} = \begin{cases} 1 & \text{if } i = j \\ 0 & \text{if } i \neq j \end{cases}$$

That is, the vectors are *unit* vectors such that they are *normal* (i.e., orthogonal) to one another. In all that follows, all vector spaces under consideration are restricted to a maximum of three dimensions.

For notational simplicity, denote the partial derivatives with respect to x, y, z, t by

$$\partial_x \equiv \tfrac{\partial}{\partial x},\ \partial_y \equiv \tfrac{\partial}{\partial y},\ \partial_z \equiv \tfrac{\partial}{\partial z},\ \partial_t \equiv \tfrac{\partial}{\partial t},\ \partial_x^2 \equiv \tfrac{\partial^2}{\partial x^2},\ \partial_y^2 \equiv \tfrac{\partial^2}{\partial y^2},\ \partial_z^2 \equiv \tfrac{\partial^2}{\partial z^2},\ \partial_t^2 \equiv \tfrac{\partial^2}{\partial t^2}$$

The definition of the *del* operator ∇ in 3-dimensional space $\mathbf{R}^3$ is:

- In rectangular coordinates, $\nabla = (\partial_x, \partial_y, \partial_z)$, and $\Delta \equiv \nabla \cdot \nabla = \partial_x^2 + \partial_y^2 + \partial_z^2$.
- In spherical coordinates, $x = r \sin\phi \cos\theta$, $y = r \sin\phi \sin\theta$, $z = r \cos\phi \Rightarrow$ $\nabla = (\partial_r, \frac{1}{r}\partial_\theta, \frac{1}{r \sin\theta}\partial_\phi)$ and $\Delta = \frac{1}{r^2}\partial_r(r^2 \partial_r) + \frac{1}{r^2 \sin\theta}\partial_\theta(\sin\theta\, \partial_\theta) + \frac{1}{r^2 \sin^2\theta}\partial_\phi^2$.

Let S be a 2-dimensional surface in $\mathbf{R}^3$ such that $\exists\{\boldsymbol{u}, \boldsymbol{v}\}$, a basis field, that spans S. That is, locally, each vector w defined at a point on S can be expressed as a linear combination of $(\boldsymbol{u}, \boldsymbol{v})$: $w = a\boldsymbol{u} + b\boldsymbol{v}$, for some real numbers a, b. Let $\boldsymbol{p}$ be a unit vector that is normal to the surface S at each point on S. Then, on the volume $S \times \mathbf{R}$ defined by S via the local coordinates given by the basis vectors $\{\boldsymbol{u}, \boldsymbol{v}, \boldsymbol{p}\}$ at each point in $S \times \mathbf{R}$,

$$\nabla \equiv \nabla_\perp + \boldsymbol{p}\,\partial_p$$

where $\nabla_\perp \equiv (\partial_u, \partial_v)$. See Figure A.1.

* A more general definition is given in a later section.

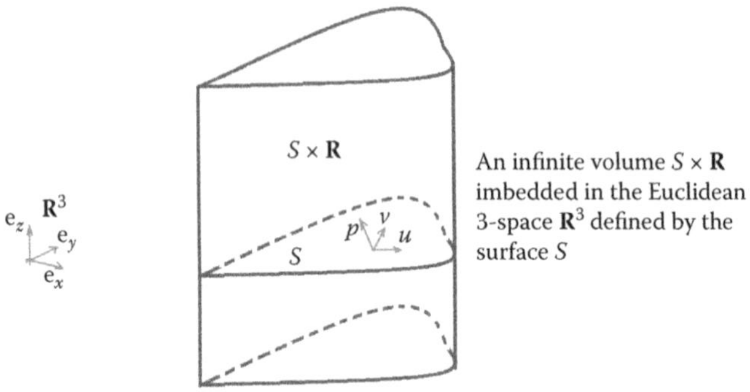

FIGURE A.1
Arbitrary coordinate system defined on $S \times \mathbf{R}$.

A.1.5 Proposition

Let $f : \mathbf{R}^3 \to \mathbf{R}^3$ be a coordinate transformation from $\{\mathbf{e}_1, \mathbf{e}_2, \mathbf{e}_3\}$-coordinates onto $\{\mathbf{e}_1', \mathbf{e}_2', \mathbf{e}_3'\}$-coordinates defined by $(x^1, x^2, x^3) \mapsto (y^1, y^2, y^3)$. Then, f has the following matrix representation $\boldsymbol{u} = f(v) \equiv D(f)v$, where $D(f) = (\partial_j y^i)$ is a 3×3 matrix called the *Jacobian* of f defined by

$$D(f) = (\partial_j y^i) = \begin{bmatrix} \partial_1 y^1 & \partial_2 y^1 & \partial_3 y^1 \\ \partial_1 y^2 & \partial_2 y^2 & \partial_3 y^2 \\ \partial_1 y^3 & \partial_2 y^3 & \partial_3 y^3 \end{bmatrix}$$

with $\partial_j y^i \equiv \partial_{x^j} y^i$ for each $i,j = 1,2,3$, $\boldsymbol{u} = (y^1, y^2, y^3)$ and $v = (x^1, x^2, x^3)$. Explicitly,

$$\begin{pmatrix} y^1 \\ y^2 \\ y^3 \end{pmatrix} = \begin{bmatrix} \partial_1 y^1 & \partial_2 y^1 & \partial_3 y^1 \\ \partial_1 y^2 & \partial_2 y^2 & \partial_3 y^2 \\ \partial_1 y^3 & \partial_2 y^3 & \partial_3 y^3 \end{bmatrix} \begin{pmatrix} x^1 \\ x^2 \\ x^3 \end{pmatrix}$$

□

A.1.6 Example

Derive the coordinate transformation between rectangular coordinates and spherical coordinates in $\mathbf{R}^3$, where in the *spherical coordinate system*, $\boldsymbol{u} = (r, \theta, \phi)$ is defined with respect to rectangular coordinates by $r = \sqrt{x^2 + y^2 + z^2}$, $\theta = \arctan \frac{y}{x}$, $\phi = \arctan \frac{\sqrt{x^2+y^2}}{z}$, and the coordinate transformation from rectangular coordinates to spherical coordinates are related by: $x = r \sin\theta \cos\phi$, $y = r \sin\theta \sin\phi$, $z = r\cos\theta$.

Let $f:(r, \theta, \phi) \mapsto (x, y, z)$, where $x = r \sin\theta \cos\phi$, $y = r \sin\theta \sin\phi$, $z = r \cos\theta$. On setting $y^1 = x, y^2 = y, y^3 = z$ and applying Proposition A.1.5:

$$\begin{pmatrix} u_x \\ u_y \\ u_z \end{pmatrix} = \begin{bmatrix} \partial_r x & \partial_\theta x & \partial_\phi x \\ \partial_r y & \partial_\theta y & \partial_\phi y \\ \partial_r z & \partial_\theta z & \partial_\phi z \end{bmatrix} \begin{pmatrix} v_r \\ v_\theta \\ v_\phi \end{pmatrix} \tag{A.1}$$

where $\partial_r y^1 = \partial_r x = \partial_r(r\sin\theta\cos\phi) = \sin\theta\cos\phi$, $\partial_\theta y^1 = \partial_\theta(r\sin\theta\cos\phi) = r\cos\theta$ $\cos\phi$, and so on. That is,

$$D(f) = \begin{bmatrix} \sin\theta\cos\phi & \cos\theta\cos\phi & -\sin\phi \\ \sin\theta\sin\phi & \cos\theta\sin\phi & \cos\phi \\ \cos\theta & -\sin\theta & 0 \end{bmatrix}$$

upon setting $\mathbf{e}_r = \hat{\mathbf{r}}, \mathbf{e}_\theta = r\hat{\theta}$ and $\mathbf{e}_\phi = r\sin\theta\hat{\phi}$ (in order for the unit vectors to have the same dimensions of length). This (invertible) linear transformation $D(f)$ leads to the transformation of vector components in spherical coordinates into rectangular coordinates:

$$v = v_r\mathbf{e}_r + v_\theta\mathbf{e}_\theta + v_\phi\mathbf{e}_\phi \quad \text{and} \quad \boldsymbol{u} = u_x\mathbf{e}_x + u_y\mathbf{e}_y + u_z\mathbf{e}_z$$

Now, the above transformation, which transforms spherical coordinates into rectangular coordinates, is in fact the inverse operation transforming rectangular coordinates into spherical coordinates. Hence,

$$D(f)^{-1} = \begin{bmatrix} \sin\theta\cos\phi & \sin\theta\sin\phi & \cos\theta \\ \cos\theta\cos\phi & \cos\theta\sin\phi & -\sin\theta \\ -\sin\phi & \cos\phi & 0 \end{bmatrix}$$

is the required linear transformation taking vector components in rectangular coordinates into components in spherical coordinates. Explicitly,

$$\begin{pmatrix} v_r \\ v_\theta \\ v_\phi \end{pmatrix} = \begin{bmatrix} \sin\theta\cos\phi & \sin\theta\sin\phi & \cos\theta \\ \cos\theta\cos\phi & \cos\theta\sin\phi & -\sin\theta \\ -\sin\phi & \cos\phi & 0 \end{bmatrix} \begin{pmatrix} u_x \\ u_y \\ u_z \end{pmatrix} \qquad \square$$

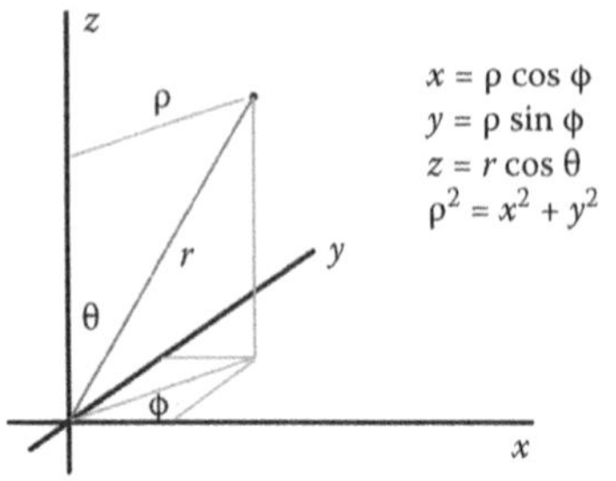

FIGURE A.2
Cylindrical coordinate system.

A.1.7 Example

In this final example, the coordinate transformation between rectangular and cylindrical coordinates is derived. First, recall that in terms of cylindrical coordinates (see Figure A.2):

$$x = \rho\ \cos\phi,\ y = \rho \sin\phi,\ z = r\ \cos\ \theta,\ \rho = \sqrt{x^2 + y^2}$$

On setting $y^1 = x, y^2 = y, y^3 = z$ and applying Proposition A.1.5, $f(v) = \boldsymbol{u} \Rightarrow$

$$\begin{pmatrix} u_x \\ u_y \\ u_z \end{pmatrix} = \begin{bmatrix} \partial_\rho x & \partial_\phi x & \partial_\theta x \\ \partial_\rho y & \partial_\phi y & \partial_\theta y \\ \partial_\rho z & \partial_\phi z & \partial_\theta z \end{bmatrix} \begin{pmatrix} v_\rho \\ v_\phi \\ v_\theta \end{pmatrix} \tag{A.2}$$

where $\partial_\rho y^1 = \partial_\rho x = \partial_r(\rho \cos\phi) = \cos\phi,\ \partial_\phi y^1 = \partial_\phi(\rho\cos\phi) = -\rho\sin\phi$, and so on.

Substituting the values yields the transformation from cylindrical coordinates to rectangular coordinates:

$$D(f) = \begin{bmatrix} \cos\phi & -\rho\sin\phi & 0 \\ \sin\phi & \rho\cos\phi & 0 \\ 0 & 0 & -r\sin\theta \end{bmatrix}$$

Inverting $D(f)$ yields the transformation from rectangular coordinates to cylindrical coordinates:

$$D(f)^{-1} = \begin{bmatrix} \cos\phi & \sin\phi & 0 \\ -\frac{1}{\rho}\sin\phi & \frac{1}{\rho}\cos\phi & 0 \\ 0 & 0 & -\frac{1}{r\sin\theta} \end{bmatrix}$$

This section concludes with a brief list of some common identities employed in vector analysis [3,12,14,18,19,21]. In what follows, scalars are denoted by lower case Greek letters, and vectors in upper case bold letters.

- $\nabla(\alpha \boldsymbol{A} + \boldsymbol{B}) = \boldsymbol{A} \cdot \nabla\alpha + \alpha\nabla \cdot \boldsymbol{A} + \nabla \cdot \boldsymbol{B}$
- $\nabla \times (\alpha \boldsymbol{A} + \boldsymbol{B}) = -\boldsymbol{A} \times \nabla\alpha + \alpha\nabla \times \boldsymbol{A} + \nabla \times \boldsymbol{B}$
- $\nabla \times (\nabla \times \boldsymbol{A}) = \nabla(\nabla \cdot \boldsymbol{A}) - \Delta \boldsymbol{A}$
- $\nabla \times \nabla\alpha = 0$
- $\nabla \cdot (\nabla \times \boldsymbol{A}) = 0$
- $\nabla \times (\boldsymbol{A} \times \boldsymbol{B}) = \boldsymbol{A}(\nabla \cdot \boldsymbol{B}) + (\boldsymbol{B} \cdot \nabla)\boldsymbol{A} - \boldsymbol{B}(\nabla \cdot \boldsymbol{A}) - (\boldsymbol{A} \cdot \nabla)\boldsymbol{B}$

A.2 Basic Point-Set Topology: A Synopsis

In this section, a sketch of the elements of general topology used in analysis, theoretical physics, and engineering is presented. Some concepts and terminologies introduced in this section are used throughout the text for mathematical convenience and to make description precise. First, let $f: U \to V$ be a mapping. Then, (a) f is *injective* if $\forall u, v \in U, f(u) = f(v) \Rightarrow u = v$; (b) f is *surjective* if $\forall v \in V, \exists u \in U$ such that $f(u) = v$; (c) f is *bijective* if f is both injective and bijective. Second, by way of motivation, recall the definition of the continuity of functions from real analysis.

A.2.1 Definition

Let f: $\mathbf{R} \to \mathbf{R}$ be a function and fix $x_0 \in \mathbf{R}$. Then, f is said to be continuous at x_0 if $\forall \varepsilon > 0, \exists \delta > 0$ such that $|x - x_0| < \delta \Rightarrow |f(x) - f(x_0)| < \varepsilon$. Moreover, if f is continuous $\forall x \in \Sigma$, where $\Sigma \subseteq R$, then f is said to be continuous on Σ.

Now, on setting $N_\delta(x_0) = (x_0 - \delta, x_0 + \delta)$ and $N_\varepsilon(f(x_0)) = (f(x_0) - \delta, f(x_0) + \delta)$, it is clear from Definition A.2.1 that the continuity of f at x_0 is equivalent to the following condition: $x \in N_\delta(x_0) \Rightarrow f(x) \in N_\varepsilon(f(x_0))$; that is, $f(N_\delta(x_0)) \subset N_\varepsilon(f(x_0))$. See Figure A.3(a). Note that by construction, $N_\delta(x_0)$ is a δ-interval of x_0 and $N_\varepsilon(f(x_0))$ is an ε-interval of $f(x_0)$.

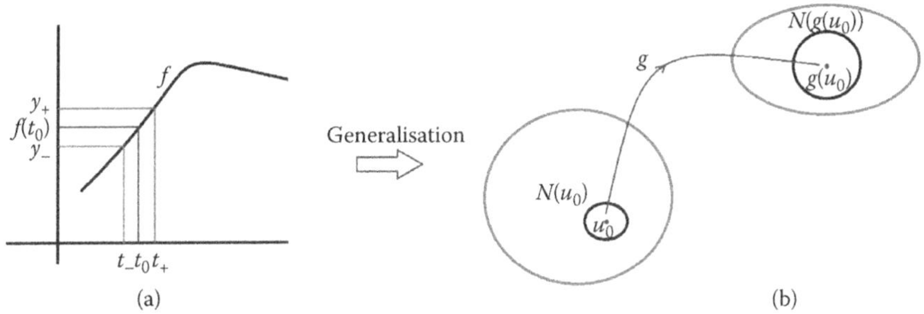

FIGURE A.3
Generalizing the concept of continuity in real analysis.

This motivates the concept of continuity on spaces that are more general than the real line. That is, the generalization of intervals, called *neighborhoods*, can be modeled on more general spaces to define continuity. Refer to Figure A.3(b), where (U,V) is a pair of topological spaces, $g:U \to V$ is a continuous map, $N(x_0)$ is a neighborhood of x_0 in U, and $N(g(x_0))$ is a neighborhood of $g(x_0)$ in V. The continuity of g guarantees that there exists a neighborhood $N(x_0)$ of x_0 such that $g(N(x_0)) \subset N(g(x_0))$. Thus, this leads to the notion of constructing neighborhoods in abstract spaces.

A.2.2 Definition

Given a nonempty set Ω, suppose $\tau \subset 2^{\Omega}$ satisfies the following criteria.

(a) $\varnothing, \Omega \in \tau$.
(b) $\forall U_k \in \tau, \bigcup_k U_k \in \tau$ (arbitrary union of subsets).
(c) $\forall U_k \in \tau, k = 1,\ldots,n < \infty, \bigcap_{k=1}^{n} U_k \in \tau$ (finite intersection of subsets).

Then, the pair (Ω, τ) is called a *topological space*, and τ defines the *topology* of Ω. Furthermore, the elements of τ are called *open subsets* (or *open neighborhoods*) of Ω. Finally, $N \subset \Omega$ is a *neighborhood* of $x \in \Omega$ if $x \in N$ and $N \in \tau$.

An example of an open subset in the real line is the open interval (a,b), where $a < b$. Observe that given an open interval (a,b) and any point $x \in (a,b)$, there exists an open interval J such that $x \in J$ and $J \subset (a,b)$. To see this, choose $c = \min\left\{x - \frac{a+x}{2}, \frac{x+b}{2} - x\right\}$. Then, by construction, $J \equiv (x - c, x + c) \subset (a,b)$ is the sought-for neighborhood about x that is contained in (a,b). More generally, the following result can be established.

A.2.3 Lemma

Given a topological space (Ω, τ), a subset $N \subset \Omega$ is open if and only if $\forall x \in N, \exists M \in \tau$ such that $x \in M$ and $M \subset N$. □

The notion of a closed interval $[a,b]$ in the real line can also be generalized to an arbitrary topological space. A subset $N \subset \Omega$ is *closed* if its complement $N^c = \Omega - N$ is open. Informally, a closed subset contains its boundary points. Thus, if $N \subset \Omega$ is open, then $N \cup \partial N$ is closed. Conversely, given a subset $N \subset \Omega$, the *interior* N° of N is the subset of all $x \in N$ such that $\exists U \in \tau$ with $x \in U \subset N$. Hence, by definition, $N^\circ \in \tau$ and in particular, $N^\circ \cap \partial N = \varnothing$.

A closely related concept defining a closed set is a sequence. Intuitively, a sequence is a set of ordered numbers: $a_1, a_2, \ldots, a_n$. A more precise definition is given below. First, an *ordered* set $(\Lambda, \leq)$ is a set such that $\forall i, j \in \Lambda$, either $i \leq j$ or $j \leq i$. In addition, if $\Lambda \subseteq \mathbf{N}$, then the pair $(\Lambda, \leq)$ is called a *countable indexing set*. Finally, a function $f\colon \mathbf{N} \to \mathbf{R}$ that assigns $k \mapsto f(k)$ defines a real *sequence*, where it is clear that $\mathbf{N}$ inherits a natural order defined by "greater than or equal to." A sequence is often represented by (a_n), $(a_n)_n$, or $\{a_n\}_n$, where $a_n = f(n)$.

In the case of a topological space Ω, a sequence is defined similarly by a mapping $f: \Lambda \rightarrow \Omega$, where Λ is a countable indexing set.

In what follows, only topological spaces whose topology can be characterized by sequences, such as the finite-dimensional Euclidean spaces, are considered.* For this class of topological spaces, continuity can be meaningfully defined via convergent sequences. In particular, all indexing sets considered herein are countable.† Again, recall from real analysis that a real sequence $(a_n)_n$ is said to converge to $a_0 \in \mathbf{R}$. if $\forall \varepsilon > 0, \exists N > 0$ such that all $n > N \Rightarrow |a_n - a_0| < \varepsilon$. This has the following obvious generalization to topological spaces.

A.2.4 Definition

Let (Ω,τ) be a nonempty topological space and $(a_n)_{n\in\Lambda}$ a sequence in Ω, where Λ is some indexing set. Then, $(a_n)_{n\in\Lambda}$ converges to the limit $a_0 \in \Omega$ if $\forall U \in \tau$ with $a_0 \in U$, $\exists n_U \in \Lambda$ such that all $n \geq n_U \Rightarrow a_n \in U$. The limit a_0 is said to be a *limit point* of the sequence $(a_n)_n$.

From this definition, it can be shown that a subset $S \subset \Omega$ is closed if and only if S contains all of its limit points. And if $U \subset \Omega$ is open, then its *closure* $\bar{U}$ is defined to be the union of U and all of its limit points. Thus, an equivalent characterization of a closed set is the following: for any sequence $(a_n)_n$ in $S \subset \Omega$ such that $a_n \rightarrow a_0 \in \Omega \Rightarrow a_0 \in S$. That is, S contains the limit point of every sequence in S that is convergent in Ω.

A.2.5 Definition

Let (Ω_i, τ_i), for i = 1,2, be nonempty topological spaces and $f: \Omega_1 \rightarrow \Omega_2$ be a map. Then, f is *continuous* at $x_0 \in \Omega_1$ if for any neighborhood $N(f(x_0)) \subset \Omega_2$ of $f(x_0)$, there exists a neighborhood $U(x_0) \subset \Omega_1$ of x_0 such that $f(U(x_0)) \subset N(f(x_0))$. In particular, f is *continuous* on Ω_1 if $\forall N \in \tau_2$, $f^{-1}(N \cap f(\Omega_1)) \in \tau_1$.

It is easy to see from Definitions A.2.4 and A.2.5 that f is *continuous* at $x_0 \in \Omega_1$ if and only if for any convergent sequence $x_n \rightarrow x_0$ in Ω_1, the sequence $(f(x_n))_n$ in Ω_2 converges to $f(x_0) \in \Omega_2$. As mentioned earlier, this technically holds for first countable spaces [4,7,11]. Finally, suppose U,V are two topological spaces and $f: U \rightarrow V$ is a continuous bijection such that its *inverse* $f^{-1}: V \rightarrow U$ is also continuous. Then, f is called a *homeomorphism* and the two spaces U, V may be identified (topologically) with each other. This is denoted by $U \cong V$.

Given a topological space (Ω, τ) and any subset $U \subseteq \Omega$, an *open cover* $C \subset \tau$ of U is defined by $U \subseteq \bigcup_{N\in C} N$. The open cover is finite if $|C| < \infty$, where the

* These spaces are called *first countable spaces*. See, for example, References [7,11] for more details.

† For an uncountable indexing set, the associated "sequence" is called a *net*. This is the generalization of a sequence (*ibid.* for further details) required for more general topological spaces; that is, spaces that do not satisfy the axiom of first countability.

cardinality |C| denotes the number of elements in C. If C is an open cover of $U \subseteq \Omega$, and $C' \subset C$ such that $U \subseteq \bigcup_{N \in C'} N$, then C' defines a *subcover* of U.

A.2.6 Definition

Given a topological space (Ω,τ), a subset $U \subseteq \Omega$ is said to be *compact* if every open cover of U has a finite subcover.

A.2.7 Proposition

Let Ω be a compact space and $K \subseteq \Omega$ Then, K is compact if and only if K is closed. □

Intuitively, a compact set is a closed set with "finite volume." To make this concept concrete, consider a finite-dimensional Euclidean space $\mathbf{R}^n$. Next, define a *metric* (or *distance function*) $\rho : \mathbf{R}^n \times \mathbf{R}^n \to \mathbf{R}$ to be a nonnegative function satisfying

(a) $\rho(x,y) \geq 0 \; \forall x,y \in \mathbf{R}^n$ with $\rho(x,x) = 0 \; \forall x \in \mathbf{R}^n$

(b) $\rho(x,y) = \rho(y,x) \; \forall x,y \in \mathbf{R}^n$ (symmetry)

(c) $\rho(x,y) \leq \rho(x,z) + \rho(z,y) \; \forall x,y,z \in \mathbf{R}^n$ (triangle inequality)

An ε-*disk* $B_\varepsilon(x_0) \subset \mathbf{R}^n$ of x_0 is defined by $B_\varepsilon(x_0) = \{x \in \mathbf{R}^n : \rho(x,x_0) < \varepsilon\}$. This defines an open subset in $\mathbf{R}^n$. An example of a metric on $\mathbf{R}^n$ is the Euclidean metric $\rho(x,y) = \|x - y\|$, where $\|x - y\| = \sqrt{(x^1 - y^1)^2 + \cdots + (x^n - y^n)^2}$.

Given a subset $U \subset \mathbf{R}^n$, if $\exists \varepsilon > 0$ such that $\forall x \in U, U \subset B_\varepsilon(x)$, then U is *bounded* in $\mathbf{R}^n$. In $\mathbf{R}^n$, a subset $U \subset \mathbf{R}^n$ is compact if and only if it is both closed and bounded. Thus, every closed ε-*disk* $\bar{B}_\varepsilon(x_0)$ is compact in $\mathbf{R}^n$. Compact sets possess some pleasant properties. An instance is given below.

A.2.8 Theorem

Let (Ω,τ) be a topological space and f:$K \to R$ be a continuous function, where $K \subset \Omega$ is compact. Then, f attains its maximum and minimum on K. That is, $\exists \bar{u}, \underline{u} \in K$ such that $f(\underline{u}) \leq f(x) \leq f(\bar{u})$ on K. □

This section closes with the concept of connectedness. Let Ω be a topological space. Then, Ω is said to be *connected* if $\forall U,V \subset \Omega$ open with $U \cap V = \varnothing$, $\Omega \neq U \cup V$. Furthermore, if $\forall u,v \in \Omega$, $\exists \gamma$: $[0,1] \to \Omega$ continuous, such that $\gamma(0) = u, \gamma(1) = v$, then Ω is *path connected*. That is, a path-connected space is a space such that any two points within the space can be connected by a (continuous) path. It can be shown that path connectedness implies connectedness; however, the converse is false. A path γ such that $\gamma(0) = \gamma(1)$ is called a *loop* or a *closed path*.

A.2.9 Definition

Let Ω be a path-connected topological space and let $x,y \in \Omega$ be a pair of arbitrary points. Suppose $\gamma, \gamma' : [0,1] \to \Omega$ are two arbitrary paths connecting x,y. That is, $\gamma(0) = x = \gamma'(0)$ and $\gamma(1) = y = \gamma'(1)$. If there exists a continuous mapping $H: [0,1] \times [0,1] \to \Omega$ such that $H(0,t) = \gamma(t)\ \forall t \in [0,1]$ and $H(1,t) = \gamma'(t)\ \forall t \in [0,1]$, then Ω is said to be *simply connected*. The mapping H is called a *homotopy*. A *multiply connected* space is a space that is not simply connected. Lastly, a loop is *null-homotopic* if there exists a homotopy mapping the loop into a point, that is, if the loop can be continuously shrunk to a point.

In short, a homotopy defines a deformation that deforms one path into another in a continuous fashion. Examples of simply connected spaces are $\mathbf{R}^n$ and the 2-sphere S^2. On the other hand, the 2-torus T^2 (i.e., a donut) is not simply connected: the presence of a hole in the torus prevents the loops λ,η encircling the torus from deforming into one another (see Figure A.4). A hole thus presents an obstruction toward the existence of a homotopy deforming λ into η, where $\lambda(0) = \eta(0) = P = \lambda(1) = \eta(1)$. On the other hand, the two paths γ,γ' in $\mathbf{R}^2$, where $\gamma(0) = \gamma'(0) = P = \gamma(1) = \gamma'(1)$, can be deformed into each other. However, observe that if the point Q should be deleted from $\mathbf{R}^2$, then γ cannot be deformed into γ' (and vice versa).

Regarding the 2-sphere $S^2 = \{(x,y,z) \in \mathbf{R}^3 : x^2 + y^2 + z^2 = 1\}$ in Figure A.4, it is also clear that γ,γ' can be deformed into each other in S^2. However, by deleting the point $Q \in S^2$, it can be intuitively seen that γ' can be deformed into γ via a homotopy that traverses around the sphere about the point $P \in S^2$. Topologically, by deleting the point $Q \in S^2$ from S^2 and stretching out the deleted point in all directions to infinity, $S^2 - \{Q\}$ can be transformed into the plane $\mathbf{R}^2$.

A.2.10 Remark

Suppose $K \subset \mathbf{R}^3$ is a compact (2-dimensional) surface such that $\partial K = \varnothing$. Then, K is called a *closed* 2-surface. Intuitively, the boundary of a boundary is empty. In what follows, a surface in $\mathbf{R}^3$ always refers to a 2-surface. Finally, a compact surface $K \subset \mathbf{R}^3$ is said to be *spanned* by a simple loop $\gamma \subset \mathbf{R}^3$ if $\gamma = \partial K$. Recall

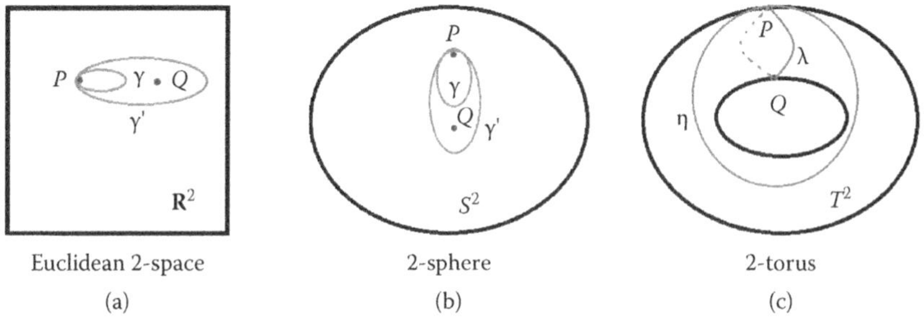

FIGURE A.4
Simple connectedness: (a) and (b); multiple connectedness: (c).

that a loop is *simple* if $\gamma(t) \neq \gamma(s) \, \forall \, t,s \in (0,1)$; that is, if the loop does not have any self-intersection.

An example of a closed surface is a 2-sphere, whereas a hemisphere is clearly a surface spanned by a loop. Suppose M is a path-connected topological space and $\gamma_i : [0,1] \to M$ are any two paths, for i = 1,2, such that $\gamma_1(1) = \gamma_2(0)$. Then, the operation $*$ can be defined on the pair (γ_1, γ_2) to construct a new path $\gamma_1 * \gamma_2$ in the following manner.

$$\gamma_1 * \gamma_2(t) = \begin{cases} \gamma_1(2t) & \text{for } 0 \leq t \leq \frac{1}{2} \\ \\ \gamma_2(2t-1) & \text{for } \frac{1}{2} \leq t \leq 1 \end{cases} \tag{A.3}$$

That is, Equation (A.3) defines a new path by joining γ_1 to γ_2.

Indeed, this process can be continued for the sequence of paths $(\gamma_i)_i$ in M if they satisfy $\gamma_i(1) = \gamma_{i+1}(0) \, \forall i : \gamma_1 * \gamma_2 * \cdots * \gamma_i * \cdots$. Finally, the *inverse* γ^{-1} of the path γ is given by $\gamma^{-1}(t) = \gamma(1-t)$. That is, the inverse path is just the reverse orientation of the original path. In particular, $\gamma * \gamma^{-1}$ defines a *degenerate* loop.

A.3 Boundary Conditions for Electromagnetic Fields

The following two important theorems play a central role in proving certain elementary properties of electromagnetism. They are stated without proof; the proofs can be found in the numerous literature on electrodynamics [3,13,18,19] or [12]. The first theorem is called Stokes' theorem, and the second theorem is called the divergence theorem.

Recall that a continuous mapping $f : \mathbf{R}^n \to \mathbf{R}^m$ is a called an *m-vector field* (or simply a *vector field*) on $\mathbf{R}^n$. If m = 1, f is called a *scalar field*. Finally, for notational convenience, let $C^k(\mathbf{R}^n, \mathbf{R}^m)$ denote the space of k-times continuously differentiable vector fields on $\mathbf{R}^n$. Intuitively, think of this space as the space of vector fields $f = (f^1(x), \ldots, f^m(x))$ satisfying

$$\frac{\partial^k f^j}{\partial x^{i1} \cdots \partial x^{ik}} = \frac{\partial^k f^j}{\partial x^{\sigma(i1)} \cdots \partial x^{\sigma(ik)}} \quad \forall \, i_1, \ldots, i_k \in \{1, \ldots, n\}, \forall j = 1, \ldots, m$$

and all permutations σ: {1,...,n} → {1,...,n}. Note that a *permutation* is just a bijection from a countable set onto itself. The space $C^k(\mathbf{R}^n, \mathbf{R}^m)$ is often abbreviated by C^k should no confusion arise, and the mappings in $C^k(\mathbf{R}^n, \mathbf{R}^m)$ are said to be of *class* C^k.

Some examples are considered below. Let $f:\mathbf{R}^3 \to \mathbf{R}$ be continuous such that $\partial_{x^i} f$ exists and is continuous on $\mathbf{R}^3$ for $i = 1,2,3$. Then, f is of class C^1. In addition, if $\partial_{x^i x^j} f$ exists and is continuous $\forall\, i,j = 1,2,3$—equivalently, $\partial_{x^i x^j} f = \partial_{x^j x^i} f \ \forall\, i,j$—on $\mathbf{R}^3$, then f is of class C^2. Finally, $f \in C^0(\mathbf{R}^3)$ denotes f is a continuous function on $\mathbf{R}^3$: this space is commonly denoted by $C(\mathbf{R}^3)$.

A.3.1 Theorem (Stokes' Theorem)

Let f be a vector field of class C^1 on a compact surface $\Sigma \subset \mathbf{R}^3$ in $\mathbf{R}^3$ spanned by $\partial\Sigma \neq \varnothing$, $\boldsymbol{l}$ a unit vector field tangent to $\partial\Sigma$ in $\mathbf{R}^3$, and N a unit vector field normal to the surface Σ in $\mathbf{R}^3$ pointing away from the interior of Σ.
Then,

$$\oint_{\partial\Sigma} f \cdot \boldsymbol{l}\, d\ell = \oiint_{\Sigma} \nabla \times f \cdot \boldsymbol{n}\, dS.$$

□

A.3.2 Corollary

If Σ is a closed compact surface in $\mathbf{R}^3$, then $\oiint_{\Sigma} \nabla \times f \cdot \boldsymbol{n}\, dS = 0$.

Proof

By assumption, $\partial\Sigma = 0$. Hence, invoking Stokes' theorem,

$$\oiint_{\Sigma} \nabla \times f \cdot \boldsymbol{n}\, dS = \oint_{\partial\Sigma} f \cdot \boldsymbol{l}\, d\ell \equiv 0$$

as asserted. □

Heuristically, Corollary A.3.2 is evident: it suffices to consider $S^2 \subset \mathbf{R}^3$ and note that $\partial S^2 = \varnothing$. That is, a sphere has no boundary. Thus, by Theorem A.3.1, the surface integral must vanish, as the integral is over an empty set.

A.3.3 Theorem (Divergence Theorem)

If a vector field f on a compact set $\Omega \subset \mathbf{R}$ is of class C^1, where Ω is simply connected and $\partial\Omega$ is closed, then $\oiint_{\partial\Omega} f \cdot \boldsymbol{n}\, dS = \oiiint_{\Omega} \nabla \cdot f\, dV$, where $\boldsymbol{n}$ is the outward, normal, unit vector field on $\partial\Omega$. □

As a side remark, Stokes' theorem and the divergence theorem are actually special cases of a more general theorem attributed to Stokes, which relates the volume integral of an n-dimensional volume to the surface integral of the associated $(n - 1)$-dimensional boundary of the n-dimensional volume. For more details, refer to any standard reference on differential geometry [4,17].

Faraday's law follows as a direct consequence of (1.1) via Stokes' theorem:

$$\oint_{\partial\Sigma} \boldsymbol{E}\cdot\boldsymbol{l}\,d\ell = \iint_{\Sigma} \nabla\times\boldsymbol{E}\cdot\boldsymbol{n}\,dS = -\partial_t \iint_{\Sigma} \boldsymbol{B}\cdot\boldsymbol{n}\,dS \equiv -\partial_t \Psi$$

where $\Psi \equiv \iint_{\Sigma} \boldsymbol{B}\cdot dS$ is the magnetic flux flowing across the surface Σ. Notice that when the compact surface Σ is closed, Faraday's law implies that $-\partial_t \Psi = 0$. That is, the magnetic flux is conserved because it does not change with time.

Furthermore, observe that the sum of magnetic flux entering and exiting a closed surface is zero. In other words, magnetic flux forms loops. To see this, it suffices to appeal to the divergence theorem: $\Psi = \iint_{\Sigma} \boldsymbol{B}\cdot dS = \iiint_M \nabla\cdot\boldsymbol{B}dV$, where $\partial M = \Sigma$. From Maxwell's equation, $\nabla\cdot\boldsymbol{B} = 0$, it follows at once that $\Psi = 0$. Thus, any magnetic flux lines that exit a compact surface must return to the same surface.

Finally, recall two important identities from vector calculus. They are stated without proof below. Suppose φ is a scalar field on $\mathbf{R}^3$ that is of class C^2, and f a vector field on $\mathbf{R}^3$ that is also of class C^2. Then, the following identities hold.

$$\nabla\times\nabla\varphi = 0 \tag{A.4}$$

$$\nabla\cdot\nabla\times f = 0 \tag{A.5}$$

The contents of the remainder of this section comprise (i) the boundary conditions for the electrostatic field, and (ii) the boundary conditions for the magnetostatic field. So, first of all, consider a perfect dielectric medium; this is an idealized medium whose conductivity is identically zero. In the presence of an applied electric field, the molecules comprising the dielectric are polarized to form dipoles.

In complete analogy with the electric field, define the *electric polarization vector* $\boldsymbol{P}$ on a dielectric medium $\Omega\subset\mathbf{R}^3$ to be the electric field induced by the dipoles via an applied external electric field; see Figure A.5. Informally,

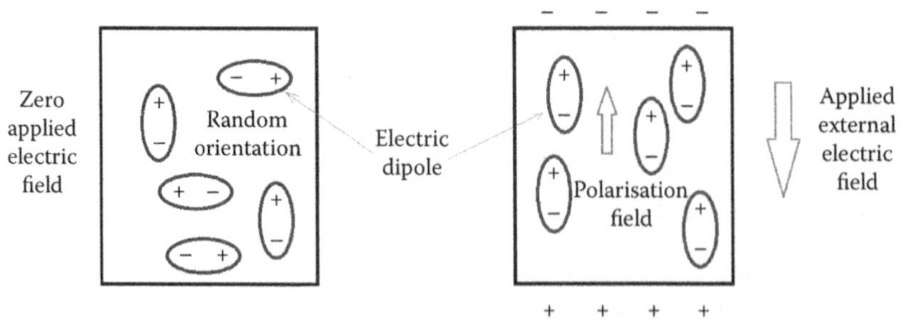

FIGURE A.5
Alignment of electric dipoles under an external electric field.

under the influence of an applied electric field, the electric dipoles of a dielectric medium align themselves according to the applied field. The aligned dipoles thus induce a polarization field of their own.

A.3.4 Definition

Let $\Omega \subset \mathbf{R}^3$ be a dielectric medium that is uncharged. Then, *bound charges* are charges within Ω such that they lie within the valence band.* An *electric polarization field* $\boldsymbol{P}$ generated by Ω is a C^1 vector field satisfying

$$\nabla \cdot \boldsymbol{P} + \tilde{\rho} = 0 \tag{A.6}$$

where $\tilde{\rho}$ is the *bound charge density* in Ω induced by the dipoles. In particular, a *bound surface charge density* $\tilde{\sigma}$ on Ω is defined by

$$\boldsymbol{P} \cdot \boldsymbol{n} = \tilde{\sigma} \tag{A.7}$$

where $\boldsymbol{n}$ is the normal vector field on $\partial\Omega$.

A.3.5 Lemma

Equations (A.6) and (A.7) are well-defined.

Proof

The total charge Q_Ω in Ω is clearly given by $Q_\Omega = \oiint_{\partial\Omega} \tilde{\sigma}\, d^2x + \oiiint_{\Omega} \tilde{\rho}\, d^3x$. By Theorem A.3.3, $\oiint_{\partial\Omega} \boldsymbol{P} \cdot \boldsymbol{n} d^2x = \oiiint_{\Omega} \nabla \cdot \boldsymbol{P} d^3x$ implies trivially that

$$0 = Q_\Omega = \oiint_{\partial\Omega} \boldsymbol{P} \cdot \boldsymbol{n} d^2x - \oiiint_{\Omega} \nabla \cdot \boldsymbol{P} d^3x = \oiint_{\partial\Omega} \tilde{\sigma}\, d^2x + \oiiint_{\Omega} \tilde{\rho}\, d^3x$$

as Ω is assumed to be uncharged, establishing the assertion. □

A more general proof can be obtained (see, e.g., Reference [3, p. 107]) via the vector identity $\nabla(\alpha \boldsymbol{A} + \boldsymbol{B}) = \boldsymbol{A} \cdot \nabla\alpha + \alpha\nabla \cdot \boldsymbol{A} + \nabla \cdot \boldsymbol{B}$ given in Section A.1.

Indeed, it is evident from Gauss' law $\nabla \cdot \varepsilon_0 \boldsymbol{E} = \rho$ and the proof of Lemma A.3.5, that the negative sign in Equation (A.6) is required in order for the polarized charge in an electrically neutral dielectric to remain zero; that is, $\nabla \cdot \boldsymbol{P} = -\tilde{\rho}$. In a sense, Lemma A.3.5 furnishes a motivation for Definition A.3.4.

Some comments are due. First, it is clear from Definition A.3.4 that the presence of a polarization field will perturb the existing applied electric field; see

---

* Charges that lie in the conduction band, such as in a conductor, can flow freely within the medium. However, charges that are in the valence band cannot flow freely throughout the medium; they are thus bound.

Figure A.5. To wit, the resultant field is the superposition of the applied field and the field induced by the aligned dipoles. Second, by Gauss' law, the total volume charge density must satisfy $\nabla \cdot \varepsilon_0 \boldsymbol{E} = \rho + \tilde{\rho}$, whence, by (A.6), $\rho = \nabla \cdot (\varepsilon_0 \boldsymbol{E} + \boldsymbol{P})$. This leads naturally to the following definition.

A.3.6 Definition

Given a dielectric medium $\Sigma \subset \mathbf{R}^3$ subject to an external electric field $\boldsymbol{E}$, the *electric displacement* $\boldsymbol{D}$ is a vector field in $\mathbf{R}^3$ given by $\boldsymbol{D} = \varepsilon_0 \boldsymbol{E} + \boldsymbol{P}$, and the *electric permittivity* ε of Ω is defined by $\boldsymbol{P} = (\varepsilon - \varepsilon_0)\boldsymbol{E}$, where ε_0 is the electric permittivity of free space. The ratio $\frac{\varepsilon}{\varepsilon_0}$ is called the *dielectric constant* (or *relative permittivity*) of Ω.

A.3.7 Proposition

Given a dielectric $\Omega \subset \mathbf{R}^3$ and an applied external electric field $\boldsymbol{E}$, the electric displacement satisfies $\boldsymbol{D} = \varepsilon \boldsymbol{E}$.

Proof

By definition, $\boldsymbol{P} = (\varepsilon - \varepsilon_0)\boldsymbol{E} \Rightarrow \boldsymbol{D} = \varepsilon_0 \boldsymbol{E} + (\varepsilon - \varepsilon_0)\boldsymbol{E} = \varepsilon \boldsymbol{E}$, as claimed. □

A.3.8 Theorem

Let $(\Omega_\pm, \varepsilon_\pm) \subset \mathbf{R}^3$ be two simply connected dielectric media such that $\partial\Omega_+ = S = \partial\Omega_-$ Suppose an electric field $\boldsymbol{E}$ is directed from Ω_+ across the boundary S into Ω_-. Then, $\boldsymbol{n} \times (\boldsymbol{E}_+ - \boldsymbol{E}_-) = 0$, where $\boldsymbol{n}$ is a unit, normal, vector field on S directed into Ω_- and $\boldsymbol{E}_\pm = \boldsymbol{E} \mid \Omega_\pm$.

Proof

Recall that for a simply connected space, $\oint_\gamma \boldsymbol{E} \cdot \mathrm{d}\boldsymbol{l} = 0$, where $\boldsymbol{l}$ is the vector field tangent to a loop γ, thus, choose a sufficiently small rectangular loop $\gamma = \gamma_\uparrow \cup \gamma_\rightarrow \cup \gamma_\downarrow \cup \gamma_\leftarrow$, where $\gamma_\rightarrow \subset \Omega_-, \gamma_\leftarrow \subset \Omega_+$ are paths parallel to a vector $\boldsymbol{v}$ tangent at some fixed point $x \in S$, and $\gamma_\uparrow, \gamma_\downarrow$ are paths normal to $\boldsymbol{v}(x)$, for a sufficiently small loop γ. That is, the path $\gamma_\rightarrow(\gamma_\uparrow)$ is of the same length and is oppositely oriented to $\gamma_\leftarrow(\gamma_\downarrow)$. Then, by construction,

$$0 = \oint_\gamma \boldsymbol{E} \cdot \mathrm{d}\boldsymbol{l} = \int_{\gamma_\uparrow} \boldsymbol{E} \cdot \mathrm{d}\boldsymbol{l} + \int_{\gamma_\rightarrow} \boldsymbol{E} \cdot \mathrm{d}\boldsymbol{l} + \int_{\gamma_\downarrow} \boldsymbol{E} \cdot \mathrm{d}\boldsymbol{l} + \int_{\gamma_\leftarrow} \boldsymbol{E} \cdot \mathrm{d}\boldsymbol{l}$$

In particular, in the limit as the lengths of $\gamma_\uparrow, \gamma_\downarrow$ go to zero, where $\gamma_\rightarrow, \gamma_\leftarrow$ are held constant,

$$0 = \int_{\gamma_\rightarrow} \boldsymbol{E} \cdot \mathrm{d}\boldsymbol{l} + \int_{\gamma_\leftarrow} \boldsymbol{E} \cdot \mathrm{d}\boldsymbol{l} = \int_{\gamma_\rightarrow} (\boldsymbol{E}_- - \boldsymbol{E}_+) \cdot \mathrm{d}\boldsymbol{l}$$

where, recall that in the limit as the lengths of $\gamma_\uparrow, \gamma_\downarrow$ tend to zero,

$$\int_{\gamma_\rightarrow} \boldsymbol{E}\cdot \mathrm{d}\boldsymbol{l} = -\int_{\gamma_\leftarrow} \boldsymbol{E}\cdot \mathrm{d}\boldsymbol{l}$$

(as they will then have the same endpoints in passing to the limit). As Ω is simply connected, this is possible; a more rigorous construction via homotopy is not pursued here.

Thus, $0=\int_{\gamma_\rightarrow}(\boldsymbol{E}_- - \boldsymbol{E}_+)\cdot \mathrm{d}\boldsymbol{l} \Rightarrow (\boldsymbol{E}_- - \boldsymbol{E}_+)\cdot \boldsymbol{l} = 0$, where $\boldsymbol{l}$ is the unit tangent vector field on $\gamma_\rightarrow$. Because the scalar product of the normal component of $\boldsymbol{E}_\pm$ with $\boldsymbol{l}$ is zero by construction, it follows from the arbitrariness of $\boldsymbol{E}$ that the $\boldsymbol{E}_+ \equiv \boldsymbol{E}_-$. Finally, as $\boldsymbol{n}$ is normal to $\boldsymbol{l}$, the result is equivalent to $\boldsymbol{n}\times(\boldsymbol{E}_- - \boldsymbol{E}_+) = 0$. That is, $\boldsymbol{E}$ is continuous across the boundary interface S, as required. □

A.3.9 Corollary

Suppose Ω_- is a perfect electrical conductor (PEC). Then, $\boldsymbol{n}\times\boldsymbol{E}_+ = 0$ on S.

Proof

In a smuch as Ω_- is PEC, it follows that the static electric field in Ω_- is zero: $\boldsymbol{E}_- = 0$. Thus, the conclusion follows. □

Now, from Proposition A.3.8, it is clear that the tangential component of $\boldsymbol{D}_\pm$ on S is thus discontinuous across S unless $\varepsilon_+ = \varepsilon_-$. Indeed this is quite evident: $\boldsymbol{n}\times\boldsymbol{E}_+ = \boldsymbol{n}\times\boldsymbol{E}_- \Rightarrow \boldsymbol{n}\times\boldsymbol{D}_+\frac{1}{\varepsilon_+} = \boldsymbol{n}\times\boldsymbol{D}_-\frac{1}{\varepsilon_-} \Rightarrow \boldsymbol{D}_+ = \frac{\varepsilon_+}{\varepsilon_-}\boldsymbol{D}_-$, as claimed. Physically, this relates to the dipoles induced on the boundary of $\Omega_\pm$. The tangential component of the electric displacement is thus reduced by the dielectric constant of the medium. Finally, from the definition of potential, $\varphi = -\int \boldsymbol{E}\cdot \mathrm{d}\boldsymbol{l}$, invoking Theorem A.2.8 leads to the continuity of the potential across S. To see this, it suffices to observe from the proof of Theorem A.2.8 that $0=\int_{\gamma_\rightarrow}\boldsymbol{E}\cdot \mathrm{d}\boldsymbol{l} + \int_{\gamma_\leftarrow}\boldsymbol{E}\cdot \mathrm{d}\boldsymbol{l} = \int_{\gamma_\rightarrow}(\boldsymbol{E}_- - \boldsymbol{E}_+)\cdot \mathrm{d}\boldsymbol{l} = \varphi_- - \varphi_+$, whence $\varphi_-|S = \varphi_+|S$. Thus, Theorem A.3.8 is equivalent to the continuity of the potential across a dielectric interface. This is formally stated below.

A.3.10 Corollary

Given any $\boldsymbol{x}\in S$, $\lim_{y\to x}\varphi_-(y) = \lim_{y'\to x}\varphi_+(y')$, where $y\in\Omega_-$ and $y'\in\Omega_+$. □

The next result establishes the normal component of the electric field across a dielectric boundary.

A.3.11 Theorem

Let $(\Omega_\pm, \varepsilon_\pm)\subset \mathbf{R}^3$ be two simply connected dielectric media such that $\partial\Omega_+ = S = \partial\Omega_-$. Suppose an electric field $\boldsymbol{E}$ is directed from Ω_+ across the boundary S into Ω_-. Then, $\boldsymbol{n}\cdot(\boldsymbol{D}_+ - \boldsymbol{D}_-) = 0$, where $\boldsymbol{n}$ is a unit vector field normal to S pointing into Ω_-, and $\boldsymbol{D}_\pm = \boldsymbol{D}|\Omega_\pm$.

Proof

Across S, fix some point $\boldsymbol{x}_0 = (x_0^1, x_0^2, x_0^3) \in S$ and consider a small cylinder $V(r,\delta) = B_2(\boldsymbol{x}_0, r) \times [x_0^3 - \delta, x_0^3 + \delta]$, where

$$B_2(\boldsymbol{x}_0, r) = \left\{ \boldsymbol{x} \in \mathbf{R}^3 : (x^1 - x_0^1)^2 + (x^2 - x_0^2)^2 \le r,\, x^3 = x_0^3 \right\}$$

is a circular surface of radius r centered about $\boldsymbol{x}_0$. Set $V_\pm(r,\delta) = \Omega_\pm \cap V(r,\delta)$. Then, by Gauss' law,

$$\int_{V(r,\delta)} \nabla \cdot \boldsymbol{D} \mathrm{d}^3 x = \int_{V(r,\delta)} \rho \mathrm{d}^3 x$$

where ρ is the charge density in $V(r, \delta)$. Invoking Theorem A.3.3,

$$\int_{V(r,\delta)} \nabla \cdot \boldsymbol{D} \mathrm{d}^3 x = \int_{\partial V_\uparrow(r,\delta)} \boldsymbol{D}_- \cdot \boldsymbol{n}_\uparrow \mathrm{d}^2 x + \int_{\partial V_\downarrow(r,\delta)} \boldsymbol{D}_- \cdot \boldsymbol{n}_\downarrow \mathrm{d}^2 x +$$
$$\int_{\partial V_0^+(r,\delta)} \boldsymbol{D}_- \cdot \boldsymbol{n}_0 \mathrm{d}^2 x + \int_{\partial V_0^-(r,\delta)} \boldsymbol{D} \cdot \boldsymbol{n}_0 \mathrm{d}^2 x$$

where $\partial V_\uparrow(r,\delta) = B_2(\boldsymbol{x}_0, \delta) \times \{x_0^3 + \delta\}$, $\partial V_\downarrow(r,\delta) = B_2(\boldsymbol{x}_0, \delta) \times \{x_0^3 - \delta\}$ are the end caps of the cylinder $V(r, \delta)$,

$$\partial V_0^+(r,\delta) = (\partial V(r,\delta) \cap \Omega_+) - \partial V_\downarrow(r,\delta) \quad \text{and} \quad \partial V_0^-(r,\delta) = (\partial V(r,\delta) \cap \Omega_-) - \partial V_\uparrow(r,\delta)$$

are the cylindrical boundary modulo the end caps of the cylinder. The unit vector field $\boldsymbol{n}_\uparrow(\boldsymbol{n}_\downarrow)$ is the outward normal on $\partial V_\uparrow(r,\delta)$ ($\partial V_\downarrow(r,\delta)$) and $\boldsymbol{n}_0$ is the outward normal unit vector field on $\partial V_0^-(r,\delta) \cup \partial V_0^+(r,\delta)$.

Then, by construction, $\lim_{\delta \to 0} \int_{\partial V_0^\pm(r,\delta)} \boldsymbol{D}_\pm \cdot \boldsymbol{n}_0 \mathrm{d}^2 x = 0$, and hence, for $r > 0$ sufficiently small, and noting that on S, $\boldsymbol{n}_\uparrow = -\boldsymbol{n}_\downarrow$,

$$\rho_S \pi r^2 \approx \lim_{\delta \to 0} \int_{V(r,\delta)} \nabla \cdot \boldsymbol{D} \mathrm{d}^3 x = \int_{\partial V_\uparrow(r,\delta)} \boldsymbol{D}_- \cdot \boldsymbol{n}_\uparrow \mathrm{d}^2 x + \int_{\partial V_\downarrow(r,\delta)} \boldsymbol{D}_+ \cdot \boldsymbol{n}_\downarrow \mathrm{d}^2 x \approx \boldsymbol{n}_\uparrow \cdot (\boldsymbol{D}_- - \boldsymbol{D}_+) \pi r^2$$

where ρ_S is the surface charge density on $\partial V(r, \delta)$. Hence, $\boldsymbol{n}_\uparrow \cdot (\boldsymbol{D}_- - \boldsymbol{D}_+) = \rho_S$. In particular, in the absence of free charges, $\rho_S = 0 \Rightarrow \boldsymbol{n}_\uparrow \cdot (\boldsymbol{D}_- - \boldsymbol{D}_+) = 0$, as desired. □

The electric field normal to a dielectric interface is thus discontinuous, and the discontinuity is the result of the formation of the induced dipoles at the interface caused by the applied electric field. Moreover, note that in a conductor, the valence electrons are not bound to the crystal lattice: they are free charges that may be approximated by a cloud of plasma in the conduction band. This observation leads to the following corollary.

A.3.12 Corollary

If Ω_- is a PEC, then $n_\downarrow \cdot D_+ = \rho_S$, where $n_\downarrow$ is the unit normal vector field on S pointing into Ω_+ and ρ_S is the surface charge induced by the applied electric field.

Proof

Because Ω_- is a PEC, it follows that the static electric field within Ω_- is zero and hence, $-n_\uparrow \cdot D_+ = n_\downarrow \cdot D_+ = \rho_S$, as required. □

Some comments regarding the proof of Theorem A.3.11 are due. An uncharged pure dielectric can only have bound charges. These charges are induced by the presence of an applied electric field that polarizes the molecules to form dipoles. However, if free charges were placed on the boundary S, then $n_\uparrow \cdot (D_- - D_+) = \rho_S$ would hold.

Conversely, a conductor has, by definition, free charges present in the conduction band in order to conduct electrons.* Hence, even if the conductor were initially neutral, in the presence of an electric field, the electric field would cause the free charges in Ω_- to migrate onto S such that the resultant static electric field in Ω_- is zero. This is encapsulated in Corollary A.3.12.

A.3.13 Example

Consider a point charge above a dielectric interface defined as follows. Given $(\Omega_\pm, \varepsilon_\pm) \subset \mathbf{R}^3$, where $\Omega_+ = \mathbf{R}^3_+$ and $\Omega_- = \mathbf{R}^3 - \Omega_+$, set $S = \partial\Omega_+$ and suppose that a charge $Q \neq 0$ is located at $x_Q = (0,0,d) \in \Omega_+$. What is the potential in $\mathbf{R}^3$? Using the method of images, consider an image charge $Q' \in \Omega_-$ located at $x_{Q'} = (0,0,-d)$. Then, the potential on Ω_+ is given by

$$\varphi_+ = \frac{1}{4\pi\varepsilon_+}\left\{\frac{Q}{\sqrt{x^2+y^2+(z-d)^2}} + \frac{Q'}{\sqrt{x^2+y^2+(z+d)^2}}\right\} \tag{A.8}$$

Next, given that Ω_- is a dielectric, the field within Ω_- is nonzero. To determine the field in Ω_-, suppose $\exists Q'' \in \Omega_+$ located at $x_Q = (0,0,d)$, where the electric permittivity of the whole space is ε_-. Then, the potential in Ω_- is given by

$$\varphi_- = \frac{1}{4\pi\varepsilon_-}\frac{Q''}{\sqrt{x^2+y^2+(z-d)^2}} \tag{A.9}$$

---

* More precisely, the intersection of the conduction band and the valence band is nonempty. Hence, the valence electrons become available for conduction as they also exist in the conduction band.

It thus remains to determine (Q', Q''). To do so, the boundary conditions must be invoked. By Theorem A.3.8, $\lim_{z\to 0^+} \varepsilon_- \boldsymbol{n}\cdot \boldsymbol{E}_-(x,y,z) = \lim_{z\to 0^-} \varepsilon_+ \boldsymbol{n}\cdot \boldsymbol{E}_+(x,y,z)$, where $\boldsymbol{n}$ is the unit normal vector field on S directed toward Ω_-. From $\boldsymbol{E} = -\nabla\varphi$, it follows that $\boldsymbol{n}\cdot\boldsymbol{E}_\pm = -\partial_z \varphi_\pm$. Thus,

$$\partial_z \varphi_+ = -\frac{1}{4\pi\varepsilon_+}\left\{\frac{Q(z-d)}{\left(x^2+y^2+(z-d)^2\right)^{\frac{3}{2}}} + \frac{Q'(z+d)}{\left(x^2+y^2+(z+d)^2\right)^{\frac{3}{2}}}\right\}$$

$$\partial_z \varphi_- = -\frac{1}{4\pi\varepsilon_+}\frac{Q''(z-d)}{\left(x^2+y^2+(z-d)^2\right)^{\frac{3}{2}}}$$

and on S, the boundary condition yields $-Q+Q' = -Q'' \Rightarrow Q'' = Q - Q'$.

Next, invoking Corollary A.3.10, $\lim_{z\to 0^+}\varphi_+ = \lim_{z\to 0^-}\varphi_-$. Thus, from (A.8) and (A.9),

$$\tfrac{1}{\varepsilon_+}(Q+Q') = \tfrac{1}{\varepsilon_-}Q'' \Rightarrow Q'' = \tfrac{\varepsilon_-}{\varepsilon_+}(Q+Q')$$

whence solving the pair of equations yields

$$Q' = \frac{\varepsilon_+ - \varepsilon_-}{\varepsilon_+ + \varepsilon_-}Q \tag{A.10}$$

$$Q'' = \frac{2\varepsilon_-}{\varepsilon_+ + \varepsilon_-}Q \tag{A.11}$$

□

It is interesting to note from Equation (A.10) that when $\varepsilon_+ > \varepsilon_-$, Q' has the same sign as Q, whereas for $\varepsilon_+ < \varepsilon_-$, Q' has the opposite sign as Q. This also explains why the mirror image induced by a conducting ground plane is oppositely charged, as the electric permittivity of a PEC is infinity. On the other hand, when a charge is embedded in a dielectric medium, its image charge induced across the boundary in a medium with a lesser electric permittivity is of the same sign as that of the original charge. This is clearly counterintuitive.

A.4 Elements of Partial Differential Equations

In this brief primer, two topics in partial differential equations (PDE) that are of primary importance to electromagnetic theory are reviewed. They are the Poisson equation and the wave equation. Readers interested in pursuing more advanced topics in PDE may refer to some excellent references [1,8,10,16,23] for a more general and abstract approach, or [9,26] for a more applied approach. An elementary knowledge of complex analysis is assumed (see Reference [4] for an excellent exposition).

Recall that the concept of distance on the real line $\mathbf{R}$ is defined by the following metric, $d:\mathbf{R}\times\mathbf{R}\to\mathbf{R}_+$, $d(x,y)=|x-y|$. More generally, on $\mathbf{R}^3$, $d_3:\mathbf{R}^3\times\mathbf{R}^3\to\mathbf{R}_+$ may be defined by the *Euclidean norm*

$$d_3(x,y)=\|x-y\|_3=\sqrt{(x^1-y^1)^2+(x^2-y^2)^2+(x^3-y^3)^2}$$

There are other equivalent definitions, however, the above suffices for the present discussion.

A.4.1 Definition

Let $f:\mathbf{R}^3\to\mathbf{R}$ be a function. If $K\subset\mathbf{R}^3$ is compact and $f=0$ on $\mathbf{R}^3-K$, then f is said to have a *compact support* $\operatorname{supp}(f)=K$.

A.4.2 Definition

Suppose $\psi: M\to R$ is of class C^2, where $M\subset\mathbf{R}^3$ is compact. Furthermore, let $f: M\to\mathbf{R}$ be continuous on M. Then, $-\Delta\psi=f$ on M is called the *Poisson* equation on M, and if $f=0$, it is known as the *Laplace* equation. The operator Δ is called the (3-dimensional) *Laplacian*. Moreover, if $-\Delta\psi=f$ satisfies

(a) $\psi=g$ on ∂M, where $g\in C(\partial M)$, then it is said to satisfy the *Dirichlet boundary condition*
(b) $\partial_n\psi=g$ on ∂M, where $\boldsymbol{n}$ is a normal vector field on ∂M (pointing outward from M) and some $g\in C(\partial M)$, then it is said to satisfy the *Neumann boundary condition*

It is crucial to note that $-\Delta\psi=f$ has no solution if both boundary conditions (a) and (b) are simultaneously satisfied on ∂M. In order for a solution to exist, it is necessary that either (a) or (b) hold. A solution ψ of the Laplace equation is called a *harmonic function*. Furthermore, Definition A.4.2 also holds for $M=\mathbf{R}^3$ if ψ has compact support in M.

A.4.3 Definition

The Poisson equation $-\Delta\psi=f$ on M subject to the Dirichlet boundary condition $\psi=g$ on ∂M is said to be *well-posed* if

(a) A unique solution ψ exists, and
(b) Given $g_i\in C(\partial M)$, $i=1,2$, and unique solutions ψ_i to the Dirichlet problem $-\Delta\psi=f$ on M subject to $\psi=g_i$ on ∂M, $\forall\varepsilon>0$, $|g_1-g_2|<\varepsilon$ on ∂M implies that $|\psi_1-\psi_2|<\varepsilon$.

Condition A.4.3(b) is known as the *continuous dependence on data* for the Dirichlet problem. Essentially, it states that if the data undergo a small perturbation, then the solution will also change by a small perturbation. That is, deforming the data continuously from g_1 into g_2 will deform the solution ψ_1 continuously into ψ_2. Finally, the solutions considered in this monograph are classical solutions. More general solutions, called weak solutions [8,23,25], are not considered here although they are briefly explored in the last section for the reader interested in applying finite element analysis [14,22,25] to electromagnetics.

A.4.4 Example

Consider $r:\mathbf{R}^2 \to \mathbf{R}$ defined by $r(x) = \sqrt{(x^1)^2 + (x^2)^2}$. Then, for $x \neq 0$,

$$\partial_{x^i} r(x) = \tfrac{x^i}{\sqrt{(x^1)^2+(x^2)^2}} = \tfrac{x^i}{r(x)} \quad \text{for} \quad i = 1,2.$$

And

$$\partial^2_{x^i} r(x) = \tfrac{x^i}{\sqrt{(x^1)^2+(x^2)^2}} = \tfrac{1}{r(x)} - \tfrac{(x^i)^2}{r^3(x)}$$

Thus, for any twice differentiable function f on $\mathbf{R}$, the composition $f \circ r : \mathbf{R}^2 \to \mathbf{R}$ is of class C^2 and hence, on setting $\psi = f \circ r$, $\Delta\psi = \psi'' + \tfrac{1}{r}\psi'$, where $\psi' = \tfrac{d}{dr}\psi$. So, when $\Delta\psi = 0$ on $\mathbf{R}^2$, setting $v = \psi' \Rightarrow \tfrac{v'}{v} = -\tfrac{1}{r} \Rightarrow v = \tfrac{1}{r}$, as $-\ln r = \ln\tfrac{1}{r}$, for $v, r \neq 0$. Thus, $\tfrac{d}{dr}\psi = \tfrac{1}{r} \Rightarrow \psi = \ln r$.

Now, as Δ is a linear operator, it follows that the most general solution is of the form $\psi = a\ln r + b$, where a,b are arbitrary constants. In particular, the solution

$$\psi = -\tfrac{1}{2\pi}\ln r \tag{A.12}$$

is called the *fundamental solution* of the 2-dimensional Laplace equation $\Delta\psi = 0$. □

A.4.5 Example

Emulating Example A.3.3, consider $r:\mathbf{R}^3 \to \mathbf{R}$ defined by $r\ (x) = \sqrt{(x^1)^2 + (x^2)^2 + (x^3)^2}$ and for any twice differentiable function f on $\mathbf{R}$, form the composition $f \circ r : \mathbf{R}^3 \to \mathbf{R}$. Then, f is of class C^2 and in view of Example A.4.4,

$$\Delta\psi = \psi'' + \tfrac{2}{r}\psi'$$

where $\psi = f \circ r$. Thus, as before, on setting $v = \psi' \Rightarrow \tfrac{v'}{v} = -\tfrac{2}{r} \Rightarrow v = \tfrac{1}{r^2}$ That is, the general solution to the 3-dimensional Laplace equation $\Delta\psi = 0$ is given by

$\psi = \frac{a}{r} + b$, for $r \neq 0$ and some arbitrary constants a,b. In particular, the fundamental solution of the 3-dimensional Laplace equation is defined by

$$\psi = \frac{1}{4\pi}\frac{1}{r} \tag{A.13}$$

□

It is clear that the solutions defined in Examples A.4.4 and A.4.5 are radial solutions. An important property satisfied by the solution of Laplace's equation is the mean value property.

A.4.6 Definition

Let $M \subset \mathbf{R}^3$ be open and $\psi: M \to \mathbf{R}$ be continuous. Suppose $\forall x \in M$, $\exists r > 0$ such that $\bar{B}(x,r) \subset M$ and $\psi(x) = \frac{1}{4\pi r^2}\int_{\partial B(x,r)} \psi(y) d^2y$. Then, ψ is said to satisfy the *mean value property*. Finally, for $M \subset \mathbf{R}^2$, the mean value property is defined by $\psi(x) = \frac{1}{2\pi r}\int_{\partial B(x,r)} \psi(y) dy$.

A.4.7 Remark

It can be established that if ψ is analytic on $M \subset \mathbf{R}^3$, then it possesses the mean value property. In particular,

$$\psi(x) = \frac{1}{4\pi r^2}\int_{\partial B(x,r)} \psi(y) d^2y = \frac{3}{4\pi r^3}\int_{B(x,r)} \psi(y) d^3y$$

Moreover, the Dirichlet problem of Definition A.4.2(a) has at the most one solution; that is, if a solution exists, it is unique. The existence of a solution is often a very difficult task to establish, and the existence depends upon the smoothness of the boundary [8,10,15]. Clearly, in this appendix, the boundary is always assumed to be sufficiently smooth (the term *regular* is often used interchangeably in the literature) to avoid pathological problems with the question of existence.

A subset $M \subset \mathbf{R}^3$ is called a *domain* if it is open, bounded, and path connected. Now, let $M \subset \mathbf{R}^3$ be a domain and consider the Dirichlet problem

$$-\Delta\psi = f \text{ on } M \tag{A.14a}$$

$$\psi = g \text{ on } \partial M \tag{A.14b}$$

where $g \in C(\partial M)$ and suppose that a solution $\psi \in C^2(M) \cap C(\partial M)$ for (A.14) exists. Is there a formal representation for the solution? This leads to the concept of Green's function.

In all that follows, the analysis is restricted to $\mathbf{R}^n$, for $n = 2,3$. Given a domain $M \subset \mathbf{R}^n$, let $\Psi_n = \Psi_n(|x' - x|)$ denote the fundamental solution on $\mathbf{R}^n$, and

$x', x \in M$, where $x' \neq x$. This is often written as $\Psi_n = \Psi_n(x', x)$. By definition, the function is symmetric: $\Psi_n(x, x') = \Psi_n(x', x)$; see Equation (A.12) and (A.13), where $r = \|x' - x\| = \|x - x'\|$. Furthermore, suppose $\exists \Phi_n = \Phi_n(x', x)$, symmetric, such that it is the solution to the following Laplace equation, for each fixed $x \in M$,

$$\Delta' \Phi_n(x', x) = 0 \; \forall x' \in M \tag{A.15a}$$

$$\Phi_n(x', x) = -\Psi_n(x', x) \; \forall x' \in \partial M \tag{A.15b}$$

where Δ' denotes the Laplacian of the primed coordinates x'. For instance, if $\Delta = \sum_i \partial^2_{x^i}$ in rectangular coordinates, then $\Delta' = \sum_i \partial^2_{x'^i}$. Then,

$$G(x', x) = \Psi_n(x', x) + \Phi_n(x', x) \tag{A.16}$$

defines the *Green function* for the Dirichlet problem (A.14). By construction, Green's function is symmetric in its variables.

It can be shown that the solution to Equation (A.14) has the representation:

$$\psi(x) = \int_M f(x') G(x', x) \mathrm{d}^n x' - \int_{\partial M} g(x') \partial_{n'} G(x', x) \mathrm{d}^{n-1} x', \tag{A.17}$$

where $n = 2,3$, and n' is the unit normal vector field on ∂M pointing outward from M; that is, exterior to M. From an electrostatic perspective, the Green function denotes an impulse function acting on a unit point charge located at x'.

A.4.8 Example

Consider the example of a charge above a grounded plane immediately following Example 1.1.5. Recasting the problem in the form given by (A.14): $M = \mathbf{R}^3_+$, $f = \delta^3(x' - x) \equiv \delta(x'^1 - x^1)\delta(x'^2 - x^2)\delta(x'^3 - x^3)$ is the Dirac-delta distribution defined by

$$\int_{-\infty}^{\infty} \delta(x' - x)\mathrm{d}x' = 1 \quad \text{and} \quad \delta(x' - x) = \begin{cases} 0 \text{ for } x' \neq x \\ \infty \text{ for } x' = x \end{cases}$$

and lastly, $g = 0$. Thus, by Equation (A.15),

$$\psi(x) = \int_M f(x') G(x', x) \mathrm{d}^n x' = G(0, x)$$

□

The fundamental solution in $\mathbf{R}^3$ is given by Equation (A.13):

$$\Psi_3 = \frac{1}{4\pi}\frac{1}{\|x'-x\|}$$

where $x' = (0,0,d)$ and $x \in$ M is any fixed arbitrary point. Returning to Example A.4.8, choose $\Phi_3 = -\frac{1}{4\pi}\frac{1}{\|x''-x\|}$, where $x'' = (0,0,-d)$. Then, on the plane $\partial\mathbf{R}^3_+$, $G(x',x) = \Psi_3(x',x) + \Phi_3(x',x) = 0 \; \forall x \in \partial\mathbf{R}^3$. This then is the essence of the method of images.

Before proceeding to the study of wave equations, some definitions must be recalled to facilitate the discussion. First, a *general linear second-order* PDE has the form:

$$\sum_{i,j=0}^{3} a_{ij}(x)\partial_{ij}\psi(x) + \sum_{j=0}^{3} b_j(x)\partial_j\psi(x) + c(y)\psi(x) = f(y) \quad \forall x \in M \subset \mathbf{R}^{3+1}$$

where x^0 of the coordinate $x = (x^0,\ldots,x^3)$ denotes time, and $a_{ij}, b_j, c \in C(M)$ such that $a_{ij} = a_{ji} \; \forall i,j$. For notational simplicity, $\partial_k = \partial_{x_k}$. Then, a nonconstant function $\xi \in C^2(M)$ defines a characteristic surface $S = \{x \in M : \xi(x) = 0\}$ if the following criterion is satisfied: $\sum_{i,j=0}^{3} a_{ij}(x)\partial_i\xi(x)\partial_j\xi(x) = 0$ on S. Otherwise, the surface is said to be noncharacteristic with respect to the linear second-order PDE. The quantity $Q_\xi(x) \equiv \sum_{i,j=0}^{3} a_{ij}(x)\partial_i\xi(x)\partial_j\xi(x)$ is called the *associated ξ-quadratic form*.

A.4.9 Remark

Consider the matrix $\mathbf{A} = (a_{ij})$ of the general second-order linear PDE. For a fixed $(x_0,t_0) \in M \times [0,\infty)$, if the set $\sigma(\mathbf{A})$ of eigenvalues of $\mathbf{A}$ satisfies

(a) $\operatorname{sgn}(\lambda_i) = \operatorname{sgn}(\lambda_j) \; \forall \lambda_i, \lambda_j \in \sigma(\mathbf{A})$, then the PDE is *elliptic* at (x_0,t_0)
(b) $\operatorname{sgn}(\lambda_i) = -\operatorname{sgn}(\lambda_{\hat{i}}) \; \forall \lambda_i \in \sigma(\mathbf{A}) - \{\lambda_{\hat{i}}\}$ for some fixed $\lambda_{\hat{i}}$, then the PDE is *hyperbolic* at (x_0,t_0)
(c) $\exists \lambda = 0$ for some $\lambda \in \sigma(\mathbf{A})$, then the PDE is *parabolic* at (x_0,t_0)

Furthermore, if the general second-order linear PDE is respectively, elliptic, hyperbolic, or parabolic on $M \times [0,\infty)$, then the PDE is said to be, respectively, *elliptic, hyperbolic,* or *parabolic.*

A.4.10 Definition

Suppose $\psi: M \times (0,\infty) \to \mathbf{R}$ is of class C^2, where $M \subset \mathbf{R}^3$ is compact. Furthermore, let $f: M \times [0,\infty) \to \mathbf{R}$ be continuous on M. Then, the hyperbolic equation

$$\partial_t^2\psi - c^2\Delta\psi = f \quad \text{on} \quad M \tag{A.18a}$$

satisfying the following initial value conditions

$$\begin{cases} \psi = g \\ \partial_t \psi = h \end{cases} \quad \text{on} \quad M \times \{0\} \tag{A.18b}$$

for some continuous functions g,h on M, is called the *inhomogeneous wave equation* on M, and if $f \equiv 0$, it is known as the *wave equation*. The coefficient c is the speed of the wave propagation.

Now, given the wave equation $\partial_t^2 \psi - c^2 \Delta \psi = 0$ on $M \times (0,\infty)$, consider the associated quadratic form $Q_\xi = (\partial_t \xi)^2 - c^2 |\nabla \xi|^2$ on $M \times [0,\infty)$, where $\nabla = (\partial_1, \partial_2, \partial_3)$ and $\partial_t = \partial_0$. Next, consider the function $\xi : M \times [0,\infty) \to \mathbf{R}$ defined by $\xi(x,t) = \frac{c^2}{2}(t-\tau)^2 - \frac{1}{2}|x-y|$, for some fixed $(y,\tau) \in M \times [0,\infty)$. Then, by definition,

$$Q_\xi = (\partial_t \xi)^2 - c^2 |\nabla \xi|^2 = c^4 (t-\tau)^2 - c^2 |x-y|^2 = 2c^2 \xi$$

whence the characteristic surface $S_\xi = \{(x,t) \in \Omega \times [0,\infty) : c|t-\tau| = |x-y|\}$, as $Q_\xi = 0$ implies at once that $c^4(t-\tau)^2 - c^2|x-y|^2 = 0 \Rightarrow c^2(t-\tau)^2 - |x-y|^2 = 0$. See Figure A.6 for an illustration of the characteristic surface based at some $(y,\tau) \in M \times (0,\infty)$.

A.4.11 Remark

Given the wave equation $\partial_t^2 \psi - c^2 \Delta \psi = 0$ on $M \times (0,\infty)$ satisfying the initial value condition $\psi|_{t=0} = g$ and $\partial_t \psi|_{t=0} = h$ on M, let $\bar{S}_\xi(x_0,t_0)$ denote the closure of the characteristic cone of the wave equation: $\bar{S}_\xi = \{(x,t) \in M \times [0,\infty) : c(t_0 - t) \geq |x - x_0|\}$ at $(x_0,t_0) \in M \times (0,\infty)$. Then, the *domain of dependence* at the point (x_0,t_0) is defined by the closed disk $\bar{B}(x_0) = \bar{S}_\xi(x_0,t_0) \cap M \times \{0\}$. The reason for this is the following: the wave solution at (x_0,t_0) only depends on its initial value on $\bar{B}(x_0)$. This is clarified by the following theorem.

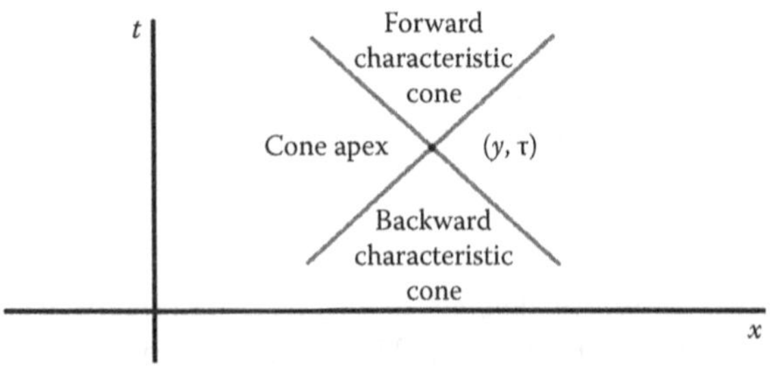

FIGURE A.6
Characteristic surface of a wave equation.

A.4.12 Theorem (Domain of Dependence Inequality)

Given the wave equation $\partial_t^2 \psi - c^2 \Delta \psi = 0$ on $M \times (0,\infty)$ satisfying the initial value condition $\psi|_{t=0} = g$ and $\partial_t \psi|_{t=0} = h$ on M, let $\bar{S}_\xi(x_0,t_0)$ denote its closed characteristic cone at $(x_0,t_0) \in M \times (0,\infty)$, and set

$$\bar{B}(x_0;\tau) = \bar{S}_\xi(x_0,t_0) \cap \Omega \times \{\tau\}$$

Then,

$$\int_{\bar{B}(x_0;\tau)} \left\{ |\nabla \psi|^2 + |\partial_t \psi|^2 \right\}_{t=\tau} \mathrm{d}^3 x \le \int_{\bar{B}(x_0)} \left\{ |\nabla \psi|^2 + |\partial_t \psi|^2 \right\}_{t=0} \mathrm{d}^3 x \quad \forall \tau \in [0,t_0] \qquad \square$$

The quantity $e(x_0;\tau) = \frac{1}{2}\int_{\bar{B}(x_0;\tau)} \left\{ |\nabla \psi|^2 + |\partial_t \psi|^2 \right\}_{t=\tau} \mathrm{d}^3 x$ is called the *energy* of the wave in $\bar{B}(x_0;\tau)$.

Now, consider (A.18) and define the average

$$\langle \psi(x;r,t) \rangle_{\partial B(x,r)} \equiv \frac{1}{|\partial B(x,r)|} \int_{\partial B(x,r)} \psi(y,t) \mathrm{d}^2 y$$

where $\partial B(x,r)$ is a sphere of radius $r > 0$ centered about x, and $|\partial B(x,r)|$ denotes the surface area of $\partial B(x,r)$. To express the solution of Equation (A.18) as a function of g,h, recall first, the following lemma [8, p.70].

A.4.13 Lemma (Euler-Poisson-Darboux)

Fix $x \in \mathbf{R}^3$ and some $r > 0$, and suppose that ψ is a solution of (A.18). Then, relative to polar coordinates, $\partial_t^2 \Psi - \partial_r^2 \Psi - \frac{2}{r}\partial_r \Psi = 0$ on $\mathbf{R}_+ \times (0,\infty)$ with $\Psi = \langle g(x;r,t) \rangle_{\partial B(x,r)}$ and $\partial_t \Psi = \langle h(x;r,t) \rangle_{\partial B(x,r)}$ defined on $\mathbf{R}_+ \times \{0\}$, where $\Psi \equiv \langle \psi(x;r,t) \rangle_{\partial B(x,r)}$. $\square$

Indeed, it can be shown via Lemma A.4.13 that the solution of Equation (A.18) for $\Omega \subseteq \mathbf{R}^3$ is given by the following *Kirchhoff formula,*

$$\psi(x,t) = \langle th(y) + g(y) + \nabla g \cdot (y-x) \rangle_{\partial B(x,t)} \tag{A.19}$$

for $t > 0$, and for $\Omega \subseteq \mathbf{R}^2$, the solution is given by the *Poisson formula*:

$$\psi(x,t) = \left\langle \frac{tg(y) + t^2 h(y) + t\nabla g \cdot (y-x)}{\sqrt{t^2 - |y-x|^2}} \right\rangle_{B(x,t)} \tag{A.20}$$

The interested reader may pursue the excellent references [8,16,23,26] for the details of the derivation.

This section closes with a cursory generalization of PDE solutions. From an application perspective (e.g., by numerical scientists and engineers) its utility arises in the formulation of the finite element method [14,22,25] for

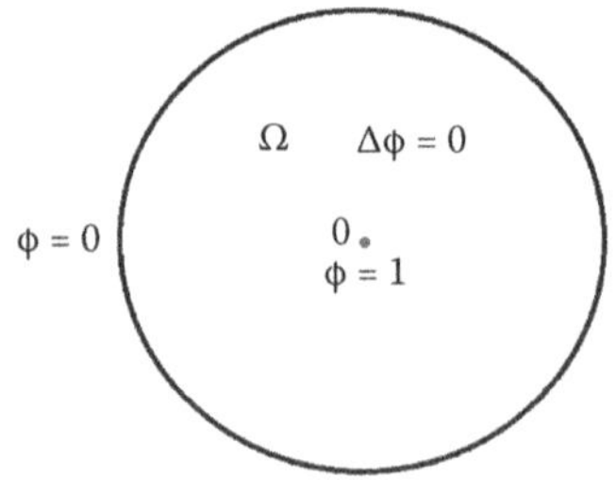

FIGURE A.7
No classical solution exists for the Dirichlet problem.

numerical computation. In essence, this final section paves the foundation for EMC engineers to apply the method of finite element analysis to solve Maxwell's equations numerically.

By way of motivation, consider Poisson's equation on some domain M:

$$\begin{cases} -\Delta\psi = f \text{ on } M \\ \\ \psi = g \text{ on } \partial M \end{cases} \tag{A.21}$$

If a solution $\psi \in C^2(M) \cap C(\bar{M})$ of Equation (A.21) exists, then it is called a *classical* (or *strong*) solution. Before proceeding to weaken the definition of the solution of (A.21), some background information is required.

Given a subset $\Omega \subseteq \mathbf{R}^3$, define its volume by $\mu(\Omega)$, where μ is the *Lebesgue measure*; see, for example, References [4,6,20] for details. Note that if $\mu(N) = 0$ for some $N \subset \mathbf{R}^3$, then N is set to be a *null* set (or a set of *measure zero*), and $\mu(\Omega \cup N) = \mu(\Omega)$ by the definition of Lebesgue measure. In particular, if $N \subset \mathbf{R}^3$ is countable, then it is a set of measure zero. Informally, the Lebesgue measure is roughly the generalisation of the Riemann integral (encountered in undergraduate engineering calculus) extended to a collection of null sets such that if N is a null set, then so is $M \subset \mathbf{R}^3$ $\forall M \subset N$. For a precise definition, consult the previously cited references.

Now, given a function $f: M \to \mathbf{R}$, the function is *Lebesgue integrable* if $\int_\Omega |f(x)|\, d\mu(x) < \infty$. Moreover, if $f, g: \Omega \to \mathbf{R}$ such that $f = g$ on $\Omega - N$ for some $N \subset \Omega$ such that $\mu(N) = 0$, then, $f = g$ *almost everywhere* (*a.e.*). Next, define the space of *square integrable* functions by $L^2(\Omega) = \left\{ f : \int_\Omega |f(x)|^2\, d\mu(x) < \infty \right\}$ (modulo functions that equal one another μ-a.e.). The space $L^2(\Omega)$ is a Hilbert space.

A.4.14 Definition

Given a real vector space V, an *inner product* $(\cdot,\cdot): V \times V \to \mathbf{R}$ is a continuous function satisfying

(a) *Positive-definiteness*: $(u,u) \geq 0\ \forall u \in V$ with $(u,u)=0$ if $u=0$.
(b) *Symmetry*: $(u,v)=(v,u)\ \forall u,v \in V$.
(c) *Linearity*: $(\alpha u+v,w)=\alpha(v,w)+(v,w)\ \ \forall u,v,w \in V$ and $\alpha \in \mathbf{R}$.

Then, V is called a *pre-Hilbert space* if it admits an inner product structure. An inner product defines a *metric* or *norm* on V in an obvious manner: $|u| = \sqrt{(u,u)}$. Furthermore, recall that a sequence $(u_n) \subset V$ is *Cauchy* if $\forall \varepsilon > 0, \exists N > 0$ such that $n,m > N \Rightarrow |u_n - u_m| < \varepsilon$; and a sequence $(u_n) \subset V$ converges $(u_n) \to u_0 \in V$, if $\forall \varepsilon > 0, \exists N > 0$ such that $n > N \Rightarrow |u_n - u_0| < \varepsilon$. The point u_0 is called a *limit point*. Then, V is *complete* with respect to the norm if every Cauchy sequence in V converges in V.

In view of Definition A.4.11, a *Hilbert space* is a pre-Hilbert space that is complete with respect to the inner product. It can be shown that the Lebesgue measure μ on $M \subseteq \mathbf{R}^3$ defines an inner product on $\mathrm{L}^2(M)$ as follows.

$$(u,v) = \int_M u(x)v(x)\mathrm{d}\mu(x) \tag{A.22}$$

It particular, $\mathrm{L}^2(M)$ endowed with (A.22) defines a Hilbert space.

A.4.15 Example

Define $\mathrm{H}^1(M) = \{u \in \mathrm{L}^2(M) : \partial_i u \in \mathrm{L}^2(M), i = 1,2,3\}$. Then, an inner product can be defined on $\mathrm{H}^1(M)$ as follows,

$$(u,v)_{\mathrm{H}^1(M)} = (u,v) + \sum_i (\partial_i u, \partial_i v) \tag{A.23}$$

with norm given by $|u|_{\mathrm{H}^1(M)} = \sqrt{(u,v) + \Sigma_i(\partial_i u, \partial_i v)}$. It can be shown that the space is complete with respect to the inner product and it is thus a Hilbert space. This is a space wherein the function together with its partial derivatives are square-integrable. □

Now, in hindsight, define $\mathrm{H}_0^1(M) = \{u \in \mathrm{H}^1(M) : u|\partial M = 0\}$. Then, in view of the homogeneous Dirichlet boundary problem, $\mathrm{H}_0^1(M)$ is constructed as a possible solution space for the homogeneous Dirichlet problem. The space $\mathrm{H}_0^1(M)$ admits an inner product inherited from $\mathrm{H}^1(M)$ via Equation (A.23). Finally, define $\mathrm{H}^{-1}(M) \equiv \mathrm{H}_0^1(M)'$ to be the *topological dual* of $\mathrm{H}_0^1(M)$; this is the space of bounded linear functionals $\xi : \mathrm{H}_0^1(M) \to \mathbf{R}$ endowed with appropriate topology that is outside the scope of this exposition [2,4].*

Returning to Equation (A.21), by way of motivation, consider two commonly cited examples due to Zaremba and Lebesgue, respectively: the former can be found in Reference [15, p. 285] and the latter in [26, p. 198]. Given a

* These spaces are more generally known as *Sobolev spaces*, the properties of which can be found in References [1,8,10].

unit 2-disk $B(0,1) \subset \mathbf{R}^2$, set $\Omega = B(0,1) - \{0\}$, illustrated in Figure A.7 no classical solution exists for the Dirichlet problem $\Delta\varphi = 0$ on Ω subject to the boundary conditions:

$$\varphi = \begin{cases} 1 \text{ for } x = 0, \\ 0 \text{ on } \partial B(0,1) \end{cases}$$

Indeed, to see the nonexistence of a classical solution, suppose $\exists\varphi \in C^2(\Omega) \cap C(\bar{\Omega})$ satisfying the above boundary conditions. By definition, φ is analytic, hence, if φ were analytic at 0, then $\varphi \equiv 0$ on $\bar{\Omega}$ by the maximum modulus principle, yielding a contradiction. Thus, the origin 0 must at most be a removable singularity by Riemann. Recall that Riemann's theorem on removable singularity states that if a function is analytic on Ω, then either the function is analytic at 0 or 0 is a removable singularity of the function [5]. Hence, a suitable value can be assigned to $\varphi(0)$, rendering φ analytic at 0. However, in order for φ to satisfy the boundary condition at 0, φ must be discontinuous at 0 as analyticity implies that $\varphi \equiv 0$ on Ω. Hence, no classical solution exists, as asserted.

The second example due to Lebesgue is more complicated, and hence, only a heuristic argument is sketched. See Figure A.8, where the exponential cusp is generated by rotating the following exponential curve about the *z*-axis,

$$x = \begin{cases} e^{-\frac{1}{z-z_0}} \text{ for } z > z_0 \\ 0 \quad \text{for } z = z_0 \end{cases}$$

for some $z_0 > 0$.

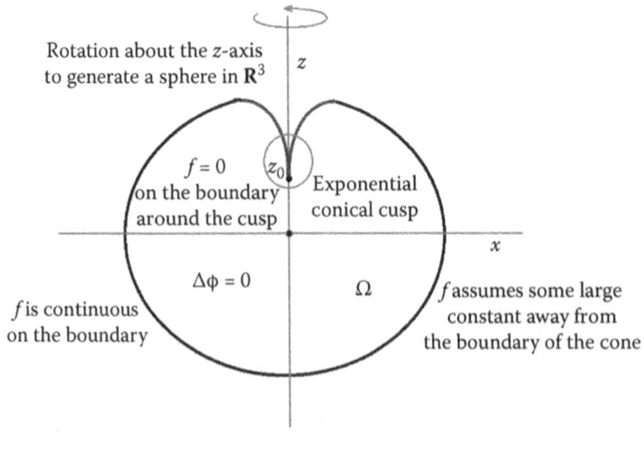

FIGURE A.8
Cross-section of a sphere with an exponential cusp.

Suppose that $f = 0$ on $\partial\Omega \cap B(z_0, \varepsilon)$, for some $\varepsilon > 0$ small, and $f = \varphi_0 >> 0$ away from the cusp, for some large constant φ_0. To see why a solution cannot exist, suppose for concreteness that φ represents steady-state temperature. Then, about $\partial\Omega \cap B(z_0, \varepsilon)$, for a sufficiently small $\varepsilon > 0$, the cusp at z_0 cannot absorb heat sufficiently fast to keep the temperature close to 0 as required by the boundary condition at the cusp due to the lack of surface area about the cusp to conduct heat away. Hence, if a solution exists, it cannot be continuous at z_0, yielding a contradiction. In short, no solution exists, as required.

Thus, the above two examples amply demonstrate that the requirement $\psi \in C^2(M) \cap C(\bar{M})$ for the existence of a solution to Equation (A.21) is too stringent. The question thus leads to the conditions under which a meaningful solution can exist when the smoothness requirement is relaxed.

First, define $L^2_{loc}(M)$ to be the set of all functions $f: M \to \mathbf{R}$ such that $\int_K |f|^2 \, d\mu < \infty$ for all compact $K \subset M$. This is the set of all locally square integrable functions, and let $C_0^\infty(M)$ denote the set of all infinitely differentiable functions on M with compact support. For $M = \mathbf{R}$, define the weak derivative $\frac{d\delta(t)}{dt}$ by

$$\int_{-\infty}^{\infty} \frac{d\delta(t)}{dt} v(t) \, d\mu(t) \equiv -\int_{-\infty}^{\infty} \delta(t) \frac{dv(t)}{dt} \, d\mu(t)$$

$\forall v \in C_0^\infty(\mathbf{R})$. The space $C_0^\infty(\mathbf{R})$ is called the *test function space.*

Motivated by this endeavor, applying Green's theorem to $-\int_M \Delta\varphi v \, d\mu$, assuming the integral exists, leads to:

$$-\int_M \Delta\varphi v \, d\mu = \int_M \nabla\varphi \cdot \nabla v \, d\mu - \int_{\partial M} \varphi \nabla v \cdot \boldsymbol{n} \, d\lambda \tag{A.24}$$

for all $v \in C_0^\infty(M)$, where λ is the Lebesgue measure on ∂M. Then, applying this to (A.14) with $g = 0$, that is, the homogeneous Dirichlet problem, yields:

$$\int_M \nabla\varphi \cdot \nabla v \, d\mu = \int_M f v \, d\mu \tag{A.25}$$

If the equality (A.25) holds $\forall v \in C_0^\infty(M)$, then φ defines a solution to Equation (A.14). In particular, φ, v can be relaxed to $\varphi, v \in H_0^1(M)$ and $f \in H^{-1}(M)$ in order for (A.25) to hold. Then, $\varphi \in H_0^1(M)$ is said to be the *weak solution* of (A.21).

A.4.16 Remark

Suppose that Equation (A.21) is inhomogeneous; that is, $g \neq 0$. This can be converted easily into a homogeneous Dirichlet problem as follows. Construct any function $\psi \in C^2(M) \cap C(\bar{M})$ such that $\psi|\partial M = g$. Set $\phi = \varphi - \psi$. Then, by construction,

$$-\Delta\phi = -\Delta\varphi - \Delta\psi = f - \Delta\psi \text{ on } M \quad \text{and} \quad \phi = \varphi - \psi = g - g = 0 \text{ on } \partial M.$$

That is, the inhomogeneous Dirichlet problem is transformed into the homogeneous Dirichlet problem:

$$\begin{cases} -\Delta\phi = f - \Delta\psi & \text{on } M \\ \phi = 0 & \text{on } \partial M \end{cases}$$

References

1. Adams, R. 1975. *Sobolev Spaces*. New York: Academic Press.
2. Berberian, S. 1974. *Lectures in Functional Analysis and Operator Theory*. New York: Springer-Verlag.
3. Cheng, D. 1992. *Field and Wave Electromagnetics*. Reading, MA: Addison-Wesley.
4. Choquet-Bruhat, Y., DeWitt-Morette, C., and Dillard-Bleick, M. 1982. *Analysis, Manifolds and Physics, Part I: Basics*. Amsterdam: North-Holland.
5. Churchill, R. and Brown, J. 1990. *Complex Variables and Applications*. New York: McGraw-Hill.
6. Cohn, D. 1980. *Measure Theory*. Boston: Birkhäuser.
7. Engelking, R. 1989. *General Topology*. Berlin: Heldermann Verlag (*Sigma Series in Pure Mathematics*, Vol. 6).
8. Evans, L. 1998. *Partial Differential Equations*. Providence, RI: American Mathematical Society (GSM Vol. 19).
9. Farlow, S. 1993. *Partial Differential Equations for Scientists and Engineers*. New York: Dover.
10. Grisvard, P. 1985. *Elliptic Problems in Nonsmooth Domains*. Boston: Pitman.
11. Hocking, J. and Young, G. 1961. *Topology*. New York: Dover.
12. Hsu, H. 1984. *Applied Vector Analysis*. San Diego: Harcourt Brace Jovanovich.
13. Jackson, J. 1962. *Classical Electrodynamics*. New York: John Wiley & Sons.
14. Johnson, C. 2009. *Numerical Solution of Partial Differential Equations by the Finite Element Method*. New York: Dover.
15. Kellogg, O. 1953. *Foundations of Potential Theory*. New York: Dover.
16. Koshlyakov, N., Smirnov, M., and Gliner, E. 1964. *Differential Equations of Mathematical Physics*. Amsterdam: North-Holland, UK.
17. Nakahara, M. 2003. *Geometry, Topology and Physics*. Bristol: IOP.
18. Neff, H. Jr. 1981. *Basic Electromagnetic Fields*. New York: Harper & Row.
19. Plonsey, R. and Collin, R. 1961. *Principles and Applications of Electromagnetic Fields*. New York: McGraw-Hill.
20. Rao, M. 1987. *Measure Theory and Integration*. New York: John Wiley & Sons.

21. Rothwell, E. and Cloud, M. 2001. *Electromagnetics.* Boca Raton, FL: CRC Press.
22. Sadiku, M. 2001. *Numerical Techniques in Electromagnetics.* Boca Raton, FL: CRC Press.
23. Sauvigny, F. 2006. *Partial Differential Equations 1: Foundations and Integral Representations.* Heidelberg: Springer-Verlag.
24. Smythe, W. 1950. *Static and Dynamic Electricity.* New York: McGraw-Hill.
25. Solin, P. 2006. *Partial Differential Equations and the Finite Element Method.* Mineola, NY: John Wiley & Sons.
26. Zachmanoglou, E. and Thoe, D. 1986. *Introduction to Partial Differential Equations with Applications.* New York: Dover.

Index

Printed and bound by CPI Group (UK) Ltd, Croydon, CR0 4YY
18/10/2024
01776256-0012